Glider Representations

Monographs and Research Notes in Mathematics

Series Editors:
John A. Burns, Thomas J. Tucker, Miklos Bona, Michael Ruzhansky

About the Series

This series is designed to capture new developments and summarize what is known over the entire field of mathematics, both pure and applied. It will include a broad range of monograph and research notes on current and developing topics that will appeal to academics, graduate students, and practitioners. Interdisciplinary books appealing not only to the mathematical community, but also to engineers, physicists, and computer scientists are encouraged.

This series will maintain the highest editorial standards, publishing well-developed monograph as well as research notes on new topics that are final, but not yet refined into a formal monograph. The notes are meant to be a rapid means of publication for current material wher the style of exposition reflects a developing topic.

Analytic Methods for Coagulation-Fragmentation Models, Volume I
Jacek Banasiak, Wilson Lamb, Philippe Laurencot

Analytic Methods for Coagulation-Fragmentation Models, Volume II
Jacek Banasiak, Wilson Lamb, Philippe Laurencot

Neutrices and External Numbers
A Flexible Number System
Dinis Bruno, Imme van den Berg

Extending Structures
Fundamentals and Applications
Ana Agore, Gigel Militaru

Inverse Scattering Problems and their Application to Non-Linear Integrable Equations
Pham Loi Vu

Spectral Methods Using Multivariate Polynomials On The Unit Ball
Kendall Atkinson, David Chien, and Olaf Hansen

Glider Representations
Frederik Caenepeel, Fred Van Oystaeyen

For more information about this series please visit: https://www.crcpress.com/Chapman--HallCRC-Monographs-and-Research-Notes-Ir Mathematics/book-series/CRCMONRESNOT

Glider Representations

Frederik Caenepeel
Fred Van Oystaeyen

CRC Press
Taylor & Francis Group
Boca Raton London New York

CRC Press is an imprint of the
Taylor & Francis Group, an **informa** business
A CHAPMAN & HALL BOOK

CRC Press
Taylor & Francis Group
6000 Broken Sound Parkway NW, Suite 300
Boca Raton, FL 33487-2742

Printed on acid-free paper

International Standard Book Number-13: 978-0-367-40657-8 (Hardback)

Visit the Taylor & Francis Web site at
http://www.taylorandfrancis.com

and the CRC Press Web site at
http://www.crcpress.com

To Matthias, the first author's brother
and to Emily and Aline, the second author's
granddaughters

Contents

Preface

This monograph is about a very new subject in mathematics. In a sense, fragments and gliders are a generalization of vector spaces over some field or modules over some ring when there is a filtration on the ring fixed. Such a filtration may be a (finite or infinite) chain of subrings. For example, in the case of a group ring of some group one may consider subrings obtained from a chain of subgroups of the group. The glider theory with respect to such a filtration is a generalized representation theory where the structure of the group with respect to the chain of subgoups intervenes. In Algebra, many filtrations of rings appear naturally and several theories are built on them. To name a few, a valuation on a field defines the valuation filtration and in algebraic (projective) geometry there are the standard filtrations on finitely generated algebras. Let us also mention the filtrations on rings of differential operators determined by the order of the differential operators or filtrations defined on enveloping algebras of Lie algebras and on Poincaré-Birkhoff-Witt deformations of commutative algebras. Finite filtrations by rings are less studied but it may be seen as a stepwise construction procedure; see for example iterative Ore extensions and quantum groups, iterated group rings already mentioned or iterated enveloping algebras of Lie algebras.

It was the aim of this book to introduce the gliders in general but then study their meaning in several different applications of the method. We believe the new structure of a glider (representation) will be useful in studying new structural properties in different areas in Algebra and Geometry, so we tried to open several doors focusing on structure results in the different situations as already mentioned above. It shows that new phenomena appear and the new theory is not a trivial generalization. Quite the opposite, intriguing mixtures of algebraic and geometric techniques appear, for example, in the theory of the glider Brauer-Severi variety of a finite dimensional central simple algebra, where the Brauer-Severi variety of the Algebra does appear but linked to the Riemann surface of the central field. More results in this direction then lead to input from classical theory of maximal orders. Throughout the book, we touch upon various branches of mathematics and we have tried to make everything as self contained as possible. Nonetheless, when setting the scene of a new chapter, we always provided references for more detailed overviews.

We present a broad spectrum of possible research in very different areas because we aimed to show gliders can be useful in many different directions. This also has the obvious consequence that a lot of doors are opened but the theories are not exhaustively developed here yet. One may continue with many research projects; for example, we included a few facts about the glider version of the Gelfand-Kirillov dimension; also a few results about the Weyl algebras are included. However, the theory of gliders over rings of differential operators with respect to filtrations given by algebraic subvarieties and subsets of differential operators defining chains of subrings, as well as the theory of holonomic gliders, is not done here. This characteristic makes the book particularly suited for advanced seminars where several participants may actually begin some research projects, as exercises in creativity. In the case of glider representations of finite groups it is evident that even chains of three or even two(!) subgroups do yield some non-trivial effects. It also remains a goal to deal, some day, with longer chains in some detail, with iterative steps along the chain transferring new structure results from the smaller to the bigger. Consider this an abstract tool box. Hopefully it may get you into constructing something with it!

FREDERIK CAENEPEEL
FRED VAN OYSTAEYEN

Acknowledgments

Research in this work was financially supported by the Research Foundation – Flanders, FWO, in terms of a PhD fellowship for the first author. We also thank the University of Antwerp, UA, and the Shanghai Center for Mathematical Sciences, Fudan University, Shanghai for support, hospitality and facilities extended to us.

The start of our collaboration that eventually led to the current manuscript can be traced back to one person in particular. We are extremely grateful to Stefaan Caenepeel, long lasting friend of the second author, father ánd mathematical brother of the first one.

We thank Ruben Henrard, Geoffrey Janssens, Eric Jespers and Adam-Christiaan Van Roosmaelen for fruitful discussions on the subject.

Last but not least, thanks to all of our friends and family members for supporting and for believing in us. We are lucky to have so many.

Acknowledgments

Authors

Frederik Caenepeel is a postdoctoral researcher at the Shanghai Center for Mathematical Sciences associated with Fudan University, Shanghai, China. He received his PhD degree from the University of Antwerp, Belgium under the supervision of Professor, Doctor Wendy Lowen and Emeritus Professor, Doctor Fred Van Oystaeyen. In his free time, Frederik enjoys doing outdoor sports activities, discovering the world and catching up with friends.

Fred Van Oystaeyen is an emeritus Professor at the University of Antwerp. His research interests include Non-commutative Algebra and Geometry, Hopf Algebras, graded rings, the Brauer group and representation theory. He has authored or co-authored over 300 scientific papers and about 25 research books and has edited several proceedings of the more than 60 international congresses. He did research evaluation in several countries, was president of the Belgian Science Foundation's math committee and served 5 years on the Academic Advisory Committee for the ERASMUS program. He was also a visiting scholar at the University of Cambridge (UK) and honorary professor at the Beijing Normal University. In his free time, he enjoys blues music and caudiciform plants, cacti and bonsai trees.

Introduction

In Mathematics it is very common to study structures over some fixed ground structure. For example in Algebra, vector spaces are viewed over some fixed ground field K and also algebras are viewed as algebras over a field. Similarly, Geometry is most often developed over the field of real or complex numbers. Number Theory starts from studying algebraic extensions of the field of rational numbers, Galois theory studies field extensions etc... On the other hand very few techniques are developed to study substructures of a given structure. We think one of the reasons for this asymmetry is that extensions or over-structures usually come with a construction description showing how it is build over the substructure, whereas substructures are not constructed from the bigger one in some concrete way. There is one basic nice example though: the valuation theory of a field. But there are many different valuation rings in a field so the field does not yield much information about the valuation rings inside it. On the other hand, knowing all or in some cases almost all valuation rings contains enough information on the field to provide a useful tool for studying the algebraic properties of it; see Number Theory or Class Field Theory. In connection with the topic of this book, the important property of a valuation ring in a field is that there exists a filtration on the field (as a ring) such that the considered valuation ring is the part of filtration degree zero.

The topic of interest will thus be rings with an ascending filtration indexed by a totally ordered group Γ, usually the integers $\mathbb{Z}$. For a totally ordered group Γ and a ring R a Γ-filtration on R is an ascending chain: $\ldots \subset F_\gamma R \subset \ldots \subset F_\tau R \subset \ldots$, for $\gamma \leq \tau$, of additive subgroups $F_\gamma R$ of R for each $\gamma \in \Gamma$, such that $1 \in F_e R$, e the neutral element of Γ, and for all $\gamma, \tau \in \Gamma$ we have $F_\gamma R F_\tau R \subset F_{\gamma\tau} R$. Moreover we usually assume $R = \bigcup_{\gamma \in \Gamma} F_\gamma R$ (FR is exhaustive) and $0 = \bigcap_{\gamma \in \Gamma} F_\gamma R$ (FR is separated). Then $S = F_e R$ is a subring of R and FR may be seen as a structural connection between R and S. Before we go on, let us state that filtrations exist on any ring; in fact many different filtrations may easily be defined, and many filtrations appear naturally and are often used in the analysis of the structure of rings. For example, standard filtrations are defined on algebras when they are viewed as epimorphic images of a free algebra $K\langle \underline{X} \rangle$ over a field K with variables from a set $\underline{X} = \{x_\alpha \mid \alpha \in \mathcal{A}\}$ for some index set $\mathcal{A}$. The filtration on the algebra is then induced by the filtration on $K\langle \underline{X} \rangle$ given by $F_{-n} K\langle \underline{X} \rangle = 0$ for $n > 0$, $F_0 K\langle \underline{X} \rangle = K, F_m K\langle \underline{X} \rangle$ is the K-vector space of sums of monomials of total degree in X less than or equal to m, for $m > 0$. Other filtrations which will be used a lot in this book are given by finite chains of subrings of a ring R: $R_0 \subset R_1 \subset \ldots \subset R_m \subset \ldots \subset R$ with $R = R_n$ for some n. The filtration FR defined by $F_{-n}R = 0$ for $n > 0$, $F_0R = R_0$ and $F_mR = R_m$ for $m > 0$ with $F_NR = R$ for $N \geq n$, is a positive filtration defining a structural connection between R and the chain of subrings.

Now in classical ring theory one uses module theory to study a ring; that is, one considers the category R-mod of (left) R-modules in order to trace properties of R. This is particularly successful in representation theory, say of groups, Lie algebras, finite dimensional algebras,... . For example, look at a subgroup H of a group G and the group algebras $KH \subset KG$ over a field K. Now we can study KH-modules and extend these to KG-modules or we can study KG-modules and restrict these to KH-modules. This provides some results in the representation theory but in fact the theory does not encode the structural detail of the pair (H, G) or (KH, KG). Thus we are prompted to consider an extension of module theory which would take into account the structural properties of the pair $H \subset G$, and this is exactly what the theory of fragments and gliders will provide for us. The idea is the following: for a ring R with filtration (here we will restrict to $\mathbb{Z}$-filtrations) FR and degree zero part $F_0R = S$ we will now define an FR-fragment as an S-module M with a descending chain $M \supset M_1 \supset M_2 \supset \ldots \supset M_n \supset \ldots$, such that for $i \leq j$: $F_iRM_j \subset M_{j-i}$, so the M_i are all S-modules. For a glider we assume $M \subset \Omega$ where Ω is an R-module and the actions of the F_iR on the M_j are all restrictions of the R-multiplication in Ω (so for general fragments we have to write down the exact compatibility conditions for the F_iR-actions on the M_j under restricting F_iR to F_kR and viewing $M_j \subset M_l$ for $k < i, l < j$; see exact definition in Chapter 1). So in general, $F_1RM_i \subset M_{i-1}$ but it is still possible that $F_1RM_i \subset M_i$ and this will happen in some fragments. For a generalized representation theory, the restriction to gliders instead of the more general fragments is very natural; the glider may be viewed as a chain of S-modules inside an R-module and in fact it may then be related to the negative part of a filtration on an R-module (the R-module generated by M inside Ω) whereas the filtration FR is usually a positive one, meaning that $F_nR = 0$ for $n < 0$.

Dealing with a new theory we have to produce more than just new structures and their properties. First we have to obtain enough structure to be able to claim the new theory is a satisfactory generalization of module theory. Secondly, we have to obtain enough feedback from the new results on existing mainstream mathematics to motivate the usefulness of the new concepts. Hence the book is presented in two parts.

The first part contains the general theory of fragments providing a useful generalization of module theory. This includes some categorical methods, induced fragments, glider morphisms, limits of fragments, finite generation and finiteness conditions, right bounded filtrations, irreducible fragments, Noetherian and Artinian fragments, decomposition results for fragments in terms of irreducible fragments. Let us only mention that the notion of an irreducible fragment or glider is somewhat more complex than in the case of modules but nevertheless we do obtain a suitable decomposition theory in semisimple situations.

The second part contains the applications to different fields of Mathematics and actually a distinction is made between the two types of filtrations we are working with, namely the filtrations corresponding to finite chains of subrings of a ring and unbounded filtrations, of which standard filtrations form an interesting example. In Chapter 4 we study glider representations of a chain of finite groups, present a Clifford theory for group algebra fragments with relations between decomposition groups and special applications to nilpotent groups. In Chapter 5 we use the generalities from fragment and glider theory to introduce the glider representation ring of a chain of groups. We clarify its ring structure and primitive central idempotents for Abelian groups with chain $1 < G$. Observe that the chains of subgroups used here are very simple; we included some comments for longer chains. To make the step towards non-Abelian groups, we include a specific example of the quaternion group Q_8, which yields insight to treat the general case. For interesting classes of groups we are able to determine the structure of the glider representation ring for $1 < G$ and we obtain a separation for some isocategorical groups. In Chapter 6 the glider theory is applied to semisimple Lie algebras with special attention to gliders in Verma modules and Verma gliders. Again, irreducible gliders are studied and we find a relation to anti-dominant weights. We provide several examples with respect to Lie algebra embeddings like $\mathfrak{sl}_2 \subset \mathfrak{sl}_3$. For $\mathfrak{sl}_n \subset \mathfrak{sl}_m$ the nilpotent orbits generated by the embedding elements correspond to inclusions of the suitable Hasse subdiagrams. We also obtain similar descriptions for the nilpotent orbits generated by the embedding elements for the inclusions $\mathfrak{sp}_{2n} \subset \mathfrak{sp}_{2m}$ and $\mathfrak{sp}_{2n+1} \subset \mathfrak{sp}_{2m+1}$.

Starting from Chapter 7 we deal with standard filtrations, as particular example of unbounded filtrations. In Chapter 7 we focus on such filtrations obtained after localization at Ore sets or more general at kernel functors. Those localization results are then applied to obtain a sheaf theory of glider representations in case of a spectrum of a Noetherian ring, a torsion-theoretic spectrum, the non-commutative site of non-commutative geometry, respectively. Many other applications are awaiting further research; we mention just a few applications to classical algebraic geometry. In Chapter 8 we remain within the geometric situation by taking inspiration from the theory of Brauer-Severi varieties of central simple algebras (csa). Being defined as the variety of irreducible left ideals of a csa A, we exploit the sets of irreducible left glider ideals in a csa A with certain unbounded filtration FA. It turns out that even in the case of fields K we obtain non-trivial results.

Finally we include further applications in the direction of coalgebras and Hopf algebras applied to iterated one-point extensions of finite dimensional elementary algebras. The Gelfand-Kirillov dimension of gliders is defined and with some remarks on gliders over Weyl algebras we point at some research directions connected to rings of differential operators.

In conclusion we may say that we have provided several applications in classical domains within Mathematics, thereby motivating the introduction of the new glider theory and we hope to have opened many doors for new future research, in particular with respect to more complex filtration chains.

Part I

General Fragment and Glider Theory

Chapter 1

Basic definitions and generalities

The first chapter encompasses the definition of fragments and glider representations, which will be defined over some filtered ring FR. At first sight, this new theory may seem very similar to the theory of filtered rings and modules. Therefore, we assume the reader is acquainted with module theory, [3], [44], and with the extended filtered module theory, [46], [54].

After giving some elementary examples, we introduce the category of (left) FR-fragments, discuss the existence of quotients, construct inductive and projective limits and discuss filtered and graded fragments.

1.1 Some notation

Throughout Γ will be a totally ordered group with positive cone Π and neutral element e. The operation of Γ will be denoted by $+$ even if it is not Abelian, the ordering is given by $:\leq$, we write $\gamma < \delta$ to indicate that $\gamma \leq \delta$ and $\gamma \neq \delta$ for $\gamma, \delta \in \Gamma$. For simplicity, the reader may always consider Γ to be $\mathbb{Z}$.

By R we shall always indicate an associative ring with unit. A Γ-filtration on R is given by an ascending chain of additive subgroups of $R : \ldots \subset F_\gamma R \subset \ldots \subset F_\delta R \subset \ldots R$, for $\gamma \leq \delta$, such that $F_\gamma R F_\delta R \subset F_{\gamma+\delta} R$ for all $\gamma, \delta \in \Gamma$ and $1 \in F_e R$. We call R with filtration FR a (Γ-)filtered ring and we will write $S = F_e R$. Observe that S is a subring of R and that every $F_\gamma R$, for $\gamma \in \Gamma$ is an S-bimodule. By convention, $A \subset B$ means that A is a subset of B and that A need not be proper. Similarly, $A \supset B$ means that A contains B, but need not be strictly bigger. To indicate a strict inclusion we will denote $A \subsetneq B$, resp. $A \supsetneq B$.

An R-module M is a filtered R-module (with respect to FR) if it has an ascending chain of additive subgroups: $\ldots \subset F_\gamma M \subset \ldots \subset M$ such that for $\gamma, \delta \in \Gamma$ we have that $F_\delta R F_\gamma M \subset F_{\delta+\gamma} M$. Note that by R-module we will mean a left R-module unless otherwise stated.

A filtration FR induces a negative filtration on S, denoted by F^eS and given by putting for $\gamma \in -\Pi, F^e_\gamma S = F_\gamma R$ and $F^e_\mu S = S$ for $\mu \in \Pi$. Note that FR also induces a positive filtration on R, denoted by F^+R and obtained by putting $F^+_\gamma R = F_\gamma R$ for $\gamma \in \Pi, F^+_e R = S$ and $F^+_\gamma R = 0$ for $\gamma < e$. A filtration FR is exhaustive whenever $R = \bigcup_{\gamma \in \Gamma} F_\gamma R$. The filtration FR is separated whenever $0 = \bigcap_{\gamma \in \Gamma} F_\gamma R$. In case FR is left bounded, i.e. there exists a $\gamma_0 \in \Gamma$ such that for all $\gamma < \gamma_0$ we have $F_\gamma R = 0$ and then we say that FR is discrete.

We assume throughout that FR is exhaustive and separated. This also implies that F^+R is exhaustive and discrete, and F^eS is exhaustive and separated too.

1.2 Definition of fragments

Fragments have been introduced in [52] in case $\Gamma = \mathbb{Z}$ and for positive filtrations only. Here we modify the definitions to the more general situation considered here.

Definition 1.2.1. Let FR be a (positive) filtration with subring $S = F_eR$. A (left) FR-fragment M is a (left) S-module together with a descending chain of subgroups

$$M_e = M \supset \ldots \supset M_\gamma \supset \ldots,$$

for $\gamma \in \Pi$ and satisfying the following properties
$\mathbf{f_1}$. For every $\gamma \in \Pi$ there exists an S-module $M \supset M^*_\gamma \supset M_\gamma$ and there is given an operation of $F_\gamma R$ on this M^*_γ by $\varphi_\gamma : F_\gamma R \times M^*_\gamma \to M,\ (\lambda, m) \mapsto \lambda.m$, satisfying $\lambda.(m+n) = \lambda.m + \lambda.n, e.m = m, (\lambda+\delta).m = \lambda.m + \delta.m$ for $\lambda, \delta \in F_\gamma R$ and $m, n \in M^*_\gamma$.

$\mathbf{f_2}$. For every $\gamma \in \Pi$ and $\delta \leq \gamma$ we have a commutative diagram

$$\begin{array}{ccccc}
M & \xleftarrow{\iota} & M_{\gamma-\delta} & \xrightarrow{\iota} & M \\
\uparrow{\scriptstyle \varphi_\gamma} & & \uparrow & & \uparrow{\scriptstyle \varphi_\delta} \\
F_\gamma R \times M_\gamma & \xleftarrow{\iota_R} & F_\delta R \times M_\gamma & \xrightarrow{\iota_M} & F_\delta R \times M_\delta
\end{array}$$

$\mathbf{f_3}$. For every $\gamma, \delta, \mu \in \Pi$ such that $F_\gamma R F_\delta R \subset F_\mu R$ we have $F_\delta R M_\mu \subset M^*_\gamma \cap M_{\mu-\delta}$. Moreover, the following diagram is commutative

$$\begin{array}{ccc}
F_\gamma R \times F_\delta R \times M_\mu & \xrightarrow{m_R \times M_\mu} & F_\mu R \times M_\mu\ , \\
\downarrow{\scriptstyle F_\gamma R \times \varphi_\mu} & & \downarrow{\scriptstyle \varphi_\mu} \\
F_\gamma R \times M_{\mu-\delta} & \xrightarrow{\overline{\varphi_\gamma}} & M
\end{array}$$

where m_R is the multiplication of R. Observe that the left vertical arrow is defined, since $1 \in F_e R$ implies that $F_\delta R \subset F_\mu R$.

In condition $\mathbf{f_3}$ above, we may in fact demand commutativity of the diagram for every $\gamma, \delta \in \Gamma$ such that $\gamma + \delta \geq e$, that is where the maps are defined. We also point to the fact that condition $\mathbf{f_2}$ implies that $F_{-\delta} R M_\gamma \subset M_{\gamma+\delta}$ for $\delta, \gamma \in \Pi$.

Corollary 1.2.2. *A left FR-fragment M also has the structure of a filtered S-module with respect to $F^e S$ by putting $F^e_\gamma M = M_{-\gamma}$ for $e \neq \gamma \in -\Pi$ and $F^e_\tau M = M$ for $\tau \in \Pi$.*

Proof. Let $\delta \leq e, \gamma \leq e$. Then we have $(F^e_\delta S) F^e_\gamma M = F_\delta R M_{-\gamma} \subset M_{-\gamma-\delta} = F^e_{\delta+\gamma} M$. □

From $\mathbf{f_2}$ it follows that for $r \in F_\delta R, m \in M_\gamma$ with $\delta \leq \gamma$, we may write $rm \in M_{\gamma-\delta}$ unambiguously. From $\mathbf{f_3}$ it follows that $(rs)m = r.(sm)$ for $r \in F_\gamma R$, $s \in F_\delta R$ and $m \in M_{\gamma+\delta}$. Hence we may simplify notation by writing sm whenever the multiplication of $m \in M$ and $s \in R$ is defined (independent of the choice of φ_γ defining the multiplication for different $\gamma \in \Gamma$).

1.3 Glider representations

Any R-module M may be viewed as an FR-fragment (unless otherwise stated, fragments, like modules, will be considered to be left fragments) in a trivial way, putting $M_\gamma = M$ for all $\gamma \in \Pi$ and all operations being induced by the scalar R-multiplication of M. If M is a filtered R-module then $F_e M$ may be viewed as an FR-fragment by putting $(F_e M)_\gamma = F_{-\gamma} M$ for $\gamma \in \Pi$, $(F_e M)_e = F_e M$. Clearly we have for $\tau \leq \gamma$: $(F_\tau R)(F_e M)_\gamma \subset F_{\tau-\gamma} M = (F_e M)_{\gamma-\tau}$. So with respect to the operations induced by the R-module structure of M, $F_e M$ is indeed an FR-fragment.

It is convenient when all operations φ_γ are induced by some operation induced from a larger object. This motivates the introduction of so-called glider representations as hereafter. Consider an R-module Ω containing an S-module M with respect to the S-structure of Ω induced by restriction of scalars from R to S. If $M \supset \ldots \supset M_\gamma \supset \ldots$ is an FR-fragment such that all the partial actions φ_γ are induced from the R-module Ω, then we call $\Omega \supset M$ a glider representation or a glider. In the definition of a fragment, the M^*_γ become in the glider case $M^*_\gamma = \{m \in M, F_\gamma R m \subset M\}$ and $\mathbf{f_3}$ is just equivalent to $F_\delta R M_\mu \subset M_{\mu-\delta}$. Let $\delta \leq \gamma, \gamma \in \Pi$ and $m \in M^*_\gamma, r \in F_\delta R$. We have that $F_{\gamma-\delta} R r \subset F_\gamma R$ and $F_{\gamma-\delta} R(rm) \subset F_\gamma R m \subset M$, thus $rm \in M^*_{\gamma-\delta}$. Hence the descending chain $\Omega \supset M \supset \ldots \supset M^*_\gamma \supset \ldots$ also defines a glider representation.

Definition 1.3.1. An FR-glider representation M is called natural if $M_\gamma = M_\gamma^* = \{m \in M, F_\gamma Rm \subset M\}$ for all $\gamma \in \Pi$.

Example 1.3.2. In case $\Gamma = \mathbb{Z}$ and FR is given as a ring filtration, i.e. each F_iR is a subring of R, the first part of the fragment condition $\mathbf{f_3}$ is equivalent to $F_\mu RM_\mu \subset M_\mu^*$ for all μ. Indeed, $F_iRF_jR \subset F_\mu R$ implies that $F_iR, F_jR \subset F_\mu R$ since $1 \in F_0R$ by definition. Then we have that $M_\mu^* \subset M_i^*, M_j^*$, so $F_jRM_\mu \subset F_\mu RM_\mu \subset M_\mu^* \subset M_i^*$. Conversely, from $F_jRM_\mu \subset M_i^*$ for all i and j such that $F_iRF_jR \subset F_\mu R$ we have in particular $F_\mu RF_\mu R \subset F_\mu R$ hence $F_\mu RM_\mu \subset M_\mu^*$. We observe that a natural glider over a ring filtration is a chain of $F_\mu R$-modules.

Glider representations will be the main objects of study in this book. Nonetheless, there are fragments that are not glider representations as the following examples show.

Example 1.3.3. Let $\mathbb{Z} \subset \mathbb{Q} \subset \mathbb{Q} \subset \ldots$ be a $\mathbb{Z}$-filtration of $\mathbb{Q}$. Then

$$\mathbb{Q} \times \mathbb{F}_2 \supset \mathbb{Q} \times \{0\} \supset 0 \supset \ldots$$

is a fragment, but $\mathbb{F}_2$ cannot be embedded in a $\mathbb{Q}$-module.

Example 1.3.4. Let S be a domain, M an S-module with non-trivial annihilator ideal $I = \mathrm{ann_S(M)}$. Denote by $S^\times$ the set of invertible elements of S. Let $s \in S \setminus (S^\times \cup I)$ and consider the filtration $FR = (S \subset S\frac{1}{s} \subset R)$ where R is the fraction field of S. Then the chain of S-modules

$$M \supset (s)M \supset 0$$

is an FR-fragment but not a glider representation since M is not embeddable in an R-vector space. For example, for $S = \mathbb{Z}$, $M = \mathbb{Z}/4\mathbb{Z}$, one can take $s = 3$ and then $\mathbb{Z}/4\mathbb{Z} \supset \mathbb{Z}/4\mathbb{Z} \supset 0$ yields a $\mathbb{Z} \subset \mathbb{Z}\frac{1}{3} \subset \mathbb{Q}$-fragment.

Also, not every fragment embedded into an R-module needs to have the induced glider structure. Even more is true! Suppose that $\Omega \supset M \supset \ldots \supset M_\gamma \supset \ldots$ is a glider representation. Then there are multiple ways of turning Ω in an FR-filtered module; e.g. one can put $F_{-\gamma}\Omega = M_\gamma$ for $\gamma \in \Pi$, $F_e\Omega = M$ and $F_\gamma\Omega = \Omega$ for $\gamma \in \Pi$. Conversely, the negative part $F^e\Omega$ of any filtered FR-module $F\Omega$ is a fragment with regard to the positive filtration F^+R on R (see the next example). At first sight it seems that we are just replacing filtered modules by some other fancy terminology, but we truly are extending the theory. Indeed, although the role of Ω seems rather redundant, it is of utter importance. Consider for example the easy algebra filtration $k \subset kG$ where k is some algebraically closed field of characteristic zero and G some finite group having two non-isomorphic 2-dimensional simple representations Ω_1 and Ω_2. Let V be a 2-dimensional vector space. We can put two non-isomorphic G-actions on V, turning it into Ω_1, Ω_2 respectively and we obtain a glider representation $\Omega = V \supset W$ where W is a 1-dimensional vector space.

For whichever action we choose, we have that $kG \cdot W = V$, but the fragment structure $V \supset W$ depends on the chosen G-action! The notion of fragment is clearly a broad one and in particular situations some types of fragments may be more useful than others.

Let us give some easy examples of fragments and glider representations.

Examples 1.3.5. 1. Consider a Γ-filtered ring R with subring $F_eR = S$. The negative filtration induced on S defines a glider for the positive filtration of R. This glider representation is natural exactly when the filtration on R is strong, that is $F_\sigma F_{\sigma^{-1}}R = F_eR = S$ holds for every $\sigma \in \Gamma$. For $\gamma \in \Pi$ we can also define $S_\gamma = \{s \in S, F_\gamma Rs \subset S\}$. If $\gamma \leq \tau$ in Γ then $S_\tau \subset S_\gamma$ is obvious. Also for $\tau \leq \gamma \in \Pi$ we have $F_\tau RS_\gamma \subset S_{\gamma-\tau}$ because $F_{\gamma-\tau}R(F_\tau RS_\gamma) \subset F_\gamma RS_\gamma \subset S$. Now by definition of the descending sequence: $S \supset \ldots \supset S_\gamma \supset \ldots$, defines a natural glider representation with respect to F^+R;

2. If FM is a filtered FR-module then the negative part of the chain on F_eM is an F^+R-fragment. Indeed, for $e \leq \delta < \gamma$ we have $F_\delta^+R(F_eM)_\gamma = F_\delta F_{-\gamma}M \subset F_{\delta-\gamma}M = (F_eM)_{\gamma-\delta}$;

3. To a $\mathbb{Z}$-filtration FR we may correspond for every $m \in \mathbb{N}$ the m-step filtration F^mR given by $F_n^mR = F_{\lfloor n/m \rfloor}R$, where $\lfloor n/m \rfloor$ denotes the largest integer smaller than n/m.

   Let us consider a positive filtration $FR, S = F_0R \subset \ldots \subset F_nR \subset \ldots \subset R$ and an FR-fragment M. Then M is also an F^2R-fragment sine $F_n^2RM_m = F_{\lfloor n/2 \rfloor}RM_m \subset M_{m-\lfloor n/2 \rfloor}$ for $n \leq m$; thus from $M_{m-\lfloor n/2 \rfloor} \subset M_{m-n}$ it follows indeed that M is an F^2R-fragment.
   If M is a natural FR-fragment such that $M_1 \neq M$ then M is not a natural F^2R-fragment since that would imply that

$$M_1 = \{m \in M, F_1^2Rm \subset M\} = \{m \in M, F_0Rm \subset M\} = M$$

   and we assumed that $M \neq M_1$. On the other hand we may look at the descending chain $M \supset M \supset M_1 \supset M_1 \supset M_2 \supset M_2 \supset \ldots$, which we denote by M^2. This obviously defines an F^2R-fragment and if M is natural as FR-fragment, then M^2 is natural as F^2R-fragment.

Now since we always assume filtrations to be exhaustive, i.e. $R = \bigcup_{\gamma \in \Gamma} F_\gamma R$, we have for any FR-fragment M that $B(M) = \bigcap_{\gamma \in \Gamma} M_\gamma$ is an R-module contained in M. We call $B(M)$ the body of M.

Lemma 1.3.6. *With notation as above, the S-module $M/B(M)$ together with descending chain $S(M)_\gamma = M_\gamma/B(M)$ for $\gamma \in \Pi$, is an FR-fragment. Moreover, $B(S(M)) = 0$. When M is a glider induced from an R-module Ω, then $S(M)$ is a glider as well and induced from the R-module $\Omega/B(M)$. If M is natural, then $S(M)$ is natural.*

Proof. The R-module structure of $B(M)$ is induced by the operations $\varphi_\gamma : F_\gamma R \times M_\gamma^* \to M$, mapping $F_\gamma R \times B(M)$ into $B(M)$; hence the φ_γ define $\overline{\varphi}_\gamma : F_\gamma R \times S(M)_\gamma \to S(M)$ unambiguously. The conditions $\mathbf{f_1}, \mathbf{f_2}, \mathbf{f_3}$ are easily checked for the $\overline{\varphi}_\gamma, \gamma \in \Pi$, consequently $S(M)$ is an FR-fragment. If $\overline{m} \in B(S(M))$ and $m \in M$ represents $\overline{m}$, then for every γ there is an $m_\gamma \in M_\gamma$ such that $\overline{m} = \overline{m_\gamma}$ in $M/B(M)$ or $m = m_\gamma + b_\gamma$ for some $b_\gamma \in B(M)$. It follows that $m \in M_\gamma$ for all γ or $m \in B(M)$. In other words, $\overline{m} = 0$, which shows that $B(S(M)) = 0$.
When $\Omega \supset M$ is a glider, then the maps in the commutative diagram

$$\begin{array}{ccc} M & \hookrightarrow & \Omega \\ \downarrow{\scriptstyle \pi} & & \downarrow{\scriptstyle \pi} \\ S(M) & \hookrightarrow & \Omega/B(M) \end{array}$$

are compatible with the fragment structures of M and $S(M)$. Hence $\Omega/B(M) \supset S(M)$ is a glider as well. If M is natural, i.e. $M_\gamma = \{m \in M, F_\gamma Rm \subset M\}$ for all $\gamma \in \Pi$, then we obtain that $S_\gamma = \{\overline{m} \in S(M), F_\gamma R\overline{m} \subset S(M)\}$. Indeed $S(M)_\gamma = \pi(M_\gamma)$ for $\pi : M \to M/B(M)$, which is an FR-fragment morphism (see next section). Hence, if $\overline{m} \in S(M)$ is such that $F_\gamma R\overline{m} \subset S(M)$ and if m represents $\overline{m}$ then

$$\begin{aligned} & \pi(F_\gamma Rm) = F_\gamma R\overline{m} \subset S(M) \\ \Rightarrow \quad & F_\gamma Rm \subset \pi^{-1}\pi(F_\gamma Rm) \subset \pi^{-1}(S(M)) = M. \end{aligned}$$

Hence $m \in M_\gamma$ and it follows that $\overline{m} \in S(M)_\gamma$. □

Definition 1.3.7. The fragment $S(M)$ is called the spirit of M. The terminology is suggested by the fact that the spirit is obtained by killing the body.

Example 1.3.8. With notation as before we have seen that for a Γ-filtered ring $R, S = F_e R$ is an $F^e S$-filtered left S-module. Moreover, the chain $\{S_\gamma, \gamma \geq e \text{ in } \Gamma\}$ defines on S the structure of a natural F^+R-fragment and also an FR-fragment since for $\tau \in \Pi$ we have $F_{-\tau}RS_\gamma \subset S_{\gamma+\tau}$ because $F_{\gamma+\tau}R(F_{-\tau}RS_\gamma) \subset F_\gamma RS_\gamma \subset S$. In case FR is a strong filtration, then $S_\gamma = \{s \in S, F_\gamma Rs \subset F_e R\} = \{s \in S, s \in F_{-\gamma}R\}$ or $S_\gamma = F_{-\gamma}R$ and the $F^e S$-filtration of S defines a natural glider representation on $R \supset S$ with respect to F^+R (and also FR). In any case, the body $B(S)$ of S with fragment structured given by the S_γ is just the largest R-submodule of S; indeed if $s \in S$ is such that $Rs \subset S$, then $s \in S_\gamma$ for all γ or $s \in B(S)$. This property holds for every natural FR-fragment.

Proposition 1.3.9. *With notation as before, let $\Omega \supset M$ be a natural FR-glider representation. Then $B(M)$ is the unique largest R-submodule of M (operation restricted from Ω).*

Proof. If N is an R-submodule of M, hence of the R-module Ω, then for $n \in N, Rn \subset M$ yields $n \in M_\gamma$ for every $\gamma \in \Gamma$, hence $N \subset B(M)$. Consequently $B(M)$ is the unique largest R-submodule of M (with R-operation induced by Ω). □

Remark 1.3.10. The proposition fails for non-natural gliders. A very easy example is obtained by looking at a positively $\mathbb{Z}$-filtered ring R and viewing R as an FR-fragment by putting $M_0 = R \supset M_1 \supset 0 \supset 0 \supset \ldots$, where M_1 is an arbitrary F_0R-submodule of R different from R. For this awkward fragment structure we have $B(R) = 0$ but R itself is a maximal R-submodule.

For a general FR-fragment N given by a chain $N = N_e \supset \ldots \supset N_\gamma \supset \ldots$ for $\gamma \in \Pi$ we may look at acceptable R-submodules of N. These are S-submodules N' of N carrying an R-module structure compatible with (induced by) the $F_\gamma R$-structure on $N' \cap N_\gamma$ for all $\gamma \in \Pi$. The body $B(N)$ is of course acceptable. Given an acceptable R-submodule N' of N, we may make a new chain:

$$N = N'_e \supset \ldots \supset N'_\gamma = N_\gamma + N' \supset \ldots$$

Obviously this defines a new FR-fragment structure on N. Now if $\{N'_j, j\}$ is a family of acceptable R-submodules of N then $\sum_j N'_j$ is an acceptable R-submodule. So we may consider the acceptable R-submodule of N generated by all acceptable R-submodules; let us write N' for this, and then look at the descending chain: $N \supset \ldots \supset N_\gamma + N' \supset \ldots$. In this new FR-fragment structure of N we have $B(N) = N'$.

Definition 1.3.11. We call an FR-fragment N saturated if $B(N) \supset N'$, where N' is as above.

To an arbitrary FR-fragment N we may associate its saturation by changing the chain of definition in the way described above. Since $B(N)$ is itself acceptable, a fragment is saturated if and only if $B(N) = N'$.

1.4 The category of FR-fragments

Let $R, FR, S = F_eR$ be as before. Consider FR-fragments M, N.

Definition 1.4.1. An S-linear morphism $f : M \to N$ is a morphism of FR-fragments if for all $\gamma \in \Pi, f(M_\gamma) \subset N_\gamma$ and $f(rm) = rf(m)$ for all $m \in M_\gamma, r \in F_\gamma R, \gamma \in \Pi$.

If $f : M \to N$ is a morphism of fragments such that f is injective, then we say that M is a subfragment of N. This means that we have chains

$$M = M_e \supset \ldots \supset M_\gamma \supset \ldots$$

and

$$N = N_e \supset \ldots \supset N_\gamma \supset \ldots$$

such that $M_\gamma \subset N_\gamma$ for all $\gamma \in \Pi$ and the action of $F_\gamma R$ on N_γ induces the action on M_γ. An epimorphism of fragments is a surjective morphism of fragments. The identity $1_M : M \to M$ is a morphism of fragments, and so is the zero map $M \to 0$. If $f : M \to N$ and $g : N \to T$ are morphisms of fragments, then the S-linear map $g \circ f : M \to T$ is again a morphism of fragments. If $f_1, f_2 : M \to N$ are morphisms of fragments then so is $f_1 - f_2$; we write $\mathrm{Hom}_{FR}(M, N)$ for the additive group of morphisms of fragments from M to N. Put H_γ equal to $\{h \in \mathrm{Hom}_{FR}(M, N) \mid h(M_\tau) \subset N_{\tau+\gamma} \text{ for all } \tau \in \Pi\}$ for $\gamma \in \Pi$. If $\delta \geq \gamma$ in Π then $H_\delta \subset H_\gamma$, hence we obtain a chain $H_e = \mathrm{Hom}_{FR}(M, N) \supset \ldots \supset H_\gamma \supset \ldots$ which defines the structure of a $\mathbb{Z}$-fragment where we consider $\mathbb{Z}$ as trivially Γ-filtered, i.e. $F_\gamma \mathbb{Z} = 0$ for $\gamma < e, F_\gamma \mathbb{Z} = \mathbb{Z}$ for $\gamma \geq e$. Note that $B_{F\mathbb{Z}}(\mathrm{Hom}_{FR}(M, N)) = \mathrm{Hom}_{FR}(M, B(N))$ ($B(N)$ is an R-module with trivial fragment structure). Therefore, if N is a spirit then $\mathrm{Hom}_{FR}(M, N)$ is a spirit as a $\mathbb{Z}$-fragment.

The class of FR-fragments together with the morphisms of fragments form a category FR-frag and it is easy to see that this category is preadditive. To see whether this is an additive category we have to establish the existence of direct sums. Consider a family of FR-fragments $\{M_\alpha, \alpha \in \mathcal{A}\}$. We may look at the S-module $\oplus_\alpha M_\alpha$ and for $\gamma \in \Pi$ consider $(\oplus_\alpha M_\alpha)_\gamma = \oplus_\alpha (M_\alpha)_\gamma$. The chain thus obtained in $\oplus_\alpha M_\alpha$ makes it into an FR-fragment as is easily verified. We write $\oplus_\alpha M_\alpha$ for the direct sum FR-fragment (beware of the difference with *fragment direct sums* to be defined later!). Note that the S-module $\oplus_\alpha M_\alpha$ can be made into an FR-fragment in many different ways. E.g one could omit any M_α deeper in the chain. When we speak of the FR-fragment $\oplus_\alpha M_\alpha$, it is understood that we mean the fragment structure defined above. We then have that $B(\oplus_\alpha M_\alpha) = \oplus_\alpha B(M_\alpha), S(\oplus_\alpha M_\alpha) = \oplus_\alpha S(M_\alpha)$. By $F^e S$-filt we denote the category of $F^e S$-filtered S-modules. We have a forgetful functor $U : FR\text{-frag} \to F^e S\text{-filt}$. To an FR-fragment M we associate the filtered S-module $F^e M$ from Corollary 1.2.2. To a morphism $f : M \to N$ of fragments, the S-linear map $F^e M \to F^e N$ is easily seen to be S-filtered.

Like $F^e S$-filt, FR-frag is not an Abelian category. The problem is that we cannot canonically define the cokernel for a fragment morphism. On the other hand, when $f : M \to N$ is a morphism of fragments then we may define the kernel $\mathrm{Ker}(f)$ and image $\mathrm{Im}(f)$ of f. Indeed, in the S-module $\mathrm{Ker}(f)$ we consider the chain given by $\mathrm{Ker}(f)_\gamma = \mathrm{Ker}(f) \cap M_\gamma$ for $\gamma \in \Pi$. This indeed defines a fragment structure since for $z \in \mathrm{Ker}(f) \cap M_\gamma$ we have that f is defined on $F_\gamma R z$ and since it is a morphism of fragments $f(F_\gamma R z) = F_\gamma R f(z) = 0$, hence $F_\gamma R \mathrm{Ker}(f)_\gamma \subset \mathrm{Ker}(f)$. We denote this FR-fragment by $\mathrm{Ker}(f)$. Similarly for the image, we define $\mathrm{Im}(f)_\gamma = f(M)_\gamma = f(M_\gamma)$ for $\gamma \in \Pi$.

1.5 Bi-fragments

Up to now we have always dealt with left (or right) FR-fragments. On an S-bimodule M we may consider the structure of a left FR-fragment, say with respect to a chain $M = M_e \supset \ldots \supset M^l_\gamma \supset \ldots, \gamma \in \Pi$, as well as a right FR-fragment structure with respect to a chain (of right S-modules this time) $M = M_e \supset \ldots \supset M^r_\gamma \supset \ldots$. Even if we assume that all M^l_γ, M^r_γ are S-bimodules, there need not exist further relations between both chains. An S-bimodule M is a bi-fragment (with respect to FR) if it is both a left and a right FR-fragment with respect to a chain $M = M_e \supset \ldots \supset M_\gamma \supset \ldots$ where each $M_\gamma, \gamma \in \Pi$, is an S-bimodule, i.e. for $\gamma \in \Pi$ we have $\sum_{\tau+\sigma=\gamma} F_\sigma R M_\gamma F_\tau R \subset M_{-\tau+\gamma-\sigma}$. Note that $B(M)$ is then an R-bimodule.

1.6 Induced and strict subfragments

If we consider an arbitrary S-module N in an FR-fragment M then the chain $\{N \cap M_\gamma, \gamma \in \Pi\}$ does not necessarily define a fragment structure on N. Indeed, we do not know whether $F_\gamma R(N \cap M_\gamma) \subset N$. Nevertheless, if we put $N_\gamma = \{n \in M_\gamma,\ F_\gamma R n \subset N\}$ for $\gamma \in \Pi$, then clearly $N_\gamma \subset N$ for all $\gamma \in \Pi$ and for $\delta < \gamma, N_\gamma \subset N_\delta$. The restrictions of the $\varphi_\gamma : F_\gamma R \times M^*_\gamma \to M$ to $F_\gamma R \times N_\gamma$ do define an FR-fragment structure on N, as is readily verified. We refer to N equipped with this particular fragment structure as the induced fragment structure (from M) on N. Observe that the fact that $M_\gamma \cap N$ does not define a fragment structure on N presents a deviation compared to the theory of filtered modules. A subfragment M' of an FR-fragment is said to be a strict subfragment if $M'_\gamma = M_\gamma \cap M'$ for all $\gamma \in \Pi$. In general, for any subfragment M' of M we may consider the S-module $\overline{M} = M/M'$. However, for $\overline{m} \in \overline{M}$ represented by $m \in M_\gamma$ for some $\gamma \in \Pi$, we would like to define for $r \in F_\gamma R$ that $r.\overline{m} = \overline{rm}$. If m_1 is another representative of $\overline{m}$, i.e. $m_1 - m \in M'$, then we do not know whether $r(m_1 - m) \in M'$ and thus the planned definition for $r.\overline{m}$ does not work here! Obviously, if M' is a strict subfragment of M then the foregoing construction does work and we obtain an FR-fragment $\overline{M} = M/M'$ with $\overline{M}_\gamma = M_\gamma/M'_\gamma = M_\gamma/(M' \cap M_\gamma)$ for $\gamma \in \Pi$. Observe that for a subfragment N of M we may look at the induced FR-fragment structure on N induced by M, denote it by $N^{(M)}$. It is obvious that N is a subfragment of $N^{(M)}$; hence the induced fragment is the maximal subfragment of M that can be defined on the given S-module underlying the subfragment.

Lemma 1.6.1. *An induced subfragment N of M is strict if and only if $F_\gamma R(M_\gamma \cap N) \subset N$ for all $\gamma \in \Pi$.*

Proof. If $F_\gamma R(M_\gamma \cap N) \subset N$ then $M_\gamma \cap N = \{n \in M_\gamma,\ F_\gamma Rn \subset N\} = N_\gamma^{(M)} = N_\gamma$. The other direction is trivial. $\square$

The proof actually shows that if N is an S-submodule of M such that $F_\gamma R(M_\gamma \cap N) \subset N$ for all $\gamma \in \Pi$, then the induced fragment $N^{(M)}$ is strict.

Lemma 1.6.2. *If $N \subset M$ has the induced fragment structure then:*

1. *If M is saturated then N is saturated;*
2. *If M is glider then N is glider;*
3. *If M is natural then N is natural.*

Proof. 1. For any FR-fragment X let us denote by ${}_RW(X)$ the maximal acceptable R-submodule in X (its scalar multiplication being induced by the fragment operations in X). If M is saturated then we have ${}_RW(M) \subset B(M)$. Look at ${}_RW(N)$ in N. Since the fragment structure of N is induced from M we have for all $\gamma \in \Pi$ that $N_\gamma = \{n \in M_\gamma,\ F_\gamma Rn \subset N\}$. The R-module structure of ${}_RW(N)$ is induced by the fragment operations of N hence by those of M. We obtain that ${}_RW(N) \subset {}_RW(M) \subset B(M)$, so if $z \in {}_RW(N)$ then $z \in M_\gamma$ and $F_\gamma Rz \subset N$ for all $\gamma \in \Pi$. Consequently, $z \in N_\gamma$ for all $\gamma \in \Pi$, or $z \in B(N)$. This proves that ${}_RW(N) \subset B(N)$; hence N is saturated.
2. Trivial
3. Suppose that $M_\gamma = \{m \in M,\ F_\gamma Rm \subset M\}$ for all $\gamma \in \Gamma$. By assumption, $N_\gamma = \{n \in M_\gamma,\ F_\gamma Rn \subset N\}$ for $\gamma \in \Pi$ and the fragment operations in N are induced by those in M. Now if $z \in N$ is such that $F_\gamma Rz \subset N$ then $F_\gamma Rz \subset M$ and thus $z \in M_\gamma$, whence $z \in N_\gamma$. This proves that $N_\gamma = \{n \in N,\ F_\gamma Rz \subset N\}$. $\square$

1.7 Glider morphisms

Let $f : M \to N$ be a morphism of FR-fragments. We say that f is a glider morphism if there exist R-modules $\Omega(M)$, resp. $\Omega(N)$, such that $\Omega(M) \supset M$ and $\Omega(N) \supset N$ are glider representations and if there exists an R-linear morphism $F : \Omega(M) \to \Omega(N)$ making the following diagram commutative

$$\begin{array}{ccc} \Omega(M) & \xrightarrow{F} & \Omega(N) \\ {\scriptstyle \iota_M}\uparrow & & \uparrow{\scriptstyle \iota_N} \\ M & \xrightarrow{f} & N \end{array}$$

where ι_M, ι_N are the canonical inclusions. Since this definition implies that $F(R\iota_M(M)) = RF(\iota_M(M)) \subset R\iota_N(M)$, and since we may identify M and

$\iota_M(M)$, resp. N and $\iota_N(N)$, we may assume without loss of generality that $\Omega(M) = RM, \Omega(N) = RN$. A sequence of fragment morphisms, say

$$P \xrightarrow{g} M \xrightarrow{f} N$$

is said to be a glider sequence if there exist R-modules $\Omega(P), \Omega(M), \Omega(N)$ and R-linear maps $G : \Omega(P) \to \Omega(M), F : \Omega(M) \to \Omega(N)$, such that we have a commutative diagram

$$\begin{array}{ccccc} \Omega(P) & \xrightarrow{G} & \Omega(M) & \xrightarrow{F} & \Omega(N) \\ {\scriptstyle \iota_P}\uparrow & & {\scriptstyle \iota_M}\uparrow & & \uparrow{\scriptstyle \iota_N} \\ P & \xrightarrow{g} & M & \xrightarrow{f} & N \end{array} \tag{1.1}$$

with conventions and notation as before. Note that in statements concerning exactness of sequences, for example if we state that (1.1) has exact rows then we may not replace $\Omega(X)$ by RX and still claim exactness of rows! A glider morphism $f : M \to N$ is said to have an (S, R)-cokernel if the S-module $\mathrm{Coker}_S(f)$ of f viewed as an S-linear morphism fits in the commutative diagram

$$\begin{array}{ccccc} \Omega(M) & \xrightarrow{F} & \Omega(N) & \xrightarrow{\pi_R} & \Omega(M)/F(\Omega(N)) \\ {\scriptstyle \iota_M}\uparrow & & {\scriptstyle \iota_N}\uparrow & & \uparrow{\scriptstyle \iota} \\ M & \xrightarrow{f} & N & \xrightarrow{\pi_S} & \mathrm{Coker}_S(f) \end{array}$$

the vertical maps being inclusions and π_R, π_S being the canonical R-, resp. S-linear maps. In a similar but dual way it is possible to define what it means to have an (S, R)-kernel.

Proposition 1.7.1. *Let $f : M \to N$ be a glider morphism, then f has an (S, R)-kernel and we have an exact diagram with exact rows*

$$\begin{array}{ccccccc} 0 & \longrightarrow & \mathrm{Ker}(F) & \longrightarrow & \Omega(M) & \xrightarrow{F} & \Omega(N) \\ & & \uparrow & & {\scriptstyle \iota_M}\uparrow & & {\scriptstyle \iota_N}\uparrow \\ 0 & \longrightarrow & \mathrm{Ker}(f) & \longrightarrow & M & \xrightarrow{f} & N \end{array} \tag{1.2}$$

If f has an (S, R)-cokernel and $\mathrm{Im}(f)$ *is a strict subfragment (we say that f is strict), then* $\mathrm{Coker}_S(f)$ *has the structure of a glider representation making the following into a commutative diagram with exact rows*

$$\begin{array}{ccccccc} \Omega(M) & \xrightarrow{F} & \Omega(N) & \xrightarrow{\pi_R} & \Omega(N)/F(\Omega(M)) & \longrightarrow & 0 \\ {\scriptstyle \iota_M}\uparrow & & {\scriptstyle \iota_N}\uparrow & & \uparrow{\scriptstyle \iota} & & \\ M & \xrightarrow{f} & N & \xrightarrow{\pi_S} & \mathrm{Coker}_S(f) & \longrightarrow & 0 \end{array} \tag{1.3}$$

where π_R, π_S and ι_M, ι_N are as before, but now π_S is also a fragment morphism.

Proof. Since f is glider, we can start from a diagram

$$
\begin{array}{ccc}
\Omega(M) & \xrightarrow{F} & \Omega(N) \\
{\scriptstyle \iota_M}\uparrow & & \uparrow{\scriptstyle \iota_N} \\
M & \xrightarrow{f} & N
\end{array}
$$

As we have seen at the end of Section 1.4, $\mathrm{Ker}(f)$ is a strict subfragment of N. Since ι_N and ι_M are injective it is clear that $\mathrm{Ker}(f) \subset \mathrm{Ker}(F)$ and moreover $\mathrm{Ker}(F) \cap M = \mathrm{Ker}(f)$ (treating ι_M and ι_N as identifications). The fragment structure of M is induced from $\Omega(M)$ and the one on $\mathrm{Ker}(f)$ from M, hence the structure from $\mathrm{Ker}(f)$ is induced from $\Omega(M)$, hence also by the R-module structure of $\mathrm{Ker}(F)$ since $R\mathrm{Ker}(f) \subset \mathrm{Ker}(F)$. Consequently f has an (S, R)-kernel and we have the commutative diagram (1.2) with exact rows.
If f has an (S, R)-cokernel then we have a commutative diagram with exact rows like (1.3) but with π_S being a priori only an S-module map. By assumption, f is strict, so by the observations in Section 1.6 we can define a quotient fragment structure on $\mathrm{Coker}_S(f)$ by taking the chain $N_\gamma / f(M_\gamma)$ for $\gamma \in \Pi$. The claims in the proposition are now obvious. □

Corollary 1.7.2. *If M is an induced subfragment of N such that the inclusion morphism is glider and has an (S, R)-cokernel, then M is strict in N and we have a commutative diagram with short exact sequences of rows*

$$
\begin{array}{ccccccccc}
0 & \longrightarrow & \Omega(M) & \longrightarrow & \Omega(N) & \longrightarrow & \Omega(N)/\Omega(M) & \longrightarrow & 0 \\
 & & {\scriptstyle \iota_M}\uparrow & & {\scriptstyle \iota_N}\uparrow & & \uparrow{\scriptstyle \iota} & & \\
0 & \longrightarrow & M & \longrightarrow & N & \longrightarrow & M/N & \longrightarrow & 0
\end{array}
$$

Proof. The assumptions lead to a commutative diagram

$$
\begin{array}{ccccccccc}
 & & \Omega_1 & \xrightarrow{F} & \Omega_2 & \longrightarrow & \Omega_2/F(\Omega_1) & \longrightarrow & 0 \\
 & & {\scriptstyle \iota_1}\uparrow & & {\scriptstyle \iota_2}\uparrow & & \uparrow{\scriptstyle \iota} & & \\
0 & \longrightarrow & M & \xrightarrow{f} & N & \longrightarrow & {}_S(N/M) & \longrightarrow & 0
\end{array}
$$

where Ω_1, Ω_2 are R-modules as in the definition of a glider morphism, f is the inclusion morphism and F induces f. Observe that a priori ${}_S(N/M)$ is only an S-module. Since f, ι_1 and ι_2 are injective, we have that $F(\Omega_1) \cap N = M$, hence for every $\gamma \in \Pi$ it holds that $F_\gamma R(M \cap N_\gamma) \subset F(\Omega_1) \cap N = M$. We obtain that M is a strict subfragment by Lemma 1.6.1, which allows us to define a quotient fragment structure on N/M. Since $\mathrm{Ker}(F) \cap M = 0$, the R-linear morphism F factorizes as

$$\Omega_1 \xrightarrow{\pi_1} \Omega_1/\mathrm{Ker}(F) \overset{F}{\hookrightarrow} \Omega_2$$

and we have an inclusion $\pi_1\iota_1 : M \to \Omega_1/\mathrm{Ker}(F) \cong F(\Omega_1)$. Putting $\Omega(M) = F(\Omega_1)$, $\Omega(N) = \Omega_2, \iota_M = \pi_1\iota_1, \iota_2 = \iota_N$ leads to the commutative diagram as claimed in the corollary by Proposition 1.7.1. □

1.8 Inductive and projective limits of fragments

In this section we focus on inductive (or direct) and projective (or inverse) limits of systems of fragments. A partially ordered set $\mathcal{A}, \leq$ is inductive if for $\alpha, \beta \in \mathcal{A}$ there is a $\gamma \in \mathcal{A}$ such that $\gamma \leq \alpha$ and $\gamma \leq \beta$. The inverse partially ordered set $\mathcal{A}^{op}$ is obtained by reversing the relation $\leq$. A projective system is a partially ordered set such that its inverse is inductive. We know that the category S-mod of S-modules has both direct (inductive) and inverse (projective) limits; i.e. if we have a family of S-modules $\{N_\alpha,\ \alpha \in \mathcal{A}\}$ and $\mathcal{A}$ is inductive, together with S-linear maps

$$N_\alpha \xrightarrow{\rho_{\alpha\beta}} N_\beta$$

when $\beta \leq \alpha$, then $\varinjlim_\alpha N_\alpha$ exists in S-mod (dually, projective limits $\varprojlim_\alpha T_\alpha$ may be defined with respect to an inverse system) and it may be characterized by a universal property: if $N \in S$-mod is such that there exist given $f_\alpha : N_\alpha \to N$, for every $\alpha \in \mathcal{A}$, fitting into commutative diagrams for $\beta \leq \alpha$:

$$\begin{array}{ccc} N_\alpha & \xrightarrow{\quad f_\alpha \quad} & N \\ & {\scriptstyle \rho_{\alpha\beta}}\searrow \qquad \nearrow{\scriptstyle f_\beta} & \\ & N_\beta & \end{array}$$

then there exists an S-linear map $f : \varinjlim_\alpha N_\alpha \to N$ making the commutative diagrams

$$\begin{array}{ccc} & N_\alpha & \\ & {\scriptstyle \rho_\alpha}\swarrow \qquad \searrow{\scriptstyle f_\alpha} & \\ \varinjlim_\alpha N_\alpha & \xrightarrow{\quad f \quad} & N \end{array}$$

commute. Here, the ρ_α derive from the $\rho_{\alpha\beta}$, making the following diagrams commutative

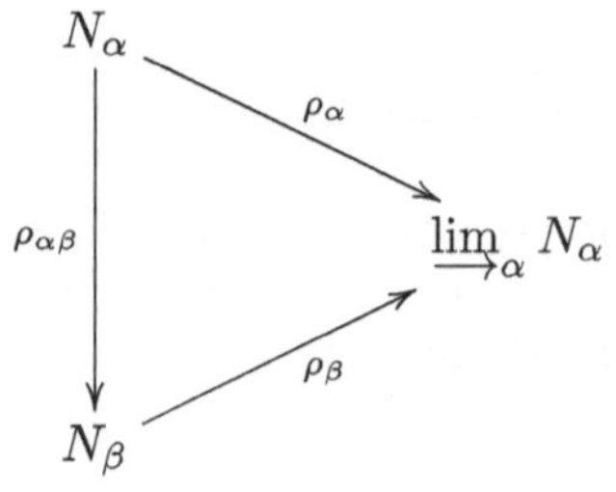

for $\beta \leq \alpha \in \mathcal{A}$. The construction (up to isomorphism) of $\varinjlim_\alpha N_\alpha$ is obtained by defining an equivalence relation $\sim$ on $\oplus_{\alpha\in\mathcal{A}} N_\alpha$ which says that $n_\alpha \in N_\alpha$ is equivalent to $n_\beta \in N_\beta$ whenever there is some $\gamma \leq \alpha, \beta$ such that $\rho_{\alpha\gamma}(n_\alpha) = \rho_{\beta\gamma}(n_\beta) \in N_\gamma$. It is not hard to check (and well-known) that $\oplus_{\alpha\in\mathcal{A}} N_\alpha / \sim$ satisfies the universal property mentioned above and therefore $\varinjlim_\alpha N_\alpha \cong \oplus_{\alpha\in\mathcal{A}} N_\alpha/\sim$ in S-mod. Dually, projective limits are defined and constructed as an S-submodule of the product $\Pi_{\alpha\in\mathcal{A}} N_\alpha$ consisting of sequences $(\ldots, n_\alpha, \ldots)_{\alpha\in\mathcal{A}}$ such that $\rho'_{\alpha\beta}(n_\beta) = n_\alpha$ for $\beta \leq \alpha$ (note that $\rho'_{\alpha\beta}$ goes from $N_\beta \to N_\alpha$ now).

We consider now an inductive system $\mathcal{A}, \leq$ and FR-fragments $\{M_\alpha,\ \alpha \in \mathcal{A}\}$ with morphisms of fragments $\rho_{\alpha\beta} : M_\alpha \to M_\beta$ for $\beta \leq \alpha$. Let $\underline{M}$ be the S-module $\varinjlim_\alpha M_\alpha$. We want to show that $\underline{M}$ has a unique, up to isomorphism, structure of an FR-fragment such that a universal property like the one mentioned above holds with respect to morphisms of fragments. On $\oplus_{\alpha\in\mathcal{A}} M_\alpha$ we have the direct sum structure as an FR-fragment, defined by $(\oplus_{\alpha\in\mathcal{A}} M_\alpha)_\gamma = \oplus_{\alpha\in\mathcal{A}} (M_\alpha)_\gamma$ for $\gamma \in \Pi$ (see Section 1.4). If we write K for the kernel of the canonical map $\oplus_\alpha M_\alpha \to \varinjlim_\alpha M_\alpha$ then we check that K can be made into a strict subfragment of $\oplus_\alpha M_\alpha$. From the remarks after Lemma 1.6.1, it suffices to verify that, for $\gamma \in \Pi$, we have that $F_\gamma R(K \cap \oplus_\alpha (M_\alpha)_\gamma) \subset K$. This follows easily from the observation that if $m_\alpha \in (M_\alpha)_\gamma, m_\beta \in (M_\beta)_\gamma$ are equivalent, i.e. $\rho_{\alpha\delta}(m_\alpha) = \rho_{\beta\delta}(m_\beta)$ in $(M_\delta)_\gamma$ for some $\delta \leq \alpha, \beta$ (note that we do end up in $((M_\delta)_\gamma)$ because the $\rho_{\alpha\beta}$ are morphisms of fragments), then for $r \in F_\gamma R$ we have that $rm_\alpha \in M_\alpha, rm_\beta \in M_\beta$ and $\rho_{\alpha\delta}(rm_\alpha) = r\rho_{\alpha\delta}(m_\alpha) = r\rho_{\beta\delta}(m_\beta) = \rho_{\beta\delta}(rm_\beta)$ by definition of fragment morphisms. Hence rm_α and rm_β are equivalent too, that is $r(m_\alpha - m_\beta) \in K$. Consequently, $(\oplus_\alpha M_\alpha)/K$ has the canonical quotient fragment structure and we arrive at $(\varinjlim_\alpha M_\alpha)_\gamma = \varinjlim_\alpha (M_\alpha)_\gamma$ as S-modules. The verification of the universal property with respect to FR-fragment morphisms $f_\alpha : M_\alpha \to M$ is now essentially the same as in the module case.

For projective limits one uses the product FR-fragment $\Pi_\alpha M_\alpha$ defined by $(\Pi_\alpha M_\alpha)_\gamma = \Pi_\alpha (M_\alpha)_\gamma$ for $\gamma \in \Pi$ and one verifies that the S-submodule given by $L = \varprojlim_\alpha M_\alpha$ obtains the structure of a strict subfragment of $\Pi_\alpha M_\alpha$, i.e. that

$$F_\gamma R(L \cap \Pi_\alpha (M_\alpha)_\gamma) \subset L$$

for every $\gamma \in \Pi$. This holds indeed, since if $(\ldots, m_\alpha, \ldots)_{\alpha \in \mathcal{A}}$ is such that $m_\alpha \in (M_\alpha)_\gamma$ for all $\alpha \in \mathcal{A}$ and $\rho'_{\alpha\beta}(m_\beta) = m_\alpha$ for $\beta \leq \alpha$, then for $r \in F_\gamma R$ all $rm_\alpha \in M_\alpha$ and $\rho'_{\alpha\beta}(rm_\beta) = rm_\alpha$ because the $\rho'_{\alpha\beta}$ are fragment morphisms. This shows that the sequence $r(\ldots, m_\alpha, \ldots)_{\alpha \in \mathcal{A}}$ is still in L as desired. The S-module L with this induced fragment structure (which is strict!) is the projective limit fragment for the system $(M_\alpha, \rho'_{\alpha\beta})$. We arrive at

Proposition 1.8.1. *The category FR-frag is an additive category with arbitrary products, inductive and projective limits.*

Observe that for any family $\{N_i,\ i \in J\}$ of subfragments of M, the S-module $\sum_{i \in J} N_i$ has a natural structure of a subfragment of M defined by $(\sum_{i \in J} N_i)_\gamma = \sum_{i \in J} (N_i)_\gamma$ for $\gamma \in \Pi$. Nevertheless, the sum need not be a strict subfragment of M even if all N_i are strict in M.

Example 1.8.2. Consider the positive ring filtration $S \subset R \subset R \subset \ldots$, where R is a Noetherian S-module. Let M be an S-module containing two R-modules M_1 and N such that $M_1 \cap N$ is simple as R-module but not simple as S-module. Let $N_1 \subsetneq M_1 \cap N$ be a non-trivial S-submodule. We have a subfragment $N \supset N_1 \supset 0 \supset \ldots$ of $M \supset M_1 \supset 0 \supset \ldots$ which is not strict. Indeed, $N \cap M_1 \supsetneq N_1$. However, we can write $N \supset N_1 \supset 0 \supset \ldots$ as the sum of strict subfragments (of $N \supset N_1 \supset 0 \supset \ldots$)

$$M_1 \cap N \supset N_1 \supset 0 \supset \ldots + \sum_i S n_i \supset 0 \supset \ldots,$$

for some $n_i \in N - M_1 \cap N$ (since R is Noetherian as S-module we can find such n_i).

For limits the situation is better:

Corollary 1.8.3. *Suppose that for all $\alpha \in \mathcal{A}, N_\alpha$ is a strict subfragment of M_α and the $\rho_{\alpha\beta} : M_\alpha \to M_\beta$, resp. $\rho'_{\alpha\beta} : M_\beta \to M_\alpha$, for $\beta \leq \alpha$ restrict to morphisms $\rho_{\alpha\beta}|N_\alpha : N_\alpha \to N_\beta$, resp. $\rho'_{\alpha\beta}|N_\beta : N_\beta \to N_\alpha$. Then $\varinjlim_\alpha N_\alpha$, resp. $\varprojlim_\alpha N_\alpha$, is a strict subfragment of $\varinjlim_\alpha M_\alpha$, resp. $\varprojlim_\alpha M_\alpha$.*

For the more particular situation of glider representations, one has to deal with limits of systems of gliders, meaning that the morphisms are glider morphisms, see Section 1.7. The construction of direct and inverse limits is, however, the same as for fragments and we obtain:

Corollary 1.8.4. *If $\Omega_\alpha \supset M_\alpha$ is a glider representation for $\alpha \in \mathcal{A}$ and $\rho_{\alpha\beta}$, resp. $\rho'_{\alpha\beta}$ are glider morphisms, then $\varinjlim_\alpha M_\alpha$, resp. $\varprojlim_\alpha M_\alpha$ is a glider representation. More precisely, if the $\rho_{\alpha\beta}$, resp. $\rho'_{\alpha\beta}$ are induced by $R_{\alpha\beta} : \Omega_\alpha \to \Omega_\beta$, resp. $R'_{\alpha\beta} : \Omega_\beta \to \Omega_\alpha$, then the fragment structure of $\varinjlim_\alpha M_\alpha$, resp. $\varprojlim_\alpha M_\alpha$ is induced from $\varinjlim_\alpha \Omega_\alpha$, resp. $\varprojlim_\alpha \Omega_\alpha$.*

Proof. The fact that $\varinjlim_\alpha M_\alpha$, resp. $\varprojlim_\alpha M_\alpha$ is glider with inducing $\varinjlim_\alpha \Omega_\alpha$, resp. $\varprojlim_\alpha \Omega_\alpha$ follows by the construction and the strictness of the subfragment in $\oplus_\alpha M_\alpha$, resp. $\Pi_\alpha M_\alpha$ describing $\varinjlim_\alpha M_\alpha$, resp. $\varprojlim_\alpha M_\alpha$. It can also be deduced from the universal properties already established. □

Limits of systems of natural gliders need probably not be natural (we have no counter-example though).

1.9 Graded and filtered fragments

In this section we introduce graded and filtered fragments, which will play an important role in Chapter 7.

Look at a positively filtered ring R with $F_0R = S$ and M an FR-fragment. Put $g(M) = \oplus_{i\geq 0}M_i/M_{i+1}$ as an S-module. Write $G_F(R) = G(R)$ where $G(R) = \oplus_n F_nR/F_{n-1}R$ is the associated graded ring. We consider on $G(R)$ the gradation filtration given by $f_nG(R) = \oplus_{p\leq n}G(R)_p$ and denote this by $fG(R)$.

Lemma 1.9.1. *If M is an FR-fragment then $g(M)$ is in a canonical way an $fG(R)$-fragment.*

Proof. Put $g(M)_p = \oplus_{i\geq p}M_i/M_{i+1}$. For $n \in \{0,\dots,p\}$ the action of F_jR with $j \leq n$ on M_i with $i \geq p \geq j$, satisfies $F_jRM_i \subset M_{i-j}$ and $F_{j-1}RM_i \subset M_{i-j+1}$. Hence that action induces in a canonical way an action of $F_jR/F_{j-1}R$ on M_i/M_{i+1} as follows: $\overline{r}_j \in G(R)_j, \overline{m}_i \in g(M)_i$ with representatives $r_j \in F_jR, m_i \in M_i$ satisfy $\overline{r}_j\overline{m}_i = r_jm_i \bmod M_{i-j+1}$; if $r_{j-1} \in F_{j-1}R, m_{i+1} \in M_{i+1}$ then $(r_j+r_{j-1})(m_i+m_{i+1}) \in r_jm_i+M_{i-j+1}$; therefore the action is well-defined. Now it is straightforward to verify that $f_nG(R)g(M)_p \subset g(M)_{p-n}$ for $n \leq p$. □

Note that $g(M)$ is not graded by the decomposition, unless M is a filtered R-module with respect to the chain reversed. Now let R be a positively graded ring $R = \oplus_{n\geq 0}R_n, R_0 = S$ and let fR be the gradation filtration given by $f_nR = \oplus_{i\leq n}R_i$.

Definition 1.9.2. An FR-fragment M is said to be a graded fragment if there exists $\{M_{(i)}, i \in \mathbb{N}\}$ of additive subgroups of M such that $M = \oplus_{i\geq 0}M_{(i)}$ and for i, j and $\nu \leq j$ we have $F_\nu RM_{i,j} \subset M_{i+\nu,j-\nu}$, where $M = M_0 \supset \dots \supset M_j \supset \dots$ is the chain defining the fragment structure of M and $M_{i,j} = M_{(i)} \cap M_j$.

It is clear that any graded R-module $X = M$ may be viewed as a graded fR-fragment by putting $M_j = M$ for every j and $M_{i,j} = M_{(i)}$, because obviously $R_\nu M_{i,j} = R_\nu M_{(i)} \subset M_{(i+\nu)} = M_{i+\nu,j-\nu}$. Now let R be filtered by FR with

$F_0R = S$. An exhaustive FR-filtration on M as an S-module is said to be an FR-fragment filtration if for $\nu \leq i$ we have: $F_\nu R F_j M_i \subset F_{\nu+j} R M_{i-\nu}$, where $F_j M_i = F_j M \cap M_i$ is the induced filtration on M_i.

Proposition 1.9.3. *Let M be an FR-fragment with FR-fragment filtration FM. Put $G_i(M)_j = F_i M_j / F_{i-1} M_j$. Then for all i: $G_i(M) = \cup_{j \in \mathbb{N}} F_i M_j / F_{i-1} M_j$; moreover $G_{FM}(M) = G(M)$ is a graded $fG(R)$-fragment.*

Proof. The first statement follows from $M = \cup_{i \in \mathbb{N}} F_i M$ and the exact commutativity of the following diagram of S-module morphisms:

$$\begin{array}{ccccccccc}
0 & \longrightarrow & F_{i-1}M & \longrightarrow & F_iM & \longrightarrow & F_iM/F_{i-1}M & \longrightarrow & 0 \\
 & & \uparrow & & \uparrow & & \uparrow & & \\
0 & \longrightarrow & F_{i-1}M \cap M_j & \longrightarrow & F_iM \cap M_j & \longrightarrow & F_iM_j/F_{i-1}M_j & \longrightarrow & 0
\end{array}$$

We can make $G(M)$ into an $fG(R)$-fragment by the chain: $G(M) \supset G(M)_1 \supset \ldots \supset G(M)_n \supset \ldots$, where $G(M)_n = \oplus_{j \geq n} G_i(M)$. We have to verify that $f_\nu G(R) G(M)_n \subset G(M)_{n-\nu}$ for $\nu \leq n$. A typical term in the first product would be $G(R)_i G_j(M)$ with $j \geq n \geq \nu \geq i$ and since $G(R)_i G_j(M) \subset G_{i+j}(M)$ with $i + j \geq i - j \geq n - \nu$, the statement is clear. Thus $G(M)$ is an $fG(R)$-fragment. We have that $G(M) = \oplus_i G_i(M)$ as additive groups, so we put $G(M)_{i,j} = G_i(M) \cap G(M)_j = G_i(M) \cap (\oplus_{k \geq j} G_k(M))$, thus $G(M)_{i,j} = 0$ if $i < j$ and if $i \geq j$ then $G(M)_{i,j} = G_i(M)$. Then $f_\nu G(R) G(M)_{i,j} \subset G(M)_{i+\nu, j-\nu}$ in both cases and thus it is clear that $G(M)$ is a graded fragment. □

In Section 2.1 we introduce finitely generated fragments, but we already give its definition here. We say that a fragment M is finitely generated if there exists a finite set $\{x_i, i = 1, \ldots, m\}$ such that $M_d = \sum_{i=1}^m F_{n_i - d} R x_i$, for some $n_i \in \mathbb{N}$. For $i_0 \in \{1, \ldots, m\}$, $M_{i_0} = \sum_{i=1}^m F_{n_i - n_{i_0}} R x_i$, i.e. $F_0 R x_{i_0} \subset M_{i_0}$ or $x_{i_0} \in M_{i_0}$ follows. So we have that $M_d = \sum_{i=1}^m F_{n_i - d} R x_i$ with $x_i \in M_{n_i}$ for all i.

Lemma 1.9.4. *If $\{x_1, \ldots, x_m\}$ with associated $n_i \in \mathbb{N}$ is a minimal set such that*

$$M_d = \sum_{i=1}^m F_{n_i - d} R x_i, \tag{1.4}$$

for all $d \in \mathbb{N}$ (i.e. no x_j can be deleted without destroying this property) then $x_i \in M_{n_i} - M_{n_i+1}$.

Proof. We have established above that $x_i \in M_{n_i}$. Suppose that $x_{i_0} \in M_{n_{i_0}+1}$. Then $M_{n_{i_0}+1} = \sum_{i=1}^m F_{n_i - n_{i_0} - 1} R x_i$ and the filtration FR being positive leads to

$$x_{i_0} \in \sum_{i \neq i_0}^m F_{n_i - n_{i_0} - 1} R x_i.$$

Write $M_d = \sum_{i \neq i_0} F_{n_i - d} R x_i + F_{n_{i_0} - d} R x_i$, then

$$\begin{aligned} F_{n_{i_0}-d} R x_{i_0} &\subset F_{n_{i_0}-d} R (\sum_{i \neq i_0} F_{n_i - n_{i_0} - 1} R x_i) \subset \sum_{i \neq i_0} F_{n_i - d - 1} R x_i \\ &\subset \sum_{i \neq i_0} F_{n_i - d} R x_i. \end{aligned}$$

Thus x_{i_0} may be deleted from $\{x_1, \ldots, x_m\}$ without destroying property (1.4) above, a contradiction. □

A fragment filtration FM is said to be a good fragment filtration if there exist $e_1, \ldots, e_m \in \mathbb{Z}$ such that for all $\nu \in \mathbb{Z}$ we have

$$F_\nu M = \sum_{j=1}^{m} F_{\nu - e_j} R x_j, \tag{1.5}$$

where $\{x_1, \ldots, x_m\}$ is a minimal set of fragment generators as in Lemma 1.9.4.

Lemma 1.9.5. *If $\{x_1, \ldots, x_m\}$ and $\{e_1, \ldots, e_m\}$ are as above then $x_i \in F_{e_i} M$. If $\{x_1, \ldots, x_m\}$ is minimal with respect to property (1.5) then $x_i \in F_{e_i} M - F_{e_i - 1} M$ for $i = 1, \ldots, m$.*

Proof. $F_{e_{i_0}} M = \sum_{i=1}^{m} F_{e_{i_0} - e_i} R x_i$ yields $F_0 R x_{i_0} \subset F_{e_{i_0}} M$. If $\{x_1, \ldots, x_m\}$ is minimal with respect to (1.5) then $x_{i_0} \in F_{e_{i_0} - 1} M$ yields $x_{i_0} \in \sum_{i=1}^{m} F_{e_{i_0} - 1 - e_i} R x_i$ and since FR is positive the x_{i_0} does not appear in the right hand member. Then $F_\nu M = \sum_{i \neq i_0} F_{\nu - e_i} R x_i + F_{\nu - e_{i_0}} R x_{i_0}$ with

$$\begin{aligned} F_{\nu - e_{i_0}} R x_{i_0} &\subset F_{\nu - e_{i_0}} R (\sum_{i \neq i_0} F_{e_{i_0} - 1 - e_i} R x_i) \\ &\subset \sum_{i \neq i_0} F_{\nu - 1 - e_i} R x_i \\ &\subset \sum_{i \neq i_0} F_{\nu - e_i} R x_i, \end{aligned}$$

contradicts the minimality of $\{x_1, \ldots, x_m\}$. □

Proposition 1.9.6. *For a good fragment filtration FM there exists a $\delta \in \mathbb{N}$, not depending on the chosen set of fragment generators of M describing the good fragment filtration, such that $F_\nu M \subset M_{\delta - \nu}$ for all ν; in other words the filtration is finer than the chain.*

Proof. From (1.5) and $F_{\nu - e_i} R x_i \subset M_{n_i + e_i - \nu}$ it follows that $F_\nu M \subset \cup_{i=1}^{m} M_{n_i + e_i - \nu}$. For some i_0 such that $n_{i_0} + e_{i_0}$ is the minimum of the $n_i + e_i$, say $\delta_{i_0}(x) = \delta_{i_0} = n_{i_0} + e_{i_0}$, we have $F_\nu M \subset M_{\delta_{i_0} - \nu}$. Now let $\{y_1, \ldots, y_t\}$ be another set of fragment generators with $y \in M_{l_j}$ and minimal with respect to (1.4); i.e. in view of Lemma 1.9.4, l_j is maximal such that $y_j \in M_{l_j}$, such that $F_\nu M = \sum_{j=1}^t F_{\nu-d_j} R y_j$. Again we have then that $y_j \in F_{d_j} M$ for all j. From the relation $F_\nu M \subset M_{\delta_{i_0}-\nu}$ we obtain that $y_j \in M_{\delta_{i_0}-d_j}$ for all j, hence $\delta_{i_0} - d_j \leq l_j$ for all j, or $\delta_{i_0} \leq d_j + l_j$ for all j. Thus $\delta_{i_0} \leq d_{j_0} + l_{j_0} = \delta_{j_0}(y) = \min\{d_j + l_j\}$. From $x_i \in F_{e_i} M$ it follows that $x_i \in M_{\delta_{j_0}(y)-e_i}$ for all i. Thus $\delta_{j_0}(y) - e_i \leq n_i$ for all i (see Lemma 1.9.4), or $\delta_{j_0}(y) \leq n_i + e_i$ for all i and $\delta_{j_0}(y) \leq \delta_{i_0}(x)$ follows. From $\delta_{i_0}(x) \leq d_j + l_j$ for all j the converse inequality $\delta_{i_0}(x) \leq \delta_{j_0}(y)$ follows, thus $\delta_{j_0}(y) = \delta_{i_0}(x) = \delta$ and $F_\nu M \subset M_{\delta-\nu}$ for all ν. □

For a good fragment filtration FM we have $F_\nu M = 0$ for $\nu < e_i$ for all i, so the filtration is necessarily discrete.

Proposition 1.9.7. *For a fragment filtration FM the completion $\hat{M} = \varprojlim_p M/F_p M$ is a fragment.*

Proof. In view of the results in Section 1.8 it will suffice to check whether the directed system

$$M/F_p M \xleftarrow[\pi_p]{} M/F_{p-1} M \longleftarrow \ldots,$$

is a directed system of fragment morphisms. Define $L(p)$ by: $M/F_p M = L_0 \supset L_1(p) = M_1/(F_p M \cap M_1) \supset \ldots \supset L_i(p) = M_i/(F_p M \cap M_i) \supset \ldots$, then $F_\nu R L_i \subset L_{i-\nu}$. It is clear that $\pi_p(L_i(p-1)) = L_i(p)$; hence the directed system satisfies the requirements mentioned in the beginning of Section 1.8, i.e.

$$\hat{M}_i = (\varprojlim_p M/F_p M)_i = \varprojlim_p M_i/(F_p M \cap M_i)$$

defines a fragment structure on $\hat{M}$. □

The technique of passing to the Rees ring of a filtration also works on the level of fragment filtrations.

Proposition 1.9.8. *Let FM be a fragment filtration; the Rees S-module $\widetilde{M} = \oplus_{i\in\mathbb{N}} F_i M \cong \sum_{i\in\mathbb{N}} F_i M X^i \subset M[X]$ has a natural graded fragment structure.*

Proof. Put $\widetilde{M}_j = \sum_{i\geq j} F_i M X^i$ and $\tilde{f}\widetilde{R}$ on $\widetilde{R} = \oplus_{n\in\mathbb{N}} F_n R \cong \sum_{n\in\mathbb{N}} F_n R X^n \subset R[X]$, by putting $\tilde{f}_\nu \widetilde{R} = \sum_{i\leq\nu} F_i R X^i$. A straightforward verification learns that

$$\tilde{f}_\nu \widetilde{R} \widetilde{M}_j \subset \widetilde{M}_{j-\nu},$$

thus $\widetilde{M}$ is an $\tilde{f}\widetilde{R}$-fragment. Furthermore $\widetilde{M}_{i,j} = F_i M \cap \widetilde{M}_j$ is zero if $i < j$ and equals $F_i M$ if $i \geq j$. So if $i < j$ then $\widetilde{R}_i \widetilde{M}_{i,j} = 0 \subset \widetilde{M}_{i+\nu,\delta-\nu}$; on the other hand if $i \geq j$ then $i+\nu \geq j-\nu$ and $\widetilde{R}_\nu \widetilde{M}_{i,j} = F_\nu R F_i M \subset F_{i+\nu} M = \widetilde{M}_{i+\nu,j-\nu}$. Consequently $\widetilde{M}$ is a graded $\tilde{f}\widetilde{R}$-fragment. □

Corollary 1.9.9. *Since X is central in $\widetilde{R}$, $\widetilde{M}/X\widetilde{M} \cong G_F(M)$ as graded $fG(R)$-fragments and $\widetilde{M}/(1-X)\widetilde{M} \cong M$ as filtered fragments.*

Proof. $X\widetilde{M}$ is strict, so $\widetilde{M}/X\widetilde{M}$ is an $f\widetilde{R}$-fragment and also an $fG(R)$-fragment. If $i \geq j$ we have:

$$(\widetilde{M}/X\widetilde{M})_{i,j} = \frac{F_iMX^i}{F_{i-1}MX^i} \cong G_i(M) = G(M)_{i,j},$$

and if $i < j$ then both $(\widetilde{M}/X\widetilde{M})_{i,j}$ and $G(M)_{i,j}$ are zero. Moreover, it is clear that the fragment structures of $fG(R)$ are compatible under this isomorphism as X is central.
Since $(1 - X)$ is central, $(1 - X)\widetilde{M}$ is also strict and

$$F_i(\widetilde{M}/(1 - X)\widetilde{M})_j \cong \frac{\sum_{k\leq i} F_kM_jX^k}{\sum_{k\leq i}(1 - X)F_kM_jX^k} \cong F_iM_j.$$

□

If M is an FR-fragment and X a variable over R then on $R[X]$ there is a filtration $FR[X]$ given by $F_n(R[X]) = \sum_{i\leq n} F_{n-i}RX^i$. The sequence on $M[X] = M \otimes_{\mathbb{R}} \mathbb{Z}[X]$ given by the chain $M_i[X]$ satisfies $F_n(R[X])M_i[X] \subset M_{i-n}[X]$, for $n \leq i$. So $M[X]$ is in a canonical way a graded fragment. This is the starting point of the change of rings problems, e.g. replacing S by a localization, integral closure, affine over rings etc... and studying the relations on fragments over the new rings with (induced) fragments on the original ones, or looking at specific fragments relating the different rings.

Chapter 2

Basic properties

2.1 Generating sets for fragments

From now on we look at totally ordered groups Γ which are Abelian. We again denote the positive cone by Π. Consider an R-fragment M with chain given by $M = M_0 \supset \ldots \supset M_\gamma \supset \ldots$ (since Γ is assumed to be Abelian, we write 0 for the neutral element). Certain subfragments of M may be constructed in a trivial way so that we may not wish to consider them as meaningful subfragments. Therefore we identify a class of neglectable subfragments in the following. A subfragment N of M is said to be trivial if either:

T_1. There is a $\gamma \in \Pi$ such that $N_\gamma = B(N)$ but $M_\gamma \neq B(M)$.
T_2. There is a $\gamma \in \Pi$ such that $N_\gamma = 0$ but $M_\gamma \neq 0$.
T_3. There exists a monotone increasing map $\alpha : \Pi \to \Pi$ such that $N_\gamma = M_{\alpha(\gamma)}$ and $\alpha(\gamma) - \tau \geq \alpha(\gamma - \tau)$ for all $\tau \leq \gamma$ in Π.

For fragments M such that $B(M) = 0$ it is obvious that T_1 and T_2 are equivalent. In that case we say that N is trivial of type I if T_2 holds and trivial of type II when T_3 holds. We will also say that N is trivial of type T_i if T_i holds for $1 \leq i \leq 3$. A fragment M is said to be irreducible if all of its subfragments are trivial; it is weakly irreducible if all strict subfragments are trivial.

Example 2.1.1. If the fragment $F^e S$ with regard to the positive filtration $F^+ R$ is irreducible, then FR is a strong filtration.

The fragment M is called separated if for $m \in M$ there exists a $\gamma \in \Pi$ such that $m \in M_\gamma$ and $m \notin M_\tau$ for all $\tau > \gamma$, unless $m \in B(M)$. For a separated fragment M we define $\delta : M \to \Pi \cup \{\infty\}$ by putting: $\delta(m) = \mu$ if $m \in M_\mu$ but $m \notin M_\tau$ for all $\tau > \mu$, and $\delta(m) = \infty$ if $m \in B(M)$. Note that $\delta(m)$ is indeed uniquely defined. We call $\delta(m)$ the degree of m in M. If a monotone increasing map $\alpha : \Pi \to \Pi$ is given, then we may look at a trivial subfragment M^α of M, where M^α is given the chain $M \supset \ldots \supset M_{\alpha(\gamma)} \supset \ldots$. Note that $F_\tau RM_{\alpha(\gamma)}$ is contained in $M_{\alpha(\gamma)-\tau}$ hence in $M_{\alpha(\gamma-\tau)} = M^\alpha_{\gamma-\tau}$ by the condition in T_3. A typical example of such an α is a translation by $\tau \in \Pi$, i.e. $\alpha(\gamma) = \gamma + \tau$ and the corresponding trivial subfragment is then just $M_\tau \supset \ldots \supset M_{\gamma+\tau} \supset \ldots$. Other examples arise when α is the restriction to Π of a group morphism of Γ mapping Π to Π. Also, any montone increasing semigroup morphism $\alpha : \Pi \to \Pi$ satisfies. Indeed, for $\tau \leq \gamma : \alpha(\gamma - \tau) + \alpha(\tau) = \alpha(\gamma)$ thus $\alpha(\gamma) - \tau \geq \alpha(\gamma) - \alpha(\tau)$ and hence $\alpha(\gamma) - \tau \geq \alpha(\gamma - \tau)$.

In case $\Gamma = \mathbb{Z}$ with cone $\Pi = \mathbb{N}$, every fragment M is separated and $\delta : M \to \mathbb{N} \cup \infty$ is well-defined. In this case we introduce the notation $\dot{M}_n = M_n \setminus M_{n+1}$. In other words, $m \in \dot{M}_{\delta(m)}$. An FR-fragment M with FR a $\mathbb{Z}$-filtration is said to have essential length $d \in \mathbb{N}$ if d is such that $M_d \neq B(M)$ but $M_{d+1} = B(M)$. We write el(M) = d. If such a $d \in \mathbb{N}$ does not exist, we write el(M) = ∞. For α as in T_3 and the associated trivial subfragment M^α we write $\delta^\alpha(m)$ for the degree of m in the fragment structure of M^α (assuming $\alpha(0) = 0$ for convenience). Observe that in this case M^α is separated too, which does not hold for arbitrary Γ. Then $B(M^\alpha) = B(M)$ and $\delta^\alpha(m) = \infty$ if and only if $\delta(m) = \infty$.

Now look at arbitrary Γ-filtrations again.

Definition 2.1.2. A fragment M is said to be generated by a set $\{m_\beta,\ \beta \in B\}$ if there exists an associated set $\{\delta_\beta \in \Gamma,\ \beta \in B\}$ such that for all $\gamma \in \Pi$:

$$M_\gamma = \sum_{\beta \in B} F_{\delta_\beta - \gamma} R m_\beta. \tag{2.1}$$

If B is finite then we say that M is finitely generated. If we restrict to $\delta_\beta - \gamma \geq e$ in (2.1), then we say that M is positively generated (this means that we may view it over the positive filtration F^+R).

Definition 2.1.3. A generated fragment M is said to be freely generated, or M is said to be a free fragment with basis $\{m_\beta,\ \beta \in B\}$ if it is generated by $\{m_\beta,\ \beta \in B\}$ with associated set $\{\delta_\beta \in \Gamma,\ \beta \in B\}$ and a relation $\sum'_{\beta \in B} r_{\delta_\beta} m_\beta = 0$ with $r_{\delta_\beta} \in F_{\delta_\beta} R$ is necessarily trivial, i.e. $r_{\delta_\beta} = 0\ \forall \beta \in B$. If M is freely generated with respect to a finite B then we say that M is free of finite rank.

The rank of a free R-module M is only unambiguously defined when R has the IBN (invariant base number) property. For fragments in particular, it is not even clear whether the rank of a free fragment is well-defined even if R is an IBN ring. In fact, even for glider representations, this has yet to be studied in detail. Moreover, for a freely generating set $\{m_\beta,\ \beta \in B\}$, there may be relations $\sum'_{\beta \in B} s_{\delta_\beta} m_\beta = 0$ with $s_\beta \in R$ and $s_\beta m_\beta$ defined, but $s_\beta \notin F_{\delta_\beta} R$. To give examples of this behavior, take some Γ-filtered ring FR. Then for any $\gamma \in \Gamma$ we may view $F_\gamma R$ as a fragment by putting $(F_\gamma R)_\tau = F_{\gamma - \tau} R$ for $\tau \in \Pi$. Note that $F_\gamma R$ need not be freely generated but this does hold if $\gamma \geq e$ because then $F_\gamma R$ is freely generated by $1 \in R$ (we usually assume that $1 \notin F_\tau R$ for $\tau < e$). A fragment M is pseudo-free if it is isomorphic to the fragment $\oplus_{\beta \in B} F_\beta R$, $F_\beta R$ having the fragment structure as above. Observe that we defined the two different but related notions of free and pseudo-free fragments. This is exactly because we do not have an unambiguously defined rank.

Proposition 2.1.4. *Let M be a separated fragment. There is a freely generated fragment E and a strict epimorphism of fragments $\pi : E \twoheadrightarrow M$.*

Proof. Since any fragment is generated by all of its elements, we can take $\{m_\beta, \beta \in B\}$ a generating set for M and suppose $\delta(m_\beta) = \delta_\beta$. Look at the free R-module generated by $\{\overline{m_\beta}, \beta \in B\}$ where $\overline{m_\beta}$ is a symbol representing m_β, $\beta \in B$. We define a pseudo-free fragment contained in $\oplus_\beta R\overline{m_\beta}$ by putting $E = \oplus_\beta F_{\delta_\beta} R\overline{m_\beta}$ with fragment structure given by $E_\gamma = \oplus_\beta F_{\delta_\beta - \gamma} R\overline{m_\beta}$. The map $E \twoheadrightarrow M$ which sends $\overline{m_\beta}$ to m_β is a fragment morphism and it is obviously strict and an epimorphism. Since all $\delta_\beta \geq 0$, the $F_{\delta_\beta} R\overline{m_\beta}$ are all freely generated, so E is also freely generated. □

For a ring R we denote the Jacobson radical of R by $J(R)$.

Lemma 2.1.5. *If $F_\tau R \subset J(F_0 R)$ for every $\tau < 0$ and M is a free fragment with basis $\{m_\beta, \beta \in B\}$ and corresponding set $\{\delta_\beta \in \Pi,\ \beta \in B\}$, then $\delta(m_\beta) = \delta_\beta$, i.e. $m_\beta \in M_{\delta_\beta} - M_{\delta_\beta + \eta}$ for every $\eta \in \Pi$.*

Proof. Since $1 \in F_0 R$, it is clear that $m_\beta \in M_{\delta_\beta}$. If $m_\beta \in M_{\delta_\beta + \eta}$, then we have

$$m_\beta \in \sum_{\gamma \in B} F_{\delta_\gamma - \eta - \delta_\beta} R m_\gamma.$$

In the right hand side of this expression, m_β appears with coefficient $c_\beta \in F_{-\eta} R$; hence $1 - c_\beta$ is a unit of $F_0 R$, leading to $m_\beta = \sum_{\gamma \neq \beta} d_\gamma m_\gamma$, with $d_\gamma \in F_{\delta_\gamma - \eta - n_\beta} R \subset F_{\delta_\gamma} R$, which contradicts the freeness of M with regard to $\{m_\gamma,\ \gamma \in B\}$. □

The condition on FR in the foregoing lemma holds for example if $\Gamma = \mathbb{Z}$ and FR is a Zariskian filtration or in case FR is a positive filtration. Zariskian filtrations will appear a lot in Chapter 7 so we recall its definition here: a filtration FR is a (left) Zariskian filtration if the Rees ring $\widetilde{R}$ associated to FR is (left) Noetherian and $F_{-1} R$ is contained in the Jacobson radical $J(F_0 R)$ of $F_0 R$.

Let us mention a useful lemma in case $\Gamma = \mathbb{Z}$.

Lemma 2.1.6. *Consider a fragment M and α as in T_3; let δ and δ^α be the degree functions corresponding to M, resp. M^α. Then*

1. *$\delta^\alpha(m)$ is the largest natural number such that $\alpha(\delta^\alpha(m)) \leq \delta(m)$;*
2. *For $\alpha(j) \leq \delta(m)$ we have $\delta(m) - \delta^\alpha(m) \geq \alpha(j) - j$.*

Proof. 1. By definition we have $M^\alpha_{\delta^\alpha(m)} = M_{\alpha(\delta^\alpha(m))}$ and the latter can only contain m if we have $\alpha(\delta^\alpha(m)) \leq \delta(m)$.

2. It is clear that $\delta(m) - \delta^\alpha(m) \geq \alpha(\delta^\alpha(m)) - \delta^\alpha(m)$. Now for any $j \in \mathbb{N}$ the number $\alpha(j) - j$ counts the number of deletions in the M-sequence, and it follows that $\alpha(j) - j \geq \alpha(i) - i$ for $j \geq i$. The assumption $\alpha(j) \leq \delta(m)$ entails that $j \leq \delta^\alpha(m)$ by the definition of δ^α. Therefore we obtain that $\alpha(\delta^\alpha(m)) - \delta^\alpha(m) \geq \alpha(j) - j$. Combined with the result from 1. we obtain for every j such that $\alpha(j) \leq \delta^\alpha(m)$ the desired inequality $\delta(m) - \delta^\alpha(m) \geq \alpha(j) - j$. □

An FR-fragment M is pseudo-(finitely) generated if there exists a set $\{m_\beta,\ \beta \in B\}$ together with $\delta_\beta \in \mathbb{N}$, such that for all $n \in \mathbb{N}, M_n = \sum'_\beta F_{\delta_\beta - \alpha(n)} R m_\beta$, for some monotone ascending $\alpha : \mathbb{N} \to \mathbb{N}$.

Corollary 2.1.7. *If α is as in T_3 and M is generated by $\{m_\beta,\ \beta \in B\}$ with respect to $\{\delta^\alpha_\beta,\ \beta \in B\}$ then M^α is as a fragment pseudo-generated with respect to $\{\delta_\beta,\ \beta \in B\}$.*

Proof. For $\alpha(j) \leq \delta(m)$ we have $\delta_\beta - \delta^\alpha_\beta \geq \alpha(j) - j$, where δ^α_β is maximal such that $\alpha(\delta^\alpha_\beta) \leq \delta_\beta$ (see Lemma 2.1.6). The claim now follows in a straightforward way. □

It is clear that problems with respect to "generating" fragments will be the basis for some phenomena appearing in a theory of "Noetherian" fragments. This will be dealt with in Section 2.3.

2.2 Projective fragments

Consider a sequence of fragment morphisms

$$L \xrightarrow{g} M \xrightarrow{f} N.$$

If the sequence is exact as a sequence of S-morphisms then $\mathrm{Im}(g) = \mathrm{Ker}(f)$ as S-modules; since $\mathrm{Ker}(f)$ is a strict subfragment the sequence will be exact as a fragment sequence only when $\mathrm{Im}(g)$ is a strict subfragment, i.e. when g is strict. Hence for a long exact sequence of fragment morphisms

$$0 \longrightarrow L \xrightarrow{g} M \xrightarrow{f} N \longrightarrow 0,$$

to be exact it is necessary and sufficient to have exactness as S-module morphisms plus f and g being strict morphisms.

Definition 2.2.1. Let $f : M \to N$ be a fragment morphism. If $m \in M$ is such that $f(m) \in N_\gamma$ entails that $m \in M_\gamma$, we say that f is degree reflecting.

For a degree reflecting fragment morphism $f : M \to N$ it follows directly that $\mathrm{Ker}(f) \subset B(M)$. In case $\Gamma = \mathbb{Z}$ we have the following proposition

Proposition 2.2.2. *Let $M \supset \ldots \supset M_n \supset \ldots$ and $N \supset \ldots \supset N_n \supset \ldots$ be fragments and $\pi : M \to N$ a strict fragment epimorphism. If M_i/M_{i+1} is a simple S-module for all $i \geq 0$, then π is degree reflecting.*

Proof. Take $m \in M$ such that $f(m) \in N_i$, but $\delta(m) < i$. Since $M_{\delta(m)+1} + Sm = M_{\delta(m)}$, we obtain $f(M_{\delta(m)}) \subset N_{\delta_m+1}$, which contradicts with π being strict epimorphic. □

Let $f : M \to N$ be a fragment morphism, and $L \subset N$ a subfragment. The chain $f^{-1}(L) \supset \ldots \supset f^{-1}(L_\gamma) \supset \ldots$ with partial actions φ_γ being the restrictions from the partial actions from M does not define a fragment structure in general. Indeed, it could be that $x \in f^{-1}(L_\gamma)$, but $x \notin M_\gamma$. Hence for some $r_\gamma \in F_\gamma R$, we do not know whether $r_\gamma f(m)$ equals $f(r_\gamma.m)$, the latter not necessarily defined! When speaking of the fragment $f^{-1}(L)$, we mean the induced fragment structure as defined in Section 1.6, i.e. $(f^{-1}(L))_\gamma = \{m \in M_\gamma \mid F_\gamma Rm \subset f^{-1}(L)\}$.

Lemma 2.2.3. *Let $f : M \to N$ be a fragment morphism.*

1. *If $L \subset N$ is a strict subfragment then $f^{-1}L \subset M$ with induced fragment structure is a strict subfragment of M. Moreover, if f is degree reflecting, then $f^{-1}(L)_\gamma = f^{-1}(L_\gamma)$.;*

2. *If $M \xrightarrow{f} N \xrightarrow{g} L$ is a sequence of fragment morphisms then $\mathrm{Ker}(gf) = f^{-1}(\mathrm{Ker}(g))$ as fragments (and these are strict subfragments);*

3. *If $g : N \to L$ is strict and $K \subset N$ a strict subfragment then if g is monomorphic, K is strict in L;*

4. *Let $f : M \to N$ be a fragment morphism and $g : M \to L$ a strict fragment epimorphism with $\mathrm{Ker}(g)$ a subfragment of $\mathrm{Ker}(f)$. There is a fragment morphism $h : L \to N$ making the diagram*

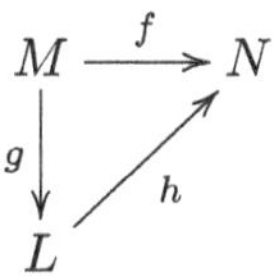

commutative, i.e. $f = hg, \mathrm{Im}(h) = \mathrm{Im}(f)$ and h is monomorphic if and only if $\mathrm{Ker}(g) = \mathrm{Ker}(f)$.

Proof. 1. Look at $f^{-1}(L) \cap M_\gamma$ and pick an x in it. Then $F_\gamma Rx \in M$ and $f(F_\gamma Rx) = F_\gamma Rf(x)$ with $f(x) \in L \cap N_\gamma = L_\gamma$ hence $f(F_\gamma Rx) \subset L$ or $F_\gamma Rx \subset f^{-1}(L)$. By Lemma 1.6.1 it follows that $f^{-1}(L)$ is strict. Assume that f is degree reflecting. Then

$$\begin{aligned} f^{-1}(L_\gamma) &= \{m \in M \mid f(m) \in L_\gamma\} = \{m \in M_\gamma \mid f(m) \in L_\gamma\} \\ &= \{m \in M_\gamma \mid F_\gamma Rm \subset f^{-1}(L)\} = f^{-1}(L)_\gamma. \end{aligned}$$

2. Since the kernel of a fragment morphism is a strict subfragment, it follows from 1. and the fact that $\mathrm{Ker}(gf) = f^{-1}(\mathrm{Ker}(g))$ as S-modules that $\mathrm{Ker}(fg) = f^{-1}(\mathrm{Ker}(g))$ as fragments.
3. Let us consider $K \subset N \subset L$, identifying N with $g(N)$ in L. Then $K \cap L_\gamma = K \cap N \cap L_\gamma = K \cap N_\gamma = K_\gamma$ for all $\gamma \in \Pi$ by the strictness assumptions.
4. We have that $L = M/\mathrm{Ker}(g)$ is the cokernel of the strict fragment morphism $\mathrm{Ker}(g) \to M$. All claims follow easily from this by defining h as follows: for $x \in L$ there is an $m \in M$ such that $g(m) = x$, then put $h(x) = f(m)$. □

Definition 2.2.4. An FR-fragment P is projective if for every strict fragment epimorphism $M \xrightarrow{\pi} N$ and given fragment morphism $f : P \to N$, there exists a fragment morphism $h : P \to M$ making the following diagram of fragment morphisms commutative:

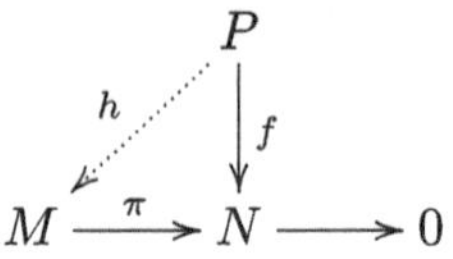

Free fragments are projective as is easily verified. Also, if M is a projective S-module, then $M \supset 0 \supset \ldots$ is a projective fragment. We have the same equivalent characterization for projective modules as being a direct summand of a free module.

Proposition 2.2.5. *A fragment P is projective if and only if it is a direct summand fragment (strict) of a freely generated fragment E.*

Proof. We have seen that there exists a freely generated fragment E and a strict fragment epimorphism $\pi : E \to P$. If P is projective then there exists a map $h : P \to E$ which makes the following diagram commutative:

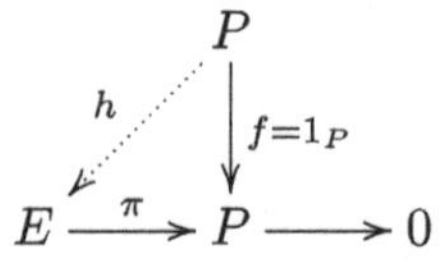

Hence $\pi h = I_P$ and $\mathrm{Ker}(h) = 0$ or h is monomorphic. Therefore we have that $h(P) \cap \mathrm{Ker}(\pi) = 0$, and $E = \mathrm{Ker}(\pi) \oplus h(P)$ as S-modules. Now look at $E_\gamma \cap h(P)$, then $\pi(E_\gamma \cap h(P)) = P_\gamma$ since π is strict. We see that $h(P_\gamma) = h(P)_\gamma$ and $E_\gamma \cap h(P)$ have the same image under π. From $\mathrm{Ker}(\pi) \cap h(P) = 0$ it then follows that $h(P) \cap E_\gamma = h(P)_\gamma$ or $h(P)$ is strict in E. We obtain that $E_\gamma \supset h(P)_\gamma \oplus (\mathrm{Ker}(\pi))_\gamma$. If $y \in E_\gamma$ then $\pi(y) \in P_\gamma$ and $h\pi(y) \in h(P)_\gamma$ is such that $y - h\pi(y) \in \mathrm{Ker}(\pi)$, thus $y - h\pi(y) \in \mathrm{Ker}(\pi) \cap E_\gamma = \mathrm{Ker}(\pi)_\gamma$. Consequently $y \in h(P)_\gamma + \mathrm{Ker}(\pi)_\gamma$, or $E_\gamma = h(P)_\gamma \oplus \mathrm{Ker}(\pi)_\gamma$.

Conversely, suppose that $P \oplus P' = E$ where P' is a subfragment of E and $E_\gamma = P_\gamma \oplus P'_\gamma$. Consider a diagram of fragment morphisms:

$$\begin{array}{ccccc} & & P & & \\ & & \big\downarrow{\scriptstyle f} & & \\ M & \xrightarrow{\ \pi\ } & N & \longrightarrow & 0 \end{array}$$

where π is a strict fragment epimorphism. By definition we have a strict exact sequence

$$0 \longrightarrow P \underset{\pi_P}{\rightleftarrows} E \longrightarrow P' \longrightarrow 0$$

with $\pi_P : E \to P$ the canonical projection which is a strict fragment epimorphism. We obtain

$$\begin{array}{ccccc} E = P \oplus P' & \underset{\pi_P}{\overset{\iota}{\rightleftarrows}} & P & & \\ {\scriptstyle h}\,\vdots & \searrow^{f\pi_P} & \big\downarrow{\scriptstyle f} & & \\ M & \xrightarrow[\pi]{} & N & \longrightarrow & 0 \end{array}$$

because E is projective. Then $h \circ \iota$ is a fragment morphism from P to M such that $\pi h \iota = f \pi_P \iota = f$. □

Corollary 2.2.6. *If a fragment $M \supset \ldots \supset M_\gamma \supset \ldots$ is projective and if $M_\gamma \neq 0$ then the fragment $M_\gamma \supset \ldots \supset M_{\gamma+\tau} \supset \ldots$ is also projective.*

Proof. From the proof of Proposition 2.1.4 and Proposition 2.2.5 we know that there is some fragment M' such that $M \oplus M' = \oplus_{b \in B} F_{\delta_b} R$, where the sum is direct on every level. If $M_\gamma \neq 0$, we have that

$$M_\gamma = \sum_{\beta \in B} F_{\delta_\beta - \gamma} R m_\beta \neq 0.$$

In particular, there is some $b \in B$ such that $F_{\delta_b - \gamma} R \neq 0$. Hence $M_\gamma \oplus M'_\gamma = \oplus_{b \in B} F_{\delta_b - \gamma} R$ and M_γ is projective. □

Corollary 2.2.7. *Let FR be a positive, right bounded filtration, say $F_n R = R$, and $M \supset M_1 \supset \ldots$ a projective FR-fragment of finite essential length and with non-zero body $B(M) \neq 0$. Then $B(M)$ is a projective R-module.*

Proof. Corollary 2.2.6 entails that $B(M) \supset B(M) \supset \ldots$ is a projective fragment, where $B(M)$ is an R-module. Let N and Q be R-modules and consider the diagram

$$\begin{array}{ccccc} & & B(M) & & \\ & & \big\downarrow{\scriptstyle f} & & \\ Q & \xrightarrow{\ \pi\ } & N & \longrightarrow & 0 \end{array}$$

in R-mod. We can associate to it the following diagram of FR-fragments

$$\begin{array}{ccccc} & & B(M) \supset B(M) \supset \dots & & \\ & & \big\downarrow f & & \\ Q \supset Q \supset \dots & \xrightarrow{\pi} & N \supset N \supset \dots & \longrightarrow & 0 \end{array}$$

Obviously, π is a strict fragment epimorphism; hence there exists a fragment morphism $h : B(M) \to Q$ which makes the second diagram commute. Since FR is right bounded and both $B(M)$ and Q are R-modules, h is R-linear, which shows that $B(M)$ is indeed projective as R-module. □

Example 2.2.8. Let A be a k-algebra with a simple non-projective module Ω which is not 1-dimensional. Then $\Omega \supset k\omega \supset 0 \supset \dots$ is a fragment with regard to the filtration $k \subset A \subset A \subset \dots$. As a fragment $k\omega \supset 0 \supset \dots$ is obviously projective. However, $A\omega = \Omega$ is not projective.

One could also ask oneself whether projectivity of $\Omega = RM$ implies the projectivity of the fragment lying below. However, there are multiple fragments lying below RM, so this does not seem plausible. We observed above that fragments $M \supset 0 \supset \dots$ with M projective as S-module are projective. The chain of the fragment does play an important role since we have:

Example 2.2.9. For $\gamma > e$ the fragment structure on $F_\gamma R$ given by $(F_\gamma R)_\tau = F_{\gamma-\tau} R$ is free, hence projective. In general however, $F_\gamma R$ is not projective as S-module. Consider for example the ring filtration $\mathbb{Z} \subset \mathbb{Q}$.

Remark 2.2.10. One can also define the notion of a projective glider representation via the lifting property of glider morphisms and relate projectivity as a glider and projectivity as a fragment.

As another corollary of Proposition 2.2.5 we may introduce the notion of a dual basis in a fragment adapted way. Recall that we consider $F_\gamma R$ as a fragment with $(F_\gamma R)_\tau = F_{\gamma-\tau} R$ for $\tau \in \Pi$. A dual basis for a fragment M is a set of pairs $\{(m_i, \varphi_i), i \in J\}$ where $m_i \in M_{\gamma_i}$ and $\varphi_i \in \mathrm{Hom}_{FR}(M, F_{\gamma_i} R)$ such that for each $m \in M$, $\varphi_i(m) = 0$ for all but a finite number of i in J and $m = \sum_i \varphi_i(m) m_i$.

Proposition 2.2.11. *Let M be a fragment; then M is projective if and only if to every generating set $\{m_\beta, \beta \in B\}$ with associated set $\{\delta_\beta \in \Gamma, \beta \in B\}$ there exists a set $\{\varphi_\beta \in \mathrm{Hom}_{FR}(M, F_{\delta_\beta} R), \beta \in B\}$ such that $\{(m_\beta, \varphi_\beta), \beta \in B\}$ is a dual basis of M.*

Proof. Assume that M is projective and $\{m_\beta, \beta \in B\}$ is a generating set with associated set $\{\delta_\beta \in \Gamma, \beta \in B\}$. By the proof of Proposition 2.1.4 we have a freely generated fragment E with basis $\{x_\beta, \beta \in B\}$ and associated set $\{\delta_\beta, \beta \in B\}$ such that we have a strict fragment epimorphism $\pi : E \to M, \pi(x_\beta) = m_\beta$ for all $\beta \in B$. Define for all $\beta \in B, \pi_\beta : E \to F_{\delta_\beta} R$ as the fragment epimorphism defined by $\pi_\beta(e) = r_\beta$ if $e = \sum_\beta r_\beta x_\beta$ in E. Since every $e \in E$ has a unique expression as in the foregoing, the π_β are well-defined. Since M is projective there exists a fragment morphism $g : M \to E$ such that $1_M = \pi g$. Put $\varphi_\beta = \pi_\beta g$ for every $\beta \in B$. This is a fragment morphism from M to $F_{\delta_\beta} R$. For $m \in M$ we have that $\varphi_\beta(m) = \pi_\beta g(m)$ and it is clear that $\varphi_\beta(m) \neq 0$ only for a finite number of $\beta \in B$. We calculate

$$m = \pi g(m) = \pi(\sideset{}{'}\sum_\beta \pi_\beta(g(m))x_\beta) = \sideset{}{'}\sum_\beta \pi_\beta(g(m))\pi(x_\beta) = \sideset{}{'}\sum_\beta \varphi_\beta(m)m_\beta.$$

Conversely, assume that there exists a dual basis $\{(m_\beta, \varphi_\beta), \beta \in B\}$ for the generated fragment M and look at the strict fragment epimorphism $\pi : E \to M$ defined by $\pi(x_\beta) = m_\beta$ for $\beta \in B$ as before. We define $g : M \to E$ by $g(m) = \sum_\beta \varphi_\beta(m)x_\beta$. Observe that the latter sum is finite. Since the φ_β are fragment morphisms, the map g is also a fragment morphism. We calculate

$$\pi g(m) = \pi(\sideset{}{'}\sum_\beta \varphi_\beta(m)x_\beta) = \sideset{}{'}\sum_\beta \varphi_\beta(m)\pi(x_\beta) = \sideset{}{'}\sum_\beta \varphi_\beta(m)m_\beta = m.$$

Thus $1_M = \pi g$ and from the proof of Proposition 2.2.5 it follows that M is a fragment direct summand of E. Therefore M is projective. □

Lemma 2.2.12. *Let* $0 \longrightarrow N \longrightarrow M \xrightarrow{\pi} P \longrightarrow 0$ *be a strict exact sequence of fragment morphisms. If* P *is a projective fragment then* $M \cong N \oplus P$ *as fragments.*

Proof. Since π is a strict fragment epimorphism there is a fragment morphism $h : P \to M$ such that $\pi h = 1_P$. If $m_\gamma \in M_\gamma$ then $\pi(m_\gamma) \in P_\gamma$ and $h\pi(m_\gamma) \in h(P)_\gamma$. Thus $m_\gamma - h\pi(m_\gamma) \in \mathrm{Ker}(\pi) \cap M_\gamma$. and the latter is $\mathrm{Ker}(\pi)_\gamma$ since $\mathrm{Ker}(\pi)$ is a strict subfragment of M. Thus $m_\gamma \in h(P)_\gamma + \mathrm{Ker}(\pi)_\gamma$, which is a direct sum since $\pi h = 1_P$. It follows that $M \cong N \oplus P$. □

The homological algebra for modules does not generalize to fragments because of the special behavior of the cokernel. Nevertheless the following proposition opens the doors to the construction of homological algebra of fragments.

Proposition 2.2.13. *Given strict exact sequences of fragment morphisms*

$$0 \longrightarrow N \xrightarrow{f} P \xrightarrow{g} M \longrightarrow 0$$

$$0 \longrightarrow N' \xrightarrow[f']{} P' \xrightarrow[g']{} M \longrightarrow 0$$

If P *and* P' *are projective, then* $P \oplus N' \cong P' \oplus N$ *as fragments (the* $\oplus$ *are as fragments).*

Proof. Since the statement is an isomorphism, we may identify N, resp. N', with $\mathrm{Ker}(g)$, resp. $\mathrm{Ker}(g')$, in P, resp. P', since f, resp. f' was strict. We obtain a diagram of fragment morphisms

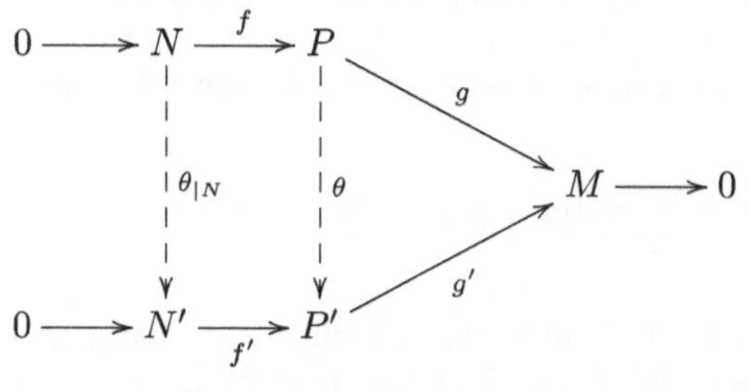

where θ is a lifting of g by projectivity of P such that $g'\theta = g$. Hence $\theta f(N)$ is in $\mathrm{Ker}(g') = N'$, so $\theta_{|N}$ maps N into N' and $\theta_{|N}$ is a fragment morphism. Now define

$$\psi : P \oplus N' \to P',\ (x, y) \mapsto \theta(x) - f'(y),$$

$$\phi : N \to P \oplus N',\ n \mapsto (f(n), \theta_{|N}(n)).$$

On $P \oplus N'$ we put the direct sum fragment structure $(P \oplus N')_\gamma = P_\gamma \oplus N'_\gamma$ for $\gamma \in \Pi$. Since θ, f, f' and $\theta_{|N}$ are fragment morphisms it is clear that ψ and ϕ are fragment morphisms. By diagram chasing we have that $\theta(P) + N' = P'$ and it follows that ψ is surjective. Also, it is obvious that ϕ is monomorphic. If $\phi(n) \in \phi(N) \cap (P \oplus N')_\gamma$, then $f(n) \in P_\gamma$ hence $n \in N_\gamma$ since f is strict. It follows that $\phi(n) \in \phi(N_\gamma) = \phi(N)_\gamma$. This shows that ϕ is strict. Concerning ψ, let $p' \in P'_\gamma$. Then from $g'(P'_\gamma) = M_\gamma = g(P_\gamma)$ it follows that there is a $p \in P_\gamma$ such that $\theta(p) - p' \in \mathrm{Ker}(g') = N'$ and in P'_γ hence $\theta(p) - p' \in N' \cap P'_\gamma = N'_\gamma$ since f' is strict. Say $\theta(p) - p' = n'_\gamma \in N'_\gamma$. Then $p' = \theta(p) - n'_\gamma \in \psi((P \oplus N')_\gamma)$. This proves that ψ is strict as well.
We arrive at a strict exact sequence of fragment morphisms

$$0 \longrightarrow N \xrightarrow{\phi} P \oplus N' \xrightarrow{\psi} P' \longrightarrow 0.$$

Since P' is projective, Lemma 2.2.12 entails that $P \oplus N' \cong P' \oplus N$. □

2.3 Fragments with finiteness conditions

For a fragment morphism $f : M \to N$ we denote by f_γ for $\gamma \in \Pi$ the map $f_\gamma = f_{|M_\gamma} : M_\gamma \to N_\gamma$. Let us reconsider the inverse image $f^{-1}(K)$. We can fix the problems made in the observations preceding Lemma 2.2.3 by defining $f^{-1}(K)_\gamma = f_\gamma^{-1}(K_\gamma) \subset M_\gamma$. This does define a fragment structure on $f^{-1}(K)$ which becomes a subfragment of the induced subfragment $f^{-1}(K)^{(M)}$. Again, if $K \subset N$ is strict, then both fragments are equal and strict as well.

Definition 2.3.1. A fragment M is said to be Noetherian if every ascending chain of subfragments $M^0 \subset M^1 \subset \ldots \subset M$ is stationary. We say that M is essentially Artinian if every descending chain of subfragments $M \supset M^0 \supset M^1 \supset \ldots \supset M^n \supset \ldots$ is such that for some $n_0 \in \mathbb{N}$ and every $n \geq n_0$, M^n is a trivial subfragment of M.

If N is a subfragment of M and M is Noetherian then N is Noetherian; however subfragments of an essentially Artinian fragment are not necessarily essentially Artinian. Counterexamples may be constructed to show that the body of an essentially Artinian fragment M need not be essentially Artinian as a subfragment even if it is obviously an Artinian (left) R-module. We can indeed have a chain of subfragments of $B(M)$, say $B(M) \supset B^1 \supset B^2 \supset \ldots \supset B^n \supset \ldots$ such that for some $\gamma_i \in \Pi$, $B(B^i) = B^i_{\gamma_i}$, then B^i is not trivial in $B(M)$ (where $B(M)_\gamma = B(M)$ for all $\gamma \in \Pi$) but it is trivial in M if $M_{\gamma_i} \neq B(M)$.

Theorem 2.3.2. *Let M be a Noetherian fragment.*

1. *For every $\gamma \in \Pi$, M_γ is a finitely generated S-module;*
2. *M is finitely generated;*
3. *If N is a strict subfragment of M then N and the canonical fragment M/N are both Noetherian fragments.*

Proof. 1. We prove that each M_γ is a Noetherian S-module. Look at an ascending sequence of S-modules $L_0 \subset L_1 \subset \ldots \subset M_\gamma$ and consider fragments L^j defined by $F_\gamma RL_j \supset \ldots \supset F_{\gamma-\tau} RL_j \supset \ldots$ for $\tau \leq \gamma$. Each L^j is a subfragment of M so the ascending chain $L^0 \subset L^1 \subset \ldots \subset L^n \subset \ldots \subset M$ is stationary, say $L^n = L^{n+1} = \ldots$ for some $n \in \mathbb{N}$. In particular, $L_n = F_0 RL_n = (L^n)_\gamma = (L^{n+1})_\gamma = L_{n+1}$, so the chain of S-submodules in M_γ is stationary. Hence M_γ is Noetherian as an S-module, thus finitely generated as an S-module.
2. Suppose that M is generated by $B = \{m_\beta,\ \beta \in B\}$ with respect to the set $\{\delta_\beta \in \Pi,\ \beta \in B\}$. Look at the fragments $F_{\delta_\beta} Rm_\beta \supset \ldots \supset F_{\delta_\beta - \tau} Rm_\beta \supset \ldots$ where for $\tau \in \Pi$, $(F_{\delta_\beta} Rm_\beta)_\tau = F_{\delta_\beta - \tau} Rm_\beta$. For ease of notation, we will write $F_\beta R$ instead of $F_{\delta_\beta} R$. Call M^1 the subfragment of M obtained in the foregoing way by m_{β_1}. If $M = M^1$ then M is generated by m_{β_1}. If $M \neq M^1$ then there is an m_{β_2} such that for some $\tau \in \Pi, F_{\beta_2-\tau} Rm_{\beta_2} \not\subset (M^1)_\tau$. Then look at the subfragment M^2 of M defined by $(M^2)_\gamma = F_{\beta_1-\gamma} Rm_{\beta_1} + F_{\beta_2-\gamma} Rm_{\beta_2}$, for $\gamma \in \Pi$. If $M^2 = M$ then M is generated by m_{β_1}, m_{β_2} with respect to $\delta_{\beta_1}, \delta_{\beta_2}$. If $M^2 \neq M$, continue with the construction. This leads to a strictly ascending chain of subfragments $M^1 \subsetneq M^2 \subsetneq \ldots \subset M$ so it cannot be stationary unless $M^n = M$ for some $n \in \mathbb{N}$. Hence the construction stops at M and M is generated by $\{m_{\beta_1}, \ldots, m_{\beta_n}\}$ with respect to $\{\delta_{\beta_1}, \ldots, \delta_{\beta_n}\}$.
3. That N is Noetherian is clear. Look at an ascending chain of subfragments $\overline{M}^0 \subset \overline{M}^1 \subset \ldots \subset \overline{M}^n \subset \ldots \subset \overline{M} = M/N$. Let $f : M \to M/N$ be the

canonical fragment morphism with $f_\gamma : M_\gamma \to (M/N)_\gamma = M_\gamma/N_\gamma$ for $\gamma \in \Pi$; this exists since N is strict. There is an ascending chain of subfragments of M, $f^{-1}(\overline{M}^0) \subset \ldots \subset f^{-1}(\overline{M}^n) \subset \ldots \subset M$ (here we use the fragment structure on the inverse image as defined at the beginning of this section!). By Noetherianity of M, there is an $n_0 \in \mathbb{N}$ such that $f^{-1}(\overline{M}^n) = f^{-1}(\overline{M}^{n_0})$ as fragments for every $n \geq n_0$. So for every $\gamma \in \Pi$ we have that $f_\gamma^{-1}(\overline{M}_\gamma^n) = f_\gamma^{-1}(\overline{M}_\gamma^{n_0})$. The surjectivity of every f_γ, $\gamma \in \Pi$, then leads to $\overline{M}^n = \overline{M}^{n_0}$ for every $n \geq n_0$ and the original chain is stationary. Thus M/N is Noetherian. □

Example 2.3.3. Let $S = K[X]$ be the polynomial ring and R its field of fractions $K(X)$, with degree filtration $\deg(X) = 1, \deg(X^{-1}) = -1$. Then $M = S$ is Noetherian but we have a non-Noetherian fragment

$$\begin{array}{ccccccc}
K[X] & & K[X] & & K[X] & & K[X]\\
\cup & & \cup & & \cup & & \cup\\
0 & & (X) & & (X) & & (X)\\
\cup & & \cup & & \cup & & \cup\\
0 & & 0 & & (X^2) & & (X^2)\\
\cup & & \cup & & \cup & & \cup\\
\vdots & \subset & \vdots & \subset & \vdots & \subset\cdots\subset & \vdots\\
\cup & & \cup & & \cup & & \cup\\
0 & & 0 & & 0 & & (X^n)\\
\cup & & \cup & & \cup & & \cup\\
\vdots & & \vdots & & \vdots & & \vdots
\end{array}$$

We see that the converse of statement 1. in the foregoing theorem does not hold in general.

Statement 3. in Theorem 2.3.2 does have a converse like in the module case. The proof, however, is an essential modification.

Theorem 2.3.4. *If N is a strict subfragment of M then M is Noetherian if and only if N and M/N are Noetherian.*

Proof. By Theorem 2.3.2 we only have to show that M is Noetherian in case N and M/N are Noetherian.
Consider an ascending chain of subfragments of M, say $M^0 \subset M^1 \subset \ldots \subset M^n \subset \ldots \subset M$. Write $f : M \to M/N$ for the canonical strict fragment epimorphism with corresponding surjective S-linear maps $f_\gamma : M_\gamma \to (M/N)_\gamma = M_\gamma/N_\gamma$ (N is strict!). Then $f(M^j)$, with $f(M^j)_\gamma = f_\gamma(M_\gamma^j)$ for every $\gamma \in \Pi$ is a subfragment of M/N. By the Noetherian assumption on M/N, the ascending chain $f(M^j)$ is stationary at some $n_0 \in \mathbb{N}$. Consider the subfragments $M^j + N$ of M, where $(M^j + N)_\gamma = M_\gamma^j + N_\gamma$, then $f(M^j + N) = f(M^j)$ as fragments and thus $f(M^n + N) = f(M^{n_0} + N)$ for all $n \geq n_0$. Thus $M_\gamma^n + N_\gamma = M_\gamma^{n_0} + N_\gamma$ for all $\gamma \in \Pi$ and thus $M^n + N = M^{n_0} + N$ for all $n \geq n_0$ as fragments in M : $M^{n_0} \subset M^{n_0+1} \subset \ldots \subset M^{n_0} + N = M^{n_0+1} + N = \ldots$ Let $K^j \subset N$

be the set of all $a \in N$ such that there are $x^j \in M^{n_0+j}$ and $x \in M^{n_0}$ such that $x^j = x + a$. It is obvious that K^j is an S-submodule of N. Similarly we define K^j_γ for $\gamma \in \Pi$ in $(M^{n_0+j})_\gamma$; thus $(M^{n_0+j})_\gamma = M^{n_0}_\gamma + K^j_\gamma$, K^j_γ is an S-module and for all γ, $K^j_\gamma \subset K^j$ and this defines a chain of S-submodules $K^j_\gamma \subset K^j_\tau \subset \ldots \subset K^j$ when $\gamma \geq \tau$.
Now it is also clear that $F_\tau RK^j_\gamma \subset K^j_{\gamma-\tau}$ follows from $F_\tau RM^{n_0}_\gamma \subset M^{n_0}_{\gamma-\tau}$ and $F_\tau RM^{n_0+j}_\gamma \subset M^{n_0+j}_{\gamma-\tau}$, hence each K^j is a subfragment of N. Observe that for all $\gamma \in \Pi$, $K^j_\gamma \subset N_\gamma$ since $K^j_\gamma \subset M_\gamma \cap N = N_\gamma$ as N is strict. Since N is a Noetherian fragment, the ascending chain of subfragments $K^1 \subset K^2 \subset \ldots \subset N$ is stationary. Hence there exists $m_0 \geq 0$ such that $K^m = K^{m_0}$ for all $m \geq m_0$ as fragments. Thus for $j \geq m_0$ we have $M^{n_0+j} = M^{n_0} + K^j = M^{n_0} + K^{m_0} = M^{n_0+m_0}$ meaning that the original chain in M is stationary and thus yielding that M is Noetherian. □

The Noetherian property, the way we defined it, is rather strong due to the fact that we allowed arbitrary chains of subfragments, even consisting of trivial subfragments. We define strict Noetherian by restricting to chains of strict subfragments. For strict Noetherian fragments we do not even have statement 1. of Theorem 2.3.2.

Example 2.3.5. Consider the ring filtration $\mathbb{Z} \subset \mathbb{Q} \subset \mathbb{Q} \subset \ldots$. The fragment $M = \mathbb{Q} \supset \mathbb{Z} \supset 0 \supset \ldots$ of essential length 1 is not Noetherian but it is strict Noetherian. Indeed, suppose that $M^0 \subset M^1 \subset \ldots$ is an ascending chain of strict subfragments. The strictness assumption implies that $M^i_1 \neq 0$ for all $i \geq 0$. Hence there exists some i_0 such that $M^i_1 = M^{i_0}_1 \lhd \mathbb{Z}$ for all $i \geq i_0$. It follows that $\mathbb{Q} = \mathbb{Q}M^i_1 \subset M^i_0 \subset \mathbb{Q}$ for all $i \geq i_0$. Of course, $M_0 = \mathbb{Q}$ is not Noetherian as $\mathbb{Z}$-module.

Which condition on a strict subfragment M will force all M_γ, $\gamma \in \Pi$ to be finitely generated S-modules (as in Theorem 2.3.2) except for the Noetherian condition of course?

Typical examples of Noetherian fragments are obtained from the valuation filtration on a field K with respect to some discrete valuation ring $\mathcal{O}_v = S$ ($K = R$ then): $\ldots \subset m_v^2 \subset m_v \subset \mathcal{O}_v \subset m_v^{-1} \subset m_v^{-2} \subset \ldots \subset K$, where m_v is the maximal ideal of $\mathcal{O}_v$ and an $\mathcal{O}_v$-module with good filtration with respect to the m_v-adic filtration. If F^eN is a filtered F^eS-module such that the filtration F^eN is good, i.e. there exist $n_1, \ldots, n_s \in N, k_1, \ldots, k_s \in \mathbb{Z}$ such that for all $r \in \mathbb{N}$:

$$F^e_{-r}N = \sum_{i=1}^{s} F^e_{-r-k_i} Sn_i,$$

then the $N_r = F^e_{-r}N$ for $r \in \mathbb{N}$ define an FR-fragment. Since F^eN is good, N is a finitely generated S-module. Since S is a DVR it is Noetherian, whence N is a Noetherian S-module. Hence, if $N^0 \subset N^1 \subset N^2 \subset \ldots \subset N$ is an ascending chain of strict subfragments, there exists an $m \geq 0$ such

that $N_0^m = N_0^{m+1} = \cdots$. By strictness it follows that for all $\gamma \in \Gamma$: $N_\gamma^m = N_0^m \cap N_\gamma = N_\gamma^{m+1} \cap N_\gamma = N_\gamma^{m+1}$. This shows that N is a strict Noetherian fragment.
Other examples are easily obtained by gliders over finite dimensional K-algebras R with positive filtration FR, $F_0R = K$ say. Indeed, if $\Omega \supset M$ is a glider, then all M_n are K-vector spaces. If $\dim_K(M) < \infty$, then M is (strict) Noetherian as a fragment. E.g. one can consider the group algebra $R = KG$ of some finite group G with $F_nR = KG_n$ for some chain of subgroups $1 \subset G_1 \subset \ldots \subset G_n \subset \ldots \subset G_m = G$. In Section 4.1 we will thoroughly study irreducible glider representations in such situations.

A fragment M is said to be (left) Artinian if any descending chain of subfragments of M is stationary; clearly an Artinian fragment is also essentially Artinian. A subfragment of an Artinian fragment is obviously again Artinian, contrary to what happens for essentially Artinian fragments.

Proposition 2.3.6. *Let M be a fragment such that M is a (left) Artinian S-module; then M is Artinian as a fragment.*

Proof. If for $\gamma_1 \in \Pi$, $M_{\gamma_1} \neq B(M)$, then we may pick $\gamma_2 \in \Pi$ such that $M_{\gamma_1} \supsetneq M_{\gamma_2}$ etc... so we find a sequence $M \supsetneq M_{\gamma_1} \supsetneq M_{\gamma_2} \supsetneq \ldots \supsetneq M_{\gamma_n} \supsetneq \ldots$ This is a sequence of S-modules, thus $M_{\gamma_n} = B(M)$. The same reasoning on chains between M_{γ_i} and $M_{\gamma_{i+1}}$, $i = 0, \ldots, n-1$, leads to the fact that there are only finitely many M_τ between M_{γ_i} and $M_{\gamma_{i+1}}$, $\gamma_{i+1} > \tau > \gamma_i$. Consequently the fragment M is given by a finite chain $M \supset \ldots \supset M_{\gamma_n} \supset \ldots \supset B(M)$ (for τ such that $\gamma_{i+1} \geq \tau > \gamma_i$, $M_\tau = M_{\gamma_{i+1}}$). Of course $M = B(M)$ is also possible.
Now consider a descending chain $M \supset M^1 \supset M^2 \supset \ldots \supset M^n \supset \ldots$. Since $B(M)$ above is also an Artinian S-module we must have that for $n \geq n_0$: $L = B(M^{n_0}) = B(M^{n_0+1}) = \ldots$. We obtain

$$\begin{array}{ccccccccc}
M^{n_0} & = & M^{n_0+1} & = & \ldots & = & M^{n_0+i} & = & \ldots \\
\cup & & \cup & & & & \cup & & \\
M^{n_0}_{\gamma_1} & \supset & M^{n_0+1}_{\gamma_1} & \supset & \ldots & \supset & M^{n_0+i}_{\gamma_1} & \supset & \ldots \\
\cup & & \cup & & & & \cup & & \\
\vdots & & \vdots & & & & \vdots & & \\
\cup & & \cup & & & & \cup & & \\
M^{n_0}_{\gamma_{l_0}} = L & = & M^{n_0+1}_{\gamma_{l_0}} = L & = & \ldots & = & M^{n_0+i}_{\gamma_{l_0}} = L & = & \ldots
\end{array}$$

Let n_1 be such that $M^{n_0+n_1}_{\gamma_1} = M^{n_0+n_1+j}_{\gamma_1}$ for all $j \geq 0$, $\ldots$, n_s such that $M^{n_0+n_s}_{\gamma_s} = M^{n_0+n_s+j}_{\gamma_s}$ for all $j \geq 0$. If $N \geq n_0 + n_1 + \ldots + n_s = m$ then $M^m = M^N$, hence the fragment is Artinian. □

Corollary 2.3.7. *Let M be a Noetherian fragment and assume that S is a (left) Artinian ring, then M is Artinian.*

Proof. In view of Theorem 2.3.2 every M_γ, $\gamma \in \Pi$, is a finitely generated S-module. Since S is an Artinian ring each M_γ is a (left) Artinian S-module. The foregoing proposition then leads to the Artinian property for M. □

The essentially Artinian condition is more closely related to the definition of irreducible fragments, which are the topic of the next chapter. It will be clear that an irreducible fragment is essentially Artinian. The Artinian condition as fragments is equivalent to the S-module Artinian condition; this is perhaps too restrictive.

Chapter 3

Irreducible fragments and gliders

In this chapter we focus on $\Gamma = \mathbb{Z}$ and we thoroughly study irreducible and weakly irreducible fragments, which are introduced in Section 2.1. First of all, both notions are indeed different. E.g. an irreducible fragment $M \supset \ldots \supset M_i \supset \ldots$ has as body $B(M)$ an irreducible R-module. For weakly irreducible fragments however, this no longer needs to be true. Indeed, if M has essential length el(M) = d and $L \subset B(M)$ is a proper R-submodule, then the subfragment $B(M) \supset \ldots \supset B(M) \supset L \supset L \supset \ldots$, with $d+1$ $B(M)$'s appearing is non-trivial, but it is also not strict! Our main goal is to describe the decomposition into irreducible subfragments for reducible ones, in fact for fragments over the situation: $S = K$ a field, $R = A$ a finite dimensional K-algebra. Arbitrary fragments would perhaps be decomposable, just like modules, under some mild conditions. In particular for group algebras $A = KG$ with G a finite group such that $(|G|, \text{char}(K)) = 1$ such a decomposition can be hoped for. So we first focus on some filtered ring R with filtration FR which is right-bounded in the sense that for some $n \in \mathbb{N}$, $F_nR = R$.

3.1 Right bounded filtrations

The filtration FR is right bounded if for some $n \in \mathbb{N}$, $F_nR = F_{n+1}R = \ldots = R$. If $M \supset M_1 \supset \ldots \supset M_n \supset \ldots$ is an FR-fragment then for $N \geq n$, $F_nRM_N = RM_N \subset M_{N-n} \subset M$. For any m, $M_m \supset M_{m+1} \supset \ldots$ is a subfragment of M and if L is a subfragment of M_m:

$$\begin{array}{ccccccc} L & \supset & L_1 & \supset & L_2 & \supset & \ldots \\ \cap & & \cap & & \cap & & \\ M_m & \supset & M_{m+1} & \supset & M_{m+2} & \supset & \ldots \end{array}$$

then $M \supset \ldots \supset M_{m-1} \supset L \supset L_1 \supset L_2 \supset \ldots$ is a subfragment of M.

Lemma 3.1.1. *If M is irreducible then $M_m \supset M_{m+1} \supset \ldots$ is also irreducible (this holds for arbitrary Γ).*

Proof. If L is a subfragment of M_m then $M(L) : M \supset \ldots \supset M_{m-1} \supset L \supset L_1 \supset \ldots$ is trivial in M by irreducibility. Suppose that $B(M(L)) = B(L) = M(L)_n$ but $M_n \neq B(M) = B(M_m)$ for some n; then if $n \geq m$ we have $L_{n-m} = B(L)$ and $(M_m)_{n-m} = M_n \neq B(M_m)$, so L is trivial in this case. If $n < m$, then $B(L) = M_n \neq B(M)$, whence $B(L) = B(M)$ and $M_n = B(M)$. In this case, M_m is the trivial chain $B(M) \supset B(M) \supset \ldots$ and $L = M_m$, so L is again trivial. If $M(L)$ is trivial of type T_2, then $M(L)_n = 0$, $M_n \neq 0$, for some n. If $n \geq m$, then $L_{n-m} = 0$, but $(M_m)_{n-m} = M_n \neq 0$. Hence L is trivial. If $n < m$, then $M_m = 0$ and L is trivial. Finally, if $M(L)_n = M_{\alpha(n)}$ for all n for some $\alpha : \mathbb{N} \to \mathbb{N}$ satisfying the conditions of T_3, then $M(L)_{m+k} = M_{\alpha(m+k)} = L_{\alpha(m+k)-m}$ for all k. The monotonic ascending function $\alpha_m(k) = \alpha(m+k)-m$ satisfies the condition of T_3 and shows that L is trivial in M_m. This covers all the cases. □

Lemma 3.1.2. *Let M be an irreducible FR-fragment where FR is right-bounded with $F_nR = R$. If $M_{m+n} \neq B(M)$ then $RM_{m+n} = M_m$ and M_m is an R-module.*

Proof. It suffices to establish the claim for $m = 0$ because we may replace M by M_m in view of the foregoing lemma. Assume that $M_n \neq B(M)$ and let k be maximal such that $M_n = M_{n+k}$; such k exists since $M_n \neq B(M)$. Now consider in M the following subfragment N:

$$\begin{array}{ccccccccccccccc}
M = M_0 & \supset & M_1 & \supset & \ldots & \supset & M_k & \supset & \ldots & \supset & M_{k+i} & \supset & \ldots & \supset & M_{k+n} & \supset & \ldots \\
\cup & & \cup & & & & \cup & & & & \cup & & & & \cup & & \\
N = RM_n & \supset & RM_{n+1} & \supset & \ldots & \supset & RM_{n+k} & \supset & \ldots & \supset & N_{k+i} & \supset & \ldots & \supset & N_{k+n} & \supset & \ldots
\end{array}$$

where $N_{k+i} = RM_{n+k+i} + F_{n-1}RM_{n+k+i-1} + \ldots + F_{n-i}RM_{n+k}$, for $i = 1, \ldots, n-1$. If $N_j = 0$ with $j < k+n$ then some $F_{n-i}RM_{n+k} = 0$ for $i \leq n$, thus $M_{n+k} = 0 = M_n$ but that contradicts $B(M) \neq M_n$. If $B(N) = N_j$ for $j \leq k+n$ then also $F_{n-i}RM_{n+k} \subset B(N) = B(M)$ for some $i \leq n$; hence $M_{n+k} = B(M)$ contradicting the choice of k. Thus N has to be trivial of type II; let $\alpha : \mathbb{N} \to \mathbb{N}$ be a monotone increasing map such that $N_i = M_{\alpha(i)}$. If $\alpha(n+k) > n+k$ then $M_{n+k} = M_{\alpha(n+k)}$ contradicts the choice of k; thus $\alpha(n+k) = n+k$ but since α is strict increasing the only possibility is $\alpha(i) = i$ for $i \leq n+k$, thus $M = RM_n, \ldots, M_k = RM_{n+k} = RM_n$. □

Corollary 3.1.3. *For a right-bounded filtration FR with $F_nR = R$, an irreducible FR-fragment M that is not an R-module has* $\mathrm{el(M)} \leq \mathrm{n}$.

If M is a fragment such that M_m is not an R-module for some $m \in \mathbb{N}$, then if M is irreducible, so is the fragment M_m by Lemma 3.1.1 and $\mathrm{el(M_m)} < \mathrm{n}$. So as far as fragment properties are concerned we may reduce the study to non R-modules which are fragments of $\mathrm{el} < \mathrm{n}$.

Lemma 3.1.4. *In the situation of Lemma 3.1.2, if M is irreducible with* $\mathrm{el(M)} = \mathrm{d}$, *then the chain of M is given by $M = F_dRM_d \supset \ldots \supset F_1RM_d \supset M_d \supsetneq B(M) \supset B(M) \supset \ldots$*

Proof. Consider the subfragment of M:

$$\begin{array}{ccccccccccccc} M=M_0 & \supset & M_1 & \supset & \dots & \supset & M_d & \supsetneq & B(M) & \supset & B(M) & \supset & \dots \\ \cup & & \cup & & & & \parallel & & & & & & \\ F_dRM_d & \supset & F_{d-1}RM_d & \supset & \dots & \supset & M_d & \supsetneq & B(M) & \supset & B(M) & \supset & \dots \end{array}$$

The latter can only be trivial in M if $M_i = F_{d-i}RM_d$ for $i = 0, \dots, d$. □

Proposition 3.1.5. *If T is a strict and trivial subfragment of M and M is irreducible then M/T is irreducible. In particular if M is irreducible then $M/B(M)$ is irreducible.*

Proof. Let $f : M \to M/T$ be the strict fragment epimorphism with $f_i = f_{|M_i} : M_i \to M_i/T_i$. Consider a subfragment $\overline{N}$ of $M/T = \overline{M}$ and let N be the inverse image of $\overline{N}$, i.e. $N_i = f_i^{-1}(\overline{N}_i)$, $i \in \mathbb{N}$. Since M is irreducible, N has to be trivial in M.
Case 1. Suppose that el(N) < el(M), say el(N) = e, e < d = el(M). Thus $B(N) = N_{e+1}$. Since $B(M)$ is an irreducible R-module (as M is an irreducible fragment), we either have $N_{e+1} = 0$ or $N_{e+1} = B(M)$. In each case we have $T_{e+1} \subset f_{e+1}^{-1}(\overline{N}_{e+1}) \subset B(M)$. Strictness of T in M yields el(T) $\leq$ el(M) = d, thus el(M/T) $\leq$ d. If $\overline{M}_m = B(\overline{M})$ then

$$\frac{M_m}{T_m} = \frac{M_{m+1}}{T_{m+1}} = \frac{M_{m+1}}{T_m \cap M_{m+1}} = \frac{M_{m+1} + T_m}{T_m},$$

hence $M_m = M_{m+1} + T_m$. Similarly: $M_{m+1} = M_{m+2} + T_{m+1}$, thus $M_m = M_{m+2} + T_m$. After a finite number of steps this leads to: $M_m = B(M) + T_m$. If el($\overline{\text{M}}$) < d we may apply the foregoing with $m = d$. But $M_d \neq B(M)$ then implies that $T_d \not\subset B(M)$. Since $e + 1 \leq d$ we have from $T_{e+1} \subset B(M)$ that $T_d \subset B(M)$, a contradiction. Consequently el($\overline{\text{M}}$) = d. Now since $N_{e+1} = N_{e+i}$ for $i \geq 1$, we have $N_{e+i} \cap T_{e+1} = T_{e+1}$. On the other hand: $N_{e+i} \cap T_{e+1} \subset M_{e+i} \cap T_{e+1} \subset M_{e+i} \cap T = T_{e+i}$. Therefore we must have that $T_{e+1} = T_{e+i}$ for $i \geq 1$. Moreover, $N_{e+1}/T_{e+1} = N_{e+i}/T_{e+i}$ for $i \geq 1$, hence el($\overline{\text{N}}$) $\leq$ e follows. From $e < d$ it then follows that el($\overline{\text{N}}$) < el($\overline{\text{M}}$) and $\overline{N}$ is trivial in $\overline{M}$.
Case 2. For some $m \in \mathbb{N}$, $N_m = 0$ but $M_m \neq 0$. Then $\overline{N}_m = 0$. If $\overline{M}_m = 0$, then $M_m = T_m$ but as $T_m \subset N_m$ this would lead to $M_m = 0$, a contradiction. Hence $\overline{N}_m = 0$ and $\overline{M}_m \neq 0$, or $\overline{N}$ is trivial in $\overline{M}$.
Case 3. If N is trivial in M because $N_i = M_{\alpha(i)}$ for some monotone increasing $\alpha : \mathbb{N} \to \mathbb{N}$, then since $T_i \subset N_i \cap T \subset M_i \cap T = T_i$ we have $T_i = N_i \cap T = M_{\alpha(i)} \cap T = T_{\alpha(i)}$ and hence N_i/T_i equals $M_{\alpha(i)}/T_{\alpha(i)}$ or $\overline{N}_i = \overline{M}_{\alpha(i)}$ for all i. Thus $\overline{N}$ is trivial in $\overline{M}$.
So $\overline{N}$ is trivial in $\overline{M}$, or $\overline{M}$ is irreducible. □

Remark 3.1.6. If M is irreducible with el(M) = d then $M/B(M)$ is irreducible with el(M/B(M)) = d.

Proposition 3.1.7. *If L is an R-submodule of an irreducible fragment M with* el(M) = d *then for the induced fragment structure on L, M/L is irreducible with* el(M/L) = d *unless $M/L = 0$.*

Proof. Since L is an R-module, putting $L_i = M_i \cap L$ defines a strict subfragment structure on L:

$$\begin{array}{ccccccccccc} M & \supset & M_1 & \supset & \dots & \supset & M_d & \supsetneq & B(M) & \supset & B(M) \\ \cup & & \cup & & & & \cup & & \cup & & \cup \\ L & \supset & L\cap M_1 & \supset & \dots & \supset & L\cap M_d & \supset & L\cap B(M) & \supset & L\cap B(M), \end{array}$$

where $L \cap B(M) = B(L)$. If $L \cap M_d = L \cap B(M)$ then el(L) < el(M) and then the foregoing proposition entails that M/L is indeed irreducible. If $L \cap M_d \neq L \cap B(M)$ then we can look at another sufragment of M:

$$\begin{array}{ccccccccccc} M & \supset & M_1 & \supset \dots \supset & M_{d-1} & \supset & M_d & \supsetneq & B(M) & \supset & \dots \\ \cup & & \cup & & \cup & & \cup & & \cup & & \\ F_dR(L\cap M_d) & \supset & F_{d-1}R(L\cap M_d) & \supset \dots \supset & F_1R(L\cap M_d) & \supset & L\cap M_d & \supsetneq & L\cap B(M) & \supset & \dots, \end{array}$$

The only way the latter subfragment can be trivial in M is via some monotone increasing α but since both chains have the same essential length, the monotone property leads to α being the identity, i.e. $M = F_dR(L \cap M_d) \subset L$, but then $M = L$ or $M/L = 0$. So in case $M \neq L$ we have $L \cap M_d = L \cap B(M)$, hence el(M/L) = d. □

We have a 'converse' for the previous proposition:

Proposition 3.1.8. *If L is an R-submodule of M such that $L \cap M_d = 0$ then L with the induced fragment structure is a trivial strict subfragment and if M/L is irreducible then the obstruction for irreducibility of M lies completely in L.*

Proof. The induced fragment structure on L is of course strict and it is trivial since $L \cap M_d = 0$. If M is irreducible then so is M/L by the previous proposition. Suppose that M/L is irreducible. Let

$$\begin{array}{ccccccccccccc} M & \supset & M_1 & \supset & \dots & \supset & M_{d-1} & \supset & M_d & \supsetneq & B(M) & \supset & \dots \\ \cup & & \cup & & & & \cup & & \cup & & \cup & & \\ P & \supset & P_1 & \supset & \dots & \supset & P_{d-1} & \supset & P_d & \supset & B(P) & \supset & \dots \end{array}$$

be a subfragment. The fragment epimorphism $\pi : M \to M/L$ sends P to the trivial subfragment

$$\frac{P}{P\cap L} \supset \frac{P_1}{P_1\cap L} \supset \dots \supset \frac{P_{d-1}}{P_{d-1}\cap L} \supset P_d \supset B(P) \supset \dots$$

If it is trivial of type T_1 then $P_d = B(P)$, hence $P \subset M$ is trivial. Triviality of type T_2 implies that $P_d = 0$, but then $P \subset M$ is also trivial since $M_d \neq 0$. For type T_3 the only possibility for $\alpha : \mathbb{N} \to \mathbb{N}$ is the identity map since both fragments have the same essential length, hence $P/(P \cap L) = M/L$ or $P + L = M$. Hence as an S-module, $M/P \cong L/(P \cap L)$ is a subquotient of L, which shows that any non-trivial subfragment P of M only differs from M by a subquotient of L.

□

An irreducible fragment M such that not every M_m is an R-module necessarily has el(M) = d $< \infty$, because of Lemma 3.1.2. If we write M_{k+1} for the first M_i that is not an R-module then the chain for M is:

$$M = RM_n \supset \ldots \supset M_k = RM_{n+k} \supset M_{k+1} \supset \ldots \supset M_{k+n} \supsetneq B(M) \supset \ldots$$

Indeed, the equality $M_{k-i} = RM_{n+k-i}$ for $i = 0, \ldots, k$ follows from the irreducibility of the fragment M_{k-i} (Lemma 3.1.1). Suppose that $M_{k+n} = B(M)$. Then $M_k = RM_{n+k} = RB(M) = B(M)$, which contradicts the choice of k. We observe that the part $M_{k+1} \supset \ldots \supset M_{k+n} \supsetneq B(M)$ is a non-module part and has $\mathrm{el}(\mathrm{M_{k+1}}) = \mathrm{n} - 1$. Indeed, if for example M_{n+k} is an R-module, then $M_k = RM_{k+n} = M_{k+n}$, contradicting the choice of k. If el(M) $\geq$ n then all M_i are embedded in R-modules, so M is a glider representation, but not necessarily a natural one. On $M_k/B(M) = \overline{M}$ we may define a separated filtration by putting $F_{-m}\overline{M}$ equal to $\overline{M}_{k+m}/B(M)$ for $m \geq 0$ and $F_m\overline{M} = M_k/B(M)$ for $m > 0$. Since $M_k \neq M_{k+1}$ (by the choice of k) it follows that $M_{k+1}, \ldots, M_{k+n}$ are not R-modules.

The above discussion shows that we may reduce the study of irreducible fragments over a right-bounded filtration to the study of R-modules and irreducible fragments of essential length lesser or equal to n. In the latter case, $M = F_nRM_n$ is an R-module, so M is a glider representation. If the right bounded filtration FR is a ring filtration, then in the former case, say el(M) = e $<$ n, Lemma 3.1.4 entails that M is a glider representation with respect to the right bounded filtration of length e on F_eR. This will be the subject of the following section.

Remark 3.1.9. Let M be an irreducible fragment with el(M) = d $< \infty$ for any filtration FR, then $M_d/B(M)$ is an irreducible S-module. Indeed, if $B(M) \subsetneq L \subsetneq M_d$ for some S-module L then we have a subfragment $M \supset M_1 \supset \ldots \supset L \supsetneq B(M) \supset \ldots$ of M and this can only be trivial in M if $L = M_d$.

Lemma 3.1.10. *Let M be an essentially left Artinian FR-fragment with* el(M) = d *and assume $B(M) \neq 0$ contains a simple proper R-submodule $L \neq 0$. Then there exists an irreducible subfragment N of M such that $B(N) = L$.*

Proof. In M we may consider the non-trivial subfragment M_L given by the chain: $M \supset M_1 \supset \ldots \supset M_d \supset L \supset L \supset \ldots$. Since M is essentially left Artinian, the family of non-trivial subfragments of M_L containing L has minimal elements. Let N be such a minimal element. If T is a subfragment of N, then either $B(T) = 0$ or $B(T) = L$ as L is simple. In case $B(T) = 0$ then T is trivial in N since $B(N) = L$. In case $B(T) = L$ then $N \supsetneq T$ would lead to T being a trivial subfragment of M; if $T_d = L$ and $M_d \neq B(M)$ then T is trivial in N since $N_d \neq L$ as N is non-trivial in M. If $T_i = M_{\alpha(i)}$ for some monotone increasing map $\alpha : \mathbb{N} \to \mathbb{N}$ then $B(M) \subset B(T) = L$, but this contradicts $L \neq B(M)$, so this is excluded. Consequently T is always trivial in N, proving that N is irreducible. □

Corollary 3.1.11. *If M is as in the previous lemma and $B(M) \neq 0$ is simple then there exists an irreducible subfragment N of M with $B(N) = B(M)$.*

Corollary 3.1.12. *If M is an essentially left Artinian FR-fragment with* el(M) = d *and $B(M) \neq 0$ is a semisimple R-module, then there exist irreducible subfragments $M^{(1)}, \dots, M^{(t)}$ with $B(M^{(i)}) = S_i$ a simple R-module such that $B(M) = S_1 \oplus \dots \oplus S_t$. Therefore M contains $M' = \sum_i M^{(i)}$ as a subfragment and $B(M') = B(M)$.*

Proof. By the previous lemma the $M^{(i)}$ exist; we have $M^{(i)}_{d_i+1} = S_i$ for $i = 1, \dots, t$. Taking $d = \max\{d_i\ ,\ i = 1, \dots, t\}$ we have $M'_{d+1} = S_1 \oplus \dots \oplus S_t$. □

Proposition 3.1.13. *Let M be an essentially left Artinian FR-fragment of* el(M) = d. *Suppose $B(M) = 0$ and M_d is a semisimple S-module, say $M_d = L_1 \oplus \dots \oplus L_t$. There exist irreducible subfragments $M^1, \dots, M^t$ of M such that $M^i_d = L_i$, $i = 1, \dots, t$ and M contains $M' = \sum_{i=1}^t M^i$ as a subfragment with $M'_d = M_d$.*

Proof. Consider the non-trivial subfragment N^i of M given by the chain:

$$M \supset M_1 \supset \dots \supset M_{d-1} \supset L_i \supset 0 \supset \dots .$$

Since M is essentially left Artinian there exists a minimal non-trivial M^i in N^i such that $M^i_d = L_i$. Look at a subfragment E of M^i; then either $E_d = 0$ or $E_d = L_i$ because L_i is a simple S-module. In case $E_d = 0$, then E is trivial in M^i since $M^i_d \neq 0$. In case $E_d = L_i$, then $E \neq M^i$ (as fragments) entails that E is a trivial subfragment of N^i. Since $E_d \neq B(E) = 0$, we must have that $E_j = M_{\alpha(j)}$ for some monotone increasing map $\alpha : \mathbb{N} \to \mathbb{N}$. Again from $E_d \neq 0$ we obtain that $E_j = M_j$ for $j < d$, $E_d = M^i_d$, hence $E = N^i = M^i$. We conclude that each M^i, $i = 1, \dots, t$ is irreducible. The other statements are obvious. □

Let us make a small digression and drop the right boundedness condition on the filtration FR. We say that an FR-fragment M, where FR is a nonpositive filtration, is F-torsion free if for $m \in M$, $F_\gamma Rm = 0$ with $\gamma \in -\Pi$ entails $m = 0$. For example, torsion free F_eR-modules are F-torsion free. If R is filtered by FR such that $S = F_eR$ is a prime ring then S with the filtration F^eS is an F-torsion free fragment.

Proposition 3.1.14. *Let FR be a Γ-filtration such that $S = F_eR$ has all $F_{-\gamma}R \neq 0$ for $\gamma \in \Pi$. Let M be an FR-fragment with $B(M) = 0$ and assume it is irreducible and F-torsion free. Then the M_γ, $\gamma \in \Pi$, are the only S-submodules of M. In case $\Gamma = \mathbb{Z}$, every M_i/M_{i+1} is a simple S-module and M is a (left) serial S-module (the lattice of S-submodules in M is linearly ordered).*

Proof. Let $L \neq 0$ be an S-submodule of M, for example $L = Sm$ for some $m \in M$. Consider the following subfragment of M

$$\begin{array}{ccccccc} M & \supset & \dots & \supset & M_\gamma & \supset & \dots \\ \cup & & & & \cup & & \\ L & \supset & \dots & \supset & F_{-\gamma}RL & \supset & \dots \end{array}$$

Since M is F-torsion free we have $F_{-\gamma}RL \neq 0$ for every $\gamma \in \Pi$. Moreover if $F_{-\gamma}RL = F_{-\delta}RL$ for all $\delta \geq \gamma$ then $F_{-\gamma}RL \subset M_\delta$ for all $\delta \geq \gamma$, thus $F_{-\gamma}RL \subset B(M) = 0$ contradicting the F-torsion freeness of M. So there must exist a monotone increasing map $\alpha : \Pi \to \Pi$ such that $L = M_{\alpha(e)}$. All claims now follow easily. □

Remark 3.1.15. This is the first time where the negative part of the filtration on R plays an important role in the definition of a fragment (Definition 1.2.1). In case $\Gamma = \mathbb{Z}$, the condition $F_{-n}R \neq 0$ for all $n \geq 0$ is satisfied for example when F_0R is a semisimple ring because then $F_{-n}R \supset (F_{-1}R)^n$ and since FR is not positive $F_{-1}R \neq 0$.

The serial property encountered above may introduce the idea that there are links with valuation theory. This link is clear from the following theorem (see also [52, Theorem 2.3] but note that we have now used a more correct notion of trivial subfragment). First we define a completely irreducible fragment as an FR-fragment M that is irreducible and if $a \in F_iR$, $m \in M_j$ are such that $am \in M_{j-i+1}$ then either $a \in F_{i-1}R$ or $m \in M_{j+1}$ (this is the $\Gamma = \mathbb{Z}$-case but it can be generalized to other Γ).

Theorem 3.1.16. *Let A be a simple Artinian ring and A' a subring of A such that $A' = F_0A$ for some positive $\mathbb{Z}$-filtration FA such that $F_iA \neq F_{i+1}A$ for all $i \in \mathbb{N}$. Then A' has a chain $A' = A'_0 \supset \dots \supset A'_i \supset \dots \neq 0$ making A' into a completely irreducible FA-fragment if and only if A is a skewfield and A' is a discrete valuation ring of A (in the sense of O. Schilling, [62]) with $F_{-n}A = A'_n$ if $n \geq 0$, $F_mA = F_mA$ with $m \geq 0$ corresponding to the valuation filtration on A.*

Proof. If A' is a discrete valuation ring of the skewfield A then A' is an FA-fragment with respect to the valuation filtration defined by $F_nA = \{a \in A,\ v(a) \geq -n\}$ where $v : A \to \mathbb{Z} \cup \{\infty\}$ is the corresponding valuation associated with A'. Obviously, $B(A') = 0$. Since left ideals of A' are linearly ordered (they are also ideals) and $F_{-n}A/F_{-n-1}A$ are simple A'-modules, in fact all isomorphic to the residue skewfield $A'/F_{-1}A = A'/A'_1$, it is easily checked that A' is an irreducible fragment. The fact that it is completely irreducible is clear from $v(ab) = v(a) + v(b)$ and $v(a)$ is the degree of $\sigma(a)$ where $\sigma(a)$ is the principal symbol in the filtration FA, i.e. $v(a) = n$ if $a \in F_{-n}A - F_{-n-1}A$.
In order to establish the converse, we start from a chain: $A' = A'_0 \supset A'_1 \supset \dots \supset A'_i \supset \dots$, making A' into a completely irreducible FA-fragment. Suppose

that $A'_i = A'_{i+1}$ for some $i \in \mathbb{N}$, then we have $F_1AA'_{i+1} \subset A'_i = A'_{i+1}$; fixing an $x \in F_1A - F_0A$ (exists by assumption) then $xA'_{i+1} \subset A'_{i+1}$ yields that $A'_{i+1} = A'_{i+2}$ in view of the completely irreducible property. Repeating this argument we find $A'_i = \ldots = A'_{i+k} = \ldots$ for all $k \geq 0$, thus $B(A') = A'_i$ and we may assume i is minimal as such, i.e. $\mathrm{el}(A') = i - 1$. Since $B(A')$ is a left A-module it follows that either A'_iA is 0 or equal to A because A is a simple ring. The first is excluded because $A'_i \neq 0$. In the latter case we have that some $z \in A'_{i-1} - A'_i$ and some unit u of A are in A'_iA; writing these elements in the form $\sum' a'_j q_j$ with $a'_j \in A'_i$, $q_j \in A$, we see that there is a $d \in \mathbb{N}$ such that $z, u \in A'_iF_dA$. Now $A'_iF_dA \cap A'_{i-1}$ is an A'-module contained in A'_{i-1} and strictly containing A'_i since $z \notin A'_i$, thus in view of Remark 3.1.9 we must have that $A'_iF_dA = A'_{i-1}$. Moreover, from Lemma 3.1.4 we obtain $F_{i-1}AA'_iF_dA = F_{i-1}AA'_{i-1} = A'$. Since $\mathrm{el}(A') = i - 1$, A'_i is a left A-module and we have that $F_{i-1}AA'_iF_dA = A'_iF_dA = A \cdot A'_iF_dA$. Then from $u \in A'_iF_dA$, $A \cdot A'_iF_dA = A$ follows and consequently $A'_iF_dA = A$ or $F_dA = A$ contradicting $F_dA \neq F_{d+1}A$. Therefore we arrive at $A'_i \neq A'_{i+1}$ for all $i \in \mathbb{N}$ and $B(A') = 0$. For every $m \in \mathbb{N}$ and $e \leq m$ we have $F_eAA'_m = A'_{m-e}$ because the subfragment $F_mAA'_m \supset \ldots \supset F_eAA'_m \supset \ldots \supset A'_m \supset \ldots$ of A' can only be trivial if $A'_{m-e} = F_eAA'_m$ for every $e \leq m$ (there are no zero terms in the sequence and its body is zero but the A'_m are nonzero).
For every $m \in \mathbb{N}$ we find: $(A'_mF_mA)A'_m = A'_m(F_mAA'_m) = A'_mA' \subset A'_m$. The last relation holds, since if $A'_mA' \not\subset A'_m$, say some $a \in A'_mA' - A'_m$, then $F_mAa \subset A' = A'_0$. Since $F_mA \neq F_{m-1}A$, there exists some $b \in F_mA - F_{m-1}A$ with $ba \subset A'_0$. By the complete irreducibility we then have that $a \in A'_{j+1}$ where $j + 1 - m = 0$, i.e. $a \in A'_m$, a contradiction. In fact, we just showed that the A'_m are also right A'-ideals. Subsequently, if A'_mF_mA would contain an $x \in F_nA - F_{n-1}A$ for some $n \geq 1$ then $xA'_m \subset A'_m$ with $m > m - n$ contradicts the completely irreducible assumption since $A'_m \neq A'_{m+1}$. Therefore $A'_mF_mA \subset A'$. Now the chain $A' \supset A'_1 \supset \ldots \supset A'_i \supset \ldots$ defines a filtration on A' because: $F_jAA'_{i+j} = A'_i$, $A'_jF_jAA'_{i+j} = A'_jA'_i$ thus $A'_jA'_i \subset A'_{i+j}$. Putting $F_{-n}A = A'_n$ for $n \geq 0$ yields a separated filtration on A again denoted by FA. It suffices to check: $A'_jF_kA \subset F_{k-j}A$ in case $k \geq j$, $A'_jF_kA \subset A'_{j-k}$ in case $j \geq k$.
First $k \geq j$: then $A'_{k-j}(A'_jF_kA) \subset A'_kF_kA \subset A'$, hence

$$A'_jF_kA = A' \cdot A'_jF_kA = (F_{k-j}AA'_{k-j})(A'_jF_kA) \subset F_{k-j}AA' = F_{k-j}A.$$

Secondly $j \geq k$: then $(A'_jF_kA)F_{j-k}A \subset A'_jF_jA \subset A'$ and now multiplying this on the right by A'_{j-k} yields $A'_jF_kA \subset A'_{j-k}$. The foregoing also implies that $A'_jF_0A = A'_j$, hence each A'_j is an ideal of A'. If $am = 0$ for $a \neq 0$ in A, $m \in A'$ then say $a \in F_iA - F_{i-1}A$, $m \in A'_j - A'_{j+1}$, but $am \in A'_{j-i+1}$ and the completely irreducible property of A' then yields that this is a contradiction, i.e. A' is F-torsion free; therefore we are in the situation where Proposition 3.1.14 applies. Thus A'_n/A'_{n+1} is a simple A'-module for every n and every $F_0A = A'$-submodule of A' is one the A'_n. Looking at the subfragment $A'_jF_jA \supset \ldots \supset A'_jF_kA \supset \ldots \supset A'_j = A'_jF_0A \supset \ldots$ of A' yields that $A'_jF_kA = A'_{j-k}$ for $k \leq j$, in particular $A'_jF_jA = A'$ for all j. It follows

that $A'_{i+j} = A'A'_{i+j} = A'_jF_jAA'_{i+j} = A'_jA'_i$ and similarly one obtains that $A'_{i+j} = A'_iA'_j$. From $A'_{n+m}F_{n+m}A = A'$ and $A'_{n+m} = A'_nA'_m$ it then follows that $F_{n+m}A \subset F_nAF_mAA'_mA'_nF_{n+m}AF_nAF_mAA'_{n+m}F_{n+m}A = F_nAF_mA \subset F_{n+m}A$ and thus $F_nAF_mA = F_{n+m}A$ for all $n, m \in \mathbb{Z}$. Note that we also have that $F_nA/F_{n-1}A$ is a simple A'-module also for positive n. Indeed if $F_nA \supsetneq L \supsetneq F_{n-1}A$ then multiplying by A'_n yields $A' \supsetneq A'_nL \supsetneq A'_1$ and that is impossible. Let $G(A)$ be the associated graded ring of FA; then $G(A)_nG(A)_m = G(A)_{n+m}$ follows from $F_nAF_mA = F_{n+m}A$. If $a \in F_nA$ represents $\overline{a} \in G(A)_n$ and $b \in F_mA$ represents $\overline{b} \in G(A)_m$ then $\overline{a}\overline{b} = 0$ in $G(A)$ means $ab \in F_{n+m-1}A$. For $0 \leq n \leq -m$ this would contradict the completely irreducible property for A'. This entails that $G(A)$ is a domain. Indeed, suppose that $\overline{a}\overline{b} = 0$ but we do not have that $0 \leq n \leq -m$. If $n, m \geq 0$, multiply on the right with some $\overline{c} \in G(A)_{-n-m}$, and then $\overline{a} = 0$ or $\overline{bc} = \overline{b}\overline{c} = 0$, but b and c satisfy $0 \leq m \leq n+m$ so $\overline{b} = 0$ or $\overline{c} = 0$ but the latter is excluded. If $n \geq 0$, $m \leq 0$, then multiply on the right with k-factors of $\overline{b}$ until $0 \leq n \leq -km$ for some $k \in \mathbb{N}$ and conclude in the same vein. If $n, m \leq 0$, multiply on the left with some $\overline{d} \in G(A)_{-n}$. Finally, if $n \leq 0$, $m \geq 0$, multiply with some $\overline{d} \in G(A)_{-n}$ on the left and a $\overline{c} \in G(A)_{-m}$ on the right. $G(A)$ being a domain entails that A is a domain too, hence A is a skewfield.
If $x \in A'_i - A'_{i+1}$ then F_iAx and xF_iA are in A' but not in A'_1 hence $F_iAx = A'$ (and $xF_iA = A'$) because F_iAx is an F_0A-submodule of A' hence one of the A'_j and it can only be $A'_0 = A'$. Thus $x^{-1} \in F_iA$. It is now clear that $v : A \to \mathbb{Z} \cup \{\infty\}$, $0 \mapsto \infty$ and $a \mapsto v(a) = n$ exactly when $a \in F_nA - F_{n-1}A$ is a valuation; it suffices to establish $v(ab) = v(a) + v(b)$ on the ring A' and this is clear from $A'_{i+j} = A'_iA'_j = A'_jA'_i$ and the fact that $G(A)$ is a domain. □

3.2 Fragments over right bounded filtrations on Artinian rings

Fragment or glider theory for finite dimensional algebras over a field provide new tools for describing a generalized representation theory; e.g. fragments of group rings KG with filtration given by a finite chain of subgroups $1 \subset G_1 \subset \ldots \subset G_n = G$ by putting $F_0KG = K$, $F_1KG = KG_1, \ldots, F_iKG = KG_i$, $F_nKG = KG$ allow us to describe links between representations of the G_i, $i = 1, \ldots, n$. In this philosophy it is normal to distinguish two types of FR-fragments: those not containing R-modules (these will be called pure fragments) and those containing a nonzero irreducible R-module. In the latter case we want to define a new fragment structure on M such that $B(M)$ is the given irreducible R-module and then reduce the study to $M/B(M)$. This is the theme of the following proposition.

Proposition 3.2.1. *Let FR be a right bounded filtration on an Artinian ring R, $F_0R \subset F_1R \subset \ldots \subset F_nR = R \subset R \subset \ldots$. Let M be a finitely generated irreducible FR-fragment with $B(M) = 0$ (in view of the statement and proof this is not really restrictive). Then either M is a pure fragment or M contains an irreducible R-module, L say. In the latter case there is a new fragment structure M' on M, i.e. $M'_0 = M_0 = M$, such that $B(M') = L$ and M' is again irreducible.*

Proof. Suppose M contains a nonzero R-submodule L'. Since R is Artinian and M is finitely generated, we may select an irreducible R-module L in L', $L \neq 0$. The assumptions imply $M_d = 0$ for some d, say minimal as such. Indeed, from Lemma 3.1.2, we have for every $m \in \mathbb{N}$: if $M_{m+2n} \neq 0$, $RM_{m+2n} = M_{m+n}$, hence $RM_{m+n} = M_{m+n} = M_m$, contradicting $B(M) = 0$.
Let e be maximal such that $M_e \not\subset L$, then $e < d$. Look at the fragment given by the chain

$$M \supset M_1 + L \supset \ldots \supset M_e + L \supsetneq L \supset \ldots \supset L \supset \ldots.$$

It is obviously an FR-fragment, say M', with $M'_0 = M$ and $B(M') = L$. Let us check that it is irreducible; consider an FR-subfragment of M' given by the chain:

$$\begin{array}{ccccccccccccc} M & \supset & M_1 + L & \supset & \ldots & \supset & M_e + L & \supsetneq & L & \supset & \ldots \supset & L & \supset \ldots \\ \cup & & \cup & & & & \cup & & \cup & & & \cup & \\ N & \supset & N_1 & \supset & \ldots & \supset & N_e & \supset & N_{e+1} & \supset & \ldots \supset & N_{e+j} & \supset \ldots \end{array}$$

If $N_{e+j} = 0$ for some j then N is trivial in M' so assume all $N_{e+j} \neq 0$. Then $0 \neq RN_{e+n+1} \subset N_{e+1} \subset L$, hence the R-irreducibility of L yields that $RN_{e+n+1} = N_{e+1} = L$ and similar $N_{e+j} = L$ for all $j \geq 1$. Since we have $M_{e+1} \subset L = N_{e+1}$ we have $M_{e+1} \subset N_{e+1}$ and similar $M_{e+j} \subset N_{e+j}$ for all $j \geq 1$. Look at the chain N' in the chain of M:

$$N' = N \supset M_1 \cap N_1 \supset \ldots \supset M_e \cap N_e \supset M_{e+1} \supset \ldots \supset M_{d-1} \supsetneq 0 \supset \ldots$$

(where $d \geq e+1$). If $N_e = L$ then N is trivial in M' so we assume $N_e \supsetneq L$. Irreducibility of M entails the existence of a monotone increasing map $\alpha : \mathbb{N} \to \mathbb{N}$ such that $M_i \cap N_i = M_{\alpha(i)}$, but then again el(M) = el(N') entails (since $M_{d-1} = M_{\alpha(d-1)} = 0$ is excluded) $M = N$, $M_i \cap N_i = M_i$ and $M_e \cap N_e = M_e$. Now from $M_i \subset N_i$ and $L \subset N_i$ we obtain $N_i = M_i + L$ for $i = 1, \ldots, e$ and $N_{e+j} = L$ for $j \in \mathbb{N}$, thus $N = M$. We conclude that M' is indeed irreducible. □

Remark 3.2.2. In the foregoing proposition it is possible to prove that $e = d-1$, i.e. $M_{e+1} = 0$. If $d - 1 > e$ then $M_{d-1} \subset L$, so if we take $x \neq 0$ in M_{d-1} then on one hand we have a subfragment $F_{d-1}Rx \supset \ldots \supset F_0Rx \neq 0$ which cannot be trivial of type I, hence $M = F_{d-1}Rx$, $M_e = F_{d-e+1}Rx$ but from $x \in L$ it follows that $M_e \subset L$, contradicting the choice of e. Thus $M_{e+1} = 0$, i.e. $e = d - 1$.

Now we restrict to the situation inspired by representation theory and the group ring KG for a finite group G. Let A be a K-algebra and consider a right bounded filtration FA given by $K = F_0A \subset F_1A = A_1 \subset \ldots \subset F_nA = A \subset \ldots$, where each A_i is a K-subalgebra of A. Then $F_iAF_jA = F_{\max\{i,j\}}A$ for every i, j. For short, we call such filtrations FA finite algebra filtrations. If each A_i is a semisimple Artinian algebra then call the filtration a semisimple filtration. For example, $K \subset KG_1 \subset \ldots \subset KG_i \subset \ldots \subset KG$ where $(\text{char}K, |G|) = 1$ is a semisimple filtration. We shall use the weaker notion of irreducibility: M is a weakly irreducible FR-fragment (for general FR) if every strict subfragment of M is trivial.

From here on we let R be a K-algebra A and FA a finite algebra filtration as defined above. Our aim is to obtain a decomposition of FA-fragments into (weakly) irreducible ones. We start with a few general remarks.

Lemma 3.2.3. *Let FA be a finite algebra filtration and M a weakly irreducible FA-fragment such that $M \neq B(M)$; then there is a $d \in \mathbb{N}$ such that $M_d \neq B(M)$ and d is maximal as such. For d as above, $M_i = F_{d-i}AM_d$, for $0 \leq i \leq d$.*

Proof. Suppose that $M_m \neq B(M)$ for all $m \in \mathbb{N}$. Then the chain $F_nAM_n \supset M_1 \cap F_nAM_n \supset \ldots \supset M_n \cap F_nAM_n = M_n \supset \ldots$ is a strict subfragment of M with body $B(M)$ ($\neq M_n$ for every n). Thus either $M_n = M_{\alpha(n)}$ with $\alpha(n) > n, M_m = M_{\alpha(m)}$ for every $m \geq n$. This yields $M_n = B(M)$, a contradiction. Hence $M = F_nAM_n = F_{n+1}AM_{n+1} = \ldots = F_{n+i}AM_{n+i} = \ldots$, for $i \in \mathbb{N}$. But $M = F_{n+i}AM_{n+i} = F_nAM_{n+i} \subset M_i$, thus $M = M_1 = \ldots = M_i = \ldots$, or $M = B(M)$, a contradiction. Consequently such a d exists and the subfragment of M given by

$$F_dAM_d \supset F_{d-1}AM_d \supset \ldots \supset M_d \supset B(M) \supset \ldots$$

is trivial, from which the last statement follows. □

Recall that for a strict epimorphism $M \xrightarrow{\pi} \overline{M}$ of fragments, a subfragment $\overline{L}$ of $\overline{M}$ is strict if and only if $\pi^{-1}(\overline{L})$ is strict in M, see Lemma 2.2.3.

Lemma 3.2.4. *If T is a strict and trivial subfragment of a weakly irreducible M then M/T is weakly irreducible. In particular, $M/B(M)$ is weakly irreducible.*

Proof. As in Proposition 3.1.5 but restricting to a strict subfragment N of M/T and $N = \pi^{-1}(\overline{N})$, where π is the strict fragmented epimorphism $\pi : M \to M/T$. Then case 2. and case 3. are identical. For case 1. we now use that $N_{e+1} = B(N)$ yields $T_{e+1} \subset B(N) \subset B(M)$ and then the proof continues completely in the same way. Also el(M/T) = el(M). □

Corollary 3.2.5 (version of Proposition 3.1.7). *If L is an R-submodule of a weakly irreducible fragment M with* el(M) = d *then for the induced fragment* *structure on L we have that M/L is weakly irreducible with* el(M/T) = d *unless $M/L = 0$.*

Proof. Since L is an R-module in M, $L \supset L \cap M_1 \supset \ldots \supset L \cap M_d \supset L \cap B(M) \supset \ldots$ is a strict subfragment of M. If $L \cap M_d = L \cap B(M)$ then this fragment is trivial in M and then the result follows from the previous lemma. If $L \cap M_d \neq L \cap B(M)$ one follows the same reasoning as in the proof of Proposition 3.1.7 to arrive at $M = L$ or $M/L = 0$. □

The foregoing, together with the observations made after Proposition 3.1.8, allows us to reduce the study of weakly irreducible fragments to weakly irreducible fragments of essential length lesser or equal to n, where $F_nA = A$. In the latter case, we deal with glider representations. In the former case, we can reduce to pure fragments.

We have a version of Proposition 3.2.1 for weakly irreducible fragments.

Proposition 3.2.6. *Let FA be a finite algebra filtration and let M be a finitely generated weakly irreducible FR-fragment with $B(M) = 0$ (in view of the statement and proof this is not really restrictive). Then either M is a pure fragment or M contains an irreducible R-module, L say. In the latter case there is a new fragment structure M' on M, i.e. $M'_0 = M_0 = M$, such that $B(M') = L$ and M' is again weakly irreducible.*

Proof. If M is not pure, we may select an irreducible A-module L in M by the Artinian property of M (M finitely generated fragment). We already know that the chain of M is given by:

$$M = M_0 \supset M_1 \supset \ldots \supset M_d \supsetneq 0 \supset 0 \supset \ldots$$

for some $d \in \mathbb{N}$. Let $e \in \mathbb{N}$ be such that $M_e \not\subset L$, e maximal as such; of course we have $e > 0$. If $e < d$ then $M_d \subset L$. We have a strict subfragment of M given by the chain $F_dAM_d \supset F_dAM_d \cap M_1 \supset \ldots \supset F_dAM_d \cap M_d = M_d \supsetneq 0 \supset \ldots$. This can only be trivial in M if $F_dAM_d = M$, hence $L = M$, a contradiction. Consequently $e = d$. Now look at M' given by:

$$M \supset M_1 + L \supset \ldots \supset M_d + L \supsetneq L \supset L \supset \ldots$$

We claim that it is weakly irreducible again and of course $B(M') = L \neq 0$. Look at a strict subfragment N of M' given by:

$$\begin{array}{ccccccc} N & \supset & N_1 & \supset \ \ldots \ \supset & N_d & \supset & L \cap N \supset L \cap N \supset \ldots \\ & & \| & & \| & & \\ & & M'_1 \cap N & & M'_d \cap N & & \end{array}$$

If $L \cap N = 0$ then N is trivial in M', if $L \cap N \neq 0$ then $L \cap N = L$ since $L \cap N = B(N)$ is an R-module and L is irreducible, i.e. we may assume $N_d \supset L$. If $N_d = L$ then N is trivial in M' so assume $N_d \supsetneq L$. Look at the chain $N \supset M_1 \cap N_1 \supset \ldots \supset M_d \cap N_d \supset 0 \supset \ldots$. Since $L \subsetneq N_d \subset M_d + L$, an $x \in N_d - L$ is of the form $m_d + \lambda$ with $m_d \in M_d$, $\lambda \in L$. Hence $m_d = x - \lambda \neq 0$ is in $M_d \cap N_d$. This shows that $M_d \cap N_d \neq 0$. The last chain on N is a strict subfragment of M hence it can only be trivial if $M = N, \ldots, M_d \cap N_d = M_d$, i.e. $M_i \subset N_i$, or $M_i + L \subset N_i$, thus $N = M'$. Consequently M' is weakly irreducible. □

If a fragment M with el(M) = d contains an irreducible fragment N also of el(N) = d then we have $N_i = F_{d-i}AN_d$ for $i = 0, \ldots, d$ and $N_d \supsetneq N_{d+1} = B(N)$. Thus $N \supset N \cap M_1 \supset \ldots \supset N \cap M_d \neq B(N) \supset B(N) \supset B(N)(= B(M) \cap N)$ is a subfragment of M since $F_iAN = N$ for $i \leq d$. We denote the latter subfragment by $N^{(M)}$; it is obviously a strict subfragment of M. We say that two irreducible subfragments N and N_1 of M are M-equivalent if $(N_1^{(M)})_0 = (N^{(M)})_0$. The set of all M-equivalent irreducible subfragments of essential length exactly d is called an M-cycle. The existence of non-singleton M-cycles is the reason for the failure of a decomposition of fragments into a direct sum of irreducible subfragments; we will establish that "decomposition" up to strictness (to be defined correctly later) will work well.

Lemma 3.2.7. *Let M be a Noetherian fragment of* el(M) = d *over an algebra filtration FA on an Artinian ring A. The weakly irreducible subfragments E of essential length exactly d are defined on the set underlying an irreducible subfragment N of M, i.e. $E_0 = (N^{(E)})_0$.*

Proof. If E is a weakly irreducible subfragment of M with el(E) = d then E contains an irreducible subfragment N of M because M is Noetherian, so we can use the Artinian properties (Corollary 2.3.7). Then E contains $N^{(E)}$ as a strict subfragment and this can only be trivial in E when $E_0 = (N^{(E)})_0 = N$. □

Remark 3.2.8. Since $F_0A = K$, Noetherian fragments are the same as finitely generated ones when we consider $\dim_K(A) < \infty$. Indeed, being finitely generated implies that M is a finite dimensional K-vector space. Since A is Artinian, the fragment chain is stationary, from which it follows that the fragment is indeed Noetherian.

Proposition 3.2.9. *Let M be a Noetherian fragment with* el(M) = d *and $B(M) = 0$ (we can reduce to this by passing to $M/B(M)$). If N is an irreducible subfragment of M of* el(N) = d *contained in a weakly irreducible $E = N^{(M)}$ then an M-cycle of N is obtained by taking irreducible fragments $N_{(\gamma)}$ corresponding to $K\gamma \neq 0$ in $M_d \cap N$.*

Proof. For $\gamma \in M_d \cap N$ nonzero look at $N_{(\gamma)} = F_dA\gamma$ with chain $(N_{(\gamma)})_i = F_{d-i}A\gamma$. These are irreducible subfragments of M with $N = (N_{(\gamma)})_0$ $(\gamma \in N)$. Clearly $N_{(\gamma)}^{(M)} = N^{(M)} = E$. □

A set $\{N^{(1)},\ldots,N^n\}$ is a generating subcycle for a subfragment E of M with el(E) = d if all $N^{(i)}$ are irreducible subfragments of M with el($N^{(i)}$) = d which are subfragments of E such that if N is any subfragment of E irreducible in M then $N \subset \sum_{i=1}^n N^{(i)}$. For example if E is weakly irreducible in M with el(E) = el(M) = d then the foregoing proposition shows that the $N_{(\gamma_i)}$ corresponding to a K-basis of E_d are a generating subcycle for $E = N^{(M)}$.

Now look at a natural glider M, for $X \subset M$ we write $X_j^* = \{x \in X,\ F_jRx \subset X\}$. The decomposition problem is reduced to two subproblems: 1. decomposition into weakly irreducible subfragments. 2. the relation between weakly irreducible subfragments E of M and their generating subcycles.
If N is an irreducible subfragment of M then $N^{(M)}$ is weakly irreducible and a strict subfragment of M; if moreover M is natural then a strict subfragment of M is also natural (Lemma 1.6.2). So if $E = N^{(M)}$ then $E = E^*$ and $E = N^*$ for every irreducible subfragment N of M such that $E = N^{(M)}$. We may look at $F_0A = K \subset F_1A = A$; the theory is then essentially a (K, A)-relative theory of K-modules contained in A-modules. The presence of other F_iA allows at least the variation of the essential length in natural irreducible fragments. The structure of arbitrary irreducible fragments (i.e. not necessarily natural) is of a much higher complexity.

Proposition 3.2.10. *Let A be a finite dimensional K-algebra with semisimple filtration FA. Let M be a natural Noetherian (hence Artinian in this case) FA-fragment. Every strict subfragment N of M is a direct summand.*

Proof. Let M be of the form $M \supset M_1 \supset \ldots \supset M_d \supset B(M) \supset \ldots$. Then N is of the form $N \supset N_1 \supset \ldots \supset N_d \supset B(N) \supset B(N) \supset \ldots$ with $N_i = M_i \cap N$ and $B(N) = B(M) \cap N$. If $B(M) \neq 0$ and $B(N) \neq 0$ then $B(M) = B(N) \oplus L$ because A is semisimple. Then $M_d \supset M_d \oplus L$ (as $N_d \cap L = N_d \cap B(M) \cap L = N \cap B(M) \cap L = B(N) \cap L = 0$), as F_dR-modules (since M is natural $M_d = \{m \in M,\ F_dAm \subset M\}$ and $F_dAF_dA = F_dA$); hence since F_dA is semisimple, $M_d = N_d \oplus L \oplus N_d''$ for some F_dA-module N_d". Put $N_d' = L \oplus N_d''$ and $M_d = N_d \oplus N_d'$, with $N_d' \supset L$. If we have obtained $M_i = N_i \oplus N_i'$ with $L \subset N_d' \subset \ldots \subset N_i'$ then $M_{i-1} \supset N_{i-1} \oplus N_i'$ because N_i' is an F_iA-module hence $F_{i-1}A$-module and $N_{i-1} \cap M_i = N_i$. Since $F_{i-1}A$ is semisimple we have $M_{i-1} = N_{i-1} \oplus N_i' \oplus L_{i-1}$; put N_{i-1}' equal to $N_i' \oplus L_{i-1}$. Hence we arrive at a chain $N' \supset N_1' \supset \ldots \supset N_d' \supset B(N') = L \supset L \supset \ldots$ such that $M_i = N_i \oplus N_i'$ for all i. The reader moreover convinces him or herself that N' is also a strict subfragment. □

For non-natural glider representations we do not have this nice property.

Example 3.2.11. Consider the semisimple filtration $K \subset KG \subset KG \subset \ldots$ and let V be an irreducible 2-dimensional G-representation. The fragment $V \supset Kv \supset 0 \supset \ldots$ is a non-natural glider and has $Kv' \supset 0 \supset \ldots$ as a strict subfragment for suitable $v' \in V$. This subfragment has no fragment direct summand.

Proposition 3.2.12. *In the situation of Proposition 3.2.10, every natural FA-fragment is a direct sum of weakly irreducible subfragments.*

Proof. If M is weakly irreducible there is nothing to prove so assume it is not. Since M is an Artinian fragment we may find a (minimal) weakly irreducible subfragment N of M which is a strict subfragment (the strictification of an irreducible subfragment). In view of Proposition 3.2.10, $M = N_1 \oplus N_1'$ and by construction N_1' is again a strict subfragment of M. Therefore N_1' is a natural, resp. glider too. By repetition of the argument for N_1', etc..., we obtain the decomposition $M = N_1 \oplus N_2 \oplus \ldots \oplus N_t$ with N_i, $i = 1, \ldots, t$ being weakly irreducible strict subfragments of M. Note that in each step a new (minimal) weakly irreducible subfragment of M is constructed in N_i by using the Artinian property on weakly irreducible subfragments of M contained in N_i. □

Now look at an irreducible glider representation N given by

$$N \supset N_1 \supset \ldots \supset N_d \supset 0 \supset \ldots$$

where N_d is a 1-dimensional K-space (A a K-algebra). If u is a unit of F_dA then we can also look at

$$F_dAuN_d = F_dAN_d = N \supset F_{d-1}AuN_d \supset \ldots \supset uN_d \supset 0 \supset \ldots$$

and we assume u is chosen such that $uN_d \neq N_d$. The latter is an irreducible fragment and we obtain a non-irreducible glider representation

$$N \supset N_1 + F_{d-1}AuN_d \supset \ldots \supset N_d \oplus uN_d \supset 0 \ldots$$

containing at least the two irreducible fragments we used in the construction. These two are not disjoint so the sum is not a direct sum. Nevertheless we have $N_d \cap N_d u = 0$. Therefore, we say that a sum of fragments $E + F$ is direct if for some $i \leq \text{el(E)}, \text{el(F)}$ we have that E_i is disjoint from F_i. A motivation for this definition is the information given by the chain of the fragment. A direct sum on every level would be too set- and module-theoretic.

Remark 3.2.13. Recall that we defined the fragment direct sum $\oplus_\alpha M_\alpha$ in Section 1.4. As we discussed above, for a nice decomposition result we have to work with a weaker notion of fragment direct sum defined here. From now on, we will refer to the direct sum as defined in Section 1.4 by the strong (or categorical) fragment direct sum. The weaker notion of fragment direct sum will be termed fragment direct sum and we will denote this by $\dot{\oplus}$.

Theorem 3.2.14. *Let FA be a finite algebra filtration on a finite dimensional K-algebra A and with $F_0A = K$. Let M be a Noetherian (hence Artinian and finite dimensional over K) fragment with $B(M) = 0$ and* el(M) = d. *Then M is the fragment direct sum of irreducible subfragments.*

Proof. A K-vector space decomposition $M_d = \oplus_{i=1}^t N_d^i$ into one dimensional K-spaces N_d^i, $i = 1, \ldots, t$, yields F_dAM_d as a fragment direct sum of irreducible subfragments $F_dAN_d^i$ of essential length exactly d. Let L_{d-1} be a K-complement of F_1AM_d in M_{d-1} and decompose it into a direct sum of 1-dimensional K-subspaces $L_{d-1} = \oplus_{j=1}^{t_1} N_{d-1}^j$; then the $F_{d-1}AN_{d-1}^j$ are irreducible subfragments of essential length exactly $d-1$ and so $F_dAM_d + F_{d-1}AM_{d-1}$ is decomposed as a fragment direct sum of irreducible subfragments of M. So we arrive at a decomposition of

$$F_dAM_d + F_{d-1}AM_{d-1} + \ldots + F_1AM_1 = N.$$

Let E be a K-complement of N in M, $E = \oplus_{k=1}^e E_k$, where the E_k are 1-dimensional K-subspaces. The subfragment $E_k \supset 0 \supset 0 \supset \ldots$ is irreducible in M, $E_k \cap N = 0$; hence we arrive at a fragmented direct sum of irreducible subfragments giving E, $F_1AM_1, \ldots, F_dAM_d$ respectively as fragmented direct sums of irreducible subfragments of M with essential length equal to $0, \ldots, d$ respectively for the new subfragments appearing in M_{i-1} not in $F_{j-i+1}AM_j$, $j = i, \ldots, d$. □

The results of Proposition 3.2.12 and Theorem 3.2.14 not only show the different nature of fragment and module theory, but they also highlight the difference of the two direct sums explained in Remark 3.2.13. Observe that in Theorem 3.2.14 we could generalize the condition $F_0A = K$ to F_0A being semisimple. For Proposition 3.2.12 we really need to have that FA is a semisimple algebra filtration, but we obtain a decomposition into strong direct sums. We will get back to these somewhat subtle observations when we introduce the glider representation ring of a chain of groups. Speaking of which, for applications to the case of the group algebra KG of a finite group with respect to a semisimple algebra filtration $K \subset KG_1 \subset \ldots \subset KG_n = KG$ the natural gliders are chains of KG_i-modules $M \supset M_1 \supset \ldots \supset M_d \supset B(M) \supset B(M) \supset \ldots$ for $d \leq n$, without connecting conditions between M_i and M_{i-1} (see Example 1.3.2). Exactly non-natural gliders of the type $M = F_{n-1}AM_{n-1}$,

$$M = M_0 \supset M_1 \supset \ldots \supset M_{n-1} \supsetneq B(M) \supset \ldots$$

contain new information linking the representation theory for the subgroups of G, $M_i = F_{n-1-i}AM_{n-1}$ is an $F_{n-1-i}A$-module for $i = 0, \ldots, n-1$ and this is sort of opposite to the situation for natural gliders (where M_i is an F_iA-module by construction). These effects are studied in what we call glider representation theory of a chain of finite groups (see Chapter 4).

3.3 Examples

We give a non-exhaustive list of situations where fragments and glider representations appear in a natural way. Many of these examples give inspiration to apply the general fragment theory in these more concrete situations. The reader will notice that Part II and III of the book are in fact based on some examples enlisted here.

a) Look at any $\mathbb{Z}$-filtration FR on a ring R; F_+R is the chain given by $F_0R \subset F_1R \subset \ldots \subset R$, F_-R is the chain given by $\ldots \subset F_{-n}R \subset \ldots \subset F_0R$, $n \geq 0$. Then $F_0R = S$ is an F_+R-fragment with its chain F_-R. We have F_0R contained in R but it need not be a natural fragment because $F_{-n}R \subset \{r \in F_0R,\ F_nRr \subset F_0R \text{ for } n \geq 0\}$ but it need not be equal. For example if $R = K$, a field and FR is the valuation filtration corresponding to a discrete valuation ring $O_v = F_0R$ then O_v becomes a natural fragment in the way explained above. For general FR the F_+R fragment F_0R is saturated exactly if $\cap_n F_{-n}R$ is the largest common ideal of F_0R and R.

b) Let K be a field, $S = K[X]$, $R = K[X, X^{-1}]$, X a variable over K. Define $F_iR = \sum_{j \leq i} K[X]X^{-j}$, $S_i = (X^i)$, $i = 0, 1, \ldots$. Then $K[X]$ is an FR-fragment. For any pair of rings $S \subset T$ and an invertible ideal I of S (i.e. there is an S-bimodule $J \subset T$ such that $IJ = JI = S$) such that $T = \cup_{n \geq 0} J^n$, we may view S as an FT-fragment where $F_0T = S$, $F_nT = J^n$ for $n \geq 0$, with respect to the chain $S_i = I^i$. On any S-module N one may define the induced negative (I-adic) filtration and view it as an FT-fragment.

c) Let V be an affine algebraic variety over a field K with coordinate ring $S = \Gamma[V] = K[\underline{X}]/I(V)$. Let $\underline{O}_V$ be the structure sheaf of V and let $\underline{O}_M$ be a coherent $\underline{O}_V$-module. To an ideal J of $\Gamma[V]$ there corresponds a Zariski open set $X(J)$ in Spec(Γ[V]). The ring of sections $\Gamma(X(J), \underline{O}_V) = Q_J(\Gamma[V]) = R(J)$ is the localization of $\Gamma[V]$ at the Gabriel filter generated by J. Define $F_nR(J) = [S : J^n] = \{x \in R(J),\ J^n x \subset S\}$. Then we have $F_0R(J) = S = \Gamma[V]$ and $FR(J)$ is a positive filtration on $R(J)$. Consider an S-module $M(J)$ with descending chain: $M(J) = M_0 \supset M_1 = JM(J) \supset \ldots \supset J^nM(J) \supset \ldots$. For $m \geq n$ we have

$$F_nR(J)M(J)_m = [S : J^n]J^mM(J) = [S : J^n]J^nJ^{m-n}M(J) \subset J^{m-n}M(J);$$

thus $M(J)$ is an $FR(J)$-fragment. So $\Gamma[V]$-modules appear as fragments over any Zariski open $X(J)$.

Coordinate rings of affine algebraic varieties come equipped with a natural, standard filtration, given by $\deg(X_i) = 1$ for the appearing variables. This observation leads to the development of a filtered localization theory and filtered structure sheaves. This will be discussed in detail in Chapter 7.

d) Consider a positively graded ring $\widetilde{A}$ containing a regular central homogeneous element of degree one, say X. Let $\widetilde{A}_X$ be the central localization of $\widetilde{A}$ at $\{1, X, X^2, \ldots\}$. Put $B = (\widetilde{A}_X)_0$ and define $F_nB = \widetilde{A}_0 + \widetilde{A}_1X^{-1} + \ldots + \widetilde{A}_nX^{-n}$, Then $F_0B = \widetilde{A}_0$ and FB is a positive filtration of B. For an $\widetilde{A}$-module M the chain $M \supset XM \supset \ldots \supset X^nM \supset \ldots$ defines an FB-fragment structure on M. This situation arises for a positively filtered ring A with filtration FA and $\widetilde{A}$ the Rees ring associated with FA. Taking for X the canonical element in $\widetilde{A}$ for which $\widetilde{A}/X\widetilde{A} = G_F(A)$, the associated graded ring corresponding to FA, and $\widetilde{A}/(1-X)\widetilde{A} = A$, then $\widetilde{A}_X \cong A[X, X^{-1}]$ and $(\widetilde{A}_X)_0 = A$. So every $\widetilde{A}$-module is realized as an FA-fragment.

e) Let R be a commutative ring, say a Noetherian domain (but the following may be generalized in the absence of this assumption). To an ideal I there is a localization $Q_I(R)$ at the Gabriel filter generated by I. The chain $R \supset I \supset I^2 \supset \ldots$ makes R into a fragment with respect to $FQ_I(R)$ where $F_0Q_I(R) = R$, $F_nQ_I(R) = [R : I^n] = \{x \in Q_I(R),\ xI^n \subset R\}$. Since $R \subset [R : I^n]$ we have that $I^n \subset I^n[R : I^n] \subset R$, hence $I^n[R : I^n]$ is in the Gabriel filter generated by I. Let $I^{(n)}$ be maximal with respect to the property that $[R : I^{(n)}] = [R : I^n]$; then we obtain that $I^{(n)}I^{(m)} \subset I^{(n+m)}$ and the chain $I^{(n)}$ is maximal with the property of defining a fragment structure on R with respect to $FQ_I(R)$ given by the $[R : I^n]$, $n \geq 0$. We obtain the symbolic I-adic filtration of R from the maximality condition on the fragment structure.

f) Let $\mathbb{A}_1(\mathbb{C})$ be the first Weyl algebra, i.e. the quotient of the free algebra $\mathbb{C}\langle X, Y\rangle$ by the two-sided ideal generated by $YX - XY - 1$. This may also be seen as the differential polynomial ring $\mathbb{C}[X][Y, \frac{\partial}{\partial X}]$ being the Ore extension of the polynomial ring $\mathbb{C}[X]$ by a variable Y with multiplication defined by $YP(X) - P(X)Y - \frac{\partial P(X)}{\partial X} = 0$, for $P(X) \in \mathbb{C}[X]$. This is an example of a ring of differential operators. Define a gradation on $\mathbb{A}_1(\mathbb{C})$ by putting $\deg(X) = -1 = -\deg(Y)$; thus $\mathbb{A}_1(\mathbb{C})_0 = \mathbb{C}[XY]$, $\mathbb{A}_1(\mathbb{C})_1 = \mathbb{C}[XY]Y$, $\mathbb{A}_1(\mathbb{C})_{-1} = \mathbb{C}[XY]X$ (note that expressions like X^mY^m can be rewritten as polynomials in XY). The filtration associated with this gradation has $F_0^g\mathbb{A}_1(\mathbb{C}) = \oplus_{n\leq 0}\mathbb{A}_1(\mathbb{C})_n$ equal to the ring generated by XY and X over $\mathbb{C}$ in $\mathbb{A}_1(\mathbb{C})$, i.e. the ring $\mathbb{C}[X][XY, X\frac{\partial}{\partial X}]$ because $(XY)X - X(XY) = X$. More generally, commuting a polynomial $P(X)$ in X with XY or a power of X will not reduce the degree of the polynomial; hence one quickly verifies that $F_0^g(\mathbb{A}_1(\mathbb{C}))$ with chain given by the induced negative filtration (in fact the X-adic filtration because $F_0^g(\mathbb{A}_1(\mathbb{C}))X$ is a an ideal of $F_0^g(\mathbb{A}_1(\mathbb{C}))$), is a fragment for the positive part of the gradation filtration on $\mathbb{A}_1(\mathbb{C})$. Similarly we have that $\mathbb{C}[X][XY, X\frac{\partial}{\partial X}]$-modules endowed with the descending chain coming from the induced X-adic filtration will be fragments for $\mathbb{A}_1(\mathbb{C})$. On the other hand we may look at the operator filtration $F^o\mathbb{A}_1(\mathbb{C})$ defined by

$$F_0^o\mathbb{A}_1(\mathbb{C}) = \mathbb{C}[X] \subset \mathbb{C}[X] + \mathbb{C}[X]Y \subset \ldots \subset \mathbb{C}[X]Y^n + \mathbb{C}[X]Y^{n-1} + \ldots \subset \ldots$$

Now consider again the ring generated by XY and X in $\mathbb{A}_1(\mathbb{C})$, $B = \mathbb{C}[X][XY, X\frac{\partial}{\partial X}]$ with descending chain given by the powers of the ideal (X) in it. A straightforward calculation learns that B is a fragment with respect to the operator filtration.

Similarly, B-modules with the chain induced by the (X)-adic filtration are also F^o-fragments. The operator filtration F^oA as well as the grading filtration F^gA (but here with $\deg(Y) = 1$ and not $\deg(Y) = -1$ as is sometimes done) allow B-modules as examples of fragments without even changing the fragment chain. Note that X is a normalizing element of B, i.e. $BX = XB$, so we may invert it by localizing at the Ore set $\{1,\ X,\ X^2, \ldots\}$ and obtain $B_X = \mathbb{A}_1(\mathbb{C})_X$ because for every $n \geq 0$ we have:

$$X^n(\mathbb{C}[X]Y^n + \mathbb{C}[X]Y^{n-1} + \ldots + \mathbb{C}[X]) \subset B.$$

It is clear that B with chain $B \supset BX \supset BX^2 \supset \ldots$ is a fragment with respect to $B_X = \mathbb{A}_1(\mathbb{C})_X$ (this is a special case of a more general localization example for non-commutative rings). The fragment structure with respect to $\mathbb{A}_1(\mathbb{C})$ then follows also by a "restriction of scalars" argument if one observes that the localized filtration on $\mathbb{A}_1(\mathbb{C})_X$ induces the operator filtration again on $\mathbb{A}_1(\mathbb{C})$. The algebra B is in fact the enveloping algebra of the two dimensional Abelian Lie algebra. Going further we may generalize the example to the case of enveloping algebras of solvable Lie algebras because these are embedded into (higher) Weyl algebras; we do not do this in more detail here.

g) Gauge algebras and quantum $\mathfrak{sl}(2)$: The following algebra appears in connection with 3-dimensional Chern-Simmons gauge field theory with associated gauge group $SU(2)$: $W_q(\mathfrak{sl}(2)) = \mathbb{C}\langle X, Y, Z\rangle/\mathcal{R}$, where the ideal of relations is generated by:

$$\begin{aligned}
&\sqrt{q}XZ - \sqrt{q}^{-1}ZX = \sqrt{q+q^{-1}}Z,\\
&\sqrt{q}^{-1}XY - \sqrt{q}YX = -\sqrt{q+q^{-1}}Y,\\
&YZ - ZY = (\sqrt{q} - \sqrt{q}^{-1})X^2 - \sqrt{q-q^{-2}}X.
\end{aligned}$$

The algebra $W_q(\mathfrak{sl}(2))$ has finite dimensional representations that are deformations of simple finite dimensional representations of $U(\mathfrak{sl}(2))$ and $W_q(\mathfrak{sl}(2))$ is a quasi-quantum group (that is: tensor products of these representations are associative). When q is moving away from 1 then "bad" representations do exist. There is a central quadratic element $c = \sqrt{q}^{-1}ZY + \sqrt{q}YZ + X^2$ in $W_q(\mathfrak{sl}(2))$ as well as a minimalizing degree one element N we do not use here.

Put $A = 1 - (\sqrt{q} + \sqrt{q}^{-1})(\sqrt{q+q^{-1}})^{-1}c$; inverting this central element will allow us to define a "nicer" algebra. Substitute:

$$
\begin{aligned}
x &= (X - \sqrt{q} + \sqrt{q}^{-1})(\sqrt{q+q^{-1}})^{-1}c\sqrt{q+q^{-2}}^{-1}A^{-1},\\
y &= Y\sqrt{q+q^{-1}}^{-1}\sqrt{A}^{-1},\\
z &= Z\sqrt{q+q^{-1}}\sqrt{A}^{-1}.
\end{aligned}
$$

In the localized algebra we may rewrite $\mathcal{R}$ as generated by:

$$
\begin{aligned}
&\sqrt{q}xz - \sqrt{q}^{-1}zx = z,\\
&\sqrt{q}^{-1}xy - \sqrt{q}yx = -y,\\
&q^{-1}zy - qyz = x,
\end{aligned}
$$

call this ideal $\mathcal{R}'$. Now $U_q^w(\mathfrak{sl}(2)) = \mathbb{C}\langle x,y,z\rangle/\mathcal{R}'$ is a Hopf algebra having only good finite dimensional representations. The relation between $W_q(\mathfrak{sl}(2))$ and the Woronowicz quantum enveloping algebra of $\mathfrak{sl}(2)$, $U_q^w(\mathfrak{sl}(2))$, may be understood by using Rees rings of filtrations. By construction, localizing $W_q(\mathfrak{sl}(2))$ at $\sqrt{A} = t$ yields the same ring as localizing $U_q^w(\mathfrak{sl}(2))$ at $\sqrt{A}^{-1} = t^{-1}$. Hence modules over one algebra become fragments over the other by using the t-adic filtration (see second part of Example b)). The examples b), d), f) and g) show some connections with the idea of coverings of projective varieties.

h) Glider representations of finite groups: Let G be a finite group and H a subgroup of G; H may be trivial, i.e. $H = \{e\}$ is allowed. Suppose we are given a chain of subgroups $G_0 = H \subset G_1 \subset \ldots \subset G$ and consider a field K. We have a positive filtration on KG given by $F_0KG = KH \subset KG_1 \subset \ldots \subset KG$. If M is an FKG-fragment with chain $M = M_0 \supset M_1 \supset \ldots \supset M_n \supset \ldots$ we actually have a chain of KH-modules with an "action" of G_i on M_j for $i \leq j$ in the sense that $G_i \subset \mathrm{Hom}_K(M_j, M_{j-i})$. In case H is central in G then $G_i \subset \mathrm{Hom}_{KH}(M_j, M_{j-i})$ for $i \leq j$. Let us take $H = \{e\}$ for example. Since G is finite, let us exclude the situation where some proper G_i is repeated (infinitely many times) in the chain for G; then there is an $n \in \mathbb{N}$ such that $KG_n = KG$. If M is a natural FKG-fragment then we have seen that each M_i is a KG_i-module but in general a nice case (M irreducible fragment) actually arises with $M_{n-j} = KG_jM_n$, $j = 0, \ldots, n$.

In fact, the name "glider representation" comes from this example. Indeed, if one writes down the matrices for the action of the G_i in some K-basis for M obtained by repeatedly extending a K-basis for M_i to M_{i-1}, $i = n, n-1, \ldots, 0$, we obtain rectangular matrices. More general representations characterized by (special) fragments over group rings have to be further investigated in a general theory of fragment representations of finite groups. The idea is to obtain information about representations of chains of subgroups of G in one theory. For example, by restricting to a pair $G_1 \subset G$, one would obtain a relative representation theory for the pair (G_1, G): existing classical results like restriction and induction of representations fit in a general fragment theory but the latter could lead to new information not expressible in module theoretic ways in a natural way. We give an elaborate survey of the obtained results so far in Chapter 4.

Part II

Right bounded algebra filtrations

Chapter 4

Glider representation theory of a chain of finite groups

In part II we work over right bounded ring or algebra filtrations FR; i.e. there exists some d such that $R = F_dR = F_{d+1}R = \ldots$ and for all $0 \leq i \leq d$ the F_iR are rings or algebras. In this chapter we focus on such filtrations that arise by considering some finite group G with fixed chain of subgroups $1 < G_1 < \ldots < G_d = G$. When the subgroups are normal in the subsequent group we denote the chain by $1 \triangleleft G_1 \triangleleft \ldots \triangleleft G_{d-1} \triangleleft G_d = G$. If K is some field, we obtain the finite algebra filtration FKG on the group algebra given by $F_{-m}KG = 0$ for $m > 0$, $F_mKG = KG_m$ for $0 \leq m \leq d$ and $F_mKG = KG$ for $m > d$. Before we start our investigation of the glider theory for chains of groups, we recall some facts which will be useful. We know that weakly irreducible fragments have finite essential length (Lemma 3.2.3) and after Proposition 3.1.7 we deduced that we may restrict the study to fragments with zero body and essential length $e \leq d$. In fact, such fragments consist of a KG_e-module M with descending chain of KG_{e-i}-modules. From Example 1.3.2 we see that this is opposite to the case of natural gliders.

Our base field K is assumed to be algebraically closed and of characteristic 0. The finite algebra filtration on KG is then semisimple and after fixing the notion of a fragment direct sum (see Remark 3.2.13) we show that every Noetherian fragment is the fragment direct sum of irreducible subfragments (Theorem 3.2.14). A first section is devoted to irreducible gliders and when restricting to finite nilpotent groups we can show some results concerning classical representation theory using the general fragment theory. In the subsequent section we prepare the second ingredient to perform a Clifford theory for glider representations. We namely give a construction to induce a fragment to a 'bigger' chain of normal subgroups. The actual Clifford theory is then performed in the third section yielding different decomposition groups at every level of the chain of groups. Section 4.4 and Section 4.5 contain a further study of these decomposition groups and their relations. Finally, we include a section on nilpotent groups of order p^kq^l and we derive when an irreducible glider is the tensor product of gliders.

We assume the reader is familiar with group theory and knows a thing or two about representation theory of finite groups. This chapter is based on [16], [17].

4.1 Irreducible gliders

Let $M \supset M_1 \supset \ldots \supset M_e \supset 0 \supset \ldots$ be an irreducible glider of essential length $e \leq d$ with regard to the group algebra filtration associated with the chain of subgroups $1 < G_1 < \ldots < G_d = G$. By Lemma 3.2.3 we have that $M_i = KG_{e-i}M_e$ for $0 \leq i \leq e$. If $e < d$ then the glider is also irreducible with regard to the filtration associated with the shorter chain $1 < G_1 < \ldots < G_e$. Even in the case of a chain of length 1, i.e. $1 < G$, the question arises regarding what the irreducible gliders of essential length 1 really are. In this simple situation, such an irreducible glider takes the form

$$\Omega = M \supset M_1 = Ka$$

such that $KGa = M$. By Maschke's theorem, the group algebra KG is semisimple and decomposes into

$$KG = M_{n_1}(K) \oplus \cdots \oplus M_{n_k}(K), \tag{4.1}$$

where k is the number of conjugacy classes of G. Moreover, there are (up to isomorphism) k irreducible representations V_i of dimension n_i and in particular the order of the group G equals $\sum_{i=1}^{k} n_i^2$. We denote by $\mathrm{Irr(G)} = \{\mathrm{V_1, \ldots, V_k}\}$ a full set of representatives of the irreducible representations of G and we write G-rep for the category of G-representations. Also recall there is an equivalence of categories between KG-mod and G-rep.

Now suppose that $M \supset Ka$ is an irreducible glider; then $M = \oplus_{i=1}^{k} V_i^{m_i}$ and $a \in M$ is such that it generates M. To answer the question for which M and a we obtain irreducible gliders, we recall the notion of (left) annihilator ideals. Let $a \in M$, then

$$\mathrm{ann(a) = \{x \in KG \mid xa = 0\}.}$$

This is a left ideal of the group algebra. The left ideal of annihilators of the whole of M,

$$\mathrm{ann(M) = \{x \in KG \mid xM = 0\}}$$

is in fact a two-sided ideal and we have that

$$\mathrm{ann(M)} = \bigcap_{\mathrm{m \in M}} \mathrm{ann(m)}.$$

If $M = V_i$ is an irreducible module, then it follows immediately from (4.1) that

$$\mathrm{ann(V_i) = M_{n_1}(K) \oplus \cdots M_{n_{i-1}}(K) \oplus M_{n_{i+1}}(K) \oplus \cdots \oplus M_{n_k}(K),}$$

whence $\dim_K(\mathrm{ann(V_i)}) = \mathrm{|G| - n_i^2}$. If U is a K-vector space, we denote by $\mathbb{P}U$ the associated projective space. The following lemma will be crucial

Lemma 4.1.1. *Let* $U \in \mathrm{Irr(G)}, \mathrm{u}, \mathrm{u}' \in \mathrm{U}$. *Then* $\mathrm{ann(u)} = \mathrm{ann(u')} \Leftrightarrow \overline{\mathrm{u}} = \overline{\mathrm{u}'} \in \mathbb{P}\mathrm{U}$.

Proof. The 'only if' direction is clear, since $K \subset Z(KG)$. Suppose that the left annihilators are equal. Since U is simple, we have that $U = KGu = KGu'$. Define a morphism

$$\varphi : KGu \to KGu', \; xu \mapsto xu'.$$

This map is well-defined exactly because the annihilators are equal. Schur's lemma then yields the result. □

We start by looking at gliders $M \supset Ka$ for which $M = V^m$, with V an n-dimensional simple representation. To begin, consider $V \oplus V$. For every $a = v_1 + v_2 \in V \oplus V$, we obtain a submodule $KGa \subset V \oplus V$. By the uniqueness of decomposition, this KGa is isomorphic to V or $V \oplus V$.

Proposition 4.1.2. *Let* $V \in \mathrm{Irr(G)}$ *and suppose that* $a = v_1 + v_2 + \ldots + v_m \in V^m$ *is such that* $KGa \cong V \subset V^m$*; then all* $\overline{v_i}$ *are equal in* $\mathbb{P}V$.

Proof. We consider first the case $m = 2$, so suppose that $\mathrm{ann(v_1)} \neq \mathrm{ann(v_2)}$ and take some $x \in \mathrm{ann(v_1)}$, which does not annihilate v_2. Then $0 \neq xa = xv_2 \subset 0 \oplus V$. But since $V = KGxv_2$, we obtain that $0 \oplus V \subset KGa$, a contradiction. So both annihilators are equal and the result follows from Lemma 4.1.1. Now consider the general case $m > 2$. If $v_1 = 0$, then $KGa \subset V^{m-1}$, so we may assume that all $v_i \neq 0$. For any $i \neq j$, consider the projection of V^m onto the i-th and j-th factor. Then KGa is projected to $KG(v_i + v_j)$. Since $v_i + v_j$ is not zero in $V \oplus V$, we have that $KG(v_i + v_j) \cong V$ and the case $m = 2$ then shows that $\overline{v_i} = \overline{v_j}$. The result follows. □

Proposition 4.1.3. *Let* $V \in \mathrm{Irr(G)}$ *be* n*-dimensional and let* $\{v_1, \ldots, v_n\}$ *be a* K*-basis for* V. *Then for* $a = v_1 + \cdots + v_n \in V^n$ *we have that* $KGa = V^n$.

Proof. We have that $\mathrm{ann(V)} = \cap_{\mathrm{i=1}}^{\mathrm{n}} \mathrm{ann(v_i)}$. If $KGa \subset V^n$ is a proper submodule, it follows that $KGa \cong V^m$ for some $m < n$. Hence its K-dimension is nm. Consider the following short exact sequence of K-vector spaces

$$0 \to \mathrm{ann(a)} \to \mathrm{KG} \to \mathrm{KGa} \to 0.$$

We have that

$$\mathrm{ann(a)} = \bigcap_{\mathrm{i=1}}^{\mathrm{n}} \mathrm{ann(v_i)} = \mathrm{ann(V)},$$

hence its K-dimension is $|G| - n^2$. It follows that the K-dimension of KGa must be n^2, a contradiction. □

Remark 4.1.4. Let $\{v_1, \ldots, v_n\}$ be a K-basis for V. The short exact sequence

$$0 \to \mathrm{ann(v_i)} \to \mathrm{KG} \to \mathrm{KGv_i} \to 0$$

shows that $\dim(\mathrm{ann(v_i)}) = |\mathrm{G}| - \mathrm{n}$. This also follows from the decomposition of the group algebra KG into a direct product of matrix algebras. Indeed, for v_1, we are searching for matrices in $M_n(K)$ that kill $(1, 0, \ldots, 0)^t$ and its dimension as a K-space is exactly $n^2 - n$. This reasoning allows us to deduce that

$$\dim(\bigcap_{i=1}^{m} \mathrm{ann(v_i)}) = |\mathrm{G}| - \mathrm{mn}.$$

As a corollary, we obtain a generalization of Proposition 4.1.3

Corollary 4.1.5. *Let $V \in \mathrm{Irr(G)}$ be n-dimensional and let $\{v_1, \ldots, v_m\}$ be $m \leq n$ linear independent vectors in V. Then for $a = v_1 + \cdots + v_m \in V^m$ we have that $KGa = V^m$.*

We denote by π_i the projection from V^n onto the i-th factor and by $\pi_{\hat{i}}$ the projection onto V^{n-1} which misses the i-th factor. Moreover, we denote by V_i the i-th component in V^n, that is, $V_i = 0 \oplus \cdots \oplus 0 \oplus V \oplus 0 \oplus \cdots \oplus 0$, where the V is in the i-th place.

Lemma 4.1.6. *Let $V \in \mathrm{Irr(G)}$ and suppose that $a = v_1 + v_2 \in V \oplus V$. Then $KGa \cong \pi_i(KGa)$ if and only if $KGa \cong V$ for $i = 1, 2$.*

Proof. If $KGa = V \oplus V$, then $\pi_i(KGa) = V$. If $KGa \cong V$, then the embedding is diagonal by Proposition 4.1.2. □

Lemma 4.1.7. *Let $V \in \mathrm{Irr(G)}$ and suppose that $a = v_1 + \cdots + v_n$. Then $v_i \in \langle v_j, j \neq i \rangle$ if and only if $\pi_{\hat{i}}(KGa) \cong KGa$.*

Proof. If $\pi_{\hat{i}}(KGa) \ncong KGa$, then $\mathrm{Ker}(\pi_{\hat{i}}) \cap KGa = V_i \cap KGa \neq 0$. Hence, we find some $x \in KG$ such that $(0, \ldots, 0, xv_i, 0, \ldots, 0) \in KGa$. If, however, v_i is a linear combination of the v_j, then any x that annihilates all v_j, also kills v_i, hence $v_i \notin \langle v_j, j \neq i \rangle$. Conversely, suppose that $v_i \notin \langle v_j, j \neq i \rangle$. We have that $0 \neq \mathrm{ann(V)} \subset \bigcap_{\mathrm{j} \neq \mathrm{i}} \mathrm{ann(v_j)}$. If $\mathrm{ann(v_i)} = \bigcap_{\mathrm{j} \neq \mathrm{i}} \mathrm{ann(v_j)}$ then we can define a morphism

$$\varphi : KGv_i \to KGv_1 \oplus \cdots \oplus \hat{KGv_i} \oplus KGv_{i+1} \oplus \cdots \oplus KGv_n,\ v_i \mapsto (v_1, \ldots, \hat{v_i}, \ldots, v_n).$$

By Schur's lemma we know that $\varphi = (\lambda_1, \ldots, \lambda_n)$ for some $\lambda_j \in K$. Hence, $v_i = \lambda_j v_j$, a contradiction. So $\mathrm{ann(v_i)} \neq \bigcap_{\mathrm{j} \neq \mathrm{i}} \mathrm{ann(v_j)}$ and we find an element $x \in KG$ that annihilates all v_j, except for v_i. The element xa then sits in $\mathrm{Ker}(\pi_{\hat{i}}) \cap KGa$. □

In particular, if V is n-dimensional, and we work in V^{n+1}, then for any choice of $n+1$ nonzero vectors in V, we have that $\pi_{\hat{1}}(KGa) \cong KGa$. Since the projection is inside V^n, KGa can at most be V^n.

Theorem 4.1.8. *Let $V \in \mathrm{Irr(G)}$ and let $v_1, \ldots, v_n \in V$. If $a = v_1 + \cdots + v_n \in V^n$, then $KGa \cong V^m$ for some $m \leq n$ if and only if $\dim(\langle v_1, \ldots, v_n \rangle) = m$.*

Proof. Suppose that V is l-dimensional. By the remarks above the theorem, we may reduce to $n \leq l$. The cases $(m,n) = (1,n)$ for all $n \leq l$ follow from Proposition 4.1.2, so we proceed by induction on m. Suppose that the result holds for $m-1$ and suppose that $KGa \cong V^m \subset V^n$. Consider the projection $\pi_{\hat{1}}(KGa) = KG(v_2 + \cdots + v_n)$. If this is isomorphic to V^{m-1}, then by induction we know that $\dim(\langle v_2, \ldots, v_n \rangle) = m-1$ and the claim follows by Lemma 4.1.7. If however, $\pi_{\hat{1}}(KGa) \cong KGa$, then $v_1 \in \langle v_2, \ldots, v_n \rangle$. Then look at $\pi_{\hat{2}}(\pi_{\hat{1}}(KGa))$. If this is not isomorphic to KGa, then it is isomorphic to V^{m-1} and by induction we have that $\dim(\langle v_1, \ldots, v_n \rangle) = \dim(\langle v_2, \ldots, v_n \rangle) = m$. If however, $\pi_{\hat{2}}(\pi_{\hat{1}}(KGa)) \cong KGa$, then we can remove v_2 as before. If $\pi_{\hat{n}} \circ \cdots \circ \pi_{\hat{1}}(KGa) \cong KGa$, then $KGa \cong V$ and Proposition 4.1.2 entails that $\dim(\langle v_1, \ldots, v_n \rangle) = 1$.
Conversely, suppose that $\dim(\langle v_1, \ldots, v_n \rangle) = m$. Up to reordering, we may assume that $\{v_1, \ldots, v_m\}$ are linearly independent. By Lemma 4.1.7, we obtain that $\pi_{\hat{n}} \circ \cdots \circ \pi_{\widehat{m+1}}(KGa) \cong KGa$, and the result follows from Corollary 4.1.5. □

Corollary 4.1.9. *A glider of the form $V^m \supset Ka$ with $V \in \mathrm{Irr(G)}$ n-dimensional and $m > n$ is never irreducible.*

The general situation where $M = \oplus_i V_i^{m_i}$ does not cause any further difficulties. Indeed, we have

Proposition 4.1.10. *Let $V_1, \ldots, V_n \in \mathrm{Irr(G)}$ be non-isomorphic irreducible G-representations. If $a = v_1 + \ldots + v_n \in \bigoplus_{i=1}^n V_i$ is such that $KGa \cong \bigoplus_{j \in I} V_j \subset \bigoplus_{i=1}^n V_i$ for some proper subset $I \subset \{1, \ldots, n\}$, then $v_k = 0$ for all $k \notin I$.*

Proof. Let φ be an isomorphism between $\bigoplus_{j \in I} V_j$ and KGa and let $w = (w_j)_{j \in I} \in \bigoplus_{j \in I} V_j$ be such that $\varphi(w) = a$. If $v_k \neq 0$ for some $k \notin I$, we obtain a non-zero morphism

$$\bigoplus_{j \in I} V_j \xrightarrow{\varphi} KGa \hookrightarrow \bigoplus_{i=1}^n V_i \xrightarrow{\pi} V_k,$$

since $w \mapsto a \mapsto v_1 + \ldots + v_n \mapsto v_k$. This is a contradiction. □

Corollary 4.1.11. *A glider of the form $V \oplus V' \supset Ka$ with $V, V' \in \mathrm{Irr(G)}$ non-isomorphic and $a = v + v', v \neq 0, v' \neq 0$ is irreducible.*

All this leads to the following

Theorem 4.1.12. *Let G be a finite group, K an algebraically closed field of characteristic zero and let $\{V_1, \ldots, V_n\}$ be a full set of irreducible G-representations of resp. dimension n_i. A glider representation*

$$M = \bigoplus_{i=1}^n V_i^{\oplus m_i} \supset Ka,$$

with $a = v_1^1 + \cdots + v_{m_1}^1 + v_1^2 + \cdots + v_{m_2}^2 + \cdots + v_1^n + \cdots + v_{m_n}^n \in M$ *is irreducible if and only if*

1. $\forall i \; m_i \leq n_i$;

2. $\forall i \; \dim(\langle v_1^i, \ldots, v_{m_i}^i \rangle) = m_i$.

When studying glider representations we would like to work with chains of bigger length, so consider $1 < G_1 < \ldots < G_d = G$. If

$$\Omega \supset M \supset \ldots \supset M_d \supset 0 \supset \ldots$$

is an irreducible glider of essential length d, then in particular, we must have that $M_d = Ka$ is a one-dimensional vector space, $M = KGa$ and Theorem 4.1.12 gives restrictions on the KG-module M. The smaller groups G_i appearing in the chain then determine the M_{d-i}. The following question now arises: If $\Omega \supset M$ is a bodyless irreducible glider of essential length d with $\dim(M_1) = n_1$, what are the possible dimensions for M and vice-versa? For starters we have

Lemma 4.1.13. *Let* $1 < G_1 < \ldots < G_d = G$ *be a chain of groups and let* $\Omega \supset M \supset \ldots \supset M_{d-1} \supset M_d = Ka$ *be an irreducible glider representation. If* $M_{d-i} = KG_i a = Ka$ *for some* i, *then* $\dim(M) \leq [G : G_i]$.

Proof. Write $G = G_i \sqcup g_2 G_i \sqcup \ldots \sqcup g_m G_i$. Since $ha \in Ka$ for all $h \in G_i$, we have that $g_j ha = k(h) g_j a$ and $g_j h' a = k(h') g_j a$ $(k(h), k(h') \in K)$ are linearly dependent for all $h, h' \in G_i$, hence $\dim(M) = \dim(KGa) \leq m = [G : G_i]$. □

Remark 4.1.14. In a similar vein one proves the more general statement that $\dim(M) \leq [G : G_i] \dim(M_{d-i})$.

In fact, the answer to the question above follows from the relation between the simple G_i-representations. If $H < G$ is some subgroup, then we have a forgetful functor $U : G\text{-rep} \to H\text{-rep}$. Let U be a irreducible G-representation. If V is a irreducible H-representation appearing in the decomposition into irreducibles of U as an H-representation, then we depict this by

and we say that U lies over V. Of course, if we decompose $U \cong U_1 \oplus \ldots \oplus U_n$ into H-representations then we obtain

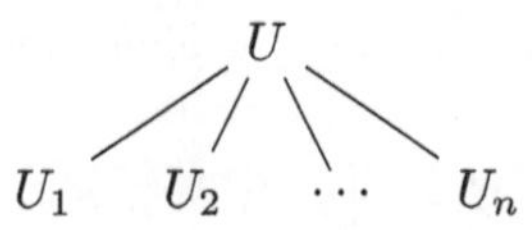

In this way we can draw a diagram relating all the irreducible representations of G and H. We wonder whether we can encounter

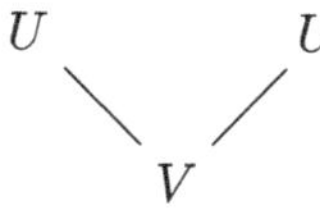

where U and U' are not of equal dimension. This happens of course when $H = 1$ so we exclude this case. We restrict to the situation of p-groups. It is well-known that they have a non-trivial center. Moreover, the index of the center in G cannot be p (otherwise G would be Abelian already), so assume that $[G : Z(G)] = p^2$ and consider a chain $1 \triangleleft Z(G) \triangleleft G_2 \triangleleft G$ where G/G_2 is cyclic of order p. Such a chain exists since a p-group is supersolvable. By the same reasoning as before, G_2 must be Abelian. Let $V_1 \in \mathrm{Irr(G_2)}$ and suppose that there exist $W_1, P \in \mathrm{Irr(G)}$ of resp. dimension 1 and p and such that $W_1 \cong V_1$ as G_2-representations and $P \cong V_1 \oplus gV_1 \oplus \ldots \oplus g^{p-1}V_1$ as G_2-representations for some $g \in G \setminus G_2$. Again, since G/G_2 is cyclic of order p, there exists $p-1$ other 1-dimensional representations $W_2, \ldots, W_p$ that lie over V_1. Let $w_i \in W_i$ for $i = 1, \ldots, p$ and let $v_1 \in V_1$. By Theorem 4.1.12, the glider representation

$$W_1 \oplus \ldots \oplus W_p \oplus P \supset KG_2(w_1, \ldots, w_p, v_1) \supset KZ(G)(w_1, \ldots, w_p, v_1) \supset K(w_1, \ldots, w_p, v_1)$$

is irreducible. However, since $KG_2(w_1, \ldots, w_p, v_1) = K(w_1, \ldots, w_p, v_1)$, this contradicts Lemma 4.1.13 since $\dim(M) = p + 1$. Because G_2 is an Abelian normal subgroup of G, the degree of each irreducible representation of G divides $[G : G_2]$. Hence we covered all the cases. A proof of the last statement mentioned can be found in Section 8.1 of [65].

Next, if $[G : Z(G)] = p^3$, then there exists a chain $1 \triangleleft Z(G) \triangleleft G_2 \triangleleft G_3 \triangleleft G$. Let $V_1 \in \mathrm{Irr(G_2)}$ be one-dimensional that has a one-dimensional G-representation W_1 lying over it. If $G/G_2 \cong C_{p^2}$ we can find p^2 non-isomorphic one-dimensional representations $W_1, \ldots, W_{p^2}$ lying over V_1. The same reasoning as above entails that there cannot lie a p-dimensional irreducible G-representation over V. If $G/G_2 \cong C_p \times C_p$, the reasoning is different. Suppose in this case that we have a one-dimensional representation W_1 and a p-dimensional representation P lying over V_1. Then we consider both W_1 and P as G_3-representations. Since $G_3/G_2 \cong C_p$, we find p non-isomorphic G_3-representations $W_1, W_2, \ldots, W_p$ lying over V_1. By induction we know that P cannot be an irreducible G_3-rep, hence $P \cong P_1 \oplus \ldots \oplus P_p$ and, up to reordering, P_1 lies over V_1. Hence we can construct the following FKG_3-glider

$$P_1 \oplus W_1 \oplus \ldots \oplus W_p \supset KG_2(p_1, w_1, \ldots, w_p) \supset KZ(G)(p_1, w_1, \ldots, w_p) \supset K(p_1, w_1, \ldots, w_p),$$

with $w_i \in W_i$ for $i = 1, \ldots, p$ and $p_1 \in P_1$. By construction we have that

$$KG_2(p_1, w_1, \ldots, w_p) = K(p_1, w_1, \ldots, w_p)$$

and Lemma 4.1.13 entails that $\dim(P \oplus W_1 \oplus \dots W_p) \leq \dim([G_3 : G_2]) = p$, a contradiction!

Theorem 4.1.15. *Let G be a p-group and let $1 \lhd Z(G) \lhd G_2 \lhd \dots \lhd G_d = G$ be a maximal chain of normal subgroups. If $V, W \in \mathrm{Irr(G)}$ are such that they lie over the same irreducible representation of $Z(G) \subsetneq G_{d-1}$, then* $\dim(V) = \dim(W)$.

Proof. If $G_2 = G$, then $G = Z(G)$ and there is nothing to prove. The case $G_3 = G$ has been dealt with above. If $G_4 = G$, suppose that V and W lie over some irreducible representation $U \in \mathrm{Irr(G_3)}$. If $\dim(U) = 1$, then $U_{|G_2} \in \mathrm{Irr(G_2)}$ and the result follows from our reasoning above. If $\dim(U) = p$, then $U \cong V_1 \oplus \dots \oplus V_p$ as G_2-representations; hence V and W lie over some irreducible $V_i \in \mathrm{Irr(G_2)}$ and the result follows again by the reasoning above. Now, let $G_5 = G$ and V, W lie over some $U \in \mathrm{Irr(G_4)}$. Suppose that $\dim(V) < \dim(W)$. Then $W_{|G_4}$ cannot be irreducible, otherwise $W, V \in \mathrm{Irr(G_4)}$ would lie over a same irreducible $U' \in \mathrm{Irr(G_3)}$. So $W \cong W_1 \oplus \dots \oplus W_p$ as G_4-representations and we may assume that $W_1 \cong U \cong V_{|G_4}$. But then it follows that $W \cong \mathrm{Ind}_{\mathrm{G_4}}^{\mathrm{G}}(\mathrm{W_1}) \cong \mathrm{Ind}_{\mathrm{G_4}}^{\mathrm{G}}(\mathrm{U}) \cong \mathrm{Ind}_{\mathrm{G_4}}^{\mathrm{G}}(\mathrm{V_{|G_4}}) \supset \mathrm{V}$, which shows that W is not irreducible, a contradiction. The general result now follows from induction. □

The theorem shows that the Hasse diagram is not connected. For example, the Hasse diagram of $\{1\} \subset \{1, -1\} \subset \mathbb{Z}_4^j \subset Q_8$ of Example 4.3.4 looks like

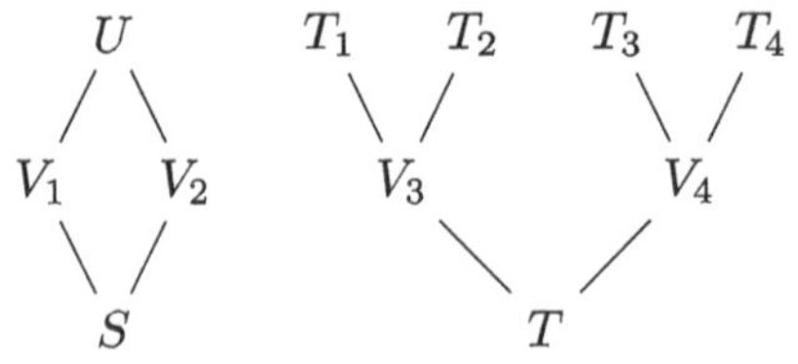

(4.2)

Since a finite nilpotent group is the direct product of its Sylow subgroups, we immediately get the same result for finite nilpotent groups.

Corollary 4.1.16. *Let $G = P_1 \dots P_n$ be a finite nilpotent group with Sylow subgroups P_i of order $p_i^{a_i}$ and let $Z(G) \subsetneq H \lhd G$ be a maximal normal subgroup. If V, W lie over the same irreducible representation U of H, then* $\dim(V) = \dim(W)$.

Proof. Suppose that $H \lhd G$ is a maximal subgroup, that is, up to reordering, $H = Q_1 P_2 \dots P_n$ with $[P_1 : Q_1] = p_1$. Since $P_1 \lhd G$ is central, $Z_{|P_1} = (Z_1)^{p_2^{a_2} \dots p_n^{a_n}}$ for $Z = V, W$ and $Z_1 \in \mathrm{Irr(P_1)}$. This implies that $\dim(V) = p_2^{a_2} \dots p_n^{a_n} \dim(V_1)$ and $\dim(W) = p_2^{a_2} \dots p_n^{a_n} \dim(W_1)$. Since V_1 and W_1 are irreducible representations of the p_1-group P_1 lying over the same irreducible representation of Q_1, the result follows from Theorem 4.1.15. □

Remark 4.1.17. We do not need the normality condition on the subgroup $H \subset G$. The only time we used the normality condition in proving Theorem 4.1.15 was when we dealt with the case $[G : Z(G)] = p^2$. To conclude that we covered all the cases we referred to a result in [65]. The condition was that $G_2 \lhd G$ is Abelian normal. The normality condition automatically follows since it is a maximal subgroup of a p-group and in any p-group the normalizer subgroup $N(G_2)$ is always strictly bigger than G_2.

Remark 4.1.18. Let G be a finite nilpotent group and $1 \neq H \lhd G$ an Abelian normal subgroup. It follows that the forgetful functor U : G-rep $\to$ H-rep is not essentially surjective. Indeed, as G is not Abelian, it has an n-dimensional irreducible representation V, which decomposes into irreducible H-representations, $V_H = \bigoplus_{i=1}^{\dim(V)} V_i$. The V_i are not isomorphic to $U(W_i)$ for W_i G-representations. This raises the question for which pair of groups $H \subset G$ the forgetful functor is essentially surjective.

Proposition 4.1.19. *Let G be a p-group and $H \lhd G$ a normal subgroup of index $[G : H] = p$. The forgetful functor U : G-rep $\to$ H-rep is essentially surjective if and only if G is Morita equivalent to the direct product $H \times C_p$.*

Proof. Suppose that the dimensions of the irreducible representations of H are $\{1, p, \ldots, p^n\}$. Essential surjectivity of U together with Theorem 4.1.15 implies that the dimensions of the irreducible representations of G also are $\{1, p, \ldots, p^n\}$. If $\mathrm{Irr(H)} = \{\mathrm{U_1}, \ldots, \mathrm{U_l}\}$, fix a set $\{W_1, \ldots, W_l\} \subset \mathrm{Irr(G)}$ such that $U(W_i) \cong U_i$, for $1 \leq i \leq l$. i Let ω be a p-th root of unity. Then we denote by $V_i : C_p \to K, \alpha \mapsto \omega^i$ $(i = 0, \ldots, p-1)$ the irreducible representation of the cyclic group $C_p =< \alpha >$. If V is a p^k-dimensional irreducible G-representation, then so is $V \otimes V_i$ for all $i = 0, \ldots, p-1$, where V_i is a G-representation via the projection $\pi : G \to G/H$. This follows from the version of [32, Theorem 3] in Isaacs book [39, Corollary 6.17], because $V_{|H}$ is irreducible. If $|H| = \sum_i' n_i^2$ then

$$\sum_i{}' p n_i^2 = p|H| = |G|,$$

and we deduce that $\mathrm{Irr(G)} = \{\mathrm{W_i \otimes V_j} \,|\, 1 \leq \mathrm{i} \leq \mathrm{l}, 0 \leq \mathrm{j} \leq \mathrm{p}-1\}$. The converse is trivial. □

4.2 Induction of fragments for filtration extensions of groups

We note that in this and the subsequent section we do everything for the more general notion of a fragment, but the examples provided are all glider representations.

Let G be a finite group and $H \triangleleft G$ some normal subgroup. Then the short exact sequence

$$1 \to H \to G \xrightarrow{\pi} G/H \to 1$$

may be viewed as defining an extension of H by G/H via the construction of a set map $\sigma : G/H \to G$ such that $\pi \circ \sigma(\overline{g}) = \overline{g}$ for $\overline{g} \in G/H$. Fixing σ defines $g = \sigma(\overline{g})h$ for a unique $h \in H$. In particular, $\sigma(\overline{g_1})\sigma(\overline{g_2}) = \sigma(\overline{g_1 g_2})h(\overline{g_1}, \overline{g_2})$ defines a map

$$h(-,-) : G/H \times G/H \to H.$$

From $(g_1 g_2)g = g_1(g_2 g)$, it follows that h is a 2-cocycle, i.e. it satisfies the following condition

$$h(\overline{g_1 g_2}, \overline{g})h(\overline{g_1}, \overline{g_2})^{\rho_{\sigma(\overline{g})}} = h(\overline{g_1}, \overline{g_2 g})h(\overline{g_2}, \overline{g}), \tag{4.3}$$

where $(-)^{\rho_{\sigma(\overline{g})}}$ denotes the conjugation by $\sigma(\overline{g})$.

Now look at a group algebra filtration of KG, K some field, given by a chain of normal subgroups $1 \triangleleft G_1 \triangleleft \ldots \triangleleft G_{d-1} \triangleleft G_d$, that is, $F_n KG = KG_n$ for $0 \leq n \leq d$. For H a normal subgroup of G, put $H_i = G_i \cap H$. We obtain a group algebra filtration of KH. By the normality of all subgroups, we have a commutative diagram

$$\begin{array}{ccccccc}
G_1/H_1 & \hookrightarrow & G_2/H_2 & \hookrightarrow & \ldots & \hookrightarrow & G/H \\
\downarrow \cong & & \downarrow \cong & & & & \downarrow id \\
G_1 H/H & \hookrightarrow & G_2 H/H & \hookrightarrow & \ldots & \hookrightarrow & G/H
\end{array}$$

which allows us to consider an ascending transversal set

$$1 \subset T_1 \subset T_2 \subset \ldots \subset T_i \subset \ldots \subset T,$$

where T_i is a set of (right) coset representations of H_i in G_i. This transversal set gives rise to a 2-cocycle h as before.

Definition 4.2.1. A 2-cocycle $h : G/H \times G/H \to H$ is said to be filtered if $h(-, \overline{g}) : G/H \to H$ is restricting to $(G/H)_i \to H_i$, where $(G/H)_i = G_i/H_i$, for every i and $g \in G$.

Example 4.2.2. If $\pi : G \to G/H$ admits a group section σ, then $G = HN$ for some subgroup N of G such that $N \cap H = \{1\}$. Then choosing N as transversal set T yields $h(-,-) = 1$, which is filtered for any chain of subgroups of G.

Example 4.2.3. Let $K \hookrightarrow L$ be a Galois extension with finite Galois group $G = \mathrm{Gal(L/K)}$ and fix a chain of normal subgroups

$$1 \triangleleft G_1 \triangleleft \ldots \triangleleft G_d = G.$$

The Galois correspondence yields a field filtration FL

$$K = K_0 \subset K_1 \subset \ldots \subset K_d = L,$$

where $K_i = L^{G_{d-i}}$. Let $A \supset A_1 \supset \ldots \supset A_d = L \supset 0 \ldots$ be a natural FL-fragment, i.e. A_i is a K_i-algebra for every $0 \leq i \leq d$. Assume moreover that every A_i is an Azumaya algebra over K_i. For $\sigma \in G_i \setminus G_{i-1}$, we obtain by the Skolem-Noether theorem a unit $u_\sigma \in A^\times$ such that for all $x \in L$

$$\sigma(x) = u_\sigma^{-1} x u_\sigma.$$

Since $L^{G_i} = K_{d-i}$, we get that $u_\sigma \in Z_A(K_{d-i}) = Z_A(Z_A(A_{d-i}))$. If A_{d-i} is a simple algebra, then the Double Centralizer Theorem yields that $u_\sigma \in A_{d-i}^\times$. Moreover, since $\sigma \notin G_{i-1}$, there exists $y \in K_{d-i+1} \setminus K_{d-i}$ such that $\sigma(y) \neq y$. This shows that $u_\sigma \notin A_{d-i+1}$. If $\dim_L(A_i) = \dim_{K_i}(L) = |G_{d-i}|$, then

$$A_i \cong \bigoplus_{\sigma \in G_{d-i}} L u_\sigma.$$

Since the Brauer group is isomorphic to the second Galois cohomology group, the isomorphism for $i = 0$ is given by a 2-cocycle $f : G \times G \to L^\times$. If f appears to be filtered, then we have in particular that f is restricting to

$$f : G_{d-i} \times G_{d-i} \to K_{d-i}^\times.$$

This implies that the Azumaya K_i-algebra A_i has a subalgebra isomorphic to

$$\bigoplus_{\sigma \in G_{d-i}} K_{d-i} u_\sigma \subset \bigoplus_{\sigma \in G_{d-i}} L u_\sigma \cong A_i.$$

Since $\dim_K(K_{d-i}) = |\mathrm{Gal(K_{d-i}/K)}| = |\mathrm{G_{d-i}}|$, this subalgebra is Azumaya over K. Hence, if an Azumaya algebra is determined by a filtered 2-cocycle for some fixed chain of normal subgroups, we obtain a chain of Azumaya algebras over the corresponding fixed fields, which all have subalgebras that are Azumaya over K

$$\begin{array}{ccccccc}
\bigoplus_{\sigma \in G} L u_\sigma & \supset & \bigoplus_{\sigma \in G_{d-1}} L u_\sigma & \supset & \ldots & \supset & L u_1 \\
\cup & & \cup & & & & \cup \\
\bigoplus_{\sigma \in G} L u_\sigma & \supset & \bigoplus_{\sigma \in G_{d-1}} K_{d-1} u_\sigma & \supset & \ldots & \supset & K u_1
\end{array}$$

In case $h(-,-)$ is filtered and in case all H_i are normal in G, e.g. when all G_i are normal in G, we can extend an FKH-fragment N into an FKG-fragment as follows. Put $M = K[G/H] \otimes_K N$ for the K-space $\oplus_{\overline{g} \in G/H} K\overline{g} \otimes N$. Similarly, for every j we put $M_j = K[G/H] \otimes_K N_j$. In this way, we obtain a descending chain of K-spaces

$$M = M_0 \supset M_1 \supset \ldots \supset M_d \supset 0 \supset \ldots$$

In order to define a KG_j multiplication on M_j, it will be enough to define $g_j(\overline{g} \otimes n_j)$ and extend this K-bilinearly. We let $\sigma : G \to T$ be as before (choice of transversal sets) with $\sigma : G_j \to T_j$ for all j. Define for $g_1 \in G_j$:

$$g_1 \cdot (\overline{g} \otimes n_j) = \overline{g_1 g} \otimes h(\overline{g_1}, \overline{g}) t^{-1} h_1 t n_j,$$

where $g_1 = t_1h_1, h_1 \in H_j, g = th,\ t_1 = \sigma(g_1)$ and $t = \sigma(g),\ n_j \in N_j$. If $n_j \in N_{l+j}$ in the foregoing, then since $h(\overline{g_1}, \overline{g}) \in H_j$ and $t^{-1}h_1t \in H_j$ we have that $g_1 \cdot (\overline{g} \otimes n_j) \in \overline{g_1g} \otimes N_l \subset M_l$. First we verify that for $g_1 \in G_j, g_2 \in G_i$ and $n \in N_{\max\{i,j\}}$, we have that

$$g_2g_1(\overline{g} \otimes n) = g_2(g_1(\overline{g} \otimes n)) \in \overline{g_2g_1g} \otimes N.$$

So consider $g_2 \in G_i, g_2 = t_2h_2$ with $t_2 = \sigma(g_2), h_2 \in H_2, g_1 = t_1h_1$ with $t_1 = \sigma(g_1)$ and $h_1 \in H_j$ and $n \in N_l$ with $l \geq \max\{i,j\}$. Then

$$g_2g_1 = t_2h_2t_1h_1 = t_2t_1(t_1^{-1}h_2t_1)h_1 = \sigma(g_2g_1)h(\overline{g_2}, \overline{g_1})(t_1^{-1}h_2t_1)h_1.$$

Therefore

$$g_2g_1(\overline{g} \otimes n) = \overline{g_2g_1g} \otimes h(\overline{g_2g_1}, \overline{g})(t^{-1}h(\overline{g_2}, \overline{g_1})t)(t^{-1}t_1^{-1}h_2t_1t)(t^{-1}h_1t)n.$$

On the other hand, we have

$$\begin{aligned} g_2(g_1(\overline{g} \otimes m)) &= g_2[\overline{g_1g} \otimes h(\overline{g_1}, \overline{g})(t^{-1}h_1t)n] \\ &= \overline{g_2g_1g} \otimes h(\overline{g_2}, \overline{g_1g})(\sigma(g_1g)^{-1}h_2\sigma(g_1g))h(\overline{g_1}, \overline{g})(t^{-1}h_1t)n. \end{aligned}$$

Both expressions are equal since $h(\overline{g_2}, \overline{g_1g})(\sigma(g_1g)^{-1}h_2\sigma(g_1g))h(\overline{g_1}, \overline{g})$ is equal to

$$\begin{aligned} &(h(\overline{g_2}, \overline{g_1g})h(\overline{g_1}, \overline{g}))(h(\overline{g_1}, \overline{g})^{-1}\sigma(g_1g)^{-1}h_2\sigma(g_1g)h(\overline{g_1}, \overline{g})) \\ = \ &h(\overline{g_2g_1}, \overline{g})(t^{-1}h(\overline{g_2}, \overline{g_1})t)(t^{-1}t_1^{-1}h_2t_1t). \end{aligned}$$

The third fragment condition $\mathbf{f_3}$ will follow from the (proof of the) following proposition.

Proposition 4.2.4. *If N is a (glider, resp. natural) KH-fragment, then $M = N^G$ is a (glider, resp. natural) KG-fragment.*

Proof. We have to establish that $KG_\mu(KT \otimes N_\mu) \subset M_\mu^*$, where $M_\mu = KT \otimes N_\mu$.

Take $m \in M_\mu$, $m = \oplus_{t \in T}\lambda_t t \otimes n_t$ with $n_t \in N_\mu$, $\lambda_t \in K$. For $g_\mu \in KG_\mu$ we have $g_\mu \cdot m = \oplus_t \lambda_t \overline{t_\mu t} \otimes h(t_\mu, t)t^{-1}h_\mu t n_t$ where $\overline{t_\mu t}$ is the representative for $t_\mu t$ in the transversal T and $h(t_\mu, t)t^{-1}h_\mu t \in H_\mu$. From $KH_\mu N_\mu \subset N_\mu^*$ we see that for every $g'_\mu \in G_\mu$ we have $g'_\mu g_\mu m \subset KT \otimes N = M$. Hence $g_\mu m \in M_\mu^*$ or $KG_\mu M_\mu \subset M_\mu^*$.

Suppose that $N \subset \Omega$ is a KH-module inducing the operations; then $M = N^G \subset KT \otimes \Omega = \Omega^G$ is a KG-module inducing the operations of G_i on $M_i = KT \otimes N_i$, and we see that M is a glider representation. If N is natural, look at $m \in M$, $m = \oplus_{t \in T}\lambda_t t \otimes n_t$, $\lambda_t \in K$, $n_t \in N$ and assume $KG_i m \subset M$, i.e. $m \in M_i^*$. If $g_i = t_ih_i$ in G_i, then $g_i m = \oplus \lambda_t \overline{t_i t} \otimes h(t_i, t)t^{-1}h_i t n_t$. Since $h(t_i, t) \in H_i$ for all t and H_i is normal in G, we obtain $H_i n_t \subset N$ or $n_t \in N_i^* = N_i$ as N is natural. Consequently $m \in KT \otimes N_i = M_i$ and M is natural. □

We call $M \supset \ldots \supset M_j \supset \ldots$ the induced fragment of N and denote it by N^G.

Lemma 4.2.5. *For $M = N^G$, we have that $M^*_\mu = KT \otimes N^*_\mu$ for all μ.*

Proof. Let $m \in M^*_\mu$, then for $g_\mu = t_\mu h_\mu$, $g_\mu m = \oplus_t \lambda_t \overline{t_\mu t} \otimes h(t_\mu, t)t^{-1}h_\mu t n_t$, where $m = \oplus_{t\in T}\lambda_t t \otimes n_t$, $\lambda_t \in K$, $n_t \in N$. So $g_\mu m = \oplus_t \lambda_t \overline{t_\mu t} \otimes h'_\mu n_t$ for some $h'_\mu \in H_\mu$. Since $g_\mu m \in M$ it follows that $g_\mu m = \oplus_t b_t t \otimes a_t$ with $b_t \in K$ and $a_t \in N$. Since $KT \otimes N$ is a direct sum $\oplus Kt \otimes N$ ($\cong \oplus_{t\in T} N$), we have, up to some permutation of T, say σ, that $Kh'_\mu n_t = Ka_{\sigma(t)} \subset N$. By the choice of h_μ in g_μ we can obtain every h^1_μ for h'_μ, hence $H_\mu n_t \subset N$, or $n_t \in N^*_\mu$. The other inclusion is trivial. □

Corollary 4.2.6. *For an FKH glider representation N, we have that N is natural if and only if N^G is natural.*

4.3 Clifford theory of group algebra fragments

In his original paper [24], A. H. Clifford elucidates the behavior of irreducible G-representations V when considered as H-representations for some normal subgroup $H \lhd G$. It appears that there are two possibilities: the induced H-representation V_H is either itself irreducible, or decomposes into irreducible components all of the same degree, which are moreover conjugate relative to G to one another. By conjugate representations V and W we mean that there is some $g \in G$ such that $h \cdot V = g^{-1}hg \cdot W$ for all $h \in H$. Clifford constructs a so-called decomposition group $H \subset G' \subset G$, which fully determines the representation V. This raises the opposite question of embedding an irreducible H-representation into some given irreducible G-representation, and a crucial role is again played by the same decomposition group G'. The answer is given by

Theorem 4.3.1. *[24, Theorem 4]*
Let $H \lhd G$ be a normal subgroup of a finite group G. A given irreducible H-representation U can be embedded in an irreducible G-representation V if and only if

1. *the subgroup $G^{dec} \subset G$ consisting of elements $g \in G$ such that the conjugate H-representation $g \cdot U$ is equivalent to U, is of finite index in G;*

2. *U can be embedded in a irreducible G^{dec}-representation U'.*

If the ground field K is algebraically closed, the latter condition is equivalent to the existence of a representation of finite degree of the group algebra $K[G^{dec}/H]$ with multiplication

$$u_S u_T = \alpha(s,t)^{-1} u_{ST},$$

where s, t are representatives of $S, T \in G^{dec}/H$ and where $\alpha : G^{dec} \times G^{dec} \to K$ is a factor-set associated with U.

We assume that the chain of groups is normal, i.e. $1 \triangleleft G_1 \triangleleft \ldots \triangleleft G_{d-1} \triangleleft G_d$ and we pick a normal subgroup $H \triangleleft G$. By putting $H_i = H \cap G_i$, we obtain an embedding of filtered group algebras $FKH \to FKG$, where $F_iKH = KH_i, F_iKG = KG_i$. It is obvious that an FKG-fragment M can be seen as an FKH-fragment, which corresponds to the usual forgetful functor $U : G$-rep $\to H$-rep. Under some suitable conditions, we constructed in the previous section an induced FKG-fragment N^G for an FKH-fragment N. This construction allows us to perform a Clifford theory for glider representations. First, we discuss the going up direction; that is, we see what happens to the induced fragment of an irreducible FKH-fragment. In this section, we do everything for fragments of essential length d and zero body. In fact, everything is analogous for smaller essential lengths. So consider an irreducible FKH-fragment $N = N_0 \supset N_1 \supset \ldots \supset N_{d-1} \supset N_d \supset 0 \ldots$ of essential length d. Irreducibility implies that N_d is one dimensional and by Lemma 3.2.3 we know that for any i, $N_{d-i} = KH_iN_d$, i.e. N_{d-i} is a KH_i-module. Remark that we do not know whether the N_i are irreducible KH_{d-i}-modules (they are semisimple of course). It is not even the case that simplicity of N_i as KH_{d-i}-module implies simplicity of N_{i+1} as KH_{d-i-1}-module.

Define M to be the induced FKG-fragment, that is

$$M = K[G/H] \otimes N.$$

Then on degree d we have a direct sum (of K-spaces) $M_d = T \otimes Ka$ where $N_d = Ka$. Any $t \in T$ generates an irreducible FKH-fragment

$$KGt \otimes a \supset KG_{d-1}t \otimes a \supset \ldots \supset Kt \otimes a.$$

Indeed, we have that $M_{d-i} = K[G/H] \otimes N_{d-i} = KT \otimes KH_iN_d = KG_iT \otimes N_d = KG_iM_d$, where the third equality follows from

$$\bar{t} \otimes h_in_d = th_it^{-1} \cdot \bar{t} \otimes n_d \text{ and } g_i\bar{t} \otimes n_d = \overline{t'} \otimes h(\overline{g_i}, \bar{t})t^{-1}h_itn_d \quad (g_i = sh_i).$$

We decomposed M into a fragment direct sum of $|T|$ irreducible FKH-fragments all of essential length d. So we have

Theorem 4.3.2. *Let N be an irreducible FKH-fragment of essential length d. Then the induced fragment $M = K[G/H] \otimes N$ decomposes into a fragment direct sum of $[G : H]$ irreducible FKH-fragments of essential length d.*

As a corollary, we obtain a Mackey decomposition theorem [47, Theorem 1]. Indeed, suppose that $E \triangleleft G$ is another normal subgroup. We have a commutative diagram

$$\begin{array}{ccccccccc} 1 & \longrightarrow & H & \longrightarrow & G & \overset{\pi}{\underset{\sigma}{\rightleftarrows}} & G/H & \longrightarrow & 1 \\ & & \uparrow & & \uparrow & & \uparrow & & \\ 1 & \longrightarrow & E \cap H & \longrightarrow & E & \overset{\pi}{\underset{\sigma}{\rightleftarrows}} & \frac{E}{E \cap H} & \longrightarrow & 1 \end{array}$$

and by putting $E_i = G_i \cap E$, we obtain two additional group algebra filtrations FKE and $FK(E \cap H)$. By the normality condition, we can begin by fixing an ascending transversal set $1 \subset S_1 \subset \ldots \subset S_d = S$, with S_i a set of right coset representations of $E_i \cap H_i$ in E_i. Consider now a subset $U_i \subset G_i$ such that $\{E_i t H_i \mid t \in U_i\}$ is a complete set of double coset representatives. Then $T_i = \{st \mid s \in S_i, t \in U_i\}$ is a complete set of representatives for H_i in G_i and $S_i \subset T_i$. In this way, we obtain an ascending transversal set $1 \subset T_1 \subset \ldots \subset T_d = T$, with T_i a set of right coset representations of H_i in G_i and the associated 2-cocylce $h : G/H \times G/H \to H$ is restricting to $h : E/(E \cap H) \times E/(E \cap H) \to E \cap H$. Moreover, for $s \in S$ and $t \in U_d = U$, we have that $h(\overline{s}, \overline{t}) = 1$. In the case that h is filtered, we can induce an FKH-fragment $N = N_0 \supset N_1 \supset \ldots \supset N_d \supset 0 \ldots$ to an FKG-fragment

$$M = N^G = K[T] \otimes N.$$

Theorem 4.3.3. *Let H, E be normal subgroups of a finite group G, with fixed ascending chain of normal subgroups*

$$1 \triangleleft G_1 \triangleleft \ldots \triangleleft G_d = G.$$

Let N be an FKH-fragment with FKA the induced group algebra filtration on $A = H, E, H \cap E$. Then the induced fragment restricted as an FKE-fragment $(N^G)_E$ is the fragment direct sum

$$(N^G)_E \cong \bigoplus_{t \in U} \left[(t \otimes N)_{H \cap E}\right]^E.$$

Proof. By construction $N^G = \bigoplus_{t \in T} t \otimes N$, and for fixed $t \in T$, the descending chain

$$t \otimes N \supset t \otimes N_1 \supset \ldots \supset t \otimes N_d$$

is easily seen to be an $FK(H \cap E)$-fragment. For $t \in U$ define

$$\varphi : \bigoplus_{s \in S} s \otimes t \otimes N \to \bigoplus_{s \in S} st \otimes N, \;\; s \otimes t \otimes n \mapsto st \otimes n.$$

Let $k = s_1 z_1 \in E_i$ with $s_1 \in S_i, z_1 \in E_i \cap H_i$ and $n \in N_i$. On the one hand we have

$$\begin{aligned} k \cdot \varphi(s \otimes t \otimes n) &= k \cdot st \otimes n = \overline{kst} \otimes h(\overline{k}, \overline{st}) t^{-1} s^{-1} z_1 stn \\ &= \overline{kst} \otimes h(\overline{k}, \overline{st}) h(\overline{s}, \overline{t}) t^{-1} s^{-1} z_1 stn. \end{aligned}$$

On the other hand, we calculate

$$\begin{aligned}\varphi(k\cdot s\otimes t\otimes n) &= \varphi(\overline{ks}\otimes h(\overline{k},\overline{s})s^{-1}z_1 s\cdot(t\otimes n))\\ &= \varphi(\overline{ks}\otimes t\otimes t^{-1}h(\overline{k},\overline{s})s^{-1}z_1 stn)\\ &= \overline{kst}\otimes h(\overline{ks},\overline{t})t^{-1}h(\overline{k},\overline{s})s^{-1}z_1 stn).\end{aligned}$$

Both expressions are equal by the 2-cocycle condition (4.3), hence φ is a morphism of FKE-fragments. The map is easily seen to be surjective and as K-spaces the domain and codomain have the same dimension, so we have an isomorphism of FKE-fragments. The result now follows. □

Let us now consider the going down direction of the Clifford theory. So suppose that M is an irreducible FKG-fragment and consider $M = M_H$ as an FKH-fragment. By irreducibility, $M_d = Ka$ is a one-dimensional K-vector space. To begin, we observe that $KH_1 a \subset M_{d-1}$ is a KH_1-submodule, which decomposes into irreducible KH_1-modules

$$KH_1 a = S_1^0 \oplus \ldots \oplus S_{e_0}^0,$$

since KH_1 is semisimple. Let $S_1^1, \ldots, S_{e_1}^1$ be irreducible KH_1-modules such that

$$M_{d-1}^H = KH_1 a \oplus S_1^1 \oplus \ldots \oplus S_{e_1}^1.$$

Subsequently, any $1 \le i \le e_1$ gives a (trivial) FKH-subfragment

$$KH_{d-1}S_i^1 \supset \ldots \supset KH_2 S_i^1 \supset S_i^1 \supset 0$$

of essential length $d-1$. Next, we find irreducible KH_2-modules $S_1^2, \ldots, S_{e_2}^2$ such that

$$M_{d-2}^H = KH_2(S_1^0 \oplus \ldots \oplus S_{e_0}^0) + KH_2(S_1^1 \oplus \ldots \oplus S_{e_1}^1) \oplus S_1^2 \oplus \ldots \oplus S_{e_2}^2.$$

Observe that the first sum no longer needs to be direct, testifying to the higher complexity of fragment structures. For every $1 \le i \le e_2$ we again obtain a (trivial) FKH-subfragment

$$KH_{d-2}S_i^2 \supset \ldots \supset S_i^2 \supset 0 \supset 0$$

of essential length $d-2$. Proceeding in this way, we arrive at a decomposition

$$\begin{aligned}M^H = M_0^H &= KH(S_1^0 \oplus \ldots \oplus S_{e_0}^0) + KH(S_1^1 \oplus \ldots \oplus S_{e_1}^1) + \ldots\\ &+ KH(S_1^{d-1} \oplus \ldots \oplus S_{e_{d-1}}^{d-1}) \oplus S_1^d \oplus \ldots \oplus S_{e_d}^d.\end{aligned}$$

Summarizing, we decomposed M^H as a fragment direct sum of "irreducible" fragments with lowest non-zero part S_j^i $(0 \le i \le d, 1 \le j \le e_i)$ and of essential length $d-i$.

Inspired by the classical Clifford theory, we can say something more. In our construction, we viewed an irreducible KH_i-module $S = S_j^i$ inside the KH_{i+1}-module $KH_{i+1}S$. Since we do not know whether the latter is irreducible, this is not entirely the classical embedding problem. Nonetheless, we can mimic the construction of [24] and use a different approach that will lead to so-called decomposition groups. We will see however, that these decomposition groups will lie between H_i and H_{i+1} and not between H_i and G_i as in the classical case.

In the decomposition of M_{d-1}^H into irreducible H_1-modules above, we fix some $S = S_1^1$. For any $h \in H_2 \backslash H_1$, hS and S are conjugate KH_1-modules relative to H_2. Moreover, hS is also irreducible. If $\forall h \in H_2$, we would have that $hS = S$, S would be an irreducible KH_2-module and thus $KH_2S = S$. Otherwise, we find a finite number of elements $h_2, \ldots, h_r \in H_2 \setminus H_1$ such that

$$S \oplus h_2 S \oplus \ldots \oplus h_r S = KH_2 S.$$

After regrouping all equivalent H_1-modules together, we get

$$R_1 \oplus R_2 \oplus \ldots \oplus R_m = KH_2 S,$$

where the R_i are the sum of equivalent modules. We define the subgroup $H'_{2,1} \subset H_2$ of elements which leave R_1 invariant. By definition, KH_2S is transitive, so all the vector spaces R_i must have the same dimension. Moreover, the $KH'_{2,1}$-module R_1 generates KH_2S. However, R_1 need not be an irreducible $H'_{2,1}$-module, since we do not have that KH_2S is irreducible (cf. [24, Observations before Section 3]). E.g. for $\mathbb{Z}_2 \subset \mathbb{Z}_4$, the two-dimensional $\mathbb{Z}_4$-representation V defined by

$$1 \mapsto \begin{pmatrix} i & 0 \\ 0 & i \end{pmatrix}$$

has $S = \mathbb{C}e_1$ an irreducible $\mathbb{Z}_2$-representation and $V = S \oplus 1 \cdot S$, both of which are isomorphic to the non-trivial irreducible $\mathbb{Z}_2$-representation. Therefore $G' = G$ and $R_1 = V$ is not irreducible. In case R_1 is not irreducible, one considers the irreducible component which contains S and introduces a (possibly) bigger decomposition group $H''_{2,1}$. Hence, after reducing to the Clifford setting and assuming K to be algebraically closed, the embedding problem (Theorem 4.3.1) gives a one-to-one correspondence between the embedding of S into an irreducible $H''_{2,1}$-module and factor sets

$$\alpha : H''_{2,1}/H_1 \times H''_{2,1}/H_1 \to \mathbb{C}$$

and modules of finite degree of the algebra $\mathfrak{a} = \mathbb{C}[H''_{2,1}/H_1]$ corresponding to α^{-1}, i.e. in which the multiplication is given by

$$\overline{h_1}\,\overline{h_2} = \alpha^{-1}(\overline{h_1}, \overline{h_2})\overline{h_1 h_2}.$$

So at stage i, we decomposed M_{d-i} into

$$KH_i a + KH_i(S_1^1 \oplus \ldots \oplus S_{e_1}^1) + \ldots + KH_i(S_1^{i-1} \oplus \ldots \oplus S_{e_{i-1}}^{i-1}) \oplus S_1^i \oplus \ldots \oplus S_{e_i}^i,$$

and the S_j^i $(1 \leq j \leq e_i)$ give rise to decomposition groups $H_i \subset H''_{i+1,j} \subset H_{i+1}$ and embeddings $S_j^i \subset V_j^i$, where the latter is an irreducible $H''_{i+1,j}$-module and generates a part of $KH_{i+1}S_j^i$.

However, we are not yet satisfied by our decomposition of M^H. In fact, we would like a more subtle relation between the H_i's and G_i's, as we already remarked above. Let us reconsider the decomposition of M_{d-1}^H into irreducible H_1-modules

$$M_{d-1}^H = S_1^0 \oplus \ldots \oplus S_{e_0}^0 \oplus S_1^1 \oplus \ldots \oplus S_{e_1}^1,$$

in which the S_i^0 add up to KH_1a. Some of these S_i^0 are (irreducible) G_1-modules. The others give rise to a $KH_1 \subset KG_1$-fragment $M_{d-1} \supset S_i^0$. In any case, since $M_{d-1} = KG_1a$ and by the irreducibility of M we know that every S_j^1 must be of the form $gS_{j_i}^0$ for some $g \in G_1 \setminus H_1$. Indeed, otherwise we would have a non-trivial FKG-subfragment by killing such an S_j^1. Therefore, the irreducible S_j^0 determine M_{d-1}. If S_j^0 is a G_1-module, there is nothing to it. If not, then

$$S_j^0 \oplus g_2 S_j^0 \oplus \ldots \oplus g_{m_j} S_j^0$$

for some $g_2, \ldots, g_{m_j} \in G_1 \setminus H_1$ appears in the decomposition of M_{d-1}. Therefore, assume that $S_1^0, \ldots, S_{f_0}^0$ are the building blocks of M_{d-1}^H, by which we mean that these S_i^0's contain all the G_1-modules and exactly enough H_1-modules such that all its conjugates relative to G_1 yield the decomposition of M_{d-1}^H. If one then fixes an $S = S_j^0$, one can introduce a decomposition group $G_1' \subset G_1$ like in the classical theory. Note that by our construction, every building block gives rise to a (possibly different) decomposition group! Moreover, if S happens to be a G_1-module, then $G_1' = G_1$.

Now we investigate what happens if we look at $M_{d-2} = KG_2a$. Recall that we considered KH_2S. If this H_2-module is even a G_2-module, then all the conjugates of S relative to G_1 become equal inside M_{d-2}^H:

$$KH_2gS = gKH_2S = KH_2S.$$

If KH_2S is not a G_2-module, then S and gS remain conjugated. By definition, there exist elements $h_2, \ldots, h_m \in H_2 \setminus H_1$ such that

$$KH_2S = S \oplus h_2S \oplus \ldots \oplus h_mS.$$

For any $g \in G_1$ we get

$$KH_2gS = gKH_2S = g(S \oplus h_2S \oplus \ldots \oplus h_mS).$$

We easily calculate

$$KH_1gh_iS = gK(g^{-1}H_1g)h_iS = gh_iK(h_i^{-1}H_1h_i)S = gh_iKH_1S,$$

since $H_1 \triangleleft G_1$ and $H_1 \triangleleft H_2$. Hence, gh_iS is an irreducible H_1-module and

$$KH_2gS = gS \oplus gh_2S \oplus \ldots \oplus gh_mS.$$

We also deduce that the irreducible H_1-factors of KH_2S and KH_2gS are either all the same, or all different. So in total, we obtain that all the conjugates of S relative to G_1 contribute to the decomposition of KG_2S into irreducible KH_1-modules. This also explains that two different building blocks S and S' do not affect one another at a higher stage.

Now, if $KG_1S = S \oplus g_2S \oplus \ldots \oplus g_lS$ as H_1-modules for some $g_2, \ldots, g_l$ in $G_1 \setminus H_1$, then by the above we obtain that

$$KG_2S = KH_2S \oplus g_2KH_2S \oplus \ldots \oplus g_nKH_2S \oplus r_1KH_2S \oplus \ldots \oplus r_tKH_2S,$$

for $n \leq l$ (up to some possible reordering) and $r_1, \ldots, r_t \in G_2 \setminus G_1H_2$. Hence we again can define $G_2' \subset G_2$ to be the subgroup of elements that leave the sum of all irreducibles equivalent to KH_2S invariant. From our discussion above, if $S \cong g_iS$ as H_1-modules, then KH_2S and g_iKH_2S have the same decomposition into irreducible H_1-components. However, this does not imply that KH_2S and g_iKH_2 are isomorphic as H_2-modules! A disappointing side effect is that there is no chance at all that $G_1' \subset G_2'$. Observe moreover, that if our normal chain of subgroups is maximal, then $G_2 = G_1H_2$ and no r's appear (see below).

Before we step up the ladder one stair further, we establish the foregoing for a concrete example.

Example 4.3.4. Look at the following graph of groups

$$\begin{array}{ccc} \mathbb{Z}_4 = \{1, j, -1, -j\} & \triangleleft & Q_8 = \langle -1, i, j \mid i^2 = j^2 = -1, ij = -ji \rangle \\ \triangledown & & \triangledown \\ \mathbb{Z}_2 = \{1, -1\} & \triangleleft & \mathbb{Z}_4 = \{1, i, -1, -i\} \end{array}$$

As transversal sets, we choose $T_1 = T_2 = \{1, j\}$ and one checks that the associated 2-cocycle h takes values in $H_1 = \mathbb{Z}_2$. This implies that h is filtered. We know that Q_8 has four 1-dimensional representations, given by

$$\begin{aligned} T_1 &: i \mapsto 1, \ j \mapsto 1, \\ T_2 &: i \mapsto -1, \ j \mapsto 1, \\ T_3 &: i \mapsto 1, \ j \mapsto -1, \\ T_4 &: i \mapsto -1, \ j \mapsto -1 \end{aligned}$$

and one irreducible 2-dimensional representation

$$U : i \mapsto \begin{pmatrix} i & 0 \\ 0 & -i \end{pmatrix}, \ j \mapsto \begin{pmatrix} 0 & -1 \\ 1 & 0 \end{pmatrix}.$$

If we consider U as a $\langle j\rangle$-representation, we diagonalize

$$\frac{1}{2}i\begin{pmatrix}-i & -1\\ -i & 1\end{pmatrix}\begin{pmatrix}0 & -1\\ 1 & 0\end{pmatrix}\begin{pmatrix}1 & 1\\ i & -i\end{pmatrix}=\begin{pmatrix}-i & 0\\ 0 & i\end{pmatrix}$$

and under base change

$$e_1 = f_1 + if_2$$
$$e_2 = f_1 - if_2$$

we get $U = V^{-i} \oplus V^i$, where V^i is the irreducible $\mathbb{Z}_4 = \langle j\rangle$-representation, defined by $j \mapsto i$ and similarly for V^{-j}. Consider now the $F\mathbb{C}Q_8$-glider

$$M = U \oplus T_3 \oplus T_2 \supset V^{-i} \oplus T_3 \oplus T_2 \supset \Delta$$

in which Δ stands for the diagonal of M_1. Let $\{f_1, f_2, e_3, e_4\}$ be a basis for M_0 establishing the direct sum decomposition. Then we will work with $\{e_1, e_3, e_4\}$ and $\{e_1 + e_3 + e_4\}$ as bases for M_1, resp. M_2. Observe that M is indeed irreducible by Theorem 4.1.12. We calculate

$$\mathbb{C}\mathbb{Z}_2\Delta = \mathbb{C}(e_3 + e_4) \oplus \mathbb{C}e_1$$

as $\mathbb{Z}_2$-modules, and $\{\mathbb{C}e_1, \mathbb{C}(e_3 + e_4)\}$ is a minimal set of building blocks. Furthermore

$$M_1 = \mathbb{C}\mathbb{Z}_2\Delta \oplus j \cdot \mathbb{C}(e_3 + e_4),$$

in which the last component equals $\mathbb{C}(-e_3 + e_4)$. The first building block $S_1 = \mathbb{C}e_1$ is a $\langle j\rangle$-module, so $G_1' = G_1$. For the second building block $S_2 = \mathbb{C}(e_3 + e_4)$, we have $S_2 \cong jS_2$, whence $G_1' = G_1 = \mathbb{Z}_4$ as well. Subsequently, we have

$$KH_2S_1 = \mathbb{C}\langle i\rangle e_1 = \mathbb{C}e_1 \oplus i \cdot \mathbb{C}e_1 = \mathbb{C}e_1 \oplus \mathbb{C}e_2 = U,$$

and

$$KH_2S_2 = \mathbb{C}\langle i\rangle(e_3 + e_4) = \mathbb{C}(e_3 + e_4) \oplus i \cdot \mathbb{C}(e_3 + e_4).$$

Since $jKH_2S_1 = KH_2S_1$ and $jKH_2S_2 = KH_2S_2$, we have that

$$\mathbb{C}Q_8e_1 = \mathbb{C}\langle i\rangle e_1 = U,\ \mathbb{C}Q_8(e_3 + e_4) = \mathbb{C}\mathbb{Z}_4^i(e_3 + e_4) = T_3 \oplus T_4.$$

Hence in both cases $G_2' = Q_8$.

Now, if $d > 2$, then we would have to look at KH_3S, but one can no longer apply the same techniques, since we do not know whether H_1 is normal in H_3. However, for every building block one can decompose KH_2S into irreducible H_2-modules

$$KH_2S = T_1 \oplus \ldots \oplus T_n,$$

extend to a decomposition of KG_2S, choose a new set of building blocks and repeat the foregoing argument. So at every stage, the relation between

G_i, H_i, G_{i+1} and H_{i+1} comes into play. Indeed, if T is an irreducible H_i-rep, then you decompose

$$\begin{aligned}
&KG_iT = T \oplus g_2T \oplus \ldots \oplus g_mT && g_2, \ldots, g_m \in G_i \setminus H_i,\\
&KH_{i+1}T = T \oplus h_2T \oplus \ldots \oplus h_tT && h_2, \ldots, h_t \in H_{i+1} \setminus H_i,\\
&KG_{i+1}T = KH_{i+1}T \oplus g_2KH_{i+1}T \oplus \ldots\\
&\quad \oplus g_nKH_{i+1}T \oplus r_1KH_{i+1}T \oplus \ldots \oplus r_uKH_{i+1}T
\end{aligned}$$

for some $n \leq m$ and $r_2, \ldots, r_u \in G_{i+1} \setminus G_iH_{i+1}$ and we obtain a decomposition group $H_{i+1} \subset G'_{i+1,T} \subset G_{i+1}$. Unfortunately, one has no hope that an ascending chain of decomposition groups arises. In the previous example, we would have

$$KH_2S_2 = \mathbb{C}\langle i \rangle S_2 = T_3 \oplus T_2$$

as decomposition into irreducible $\langle i \rangle$-modules, but $S_2 = \mathbb{C}(e_3 + e_4)$ does not fits nicely in one of the two irreducible components.

We summarize the foregoing in

Theorem 4.3.5. *Let K be an algebraically closed field of characteristic zero and let $H \triangleleft G$ be a normal subgroup of some finite group G, with fixed ascending chain of normal subgroups*

$$1 \triangleleft G_1 \triangleleft \ldots \triangleleft G_d = G.$$

If M is an irreducible FKG-fragment of essential length d, then M_H is either an irreducible FKH-fragment or decomposes into a fragment direct sum

$$M_H = \bigoplus_S M(S),$$

where $M(S) = KGS \supset KG_{d-1}S \supset \ldots \supset KG_1S \supset M_d \cap KG_1S \supset 0 \ldots$, and where the sum runs over a set of building blocks for KG_1M_d, consisting of H_1-modules. Moreover, to every S, we have a decomposition group $H_1 \subset G'_{1,S} \subset G_1$.
In the latter case, to every such S there is associated a set of building blocks of KG_2S consisting of H_2-modules and we obtain a fragment direct sum

$$M(S) = \bigoplus_T M(T),$$

where

$$M(T) = KGT \supset KG_{d-1}T \supset \ldots \supset KG_2T \supset KG_1S \cap KG_2T \supset M_d \cap KG_2T \supset 0 \ldots$$

Again the sum runs over the set of building blocks and we obtain decomposition groups $H_2 \subset G'_{2,T} \subset G_2$. We obtain similar fragment decompositions of the $M(T)$ and so we arrive at decomposition groups $H_i \subset G'_i \subset G_i$ at every stage $1 \leq i \leq d$.

Example 4.3.6. We recover the classical Clifford theory for a normal subgroup $H \triangleleft G$, if we consider the trivial filtration

$$\begin{array}{ccc} K & \subset & KG \\ \| & & \triangledown \\ K & \subset & KH. \end{array}$$

Indeed, suppose that $M \supset M_1$ is an irreducible FKG-fragment, with M an irreducible G-module. We have that $M_1 = Ka$ is one-dimensional and $M = KGa$. If $M \supset M_1$ is also irreducible as an FKH-fragment, then $M = KHa$ and it follows that M is an irreducible KH-module. If the FKH-fragment is not irreducible, we have that $KHa \neq KGa = M$. According to our approach, we decompose KHa into irreducible H-modules $KHa = S_1 \oplus \ldots \oplus S_n$. Then we deduced that $M = KGa$ can be decomposed as a sum of irreducible H-modules which are all conjugate relative to G to one of the S_i. A set of building blocks that contains some S_i and S_j would entail that these irreducible KH-modules are not conjugate relative to G to one another. But then we would have that $KGS \subset M$ would be a proper G-submodule, a contradiction. Therefore, $\{S_1\}$ is a set of building blocks and we find that M can be decomposed as

$$M = S_1 \oplus g_2 S_1 \oplus \ldots \oplus g_m S_1$$

for some $g_2, \ldots, g_m \in G \setminus H$.

4.4 Geometric aspect of decomposition groups

Suppose that the normal chain of subgroups is maximal. Observe that an irreducible FKG-fragment is completely determined by a KG-module $M = M_0$ and a one-dimensional K-subspace $Ka \subset M$. However, as we observed in Theorem 4.1.12, there are some constraints on the element $a \in M$; suppose that

$$M = S_1^{n_1} \oplus \ldots \oplus S_k^{n_k} \oplus T_1^{m_1} \oplus \ldots \oplus T_l^{m_l}, \tag{4.4}$$

is a decomposition of M into irreducible KG-modules and $(e_i)_{i \in I}$ an ordered basis establishing this decomposition. Moreover, we assume that all the T_j are one dimensional and the S_i are s_i-dimensional with $s_i > 1$. Let $a = \sum a_i e_i$. If $Ke_i = T_j$, one of the one dimensional G-representations occurring in (4.4), then since M must equal KGa, the coefficient a_i of a is nonzero. For every S_i-module, we obtain a point $[a_0 : a_1 : \ldots : a_{s_i - 1}] \in \mathbb{P}^{s_i - 1}$ and it is clear that the choice of a is indeed independent up to a scalar multiplication of the coefficients per irreducible component of M. In total, we obtain a point in the projective variety

$$\mathbb{P}^{s_1-1} \times \ldots \times \mathbb{P}^{s_1-1} \times \mathbb{P}^{s_2-1} \times \ldots \times \mathbb{P}^{s_2-1} \times \ldots \times \mathbb{P}^{s_k-1} \times \ldots \times \mathbb{P}^{s_k-1},$$

with n_1 factors $\mathbb{P}^{s_1-1}, \ldots, n_k$ factors $\mathbb{P}^{s_k-1}$.

Since $KH_iT = KG_iT$ for all T and $0 \leq i \leq d$, we do not get any non-trivial decomposition groups from the one dimensional irreducibles. Concerning the higher dimensional irreducibles, we fix $S = S_1$, which is $s = s_1$-dimensional and the part $[a_0 : \ldots : a_{s-1}]$ of a. We can represent KH_1a as an $s \times |H_1|$-matrix A_1, of which the i-th column gives the action of h_i on a in the basis $\{e_0, \ldots, e_{s-1}\}$ of S. Similarly, we introduce a matrix B_1 for KG_1a. Clearly, if $\mathrm{rk}(\mathrm{A}_1) = \mathrm{rk}(\mathrm{B}_1)$, then $KG_1a = KH_1a$ and $G_1' = G_1$ follows. Hence we obtain a Zariski open or closed set $X \subset \mathbb{P}^{s-1}$ on which no non-trivial decomposition groups occur. In the situation from Example 4.3.4, for $S = U$ a 2-dimensional representation with basis $\{f_1, f_2\}$ and $a = a_0f_1 + a_1f_2$, we calculate (for ordering $H_1 = \mathbb{Z}_2 = \{1, -1\}$ and $G_1 = \mathbb{Z}_4 = \{1, j, -1, -j\}$)

$$A_1 = \begin{pmatrix} a_0 & -a_0 \\ a_1 & -a_1 \end{pmatrix}, \quad B_1 = \begin{pmatrix} a_0 & -a_1 & -a_0 & a_1 \\ a_1 & a_0 & -a_1 & -a_0 \end{pmatrix}.$$

The rank of A_1 is always 1, so $X = \mathbb{V}(x_0^2 + x_1^2)$ if $\mathbb{P} = \mathrm{Proj}\ \mathbb{C}[\mathrm{x}_0, \mathrm{x}_1]$, which consists of two points $\{[1 : i], [1 : -i]\}$. Remark that these points correspond to the base change which establishes the decomposition of U into two irreducible $K\langle j \rangle$-modules. We considered the point $[1 : i]$ and obtained indeed that $G_1' = G_1$.

For H_2 and G_2 we introduce the (extended) matrices A_2 and B_2 as well as the Zariski open X_2. Now on $X_1 = X$, we have $KH_1a = KG_1a$, whence

$$KH_2a = KH_2KH_1a = KH_2KG_1a = KG_2a.$$

The last equality follows since our normal chain is maximal. Indeed, since $h_2g_1h_2'g_1' = h_2h_2'h_2'^{-1}g_1h_2'g_1' \in H_2G_1$ and $(hg)^{-1} = g^{-1}h^{-1} = h^{-1}hg^{-1}h^{-1}$ in H_2G_1, we see that H_2G_1 is a group and since

$$g_2^{-1}h_2g_1g_2 = h_2h_2^{-1}g_2^{-1}h_2g_2g_2^{-1}g_1g_2 \in H_2G_1,$$

we have that $G_1 \subset H_2G_1 \triangleleft G_2$, whence $H_2G_1 = G_2$.
Therefore, $X_1 \subset X_2 \subset \mathbb{P}^{s-1}$. In the example

$$A_2 = \begin{pmatrix} a_0 & ia_0 & -a_0 & -ia_0 \\ a_1 & -ia_1 & -a_1 & ia_1 \end{pmatrix},$$

which has rank 2 on $\mathbb{X}(x_0) \cap \mathbb{X}(x_1) = \mathbb{X}(x_0x_1)$. So the possible interesting cases are $[a_0 : a_1] \in \mathbb{P} \setminus \mathbb{X}(x_0x_1) = \{[1 : 0], [0 : 1]\}$. Take $a = [1 : 0]$, the other case being analogous. We have

$$M_1 = \mathbb{C}\langle j \rangle f_1 = \mathbb{C}f_1 \oplus j \cdot \mathbb{C}f_1 = \mathbb{C}f_1 \oplus \mathbb{C}f_2.$$

Both $\mathbb{Z}_2$-modules are equivalent, so $G_1' = G_1$. One stair further, we have

$$\mathbb{C}Q_8f_1 = \mathbb{C}f_1 \oplus j \cdot \mathbb{C}f_1 = \mathbb{C}f_1 \oplus \mathbb{C}f_2,$$

but both components are no longer isomorphic as $K\langle i \rangle$-modules. Therefore, $G_2' = \langle i \rangle$. Observe that $G_1' \not\subset G_2'$!

Example 4.4.1. Look at the following graph of groups

$$\begin{array}{ccc} \langle a \rangle & \triangleleft & D_8 = \langle a, x | a^4 = x^2 = 1,\ xax^{-1} = a^{-1} \rangle \\ \triangledown & & \triangledown \\ \{e, a^2\} & \triangleleft & \{e, x, a^2, a^2x\} \end{array}$$

and consider the two-dimensional irreducible representation S defined by

$$a \mapsto \begin{pmatrix} i & 0 \\ 0 & -i \end{pmatrix}, \quad x \mapsto \begin{pmatrix} 0 & 1 \\ 1 & 0 \end{pmatrix}.$$

Then one calculates that $X_1 = \mathbb{V}(x_0x_1) = \{[1:0],[0:1]\} \subset X_2 = \mathbb{X}(x_0^2 - x_1^2)$.

4.5 Relation between decomposition groups

For a finite group G with chain of normal subgroups $1 \triangleleft G_1 \triangleleft \ldots \triangleleft G_d = G$ and $H \triangleleft G$ another normal subgroup we obtain for $0 \leq i \leq d$ the following group algebra filtrations

$$\begin{array}{ll} (f1): & K \subset KH_1 \subset \ldots \subset KH_{i-1} \subset KH_{i-1} \\ (f2): & K \subset KH_1 \subset \ldots \subset KH_{i-1} \subset KH_i \\ (f3): & K \subset KG_1 \subset \ldots \subset KG_{i-1} \subset KG_{i-1} \\ (f4): & K \subset KG_1 \subset \ldots \subset KG_{i-1} \subset KG_i \end{array} \tag{4.5}$$

An FKG-glider representation M is obviously also a glider representation with regard to each of these four filtrations. When performing the Clifford theory in Section 4.3 we only considered the filtrations $(f2) \subset (f4)$ and obtained decomposition groups $H_i \subset G_i' \subset G_i$ depending on some building block S, which is an irreducible H_{i-1}-representation. What we want to do now is to consider the inclusions

$$\begin{array}{ccc} (f2) & \subset & (f4) \\ \cup & \nearrow & \cup \\ (f1) & \subset & (f3) \end{array}$$

In total there are five inclusions, so for any building block S, say an irreducible KH_i-module, we obtain five decomposition groups:

$$\begin{array}{ccccc} H_{i+1} & \subset & G'_{i+1} & \subset & G_{i+1} \\ \cup & & & \nearrow & \cup \\ H'_i & & H_i^2 & & G''_i \\ \cup & \nearrow & & & \cup \\ H_i & \subset & G'_i & \subset & G_i \end{array}$$

We wonder whether there are some relations between these 5 decomposition groups. We assume that the chains on G and H are maximal, i.e. the subsequent factor groups are all irreducible and $H_{i+1}G_i = G_iH_{i+1} = G_{i+1}$.

For the convenience of the reader, we briefly recall from [24] how the classical decomposition groups are constructed: let $H \triangleleft G$ be a normal subgroup and let V be some irreducible G-representation. If V_H is no longer irreducible, there is some irreducible H-subrepresentation W and we can find elements $g_2, \ldots, g_r \in G \setminus H$ such that

$$V = W \oplus g_2W \oplus \ldots \oplus g_rW.$$

All of the g_iW are irreducible H-representations and of the same degree. Group together the ones that are isomorphic to yield a decomposition

$$V = R_1 \oplus \ldots \oplus R_m,$$

with $W \subset R_1$ and $m|r$. Moreover, all the R_i are of equal dimension and the elements of G permute the vector spaces R_i among each other ([24, Theorem 2]). The decomposition group $H \subset G' \subset G$ is then the group of elements g that leave R_1 invariant, i.e. $gR_1 = R_1$.

Lemma 4.5.1. *The decomposition group G' is exactly the group of elements $g \in G$ such that $gW \cong W$ as H-modules.*

Proof. If $g \in G'$, then (up to reordering the g_i from above) $gR_1 = g(W \oplus g_2W \oplus \ldots \oplus g_{r'}W) = R_1 = W \oplus g_2W \oplus \ldots \oplus g_{r'}W$ (where $r' = r/m$). Hence $gR_1 = gW \oplus gg_2W \oplus \ldots \oplus gg_{r'}W$ and because the decomposition into H-components is unique, we must have that $gW \cong g_iW$ for some $1 \leq i \leq r'$. But by construction, $g_iW \cong W$. Conversely, suppose that $gW \cong W$ as H-modules; then for all $1 \leq i \leq r'$, $gg_iW \cong gW \cong W$. Because $gR_1 = R_j$ for some $1 \leq j \leq m$, j must be 1, since the irreducible H-components of gR_1 are all isomorphic to W. □

The previous lemma allows us to obtain a first result.

Proposition 4.5.2. *In the situation above, we have the following*

- $H_i^2 = G_i \Leftrightarrow G_i' = G_i$ and $H_i' = H_i$;
- $H_i^2 = H_{i+1} \Leftrightarrow G_i' = H_i$ and $H_i' = H_{i+1}$;
- $H_i^2 = G_{i+1} \Leftrightarrow G_i' = G_i$ and $H_i' = H_{i+1}$.

Proof. Observe that $H_i'G_i' \subset H_i^2$. Suppose that $G_i' = G_i$ and $H_i' = H_{i+1}$, then $G_{i+1} = G_iH_{i+1} \subset H_i^2$. Conversely, if $H_i^2 = G_{i+1}$, then for all $g \in G_i$, $gS \cong S$ so $G_i' = G_i$ and similarly we have $H_i' = H_{i+1}$. This covers the last equivalence. The first two equivalences are symmetric, so we only prove the first one. If $H_i^2 = G_i$ then automatically $G_i' = G_i$ and H_i' must be H_i

by the third equivalence and our maximality condition. Conversely, we have $H_i'G_i' = G_i \subset H_i^2$ and there are no proper subgroups in between G_i and G_{i+1} so $H_i^2 = G_i$. □

The previous proposition indicates that the two smaller decomposition groups G_i' and H_i' almost determine the group H_i^2. The interesting situation we have to study deeper is when H_i^2 equals H_i or some group F lying strictly between $H_i \subsetneq F \subsetneq G_{i+1}$. In many cases, however, this situation does not arise:

Lemma 4.5.3. *In the situation above, if $H_i \subset Z(G_{i+1})$ then the three decomposition groups H_i', G_i' and H_i^2 are all maximal, that is, they are H_{i+1}, G_i and G_{i+1} respectively.*

Proof. Let us prove this for H_i', the other two cases being completely analogous. Decompose $KH_{i+1}S = S \oplus h_2 S \oplus \ldots \oplus h_r S$ into irreducible KH_i-modules for some $h_j \in H_{i+1} \setminus H_i$. Since $H_i \subset Z(G)$, H_i commutes with the appearing h_j so for $h \in H_i$ we have $h \cdot h_j S = hh_j S = h_j(h \cdot S)$, which shows that S and $h_j S$ are isomorphic H_i-modules. Hence $KH_{i+1}S = R_1$ and it follows that $H_i' = H_{i+1}$. □

Example 4.5.4. Consider again the graph of groups from Example 4.3.4

$$\begin{array}{ccc} \mathbb{Z}_4 = \{1, j, -1, -j\} & \triangleleft & Q_8 = \langle -1, i, j | i^2 = j^2 = -1, ij = -ji \rangle \\ \triangledown & & \triangledown \\ \mathbb{Z}_2 = \{1, -1\} & \triangleleft & \mathbb{Z}_4 = \{1, i, -1, -i\} \end{array}$$

We studied the following irreducible glider representation

$$\Omega = M = U \oplus T_3 \oplus T_2 \supset V^{-i} \oplus T_3 \oplus T_2 \supset \Delta \supset 0 \supset \ldots.$$

We showed that $\{\mathbb{C}e_1, \mathbb{C}(t_3 + t_2)\}$, where $t_3 \in T_3$, $t_2 \in T_2$, is a minimal set of building blocks and after some calculations we arrive at the following decomposition groups:

$$\mathbb{C}e_1 : \quad \begin{array}{ccc} & Q_8 & \\ \langle i \rangle & Q_8 & \langle j \rangle \\ & \langle j \rangle & \end{array} \qquad \mathbb{C}(t_3 + t_2) : \quad \begin{array}{ccc} & Q_8 & \\ \langle i \rangle & Q_8 & Q_8 \\ & \langle j \rangle & \end{array}$$

$H_1 = \{1, -1\} = Z(Q_8)$, so the three lower decomposition groups are fixed. But we see that for the two others, there are some differences.

Let us look at the following situation for our diagram

$$\begin{array}{ccccc} H_2 & \subset & H_2 & \subset & G_2 \\ \cup & & & \subsetneq & \cup \\ H_1 & & F & & G_i'' \\ \cup & \subsetneq & & & \cup \\ H_1 & \subset & H_1 & \subset & G_1 \end{array} \tag{4.6}$$

Start with an irreducible H_1-module S in some irreducible glider representation M and decompose $KH_2S = S \oplus h_2S \oplus \ldots \oplus h_rS$. Since $H_1' = H_1$, we have that $h_iS = R_i$, $i = 1, \ldots, r$, whence $[H_2 : H_1] = r$. We do the same for

$$\begin{aligned} KG_2S &= KH_2S \oplus g_2KH_2S \oplus \ldots \oplus g_mKH_2S \\ &= S \oplus h_2S \oplus \ldots \oplus h_rS \oplus g_2h_2S \oplus \ldots \oplus g_mh_rS. \end{aligned}$$

Again, since $G_2' = H_2$ we have $[G_2 : H_2] = m$ and as $H_1^2 = F$ is proper, there is some $m' \leq m$ such that (up to reordering)

$$R_1^{H_1^2} = S \oplus g_2h_{j(2)}S \oplus \ldots \oplus g_{m'}h_{j(m')}S.$$

Hence we have that

$$[G_2 : F] = \frac{mr}{m'} = \frac{[G_2 : H_2][H_2 : H_1]}{m'} = \frac{[G_2 : H_1]}{m'}.$$

Assume now that G_2 is a p-group. Then $p = [G_2 : F] = \frac{p^2}{m'}$, hence $m' = p = m$. Moreover, since $G_2' = H_2$, we have that $g_iH_2 \cap g_jH_2 = \emptyset$ for $1 \leq i \neq j \leq p$, so $\{H_2, g_2H_2, \ldots, g_pH_2\}$ is a full set of left cosets of H_2 in G_2. We have that $[F : H_1] = p$, hence we can find f_i, $i = 1, \ldots, p$ forming a full set of representatives. Write $f_i = g_{\alpha(i)}h_{\beta(i)}$. Suppose that $\alpha(i) = \alpha(j)$ for some $i \neq j$. Because $f_iS \cong f_jS$, this would imply that $h_{\beta(i)}S \cong h_{\beta(j)}S$, or $h_{\beta(i)}h_{\beta(j)}^{-1}$ in H_1. However, $f_i^{-1}f_j = h_{\beta(i)}^{-1}h_{\beta(j)} \notin H_1$, a contradiction. Therefore all the g_iKH_2S have one component isomorphic to S, whence they have the same decomposition as H_1-modules (since H_2 is normal in G_2). So we have proven the following.

Proposition 4.5.5. *Let G_2 be a p-group and S some building block such that its associated decomposition graph takes the form (4.6). Then H_1 has at least p non-isomorphic irreducible representations of degree* $\dim(S)$, *call these $S_1, \ldots, S_p$. Moreover, there are at least p non-isomorphic H_2-modules that decompose into $\oplus_{i=1}^p S_i$.*

Next, we consider the situation

$$\begin{array}{ccccc} H_2 & \subset & G_2 & \subset & G_2 \\ \cup & & & \subsetneq & \cup \\ H_1 & & F & & G_1'' \\ \cup & \subsetneq & & & \cup \\ H_1 & \subset & H_1 & \subset & G_1 \end{array} \qquad (4.7)$$

To begin with, $H_1' = H_1$ means that we can find p elements $h_i \in H_2 \setminus H_1$ such that $KH_2S = S \oplus h_2S \oplus \ldots \oplus h_pS$, and all h_iS are non-isomorphic H_1-modules. In fact, since $|H_2/H_1| = p$, the factor group is cyclic and we can choose the elements h_i such that $h_i = h_2^{i-1}$ for $i = 2, \ldots, p$ and $h_1 = 1$. There are two situations when $G_2' = G_2$ occurs. Indeed, either $KG_2S = KH_2S$, saying that KH_2S is already a G_2-representation, or either

$$KG_2S = KH_2S \oplus g_2KH_2S \oplus \ldots \oplus g_pKH_2S$$

for p elements $g_i \in G_2 \setminus H_2$. In the first situation, however, $KG_2S = S \oplus h_2S \oplus \ldots \oplus h_pS$ and $H_1 \subsetneq F \subsetneq G_2$ contradicts H_1' being equal to H_1. So

$$KG_2S = (S \oplus h_2S \oplus \ldots \oplus h_pS) \oplus g_2(S \oplus h_2S \oplus \ldots \oplus h_pS) \oplus \ldots$$

Clearly, all the g_iKH_2S have the same decomposition into H_1-modules (as was the case in situation (4.6)), so the only difference between situations (4.6) and (4.7) lies in the fact the g_iKH_2S are isomorphic H_2-modules or not. Let us treat both cases simultaneously and see what happens.

For some $\alpha : \{1, \ldots, p\} \to \{1, \ldots, p\}$, we have:

$$S \cong g_2h_{\alpha(2)}S \cong \ldots \cong g_ph_{\alpha(p)}S.$$

Since the factor group G_2/H_1 is of order p^2, it is isomorphic to either C_{p^2} or $C_p \times C_p$. However, clearly $F/H_1, H_2/H_1$ and G_1/H_1 are different subgroups of order p in G_2/H_1, so since C_{p^2} has only one subgroup of order p, G_2/H_1 must be isomorphic to $C_p \times C_p$.

Of course we have that $\alpha(1) = 1$. Suppose that $\alpha(i) = 1$ for some $2 \leq i \leq p$, then $g_i = g_2^{i-1} \in F$, whence $g_j \in F$ for all $1 \leq j \leq p$ and it follows that α maps to 1. If $\alpha(i) \neq 1$ for all $2 \leq i \leq p$, then α must be bijective; otherwise $g_2^i \in F$ for some i, meaning that $\alpha(i) = 1$. Up to changing g_2 by $g_2h_{\alpha(2)}$ we may assume that $\alpha = 1$ and we see that $F/H_1 = \langle \overline{g_2} \rangle$. Going the other way, that is, via G_1, we find a $z_2 \in G_2 \setminus G_1$ such that $F/H_1 = \langle \overline{z_2} \rangle$. Hence $z_2H_1 = g_2H_1$ and we have that $z_2 \notin H_2, g_2 \notin G_1$.

Proposition 4.5.6. *Let G_2 be a p-group and S some building block such that its associated decomposition graph takes the form* (4.6) *or* (4.7). *Then G_2/H_1 is isomorphic to $C_p \times C_p$ and we find elements $g, z \in G_2 \setminus (G_1 \cup H_2)$ such that $F/H_1 = \langle \overline{g} \rangle = \langle \overline{z} \rangle$.*

Example 4.5.7. Assume $p = 5$ and H_1 some 5-group. Consider $G_2 = H_1 \rtimes (C_5 \times C_5)$, a semi-direct product defined by some group morphism $\varphi : C_5 \times C_5 \to \mathrm{Aut(H_1)}$, i.e.

$$(h, a^i, a^j) \cdot (h', a^k, a^l) = (h\varphi((a^i, a^j))(h'), a^{i+k}, a^{j+l}),$$

where $C_5 = \langle a \rangle$. Then we look at

$$\begin{array}{ccccc}
H_1 \rtimes \langle (a^2, a) \rangle & \subset & G_2' & \subset & H_1 \rtimes (C_5 \times C_5) \\
\cup & & & \subsetneq & \cup \\
H_1 & & F & & G_1'' \\
\cup & \subsetneq & & & \cup \\
H_1 & \subset & H_1 & \subset & H_1 \rtimes \langle (a, a^2) \rangle
\end{array}$$

Suppose that if we take $g = g_2 = (e, a^3, a^2)$ that $\alpha = 1$. Then $F/H_1 = \langle \overline{g_2} \rangle$, or $F = H_1 \sqcup H_1 g \sqcup \ldots \sqcup H_1 g^{p-1}$. Define a map

$$\psi : F \to H_1 \rtimes \langle \overline{g} \rangle, \ f \mapsto (h, \overline{g}^i),$$

if $f = hg^i$ and where the semi-direct product structure $H_1 \rtimes \langle \overline{g} \rangle$ is defined by $\varphi_{|\langle (a^3,a^2) \rangle}$, i.e. $\varphi(\overline{g})(h) = \varphi((a^3, a^2))(h)$. We calculate

$$\begin{aligned}
\psi(hg^i)\psi(h'g^j) &= (h, \overline{g}^i)(h', \overline{g}^j) \\
&= (h\varphi(\overline{g}^i)(h'), \overline{g}^{i+j}) \\
\psi(hg^i h' g^j) &= \psi(h\varphi((a^3, a^2)^i)(h')g^{i+j}) \\
&= (h\varphi(\overline{g}^i)(h'), \overline{g}^{i+j}),
\end{aligned}$$

which shows that ψ is a group morphism, which is easily seen to be surjective, hence bijective. Moreover,

$$H_1 \rtimes \langle \overline{g} \rangle = H_1 \rtimes F/H_1 \subset H_1 \rtimes G_2/H_1 = H_1 \rtimes G_2/H_2 \times G_2/G_1,$$

so $F = H_1 \rtimes \langle (a^3, a^2) \rangle$.

Observe that, in the previous example, if we drop the semi-direct product with H_1, then everything is Abelian and all decomposition groups would be maximal by Lemma 4.5.3. In fact, the only possibility to have situation (4.6) or (4.7) is when $G_2 \cong H_1 \rtimes (C_p \times C_p)$. Indeed

Theorem 4.5.8. *Let G_2 be a p-group and S some building block such that its associated decomposition graph takes the form (4.6) or (4.7). Then G_2 is isomorphic to $H_1 \rtimes (C_p \times C_p)$ where the semi-direct product is not a direct product; Moreover $F = H_1 \rtimes \langle \overline{g} \rangle = H_1 \rtimes \langle \overline{z} \rangle$ for some $g, z \in G_2 \setminus (H_2 \cup G_1)$ such that $F \cong H_1 \rtimes (\langle \overline{g}, \overline{z} \rangle) \subset H_1 \rtimes G_2/H_2 \times G_2/G_1$.*

Proof. By our observations above, we can find elements $z, g \in G_2 \setminus (H_2 \cup G_1)$ such that $F/H_1 = \langle \overline{g} \rangle = \langle \overline{z} \rangle$. This shows that

$$F = H_1 \rtimes \langle \overline{g} \rangle = H_1 \rtimes \langle \overline{z} \rangle.$$

We also have that $G_2/H_1 \cong G_2/H_2 \times G_2/G_1$ and the isomorphism maps $\overline{g}$ to $(\overline{g}, \overline{z})$. Hence

$$F = H_1 \rtimes \langle \overline{g} \rangle \cong H_1 \rtimes \langle (\overline{g}, \overline{z}) \rangle \subset H_1 \rtimes (\langle \overline{g} \rangle \times \langle \overline{z} \rangle) = H_1 \rtimes (G_2/H_2 \times G_2/G_1) \cong G_2.$$

This also shows that the semi-direct product is not a direct product; otherwise the decomposition group $H_1^2 = G_2$, because in this case H_1 commutes with $\{1\} \times \langle \overline{g} \rangle \times \langle \overline{z} \rangle$. □

Corollary 4.5.9. *Let G_2 be a p-group and S some building block such that its associated decomposition graph takes the form* (4.6) *or* (4.7). *Then $H_1 \neq C_p$.*

Proof. The order of the automorphism group

$$|\mathrm{Aut}(\mathrm{C_p})| = \varphi(\mathrm{p}) = \mathrm{p} - 1,$$

where φ is the Euler characteristic. Hence there are no non-trivial group morphisms $f : C_p \times C_p \to \mathrm{Aut(C_p)}$, so G_2 would not be a semi-direct product. □

The last situation we have to consider is when $H_i^2 = H_i$, i.e.

$$\begin{array}{ccccc} H_2 & \subset & G_2' & \subset & G_2 \\ \cup & & & \rotatebox{45}{$\subset$} & \cup \\ H_1 & & H_1 & & G_1'' \\ \cup & \rotatebox{45}{$\subset$} & & & \cup \\ H_1 & \subset & H_1 & \subset & G_1 \end{array} \tag{4.8}$$

We have the decomposition of KG_2S into p^2 irreducible H_1-representations. This shows for example that $p^2 \dim(S) \leq |H_1|$. In particular, we have that $KG_2S \supsetneq KH_2S$ and $KG_2S \supsetneq KG_1S$. So we can write

$$KG_2S = KH_2S \oplus g_2KH_2S \oplus \ldots \oplus g_pKH_2S,$$

$$KG_2S = KG_1S \oplus z_2KG_1S \oplus \ldots \oplus z_pKG_1S,$$

for some $g_i \in G_2 \setminus H_2$ and $z_i \in G_2 \setminus G_1$. If for example $KH_2S \cong g_2KH_2S$ as H_2-representations, then they would have the same decomposition into H_1-representations, which is a contradiction. Hence we arrive at

Proposition 4.5.10. *Let G_2 be a p-group and S some building block such that its associated decomposition graph takes the form* (4.8). *Then $G_2' = H_2$ and $G_1'' = G_1$.*

4.6 Nilpotent groups of order $p^k q^l$

Let G be a p-group, H a q-group. We look at the nilpotent group GH. If the order of GH is pq and $p-1$ does not divide q, then the group is cyclic, and it is just isomorphic to $C_p \times C_q$, which is not interesting regarding chains of subgroups. So assume that $|GH| = p^k q^l$ with $k, l > 1$. We consider the following chains of subgroups

$$\begin{array}{ccccccccc}
1_G & \subset & G_1 & \subset & \dots & \subset & G_{d-1} & \subset & GH \\
 & & & & & & & \hookrightarrow & \cup \\
 & & & & & & G_{d-1}H_{d-1} & & H_{d-1} \\
 & & & & & \hookrightarrow & & & \cup \\
 & & & & \iddots & & & & \vdots \\
 & & & \hookrightarrow & & & & & \cup \\
 & & G_1H_1 & & & & & & H_1 \\
 & \hookrightarrow & & & & & & & \cup \\
1 & & & & & & & & 1_H
\end{array}$$

The Frobenius divisibility theorem, see [28, Theorem 4.16], states that the dimension of an irreducible representation V divides the order of the group, so we enlist the irreducible representations of GH by

$$V_i \ (i \in I_1), \ P_i \ (i \in I_2), \ Q_i \ (i \in I_3), \ W_i \ (i \in I_4),$$

where V_i is 1-dimensional, P_i is p^α-dimensional, Q_i is q^β-dimensional and W_i is $p^\alpha q^\beta$-dimensional. Let $\Omega \supset M$ be an irreducible $FKGH$-glider of essential length d. Theorem 4.1.12 shows that

$$M = \bigoplus_i V_i \oplus \bigoplus_i P_i^{m_i} \oplus \bigoplus_i Q_i^{k_i} \oplus \bigoplus_i W_i^{l_i}. \tag{4.9}$$

We can consider $M \supset M_1 \supset \dots$ both as an FKG- and FKH-glider. We denote these gliders by M_G, resp. M_H.

Lemma 4.6.1. *If M_H is an irreducible FKH-glider, then the m_i in the decomposition* (4.9) *are all zero.*

Proof. In the decomposition of the GH-module M as an H-module, there appears the decomposition of P_i into H-modules (if $m_i \neq 0$). Since P_i is p^α-dimensional, its decomposition into irreducible H-representations must be

$$P_i = \oplus_{j \in J} U_j^{n_j},$$

where all the U_j are one-dimensional. In fact, all the U_j must be isomorphic, since we have that $KGHU_1 = P_i$ and in the procedure of determining the decomposition group, we can take for elements in $GH \setminus H$ elements of the form $(g, 1_H)$. Since $(g, 1_H)$ commutes with $1_G \times H$, the decomposition group is the whole group GH, whence all the U_i are isomorphic. But this contradicts with Theorem 4.1.12, so no factors P_i appear in (4.9). □

Lemma 4.6.2. *If M_H is an irreducible FKH-glider and $W_i^{l_i}$ appears in* (4.9) *with $\dim(W_i) = p^\alpha q^\beta$ then $l_i p^\alpha \leq q^\beta$.*

Proof. In the decomposition of M as an H-module there appears the decomposition of $W_i^{l_i}$ into H-modules. By the same argument in the proof of the previous lemma, the decomposition group must be GH, hence $W_i = U^k$ for some irreducible H-rep U and some $k \in \mathbb{N}$. U cannot be one-dimensional, because then M_H would not be irreducible. Thus $\dim(U) = q^j$ for some j and $k = p^\alpha q^{\beta-j}$. In particular we have that $\beta \geq j$. Theorem 4.1.12 then entails that $l_i k = l_i p^\alpha q^{\beta-j} \leq q^j$. Hence

$$l_i p^\alpha \leq q^{2j-\beta} = q^{2j-2\beta} q^\beta \leq q^\beta,$$

where the last inequality follows since $j - \beta \leq 0$. □

Proposition 4.6.3. *Let $\Omega \supset M$ be an irreducible $FKGH$ glider of essential length d. If M_H and M_G are still irreducible, then $M = \oplus_i V_i$.*

Proof. Irreducibility as FKH-, resp. FKG-glider shows that no P_i's, resp. Q_i's appear. Suppose that some $W_i^{l_i}$ of dimension $p^\alpha q^\beta$ appears. From Lemma 4.6.2 applied for KH and KG, it follows that

$$l_i p^\alpha \leq q^\beta \leq \frac{p^\alpha}{l_i},$$

hence $l_i = 1$, but then $p^\alpha = q^\beta$, a contradiction. □

Recall that the irreducible representations of a nilpotent group GH of order $p^\alpha q^\beta$ arise as tensor products $U \otimes V$ of irreducible representations of G, resp. H. We clearly have that for an irreducible GH-representation W, W_G and W_H are irreducible if and only if W is one-dimensional. Proposition 4.6.3 gives one direction of this statement in the situation of glider representations. Let $M = \oplus_i V_i \supset \ldots \supset Ka$ be an irreducible $FKGH$-glider of essential length d. If some of the appearing V_i, say V_1 and V_2, are isomorphic as G-representations, then M_G is definitely no longer irreducible by Theorem 4.1.12. So we have the glider analogue:

Proposition 4.6.4. *Let $\Omega \supset M$ be an irreducible $FKGH$ glider of essential length d. Then M_H and M_G are irreducible if and only if $M = \oplus_i V_i$, where all the V_i are non-isomorphic both as G- and as H-representations.*

Suppose now that $M = \oplus_i V_i$, where the V_i are non-isomorphic one-dimensional GH-representations, but we drop the condition of M_H and M_G remaining both irreducible gliders. We can consider $\Omega \supset M$ as an $FKG_{d-1}H_{d-1}$-glider with the second filtration of (4.5) on $G_{d-1}H_{d-1}$. Then either M remains irreducible, which says that $K_{d-1}H_{d-1}a = \oplus_i V_i$, or either $K_{d-1}H_{d-1}a \subsetneq \oplus_i V_i$. The first situation occurs exactly when all the V_i remain non-isomorphic as $G_{d-1}H_{d-1}$-representations. If we are in the latter case, suppose that $V_1 \cong V_2$ as $G_{d-1}H_{d-1}$. This means that, up to renumbering, we can take $S \cong V_1$ as a $G_{d-1}H_{d-1}$-building block. However, looking at the decomposition group $G_{d-1}H_{d-1} \subset F \subset GH$ does not give us more information.

Indeed, in the decomposition of M we have S and some conjugate $(g,h)S$ for some $(g,h) \in GH \backslash G_{d-1}H_{d-1}$, but these are equivalent as $G_{d-1}H_{d-1}$-modules by assumption. Hence $F = GH$. So we have to take a different approach here. Rewrite

$$M = \bigoplus_i U_i \otimes U_i',$$

where the U_i, resp. U_i' are one-dimensional G-, resp. H-representations. The following statements are obvious

M_G is irreducible as an FKG–fragment $\Leftrightarrow$ all the appearing U_i are non–isomorphic;
M_H is irreducible as an FKH–fragment $\Leftrightarrow$ all the appearing U_i' are non–isomorphic.

Example 4.6.5. If $M = (U_1 \otimes U_1' \oplus U_1 \otimes U_2') \supset \ldots \supset Ka$ is an irreducible $FKGH$-glider, then

$$M_G \cong U_1^{\oplus 2}, \quad M_H \cong U_1' \oplus U_2',$$

as G-, resp. H-modules (since $U_i, U_i', i = 1, 2$ are one-dimensional!). One verifies that the decomposition groups in both cases are just $F = GH$. The difference, however, lies in the number of building blocks! Indeed, for M_G we have that $KGa \cong U_1 \subset U_1^{\oplus 2}$, where the embedding is of diagonal type. Hence there is only one building block $S = KGa$ and $M_G = S \oplus (1_G, h)S$. On the other hand, $KHa \cong U_1' \oplus U_2'$, so we have two buidling blocks $S_i = U_i'$, $i = 1, 2$.

In fact, the previous example easily generalizes to

Lemma 4.6.6. *Let $M = \oplus_i V_i$ be an irreducible $FKGH$-glider; then the number of building blocks as FKG-, resp. FKH-glider corresponds to the number of non-isomorphic G-, resp. H-representations among the appearing V_i in the decomposition of M.*

The category KG-mod is monoidal, which allows us to define the tensor product of fragments and glider representations.

Proposition 4.6.7. *Let $\Omega \supset M$, $\Omega' \supset M'$ be FKG-glider representations of essential length $\leq d$. Then the descending chain*

$$\Omega \otimes \Omega' \supset M \otimes M' \supset M_1 \otimes M_1' \supset \ldots \supset M_d \otimes M_d' \supset 0 \ldots$$

is an FKG-glider representation.

Proof. All tensor products are over K, which explains the inclusions. The fragment conditions are satisfied since the comultiplication

$$\Delta : KG \to KG \otimes KG$$

is given by $\Delta(g) = g \otimes g$, extended linearly. □

Definition 4.6.8. Let $M \supset M_1 \supset \ldots$, $M' \supset M_1' \supset \ldots$ be FKG-fragments. Their tensor product is the fragment with chain

$$M \otimes M' \supset M_1 \otimes M_1' \supset \ldots.$$

If $M \supset M_1 \supset \ldots$ is an FKG-fragment, then we can consider it as an $FKGH$-fragment by defining $(g_i, h_i) \cdot m_i = g_i \cdot m_i$ for $g_i \in G_i, h_i \in G_i, m \in M_i$. Similarly, one defines an $FKGH$-fragment structure on an FKH-fragment, whence we can define the tensor product of an FKG-fragment and an FKH-fragment.

To any irreducible $FKGH$-glider of essential length d and of the form $M = \oplus_i V_i$ we now associate a triple (a, b, c), where

$$\begin{aligned} a &= \text{ number of } GH\text{−components of } M \\ b &= \text{ number of building blocks as } FKG\text{−glider,} \\ c &= \text{ number of building blocks as } FKH\text{−glider.} \end{aligned}$$

Proposition 4.6.9. *In the situation above, the glider M is the tensor product $M = M_1 \otimes M_2$ of an FKG- and an FKH-glider if and only if the associated triple (a, b, c) satisfies $a = bc$.*

Proof. Suppose that $M = M_1 \otimes M_2$. If $M_1 = U_1^{\oplus n_1} \oplus \ldots \oplus U_b^{\oplus n_b}$ and $M_2 = (U_1')^{\oplus m_1} \oplus \ldots \oplus (U_c')^{\oplus m_c}$ with at least one of the $n_i > 1$ or $m_1 > 1$, say $n_1 > 1$, then M would contain two distinct isomorphic components $U_1 \otimes U_1'$, which contradicts the irreducibility of M since these components are one-dimensional. Hence $M_1 = U_1 \oplus \ldots \oplus U_b$ and $M_2 = U_1' \oplus \ldots \oplus U_c'$, from which $a = bc$ follows. Conversely, assume $a = bc$. Up to reordering, we can write

$$M = (V_1 \oplus \ldots \oplus V_b) \oplus (V_{b+1} \oplus \ldots \oplus V_{2b}) \oplus \ldots \oplus (V_{b(c-1)+1} \oplus \ldots \oplus V_{bc}),$$

such that for every $0 \leq i \leq b$: $V_{jb+i} \cong V_{kb+i}$ as G-representations for all $0 \leq j, k \leq c-1$. Hence we have that

$$U_i \otimes U_{jb+i}' \cong V_{jb+i} \cong V_{kb+i} \cong U_i \otimes U_{kb+i}'$$

and it follows that

$$M \cong U_1 \otimes (U_1' \oplus U_{b+1}' \oplus \ldots U_{b(c-1)+1}') \oplus \ldots \oplus U_b \otimes (U_b' \oplus U_{2b}' \oplus \ldots \oplus U_{bc}').$$

If for example $U_1' \cong U_{b+1}'$ as H-representations, then $V_1 \cong V_{b+1}$ as GH-rep which contradicts the irreducibility of M. Since only c non-isomorphic H-representations occur in the above decomposition we have that

$$U_1' \oplus U_{b+1}' \oplus \ldots \oplus U_{b(c-1)+1}' \cong \ldots \cong U_b' \oplus U_{2b}' \oplus \ldots \oplus U_{bc}'$$

as H-representations, which entails that

$$M \cong (U_1 \oplus \ldots \oplus U_b) \otimes (U_1' \oplus \ldots \oplus U_{b(c-1)+1}').$$

This is a priori an isomorphism as modules. If $M_d = Km$, then since the U_i and U_j' are one-dimensional, we can write

$$m = \sum_{i=1}^{b} \sum_{j=1}^{c} \lambda_{i,j} u_i \otimes u_j' = (\sum_{i=1}^{b} \sum_{j=1}^{c} \lambda_{i,j} u_i) \otimes (\sum_{j=1}^{c} u_j'),$$

which yields an isomorphism as glider representations. □

Corollary 4.6.10. *If $\Omega_1 \supset M_1$, resp. $\Omega_2 \supset M_2$ are FKG-, resp. FKH-gliders such that all the appearing components of Ω_1 and Ω_2 are one-dimensional, then we have that $(\Omega_1 \supset M_1) \otimes (\Omega_2 \supset M_2)$ is an irreducible $FKGH$-glider if and only if $\Omega_1 \supset M_1$ and $\Omega_2 \supset M_2$ are irreducible.*

If we drop the condition that the appearing components are all one-dimensional, then we still have that irreducibility as FKG- and as FKH-glider implies irreducibility of the tensor product. The converse is no longer true.

Example 4.6.11. Consider the trivial chain $K \subset KGH$. Let P be a p-dimensional irreducible G-representation with $t, t' \in P$ linear independent. If U is a one-dimensional H-rep then $M \supset Ka$ with $M = P \otimes U_1 \oplus P \otimes U_1$, $a = t \otimes u + t' \otimes u$ is an irreducible $FKGH$-glider. There is an isomorphism of glider representations

$$(M \supset Ka) \cong (P \supset K(t+t')) \otimes (U^{\oplus 2} \supset K(u,u)),$$

but $U_1^{\oplus 2} \supset K(u,u)$ is not irreducible.

As another corollary we have a nice characterization of Abelian nilpotent groups in terms of glider representations:

Theorem 4.6.12. *Let GH be a nilpotent group of order $p^\alpha q^\beta$. TFAE*

1. *GH is Abelian;*
2. *Every irreducible $FKGH$ glider M is isomorphic to the tensor product $M_1 \otimes M_2$ of an FKG- and an FKH-glider if and only if the associated triple (a, b, c) of M satisfies $a = bc$.*

Proof. If GH is Abelian, then all irreducible representations are one-dimensional and the result follows from Proposition 4.6.9. Conversely, suppose that GH is not Abelian, so there exists some p-dimensional irreducible representation $P \in \mathrm{Irr(G)}$ (up to interchanging the role of G and H) and from Example 4.6.11 we get a counterexample. □

Example 4.6.13. Suppose that $d = 3$, i.e. both G_2 and H_2 are Abelian subgroups and let $\Omega \supset M \supset M_1 \supset M_2 \supset Km$ be an irreducible $FKGH$-glider with $M_1 = Km$. Theorem 4.1.15 entails that any irreducible G_2-representation (which is one-dimensional as G_2 is Abelian) lifts to either a p-dimensional irreducible G-representation or to p non-isomorphic one-dimensional irreducible G-representations. Moreover, since $p^2 + (|G_2| - p)p = p|G_2| = |G|$, we know that G has only one p-dimensional irreducible, with p non-isomorphic G_2-components. We can do the same for H (replacing p by q) and since $M_1 \cong V \otimes U$, with $V \in \mathrm{Irr(G_2)}, \mathrm{U} \in \mathrm{Irr(H_2)}$ ($M_1 = KG_2H_2m = Km$, with G_2H_2 Abelian), we have four cases. We introduce the following notation: if W is an irreducible G_2- (or H_2-)rep that lifts to a p- (or q-)dimensional irreducible, we denote it by $\bigwedge$, if W lifts to p one-dimensional irreducibles, we denote it by $\bigvee$. So we have the following four cases:

$$\begin{array}{ll} (\bigvee,\bigvee): & \text{the situation we had before,} \\ (\bigwedge,\bigwedge): & M \cong P \otimes Q, \\ (\bigwedge,\bigvee): & M \cong P \otimes (U_1^\epsilon \oplus \ldots \oplus U_q^\epsilon), \\ (\bigvee,\bigwedge): & M \cong (V_1^\epsilon \oplus \ldots \oplus V_p^\epsilon) \otimes Q, \end{array}$$

where $\epsilon \in \{0, 1\}$. In the first case, we can apply Proposition 4.6.9 to determine whether there is an isomorphism as tensor product of gliders. In the other three cases, we have an isomorphism $M \cong N_1 \otimes N_2$ as modules, but we cannot determine in general whether there exists an isomorphism as glider representations!

In fact, all results from this last section equally hold for more general nilpotent groups $G = P_1 \ldots P_n$ of order $p_1^{a_1} \ldots p_n^{a_n}$.

Chapter 5

Glider representation rings of finite groups and glider character theory

Let G be a finite group and let K be an algebraically closed field of characteristic zero. Character theory of finite groups starts with the definition of class functions, which are maps $f : G \to \mathbb{C}$ constant on the conjugacy classes of G. The trace maps of G-representations are particular examples of such class functions and one terms them characters of the group G. A special role is deserved for the trace maps of the irreducible representations of G, the so called irreducible characters, as it turns out that a class function is a character exactly when it is a linear combination of the irreducible characters with non-negative integer coefficients. Therefore, one introduces the additive group $R(G)$ as the free $\mathbb{Z}$-module generated by the irreducible characters. The fact that the trace map of the tensor product corresponds to the component-wise multiplication of the trace maps allows us to define a multiplication on $R(G)$, turning it into the representation ring of the group G. We refer the reader to [65] for a detailed overview.

In the previous chapter we studied glider theory for chains of finite groups and we would like to investigate whether it is possible to obtain a so called glider character theory. In [17] we introduced the notion of a glider character associated with a glider representation and a deeper study was carried out in [18]. However, thanks to fruitful conversations with Ruben Henrard, Geoffrey Janssens and Adam-Christiaan van Roosmaelen, we take a different approach in this monograph. As will become clear in Section 5.8, there is in general no one-to-one correspondence between irreducible glider representations over a fixed chain of finite groups and the associated glider characters as defined in loc. cit. Therefore there is no direct generalization of the representation ring defined in terms of glider characters, which is why we start with the introduction of so called glider representation rings.

In Section 5.1 we fix a chain of subgroups $G_0 < G_1 < \ldots < G_d = G$ and introduce the additive group $R(\widetilde{G})$ as the free $\mathbb{Z}$-module generated by the irreducible glider representations (the $\sim$ referring to the chain of subgroups considered). We use the tensor product of glider representations defined in Section 4.6 to define a multiplication on $R(\widetilde{G})$, but one has to be very careful.

In fact, there will be some restrictions on the chains of subgroups considered, but these conditions are not too severe. Working modulo irreducible gliders of essential length $< d$ allows us to consider the glider representation ring $R(G_0 < G_1 < \ldots < G_d)$, also denoted $R(\widetilde{G})$. The section ends with a first description for the chain $1 < G$, G finite Abelian, which is further elaborated in Section 5.2. We are able to give an explicit description of $\mathbb{Q}(\widetilde{G}) = \mathbb{Q} \otimes_{\mathbb{Z}} R(\widetilde{G})$, Theorem 5.2.8, and classify its primitive idempotents. Before tackling the structure of $\mathbb{Q}(1 < G)$ for non-Abelian groups, we explicitly describe the glider representation ring for $G = Q_8$ the quaternion group. This is done in Section 5.3. Section 5.4 is a rather technical section, needed to derive a short exact sequence of KG^{ab}-modules, see Theorem 5.5.10. In fact, we are forced to introduce three KG^{ab}-modules P, Q and R, the structure of which is studied in more depth in Section 5.6. With the results obtained there, we are then able to deduce a complete description of $\mathbb{Q}(\widetilde{G})$ for groups of nilpotency class ≤ 2 having $Q(G) = 0$. We explain this in Section 5.7 and arrive at a distinction between some isocategorical groups such as D_8 and Q_8.

In Section 5.8 we turn back to our initial goal of developing a glider character theory. We introduce glider class functions and glider characters associated with glider representations. From Section 4.2 we know how to induce glider representations from a normal subgroup $H \lhd G$ to the bigger group G, by using a chain of transversal sets. We argue how to obtain an irreducible glider out of the induced FKG-glider, which then yields an induction morphism of additive groups $S_{\widetilde{H}}^{\widetilde{G}} : R(\widetilde{H}) \to R(\widetilde{G})$. In fact, this brings us into the right setting to prove a generalization of Artin's theorem, Theorem 5.8.7. Next, we use the classical inproducts on the rings of class functions $R(G_i)$ to define a glider inproduct on the set of glider class functions and end the section with some results using this glider inproduct. Finally, in Section 5.9 we consider chains $1 < H < G$ of length 2 with H Abelian. These conditions namely allow us to define another multiplication on $R(1 < H < G)$.

The definition of a glider character appeared for the first time in [17] and the character theory is further developed in [18]. In [15] we started the study of $\mathbb{Q}(\widetilde{G})$ for non-Abelian groups. We mention here that in [17], [18] the authors used different terminology as we were primarily working with chains $1 < G$. In particular, *generalized character* and *generalized class function* are renamed here by *glider character* and *glider class function*, respectively. Also, the term *generalized character ring* is now termed *glider representation ring*. The reader is assumed to be familiar with character theory of finite groups.

5.1 Glider representation rings

Throughout the entire chapter we will work over an algebraically closed field of characteristic 0, say K. Let G be a finite group and let $G_0 < G_1 < \ldots < G_d = G$ be a chain of subgroups, leading to the algebra filtration FKG as in the previous chapter. We denote by $R(G_0 < G_1 < \ldots < G_d)$ the free $\mathbb{Z}$-module generated by all the irreducible FKG-glider representations with zero body. When the algebra filtration FKG is clear from the context, we will also denote this additive group by $R(\widetilde{G})$. Moreover, for an irreducible glider $M \supset M_1 \supset \ldots \supset M_d \supset 0 \supset \ldots$ we will denote its generator in $R(\widetilde{G})$ by $\chi_{M \supset M_1 \supset \ldots \supset M_d}$ or simply χ_M when no confusion arises.

In the classical group case, the additive group $R(G)$ carries a natural multiplication and it holds that the product of characters corresponds to the character of the tensor product of the representations. Recall the tensor product of glider representations, Definition 4.6.8, which will be used to define a multiplication on $R(\widetilde{G})$. In order to obtain a well-defined product we refer again to Remark 3.2.13. Although the filtration FKG is semisimple Artinian we can use Proposition 3.2.12, but the decomposition into a strong fragment direct sum is only valid for natural gliders. For this reason we work with the weaker notion of the fragment direct sum and we can use Theorem 3.2.14 to decompose the tensor product $M \otimes N$ of two irreducible gliders. However, for general groups G_0 it is still unclear that the decomposition into irreducible gliders is unique. For example, let M, N be two irreducible gliders of essential length d and suppose that G_0 has an irreducible representation M_d of dimension strictly bigger than 1 and such that $M_d \otimes N_d$ decomposes as G_0-representations

$$M_d \otimes N_d \cong \bigoplus_{i=1}^{n} V_i.$$

Although the decomposition into G_0-components is unique, there can exist more than one K-basis furnishing such a decomposition, which could then lead to non-isomorphic glider representations.

In case G_0 is Abelian, the KG_0-representation $M_d \otimes N_d$ is one-dimensional and the glider representation of essential length d

$$KG(M_d \otimes N_d) \supset KG_{d-1}(M_d \otimes N_d) \supset \ldots \supset M_d \otimes N_d$$

splits off from $M \otimes N$. However still, when $N_{d-1} \otimes M_{d-1}$ has dimension > 2 the same problem mentioned above can occur. Therefore we make the convention that G_0 is always Abelian and we work modulo gliders of essential length $< d$. The question of whether this is a necessary restriction is still open. In any case, in this way, the group $R(\widetilde{G})$ becomes the free $\mathbb{Z}$-module generated by the irreducible gliders of essential length exactly d and zero body, and the product of χ_M and χ_N becomes

$$\chi_M \cdot \chi_N := \chi_{KGM_d\otimes N_d \supset KG_{d-1}M_d\otimes N_d \supset \ldots \supset M_d \otimes N_d}, \tag{5.1}$$

which we will denote by $\chi_{M\otimes N}$. It is clear that the unit element is given by $T \supset T \supset \ldots \supset T \supset 0 \supset \ldots$ where T denotes the trivial G-representation. The associativity just follows from the associativity of the tensor product. We call $R(\widetilde{G})$ together with the multiplication just defined the glider representation ring associated with the chain $G_0 < G_1 < \ldots < G_d = G$.

Remark 5.1.1. In Section 5.9 we will define another multiplication on the glider representation ring for chains of length two, that is $G_0 < G_1 < G_2$ with G_0 and G_1 Abelian.

Let us deduce the ring structure of $R(\widetilde{G})$ for G a finite Abelian group and with chain $e < G$. Finite Abelian groups are isomorphic to their character group $\widehat{G}$, but the isomorphism is not canonical. By choosing such an isomorphism, one establishes a one-to-one correspondence between the irreducible gliders of essential length ≤ 1 and zero body and the subsets of $\{g \mid g \in G\}$. To see this, we first prove the following lemma.

Lemma 5.1.2. *Let $S, T \in \mathrm{Irr(G)}$ be non-isomorphic and let $s, s' \in S, t, t' \in T$. The irreducible gliders $S \oplus T \supset K(s+t)$ and $S \oplus T \supset K(s'+t')$ are isomorphic.*

Proof. By Schur's lemma, $\mathrm{End_{KG}(S \oplus T)} \xrightarrow{\phi} \mathrm{K}^2$ is an isomorphism. Since S and T are one-dimensional, there exists $(\lambda, \mu) \in K^2$ such that $s' = \lambda s, t' = \mu t$. Hence $\phi^{-1}(\lambda, \mu)$ defines a glider isomorphism. □

Theorem 4.1.12 then shows that the aforementioned one-to-one correspondence is given by

$$A \in \mathcal{P}(G) = \mathcal{P} \longleftrightarrow \bigoplus_{g \in A} T_g \supset Ka,$$

where T_g denotes the irreducible G-representation associated to $g \in G$ and $a = \sum_{g \in A} t_g, t_g \in T_g$. We denote by χ_A the character associated with the corresponding irreducible glider. For example $\chi_{\{g\}} = \chi_{T_g \supset Kt_g}$. If $A = \emptyset$, then the associated glider is $K \supset 0$, but we work modulo the gliders of essential length 0. Let $A, B \subset G$ be non-empty subsets. From the definition of the multiplication in $R(\widetilde{G})$ it easily follows that

$$\chi_A \chi_B = \chi_C, \tag{5.2}$$

where $C = \{gh \mid g \in A, h \in B\}$.

Proposition 5.1.3. *Let G be a finite Abelian group. The ring morphism $\iota : R(G) \to R(\widetilde{G})$, $\chi_g \mapsto \chi_{\{g\}}$ defines an integral ring extension.*

Proof. It is clear the ι defines a ring extension. Let $A \in \mathcal{P}$ and write $\chi_A^n = \chi_{A_n}$ for $n \geq 1$. If $e \in A$, then

$$A = A_1 \subset A_2 \subset A_3 \subset \ldots$$

Since G is finite, this chain stabilizes at some A_n, i.e. χ_A^n is an idempotent, so χ_A is indeed integral over $R(G)$. If $1 \notin A$, take some $g \neq 1 \in A$. It follows that $A_{o(g)}$ contains 1, where $o(g)$ denotes the order of g. We know that $\chi_A^{o(g)}$ is integral, whence so is χ_A. □

5.2 Primitive central idempotents

In this section, G is a finite Abelian group with chain $1 < G$. From the previous section we know that we can choose an isomorphism between a finite Abelian group G and its character group $\widehat{G}$. This isomorphism extends automatically to an isomorphism of the group ring $\mathbb{Z}G$ and the representation ring $R(G)$. By tensoring with the rational numbers, we then obtain an isomorphism $\mathbb{Q}G \cong \mathbb{Q} \otimes_{\mathbb{Z}} R(G)$ of $\mathbb{Q}$-algebras. We also provided the ring structure on $R(\widetilde{G})$ and we observe that $R(\widetilde{G})$ is the semigroup algebra $\mathbb{Z}S$ of the semigroup S given by

$$S = \{A \in \mathcal{P}(G) \setminus \{\emptyset\}\}$$

with multiplication as defined in (5.2). For generalities on semigroups, we refer to [57]. Also, from now on, when we write $\mathcal{P}(G)$, we mean all non-empty subsets of G.

In the previous section, we derived an integral ring extension $R(G) \hookrightarrow R(\widetilde{G})$. This allows us to consider the $\mathbb{Q}$-algebra extension $\mathbb{Q}G \hookrightarrow \mathbb{Q} \otimes_{\mathbb{Z}} R(\widetilde{G})$ ($\mathbb{Q}$ is flat as $\mathbb{Z}$-module). We denote $\mathbb{Q}(\widetilde{G}) := \mathbb{Q} \otimes_{\mathbb{Z}} R(\widetilde{G})$. Since G is finite, Maschke's theorem entails that the rational group algebra $Q(G)$ is semisimple, hence our first task is to determine the Jacobson radical of $Q(\widetilde{G})$. Secondly, we will determine the primitive central idempotents. To this extent, we mention the results obtained in [41]. In loc. cit. the authors describe the primitive central idempotents of the group algebra $\mathbb{Q}G$ for G a finite nilpotent group. Let us introduce some notation. Let $H \leq G$ be a subgroup. Then $\hat{H} = \frac{1}{|H|} \sum_{h \in H} h$ is an idempotent, which is central if and only if H is normal. Denote by $\mathcal{M}(G)$ the set of all minimal normal subgroups and define

$$\epsilon(G) = \Pi_{L \in \mathcal{M}(G)}(1 - \hat{L}).$$

For a normal subgroup $N \lhd G$, we consider $\mathcal{M}(G/N)$ and write L for the associated subgroup to $\overline{L}$ of G/N containing N. It appears that

$$\epsilon(G, N) = \hat{N} \Pi_{\overline{L} \in \mathcal{M}(G/N)}(1 - \hat{L})$$

is a primitive central idempotent if and only if G/N is cyclic and these are all the primitive central idempotents, see [41, Corollary 2.1].

In proving their results in [41], the authors used the isomorphism $\mathbb{Z}(G/H) \cong (\mathbb{Z}G)\hat{H}$, for H a normal subgroup in G. In our setting, we can extend the canonical homomorphism $\omega : G \to G/H$ to a $\mathbb{Z}$-linear morphism $\omega_H : R(\widetilde{G}) \to R(\widetilde{G/H})$, defined by

$$\omega_H(\sum_{A\in\mathcal{P}(G)} a_A\chi_A) = \sum_{A\in\mathcal{P}(G)} a_A\chi_{\omega(A)}.$$

It is clear ω_H is a ringmorphism.

To get some feeling with these semigroup algebras, we include the following example:

Example 5.2.1. For $G = C_p = \langle a\rangle$, the cyclic group of order p, we have that

$$1 = \epsilon(G) + \hat{G} \in \mathbb{Q}(C_p).$$

In $\mathbb{Q}(\widetilde{C_p})$ however, one verifies that

$$\psi(C_p) := \sum_{i=0}^{p-1} \frac{1}{p}\chi_{\{a^i\}} - \chi_{C_p}$$

is idempotent and

$$1 = \chi_{\{1\}} = \epsilon(C_p) + \psi(C_p) + \chi_{C_p},$$

is a decomposition into orthogonal idempotents. For now, it is not clear whether $\psi(C_p)$ is primitive or not.

Let us compute the kernel $\mathrm{Ker}(\omega_H)$. Choose a transversal set T for H in G. Every element g in G can be written uniquely as the product th for some $t \in T$, $h \in H$; hence we can associate to any $A \subset G$, the subset $T_A \subset T$ consisting of all the t's appearing in these products. Define

$$\mathcal{P}(G,B) = \{A \in \mathcal{P}(G) \mid T_A = B\}, \quad B \in \mathcal{P}(T).$$

Clearly, it holds that $\mathcal{P}(G) = \sqcup_{B\in\mathcal{P}}\mathcal{P}(G,B)$ and in particular the element $\chi_{BH} \in \mathcal{P}(G,B)$. Let $\alpha = \sum_{A\in\mathcal{P}(G)} a_A\chi_A \in R(\widetilde{G})$. Define

$$\alpha' = \sum_{B\in\mathcal{P}(T)} (\sum_{A\in\mathcal{P}(G,B)} a_A)\chi_{BH}.$$

We have that $\omega_H(\alpha) = \omega_H(\alpha')$. Suppose that $\omega_H(\alpha) = 0$, then for any $B \in \mathcal{P}(T)$ we obtain

$$a_{BH} = -\sum_{A\in\mathcal{P}(G,B)\setminus\{BH\}} a_A.$$

It follows that

$$\begin{aligned}
\alpha &= \sum_{B\in\mathcal{P}(T)} \left[\sum_{A\in\mathcal{P}(G,B)\setminus\{BH\}} -a_A\chi_{BH} + a_A\chi_A \right] \\
&= \sum_{B\in\mathcal{P}(T)} \left[\sum_{A\in\mathcal{P}(G,B)\setminus\{BH\}} -a_A\chi_A\chi_H + a_A\chi_A \right] \\
&= \sum_{B\in\mathcal{P}(T)} \left[\sum_{A\in\mathcal{P}(G,B)\setminus\{BH\}} a_A\chi_A(1-\chi_H) \right].
\end{aligned}$$

Hence we have just proven the following

Proposition 5.2.2. *The kernel of ω_H equals* $\mathrm{Ker}(\omega_H) = (\chi_H - 1)$.

As a corollary we obtain that $R(\widetilde{G/H}) \cong R(\widetilde{G})\chi_H$, hence $\mathbb{Q}(\widetilde{G/H}) \cong \mathbb{Q}(\widetilde{G})\chi_H$.

Proposition 5.2.3. *The element $\chi_G \in \mathbb{Q}(\widetilde{G})$ is a primitive central idempotent.*

Proof. The kernel of the ringmorphism $\cdot\chi_G : \mathbb{Q}\otimes_{\mathbb{Z}} R(\widetilde{G}) \to \mathbb{Q}\otimes_{\mathbb{Z}} R(\widetilde{G})\chi_G$ is the annihilator of χ_G. Hence by Proposition 5.2.2 we have that

$$\mathbb{Q}\otimes_{\mathbb{Z}} R(\widetilde{G})\chi_G \cong \mathbb{Q}\otimes_{\mathbb{Z}} R(\widetilde{G/G}) \cong \mathbb{Q}\otimes_{\mathbb{Z}} \mathbb{Z} \cong \mathbb{Q}. \quad \square$$

It is known that any cyclic semigroup $\langle s\rangle$ in a finite semigroup S contains an idempotent. There is also a natural order on the set $E(S)$ of idempotents given by $e \leq f$ if and only if $e = ef = fe$ for $e, f \in E(S)$.

Lemma 5.2.4. *Let $A \in \mathcal{P}(G)$, then the cyclic semigroup $\langle\chi_A\rangle$ contains a unique minimal idempotent.*

Proof. It is clear that $\chi_B^2 = \chi_B$ if and only if $B \leq G$ is a subgroup. Suppose that $A^n = H$ and $A^m = H'$ are both idempotents, i.e. subgroups of G and suppose that $n \leq m$. We can write $m = kn + r$ for some $0 \leq r \leq m-1$. It follows that $H' = A^m = HA^{m-n} = HA^{(k-1)n}A^r = HA^r$. Hence $(HA^r)^2 = HA^{2r} = HA^r$. From this it follows that

$$A^m = (HA^r)^2 = (HA^r)^n = H = A^n.$$

It follows that there is only one idempotent in $\langle A\rangle$ which is therefore minimal. $\square$

Definition 5.2.5. For $A \in \mathcal{P}(G)$ we define $n(A)$ to be the unique minimal idempotent in $\langle A\rangle$.

Lemma 5.2.6. *Let $A \in \mathcal{P}(G)$ and $n(A) = H$. Then either $A \subset H$ or $A \cap H = \emptyset$.*

Proof. Suppose that $b \in A \cap H$ and let $n \geq 0$ be such that $A^n = H$. Let $a \in A$, then $ab^{n-1} \in H$. Since $b \in H$, so is b^{1-n}, whence $a \in H$. This shows that $A \subset H$. □

Lemma 5.2.7. *Let $A \in \mathcal{P}(G)$ and $n(A) = H$. Then $A \subset gH$ for some $g \in G$.*

Proof. If $A \cap H \neq \emptyset$ then $g = 1$ satisfies by the previous lemma, so suppose that $A \cap H = \emptyset$. Let $n \geq 0$ be such that $A^n = H$, then we have that $A \subset a^{1-n}H$ for some $a \in A$. □

We can now construct nilpotent elements in the semigroup algebra $\mathbb{Q}(\widetilde{G})$. Let $A \in \mathcal{P}(G)$ with $n(A) = H$. By Lemma 5.2.6 there exists some $g \in G$ such that $A \subset gH$. The element $\chi_{gH} - \chi_A \in J$ where $J = J(\mathbb{Q}(\widetilde{G}))$ denotes the Jacobson radical. Indeed, suppose first that $g = 1$. For m the smallest integer such that $A^m = H$ we have

$$\begin{aligned}(\chi_H - \chi_A)^m &= \sum_{i=1}^{m} \binom{m}{i} (-1)^{m-i} \chi_H^i \chi_A^{m-i} \\ &= \sum_{i=1}^{m} \binom{m}{i} (-1)^{m-i} \chi_H \\ &= (1-1)^m \chi_H = 0.\end{aligned}$$

Now if $g \notin H$, then $g^{-1}A \subset H$ and $\chi_H - \chi_{g^{-1}A} \in J$. It follows that

$$\chi_{gH} - \chi_A = \chi_g(\chi_H - \chi_{g^{-1}A}) \in J.$$

Denote by I the ideal generated by the elements $\chi_{gH} - \chi_A$ where A runs over the subsets of G. Observe that if $A = \{g\}$ is a singleton, that $n(A) = 1$ and $\chi_{\{g\}1} - \chi_A = 0$.

Theorem 5.2.8. *The Jacobson radical J equals I and we have a ring isomorphism*

$$\varphi : \bigoplus_{H \leq G} \mathbb{Q}(G/H) \to \mathbb{Q}(\widetilde{G})/J,$$

where the unit element e_H of $\mathbb{Q}(G/H)$ is sent to $\Pi_{\overline{H'} \in \mathcal{M}(G/H)}(\chi_H - \chi_{H'})$.

Proof. In order for φ to be well-defined, it suffices to check that $\varphi(e_H)\varphi(e_Z) = 0$ for two different subgroups H, Z of G. To see this, choose subgroups $\overline{H'} \in \mathcal{M}(G/H), \overline{Z'} \in \mathcal{M}(G/Z)$ such that $H' \leq HZ$ and $Z' \leq HZ$. In the product $\varphi(e_H)\varphi(e_Z)$ there appear the factors $(\chi_H - \chi_{H'})$ and $(\chi_Z - \chi_{Z'})$. We calculate that their product equals

$$\chi_{HZ} - \chi_{H'Z} - \chi_{HZ'} + \chi_{H'Z'}. \tag{5.3}$$

We have the inclusions $H'Z \leq HZ \leq H'Z$ and $HZ' \leq HZ \leq HZ'$, or $HZ = H'Z = HZ'$. From this it follows that

$$H'Z' = H'HZ' = H'HZ = H'Z = HZ,$$

and we see that (5.3) is indeed 0. To show that φ is surjective, we observe that $\varphi(e_H) = \chi_H - \chi_G$ for $H < G$ a maximal subgroup. Since $\varphi(e_G) = \chi_G$ it follows that $\chi_H \in \mathrm{Im}(\varphi)$. If $n = l(G)$ is the Jordan-Hölder length of G, then for an H of length $n-2$ we have that

$$\varphi(e_H) = \chi_H + \sum_{\overline{H'} \in \mathcal{M}(G/H)} a_{H'}\chi_{H'} \in \mathrm{Im}(\varphi).$$

Surjectivity now follows from a downwards induction on the Jordan-Hölder lengths of the subgroups. That φ is an isomorphism finally follows from a dimension argument and the claim about the Jacobson radical is a direct consequence of Wedderburn-Artin. □

As a corollary we can give a complete list of the primitive central idempotents in $\mathbb{Q}(\widetilde{G})$. To this extent we denote

$$\psi(G,H) := \varphi(e_H) = \Pi_{\overline{H'} \in \mathcal{M}(G/H)}(\chi_H - \chi_{H'}).$$

Also, for $A \subset G$ we have the associated element $\hat{A} = \frac{1}{|A|}\sum_{g\in A} g$ in $\mathbb{Q}(G)$. We denote the element $\frac{1}{|A|}\sum_{g\in A}\chi_{\{g\}}$ in $\mathbb{Q}(\widetilde{G})$ by $\check{A}$. Of course, there is an embedding $\mathbb{Q}(G) \subset \mathbb{Q}(\widetilde{G})$, but this does not correspond to the embedding given by φ. Finally, by $\check{\epsilon}(G,N)$ we denote the element

$$\check{\epsilon}(G,N) = \check{N}\Pi_{\overline{L}\in\mathcal{M}(G/N)}(1-\check{L}) \in \mathbb{Q}(\widetilde{G}).$$

Let $H < G$ be a subgroup and take a subgroup $\overline{N}$ in G/H. The group $\overline{N}$ is isomorphic to some N/H with $H \leq N \leq G$ a subgroup. Pick a transversal set T_H^N for H in N. Inside the group algebra $\mathbb{Q}(G/H)$ we have the element

$$\hat{\overline{N}} = \frac{1}{|T_H^N|}\sum_{t\in T_H^N} e_{tH}.$$

It follows that

$$\varphi(\hat{\overline{N}}) = \Big(\frac{1}{|T_H^N|}\sum_{t\in T_H^N}\chi_{\{t\}}\Big)\psi(G,H) = \check{T_H^N}\psi(G,H)$$

Let e be a primitive idempotent in $\mathbb{Q}(\widetilde{G})$. By the results in [41] and Theorem 5.2.8

$$\overline{e} = \varphi(\epsilon(G/H,\overline{N}))$$

for subgroups $H \leq N \leq G$ such that G/N is cyclic. We calculate

$$\begin{aligned}
\varphi(\epsilon(G/H,\overline{N})) &= \varphi(\hat{\overline{N}})\Pi_{[\overline{L}]\in\mathcal{M}(\frac{G/H}{\overline{N}})}(\varphi(e_H) - \varphi(\hat{\overline{L}})) \\
&= \check{T_H^N}\psi(G,H)\Pi_{[\overline{L}]\in\mathcal{M}(\frac{G/H}{\overline{N}})}(\psi(G,H) - \check{T_H^L}\psi(G,H)) \\
&= \psi(G,H)\check{T_H^N}\Pi_{\overline{L}\in\mathcal{M}(G/N)}(1-\check{T_H^L}) \\
&= \psi(G,H)\check{\epsilon}(G,N).
\end{aligned}$$

Theorem 5.2.9. *Let G be a finite Abelian group. The primitive central idempotents of $\mathbb{Q}(\widetilde{G})$ are precisely the elements $\psi(G,H)\check{\epsilon}(G,N)$ for subgroups $H \leq N \leq G$ such that G/N is cyclic.*

5.3 The quaternion group Q_8

To get some more insight in the structure of glider representation rings, we will compute $R(\widetilde{Q_8})$ for the quaternion group

$$Q_8 = \langle i, j, k \mid i^2 = j^2 = k^2 = ijk \rangle$$

with chain $1 \subset Q_8$. The character table of Q_8 is given by

	$\{1\}$	$\{-1\}$	$\{i,-i\}$	$\{j,-j\}$	$\{k,-k\}$
T_1	1	1	1	1	1
T_i	1	1	1	−1	−1
T_j	1	1	−1	1	−1
T_k	1	1	−1	−1	1
U	2	−2	0	0	0

The irreducible characters associated with the one-dimensional representations generate an Abelian subgroup of the representation ring $R(Q_8)$ isomorphic to the Klein four group $V_4 = \langle a, b \mid a^2 = b^2 = 1 \rangle$. We write $ab = c$ and fix the isomorphism

$$1 \mapsto T_1 \quad a \mapsto T_i \quad b \mapsto T_j \quad c \mapsto T_k$$

To discuss the behavior of the two-dimensional irreducible representation U we fix a basis $\{e_1, e_2\}$ such that U has the following presentation:

$$i \mapsto \begin{pmatrix} 0 & i \\ i & 0 \end{pmatrix}, \; j \mapsto \begin{pmatrix} -i & 0 \\ 0 & i \end{pmatrix}.$$

Theorem 4.1.12 together with Schur's lemma shows that every point $[\lambda : \mu] \in \mathbb{P}^1$ determines an irreducible glider representation $U \supset K(\lambda e_1 + \mu e_2)$ and two different points in $\mathbb{P}^1$ yield non-isomorphic gliders. Since U is two-dimensional, there also exist irreducible gliders of the form $U^2 \supset K(u_1 + u_2)$ such that $\dim_K(\langle u_1, u_2 \rangle) = 2$. By the following lemma, there is, up to isomorphism, only one such irreducible glider.

Lemma 5.3.1. *Let $M \supset M_1 \supset 0 \supset \ldots, N \supset N_1 \supset 0 \supset \ldots$ be irreducible gliders with $M \cong U^2 \cong N$. Then both gliders are isomorphic.*

Proof. There exist $u_i, u_i' \in U$ $(i = 1, 2)$ such that $M_1 = K(u_1 + u_2), N_1 = K(u_1' + u_2')$. Let B be a base change matrix for $\{u_1, u_2\}$ and $\{u_1', u_2'\}$. Since $\mathrm{End}_{Q_8}(U^2) \cong M_2(K)$, we have that B yields an isomorphism between both gliders. □

As a corollary we deduce that the irreducible gliders of essential length 1 and zero body are labeled by

$$\mathcal{P}(V_4) \times (\mathbb{P}^1 \sqcup *),$$

where $*$ indicates that $U^{\oplus 2}$ appears in the decomposition of M. If an irreducible glider has label $(A, [\lambda : \mu])$, we denote the associated generator in $R(\widetilde{Q_8})$ by $\chi_{(A,[\lambda:\mu])}$ as we did in the Abelian case. For example, the corresponding element of

$$T_1 \oplus T_b \oplus U^2 \supset \mathbb{C}(t_1 + t_b + u_1 + u_2)$$

is denoted by $\chi_{(\{1,b\},*)}$. Let us now compute the multiplication rules in $R(\widetilde{Q_8})$, by looking at the tensor product

$$\big[U \supset K(\lambda_1 e_1 + \mu_1 e_2)\big] \otimes \big[U \supset K(\lambda_2 e_1 + \mu_2 e_2)\big]. \tag{5.4}$$

We have the following classical result:

Proposition 5.3.2. *Let G be a finite group, V an irreducible representation. Then $V \otimes V$ is irreducible if and only if V is one-dimensional.*

Proof. If $\dim_{\mathbb{C}}(V) > 1$, then the switch map $\tau : V \otimes V \to V \otimes V$, $v \otimes w \mapsto w \otimes v$ is not a scalar multiplication of the identity morphism, hence $\dim_K \mathrm{End}_{\mathrm{G}}(\mathrm{V} \otimes \mathrm{V}) > 1$, i.e. $V \otimes V$ is reducible. □

We know that $U \otimes U = S(U \otimes U) \oplus A(U \otimes U)$ decomposes into a symmetric and antisymmetric part, from which it follows that $\chi_U = \chi_S + \chi_A$, see e.g. [40, Chapter 19]. Here, χ_U denotes the associated character of U, i.e. $\chi_U \in R(Q_8)$. One checks that $\chi_A = \chi_{T_1}$ and $\chi_S = \chi_{T_i} + \chi_{T_j} + \chi_{T_k}$. A decomposition is given by

$$\begin{aligned} U \otimes U \;=\; & \underbrace{K(e_1 \otimes e_2 - e_2 \otimes e_1)}_{T_1} \oplus \underbrace{K(e_1 \otimes e_2 + e_2 \otimes e_1)}_{T_j} \\ & \oplus \underbrace{K(e_1 \otimes e_1 + e_2 \otimes e_2)}_{T_k} \oplus \underbrace{K(e_1 \otimes e_1 - e_2 \otimes e_2)}_{T_i}. \end{aligned} \tag{5.5}$$

Let $[\lambda_i : \mu_i] \in \mathbb{P}^1$, $i = 1, 2$, then $(\lambda_1 e_1 + \mu_1 e_2) \otimes (\lambda_2 e_1 + \mu_2 e_2)$ decomposes into

$$a(e_1 \otimes e_2 - e_2 \otimes e_1) + b(e_1 \otimes e_2 + e_2 \otimes e_1) + c(e_1 \otimes e_1 + e_2 \otimes e_2) + d(e_1 \otimes e_1 - e_2 \otimes e_2).$$

and we obtain that (5.4) remains irreducible if and only if $abcd \neq 0$. Since

$$\begin{pmatrix} \lambda_1\lambda_2 & \lambda_1\mu_2 & \mu_1\lambda_2 & \mu_1\mu_2 \end{pmatrix} = \begin{pmatrix} c+d & a+b & b-a & c-d \end{pmatrix} \tag{5.6}$$

this is equivalent to

$$\begin{vmatrix} \lambda_1 & \pm\lambda_2 \\ \mu_1 & \mu_2 \end{vmatrix} \neq 0 \text{ and } \begin{vmatrix} \lambda_1 & \pm\mu_1 \\ \mu_2 & \lambda_2 \end{vmatrix} \neq 0.$$

Remark 5.3.3. The above is independent of the choice of base. Indeed, suppose that $\begin{pmatrix} e_1 & e_2 \end{pmatrix} = \begin{pmatrix} f_1 & f_2 \end{pmatrix} B$ is a base change, then

$$\begin{pmatrix} e_1 \otimes e_1 & e_1 \otimes e_2 & e_2 \otimes e_1 & e_2 \otimes e_2 \end{pmatrix} = \begin{pmatrix} f_1 \otimes f_1 & f_1 \otimes f_2 & f_2 \otimes f_1 & f_2 \otimes f_2 \end{pmatrix} B \otimes B,$$

where $B \otimes B$ denotes the Kronecker tensor product of two matrices. With regard to the basis $\{f_1, f_2\}$ equation (5.6) becomes

$$\begin{pmatrix} \lambda_1\lambda_2 & \lambda_1\mu_2 & \mu_1\lambda_2 & \mu_1\mu_2 \end{pmatrix} (B \otimes B)^T = \begin{pmatrix} c+d & a+b & b-a & d-c \end{pmatrix} (B \otimes B)^T,$$

and $B \otimes B$ is invertible.

Generically, we have

$$\chi_{(\emptyset,[\lambda_1:\mu_1])}\chi_{(\emptyset,[\lambda_2:\mu_2])} = \chi_{(V_4,\emptyset)}.$$

If, for example, $\lambda_1\mu_2 = \mu_1\lambda_2$, but $\lambda_1\lambda_2 \neq \pm\mu_1\mu_2$ then $\chi_{(\emptyset,[\lambda_1:\mu_1])}\chi_{(\emptyset,[\lambda_2:\mu_2])} = \chi_{(\{a,b,c\},\emptyset)}$ and similar for the other non-generic cases. In particular, $\chi^2_{(\emptyset,[\lambda:\mu])} = \chi_{(\{a,b,c\},\emptyset)}$ for $[\lambda : \mu] \neq [1:1], [1:-1], [1:0], [0:1], [1:i], [1:-i]$.

To calculate the product $\chi_{(\emptyset,*)}\chi_{(\emptyset,[\lambda:\mu])}$ we may present the glider associated to the first factor by

$$U \oplus U \supset K(e_1 + e_2).$$

Hence the tensor product with $U \supset K(\lambda e_1 + \mu e_2)$ becomes

$$(U \otimes U) \oplus (U \otimes U) \supset K(\underbrace{(\lambda e_1 \otimes e_1 + \mu e_1 \otimes e_2)}_{v_1} + \underbrace{(\lambda e_2 \otimes e_1 + \mu e_2 \otimes e_2)}_{v_2}).$$

We denote the coefficients of the decomposition of v_i with regard to (5.5) by a_i, b_i, c_i, d_i, $i = 1, 2$. The vector v_1 yields the equation

$$\begin{pmatrix} \lambda & \mu & 0 & 0 \end{pmatrix} = \begin{pmatrix} c_1 + d_1 & a_1 + b_1 & b_1 - a_1 & c_1 - d_1 \end{pmatrix},$$

from which we obtain that $a_1b_1c_1d_1 \neq 0$ unless $\lambda\mu = 0$. If, say, $\lambda = 0$, then $c_1 = d_1 = 0$. In this case, the decomposition of the v_2 yields that $c_2d_2 \neq 0$. So we arrive at

$$\chi_{(\emptyset,*)}\chi_{(\emptyset,[\lambda:\mu])} = \chi_{(V_4,\emptyset)}, \quad \forall\, [\lambda : \mu] \in \mathbb{P}^1.$$

A similar reasoning entails that

$$\chi^2_{(\emptyset,*)} = \chi_{(V_4,\emptyset)}.$$

Next, to calculate the products of the form $\chi_{(A,\emptyset)}\chi_{(\emptyset,[\lambda:\mu])}$, we need to write explicit isomorphisms $T_r \otimes U \cong U$ for $r = 1, i, j, k$. These are given by

$U \xrightarrow{\varphi_1} T_1 \otimes U$	$U \xrightarrow{\varphi_i} T_i \otimes U$
$e_1 \mapsto t_1 \otimes e_1$	$e_1 \mapsto t_i \otimes e_2$
$e_2 \mapsto t_1 \otimes e_2$	$e_2 \mapsto t_i \otimes e_1$
$U \xrightarrow{\varphi_j} T_j \otimes U$	$U \xrightarrow{\varphi_k} T_k \otimes U$
$e_1 \mapsto -t_j \otimes e_1$	$e_1 \mapsto -t_k \otimes e_2$
$e_2 \mapsto t_j \otimes e_2$	$e_2 \mapsto t_k \otimes e_1$

$$\Rightarrow \begin{cases} \chi_{(\{1\},\emptyset)}\chi_{(\emptyset,[\lambda:\mu])} = \chi_{(\emptyset,[\lambda:\mu])} \\ \chi_{(\{a\},\emptyset)}\chi_{(\emptyset,[\lambda:\mu])} = \chi_{(\emptyset,[\mu:\lambda])} \\ \chi_{(\{b\},\emptyset)}\chi_{(\emptyset,[\lambda:\mu])} = \chi_{(\emptyset,[-\lambda:\mu])} \\ \chi_{(\{c\},\emptyset)}\chi_{(\emptyset,[\lambda:\mu])} = \chi_{(\emptyset,[-\mu:\lambda])} \end{cases}$$

For subsets $A \subset V_4$ of two elements, we generically have

$$\chi_{(A,\emptyset)}\chi_{(\emptyset,[\lambda:\mu])} = \chi_{(\emptyset,*)}, \tag{5.7}$$

but there are a few special cases. For example, for $A = \{1, b\}$ we have that $\chi_{(\{1,b\},\emptyset)}\chi_{(\emptyset,[0:1])} = \chi_{(\emptyset,[0:1])}$, because $[0 : 1] = [\lambda : \mu] = [-\lambda : \mu] = [1 : 0]$. We enlist all these special cases in the following table

A	non − generic points $[\lambda : \mu]$	$\chi_{(A,\emptyset)}\chi_{(\emptyset,[\lambda:\mu])}$
$\{1, a\}$	$[1 : 1], [1 : -1]$	$\chi_{(\emptyset,[1:1])}, \chi_{(\emptyset,[1:-1])}$
$\{b, c\}$	$[1 : 1], [1 : -1]$	$\chi_{(\emptyset,[-1:1])}, \chi_{(\emptyset,[1:1])}$
$\{1, b\}$	$[0 : 1], [1 : 0]$	$\chi_{(\emptyset,[0:1])}, \chi_{(\emptyset,[1:0])}$
$\{a, c\}$	$[0 : 1], [1 : 0]$	$\chi_{(\emptyset,[1:0])}, \chi_{(\emptyset,[0:1])}$
$\{1, c\}$	$[1 : i], [1 : -i]$	$\chi_{(\emptyset,[1:i])}, \chi_{(\emptyset,[1:-i])}$
$\{a, b\}$	$[1 : i], [1 : -i]$	$\chi_{(\emptyset,[1:-i])}, \chi_{(\emptyset,[1:i])}$

For $|A| \geq 3$, formula (5.7) holds for all $[\lambda : \mu] \in \mathbb{P}^1$. Finally, it easily follows that

$$\chi_{(A,\emptyset)}\chi_{(\emptyset,*)} = \chi_{(\emptyset,*)}, \quad \forall A \in \mathcal{P}(V_4).$$

All the other products in $R(\widetilde{Q_8})$ can now be deduced easily. For example

$$\chi_{(\{a\},[1:i])}\chi_{(\{b\},[1:i])} = \chi_{(\{a,b,c\},[1:-i])},$$

and

$$\chi_{(\{a,b\},*)}\chi_{(\emptyset,[\lambda:\mu])} = \chi_{(V_4,\emptyset)}.$$

We also include the exponents of $\chi_{(\emptyset,[1:1])}$

n	$\chi^n_{(\emptyset,[1:1])}$
1	$\chi_{(\emptyset,[1:1])}$
2	$\chi_{(\{b,c\},\emptyset)}$
3	$\chi_{(\emptyset,[-1:1])}$
4	$\chi_{(\{1,a\},\emptyset)}$
5	$\chi_{(\emptyset,[1:1])}$
$\vdots$	$\vdots$

Now that we know how to multiply in $R(\widetilde{Q}_8)$, whence also in $\mathbb{Q}(\widetilde{Q}_8)$, we wonder what the Jacobson radical $J = J(\mathbb{Q}(\widetilde{Q}_8))$ looks like. To begin with, the element $\chi_{(V_4,*)}$ is easily seen to be a 0-element, that is, the ideal generated by it is just $\mathbb{Q}\chi_{(V_4,*)}$.

Lemma 5.3.4. *For any non-empty subset $A \subset V_4$ and $[\lambda : \mu] \in \mathbb{P}^1$ there exists an $n > 0$ such that $\chi^n_{(A,[\lambda:\mu])} = \chi_{(V_4,*)}$.*

Proof. From (5.7) it follows that $n = 2$ suffices for $|A| \geq 3$. Generically we have that $\chi^2_{(\emptyset,[\lambda:\mu])} = \chi_{(\{a,b,c\},\emptyset)}$, hence generically the result also follows for $|A| = 2$. For the non-generic points one has to check all the cases. For example

$$\chi^2_{(\{b,c\},[1:1])} = \chi_{(V_4,[1:-1])} \Rightarrow n = 3 \text{ satisfies.}$$

Finally if $|A| = 1$, then $\chi^2_{(A,[\lambda:\mu])} = \chi_{(A',[\lambda':\mu'])}$ with $|A'| \geq 2$ and the result again follows. □

Corollary 5.3.5. *The inclusion of* $\mathbb{Q}$*-algebras*

$$\mathbb{Q}(\widetilde{V}_4) \hookrightarrow \mathbb{Q}(\widetilde{Q}_8), \ \chi_A \mapsto \chi_{(A,\emptyset)}$$

is an integral extension.

Proof. The 0-element $\chi_{(V_4,*)}$ is idempotent, hence integral. The elements $\chi_{(A,[\lambda:\mu])}$ for A non-empty are integral by Lemma 5.3.4. The elements $\chi_{(A,*)}$ are integral since $\chi^2_{(A,*)} = \chi_{(V_4,*)}$. Finally, the elements $\chi_{(\emptyset,[\lambda:\mu])}$ are integral since their squares sit in $\mathbb{Q}(\widetilde{V}_4)$. □

As another corollary of Lemma 5.3.4 we know that $\chi_{(A,[\lambda:\mu])} = \chi_{(V_4,*)}$ modulo the Jacobson radical J for all A non-empty and $[\lambda : \mu] \in \mathbb{P}^1$. For $A = \emptyset$ and $[\lambda : \mu]$ a generic point one calculates that $(\chi_{(\emptyset,[\lambda:\mu])} - \chi_{(\emptyset,*)})^3 = 0$. Denote by I' the ideal generated by

$$\chi_{(\emptyset,[\lambda:\mu])} - \chi_{(\emptyset,*)}, \ [\lambda : \mu] \in \mathbb{P}^1 \text{ generic,}$$

and

$$\chi_{(A,[\lambda:\mu])} - \chi_{(V_4,*)}, \ A \subset V_4 \text{ non}-\text{empty}, [\lambda : \mu] \in \mathbb{P}^1.$$

By Corollary 5.3.5 $J(\mathbb{Q}(\widetilde{V}_4)) = J \cap \mathbb{Q}(\widetilde{V}_4)$, hence we have an inclusion

$$\iota : \mathbb{Q}(\widetilde{V}_4)/J(\mathbb{Q}(\widetilde{V}_4)) \hookrightarrow \mathbb{Q}(\widetilde{Q}_4)/J.$$

From Theorem 5.2.8 we know how the left hand side decomposes, that is

$$\varphi : \bigoplus_{H \leq V_4} \mathbb{Q}(V_4/H) \xrightarrow{\cong} \mathbb{Q}(\widetilde{V}_4)/J(\mathbb{Q}(\widetilde{V}_4)),$$

where the element e_{V_4} is mapped to χ_{V_4}.

Theorem 5.3.6. *The Jacobson radical* J *equals* $I' + J(\mathbb{Q}(\widetilde{V}_4))$ *and we have a ring isomorphism*

$$\mathbb{Q}(\widetilde{Q}_8)/J \cong \mathbb{Q}(V_4) \oplus \mathbb{Q}(\mathbb{Z}_4)^{\oplus 3} \oplus \mathbb{Q}^{\oplus 2} \oplus \mathbb{Q}.$$

Proof. By definition of I' we have that

$$\dim_{\mathbb{Q}}\left(\frac{\mathbb{Q}(\widetilde{Q}_8)}{I' + J(\mathbb{Q}(\widetilde{V}_4))}\right) = \dim_{\mathbb{Q}}\left(\frac{\mathbb{Q}(\widetilde{V}_4)}{J(\mathbb{Q}(\widetilde{V}_4))}\right) + 8.$$

Further, recall from Theorem 5.2.8 the decomposition of $\mathbb{Q}(\widetilde{V}_4)/J(\mathbb{Q}(\widetilde{V}_4))$ with resp. unit elements

$$\begin{array}{l|l} \mathbb{Q}(V_4/\{1\}) & (1-\chi_{\{1,a\}})(1-\chi_{\{1,b\}})(1-\chi_{\{1,c\}}) \\ \mathbb{Q}(V_4/\{1,a\}) & (\chi_{\{1,a\}}-\chi_{V_4}) \\ \mathbb{Q}(V_4/\{1,b\}) & (\chi_{\{1,b\}}-\chi_{V_4}) \\ \mathbb{Q}(V_4/\{1,c\}) & (\chi_{\{1,c\}}-\chi_{V_4}) \\ \mathbb{Q}(V_4/V_4) & \chi_{V_4} \end{array} \tag{5.8}$$

Inside $\mathbb{Q}(\widetilde{Q}_8)$ we have that

$$\chi_{(V_4,\emptyset)} = (\frac{1}{2}\chi_{(V_4,\emptyset)} + \frac{1}{2}\chi_{(\emptyset,*)}) + (\frac{1}{2}\chi_{(V_4,\emptyset)} - \frac{1}{2}\chi_{(\emptyset,*)}),$$

and the elements $\frac{1}{2}\chi_{(V_4,\emptyset)} \pm \frac{1}{2}\chi_{(\emptyset,*)}$ are orthogonal idempotents, which are also orthogonal to the first four idempotents of (5.8). Next, we consider the inclusion $\iota(\varphi(\mathbb{Q}(V_4/\{1,a\}))$. The element $\iota\varphi(e_{b\{1,a\}}) = \iota(\chi_{\{b,c\}} - \chi_{V_4})$ is the square of

$$\chi_{(\emptyset,[1:1])} - \chi_{(\emptyset,*)},$$

and we see that $\iota(\varphi(\mathbb{Q}(V_4/\{1,a\}))$ is the subgroup algebra $\mathbb{Q}(\mathbb{Z}_2)$ of

$$\mathbb{Q}(\langle\chi_{(\emptyset,[1:1])} - \chi_{(\emptyset,*)}\rangle) \cong \mathbb{Q}(\mathbb{Z}_4).$$

The inclusions $\iota(\varphi(\mathbb{Q}(V_4/\{1,b\}))$ and $\iota(\varphi(\mathbb{Q}(V_4/\{1,c\}))$ behave analogously and they use the non-generic points $[1:0],[0:1]$ and $[1:i],[1:-i]$ respectively. The isomorphism now follows since the elements

$$\begin{array}{ll} (1-\chi_{(\{1,a\},\emptyset)})(1-\chi_{(\{1,b\},\emptyset)})(1-\chi_{(\{1,c\},\emptyset)}) & \frac{1}{2}\chi_{(V_4,\emptyset)} \pm \frac{1}{2}\chi_{(\emptyset,*)} \\ \chi_{(\{1,a\},\emptyset} - \chi_{(V_4,\emptyset)} & \chi_{(\{1,b\},\emptyset)} - \chi_{(V_4,\emptyset)} \\ \chi_{(\{1,c\},\emptyset)} - \chi_{(V_4,\emptyset)} & \chi_{(V_4,*)} \end{array}$$

are all orthogonal idempotents. □

5.4 Induced morphisms between glider representation rings

To obtain a deeper understanding of $R(1 < G)$ for G a general finite group, we try to obtain some functorial relations. To this extent, let $\varphi : H \to G$ be a group morphism between finite groups G and H and denote the irreducible representations of H, resp. G by

$$\mathrm{Irr(H)} = \{\mathrm{W}_1, \ldots, \mathrm{W}_n\}, \quad \mathrm{Irr(G)} = \{\mathrm{V}_1, \ldots, \mathrm{V}_m\}.$$

We would like to define a morphism between the glider representation rings $R(1 < H)$ and $R(1 < G)$ which preserves multiplication. In order to arrive at such a morphism, we have to associate with any H-representation W a G-representation. The idea is to include all the irreducible G-representations that are connected to through φ to W. We proceed as follows: let $V = V_i \in \mathrm{Irr(G)}$. The group morphism φ allows us to consider $V = V_\varphi$ as an H-representation and by our assumption on the ground field, we have a decomposition into irreducible H-representations

$$V_\varphi = \bigoplus_{i=1}^{n} W_i^{e(V,W_i)}.$$

We fix bases $\{w_{i,1}, \ldots, w_{i,d_i}\}$ for the irreducible H-representations W_i, $1 \leq i \leq n$ and for every V_j, $1 \leq j \leq n$ we denote and fix bases in $(V_j)_\varphi$

$$\begin{array}{lr}
w_{1,1}^{j,1}, \ldots, w_{1,d_1}^{j,1} & (\text{first } W_1) \\
\vdots & \\
w_{1,1}^{j,e(V_j,W_1)}, \ldots, w_{1,d_1}^{j,e(V_j,W_1)} & (e(V_j,W_1)^{\text{th}}\ W_1) \\
w_{2,1}^{j,1}, \ldots, w_{2,d_2}^{j,1} & (\text{first } W_2) \\
\vdots & \\
w_{2,1}^{j,e(V_j,W_2)}, \ldots, w_{2,d_2}^{j,e(V_j,W_2)} & (e(V_j,W_2)^{\text{th}}\ W_2) \\
\vdots & \\
w_{n,1}^{j,e(V_j,W_n)}, \ldots, w_{n,d_n}^{j,e(V_j,W_n)} & (e(V_j,W_n)^{\text{th}}\ W_n)
\end{array} \tag{5.9}$$

which establishes the decomposition of $(V_j)_\varphi$ into H-components. A basis element $w_{i,j}^{k,l}$ denotes the associated basis element $w_{i,j}$ from W_i embedded into the l-th component W_i of V_k. In particular, we have $KHw_{i,j}^{k,l} = W_i$ for all $1 \leq i \leq n, 1 \leq l \leq e(V_k, W_i), 1 \leq j \leq d_i$.

Let $W = \bigoplus_{i=1}^{n} W_i^{f_i}$ be an H-representation. We call the G-representation

$$\overline{\varphi}(W) = \bigoplus_{j=1}^{m} V_j^{\sum_{i=1}^{n} f_i e(V_j,W_i)},$$

the *associated G-representation* of W via φ.

Lemma 5.4.1. *For W, W' H-representations, we have $\overline{\varphi}(W \oplus W') = \overline{\varphi}(W) \oplus \overline{\varphi}(W')$.*

Proof. By construction. □

Next, for every $1 \leq i \leq n, 1 \leq j \leq m$ we define the H-module morphism

$$\iota_{W_i,V_j} : W_i \xrightarrow{\Delta} W_i^{\oplus e(V_j,W_i)} \hookrightarrow V_j$$

where Δ denotes the diagonal embedding and $W_i^{\oplus e(V_j,W_i)}$ is embedded in V_j according to the choice of bases (5.9). Specifically, $w = \sum_{l=1}^{d_i} \lambda_l w_{i,l}$ is mapped to

$$\iota_{W_i,V_j}(w) = \sum_{k=1}^{e(V_j,W_i)} \sum_{l=1}^{d_i} \lambda_i w_{i,l}^{j,k}.$$

Let W be an H-representation with decomposition $W \cong \bigoplus_{i=1}^n W_i^{f_i}$; then for $1 \leq j \leq m$, we can embed every component W_i into V_j (if $e(V_j, W_i) \neq 0$) and if $f_i \geq 2$, and then we embed the f_i components W_i into different copies of V_j. Using matrix notations, we define

$$\iota_{W,V_j} : W \hookrightarrow V_j^{\sum_{i=1}^n f_i e(V_j,W_i)}$$

by

$$\operatorname{diag}(\iota_{W_1,V_j}, \ldots, \iota_{W_1,V_j}, \iota_{W_2,V_j}, \ldots, \iota_{W_2,V_j}, \ldots, \iota_{W_n,V_j}),$$

where there are f_1 ι_{W_1,V_j}'s, f_2 ι_{W_2,V_j}'s, $\ldots$, f_n ι_{W_n,V_j}'s.
Finally, for $W \cong \bigoplus_{i=1}^n W_i^{f_i}$ we define the H-module morphism

$$\iota_{W,\overline{\varphi}(W)} : W \hookrightarrow \overline{\varphi}(W) = \bigoplus_{j=1}^m V_j^{\sum_{i=1}^n f_i e(V_j,W_i)},$$

by

$$\begin{pmatrix} \iota_{W,V_1} \\ \iota_{W,V_2} \\ \vdots \\ \iota_{W,V_m} \end{pmatrix}$$

up to reordering the components in $\overline{\varphi}(W)$. To summarize, per component W_i of W, we add V_j if W_i appears in the decomposition of $(V_j)_\varphi$.

Because the multiplication in glider representation rings is based upon tensor products of group representations, we now elucidate the behavior of the morphisms $\iota_{W,\overline{\varphi}(W)}$ under taking tensor products. To this extent, let $w = w_{1,1} \in W_1, w' = w_{2,1} \in W_2$ and suppose that $W_1 \otimes W_2 \cong \oplus_{i=1}^n W_i^{\oplus k_i}$. Accordingly, there exist $c_{ikj} \in K$ such that $w \otimes w' = \sum_{i=1}^n \sum_{k=1}^{k_i} \sum_{j=1}^{d_i} c_{ikj} w_{i,j,k}$. The extra subindex k of $w_{i,j,k}$ denotes the k-th copy of W_i. By construction we have

$$\iota_{W_1,V_1}(w) = \sum_{l=1}^{e(V_1,W_1)} w_{1,1}^{1,l}, \quad \iota_{W_2,V_1}(w') = \sum_{r=1}^{e(V_1,W_2)} w_{2,1}^{1,r},$$

$$\begin{aligned} \iota_{W_1\otimes W_2,V_1}(w\otimes w') &= \sum_{i=1}^{n}\sum_{k=1}^{k_1}\iota_{W_i,V_1}(\sum_{j=1}^{d_i} c_{ikj}w_{i,j,k}) \\ &= \sum_{i=1}^{n}\sum_{k=1}^{k_1}\sum_{t=1}^{e(V_1,W_i)}\sum_{j=1}^{d_i} c_{ikj}w_{i,j,k}^{1,t} \end{aligned}$$

and

$$\iota_{W_1,V_1}(w)\otimes\iota_{W_2,V_1}(w') = \sum_{l=1}^{e(V_1,W_1)}\sum_{r=1}^{e(V_1,W_2)} w_{1,1}^{1,l}\otimes w_{2,1}^{1,r}.$$

In general, the last two expressions are not equal. However, the tensor product of $w_{1,1}^{1,l}\otimes w_{2,1}^{1,r}$ decomposes in $V_1\otimes V_1$ as does $w\otimes w'$ in $W_1\otimes W_2$. By this we mean that the coefficients of the basis vectors $w_{i,j,k}^{u,v}$ are $\lambda_{l,r}c_{ikj}$, for some $\lambda_{l,r}\in K$. This observation allows us to prove the following.

Lemma 5.4.2. *For W, W' H-representations such that both $\overline{\varphi}(W), \overline{\varphi}(W')\neq 0$, there exists a G-linear embedding $\overline{\varphi}(W\otimes W')\subset\overline{\varphi}(W)\otimes\overline{\varphi}(W')$ that maps*

$$\iota_{W\otimes W',\overline{\varphi}(W\otimes W')}(w\otimes w')\mapsto\iota_{W,\overline{\varphi}(W)}(w)\otimes\iota_{W',\overline{\varphi}(W')}(w').$$

Proof. It suffices to prove this for $W = W_i, W' = W_k$ irreducible H-representations. We use coefficients b_s^{jl} to indicate the decomposition $V_j\otimes V_l\cong\bigoplus_{s=1}^m V_s^{b_s^{jl}}$ and we use a_s^{jl} for the decomposition of $W_j\otimes W_l$. Suppose that $W_j\otimes W_l\cong\bigoplus_{s=1}^n W_s^{a_s^{jl}}$. By definition, we have

$$\begin{aligned} \overline{\varphi}(W_i)\otimes\overline{\varphi}(W_k) &\cong \bigoplus_{j,l=1}^{m} V_j^{e(V_j,W_i)}\otimes V_l^{e(V_l,W_k)} \\ &\cong \bigoplus_{j,l=1}^{m}(V_j\otimes V_l)^{e(V_j,W_i)e(V_l,W_k)} \\ &\supset \bigoplus_{j,l=1}^{m}(W_i\otimes W_k)^{e(V_j,W_i)e(V_l,W_k)}. \end{aligned}$$

It follows that the amount of V_t-components of $\overline{\varphi}(W_i)\otimes\overline{\varphi}(W_k)$ is bigger than or equal to

$$\begin{aligned} &\sum_{j,l=1}^{m}\sum_{s=1}^{n} e(V_j,W_i)e(V_l,W_k)a_s^{ik}e(V_t,W_s) \\ \geq\ &\sum_{s=1}^{n} a_s^{ik}e(V_t,W_s) \\ =\ &\overline{\varphi}(W_i\otimes W_k)_t, \end{aligned}$$

where $\overline{\varphi}(W_i \otimes W_k)_t$ denotes the number of V_t components of $\overline{\varphi}(W_i \otimes W_k)$. Together with the observations made before the lemma, this shows that it is possible to embed $\overline{\varphi}(W \otimes W')$ in $\overline{\varphi}(W) \otimes \overline{\varphi}(W')$ such that

$$\iota_{W\otimes W',\overline{\varphi}(W\otimes W')}(w \otimes w') \mapsto \iota_{W,\overline{\varphi}(W)}(w) \otimes \iota_{W',\overline{\varphi}(W')}(w'). \quad \square$$

Example 5.4.3. We consider the example $\varphi : \mathbb{Z}_4 = \langle j \rangle \hookrightarrow Q_8$ and use notation from Section 5.3. The irreducible representations of $\mathbb{Z}_4 = \langle j \rangle$ are denoted by V_1, V_{-1}, V_i, V_{-i}. One checks that $KQ_8 e_1 \otimes e_2 = T_1 \oplus T_j \subset U \otimes U$. The irreducible Q_8-representations decompose as $\langle j \rangle$ in the following way

$$(T_1)_\varphi = (T_j)_\varphi = V_1, \quad (T_i)_\varphi = (T_k)_\varphi = V_{-1}, \quad U_\varphi = V_i \oplus V_{-i}$$

By definition it follows that

$$\overline{\varphi}(V_{-i}) = U, \quad \overline{\varphi}(V_1 \oplus V_i) = T_1 \oplus T_j \oplus U.$$

So on the one hand we have

$$\overline{\varphi}(V_{-i} \otimes (V_1 \oplus V_i)) = \overline{\varphi}(V_1 \oplus V_{-i}) = T_1 \oplus T_j \oplus U,$$

and on the other hand we have

$$\overline{\varphi}(V_{-i}) \otimes \overline{\varphi}(V_1 \oplus V_i) = U \oplus U \oplus T_1 \oplus T_i \oplus T_j \oplus T_k$$

and the embedding will map the component U diagonally into $U \oplus U$. Adding in vectors, one checks that

$$\iota_{V_1 \oplus V_{-i}, T_1 \oplus T_j \oplus U}(v_1 + v_{-i}) = t_1 + t_j + e_1$$

and

$$\begin{aligned} \iota_{V_{-i},U}(v_{-i}) \otimes \iota_{V_1 \oplus V_i, T_1 \oplus T_j \oplus U}(v_1 + v_i) &= e_1 \otimes (t_1 + t_j + e_2) \\ &= e_1 + e_1 + t_1 + t_j, \end{aligned}$$

where we choose the isomorphism $U \otimes T_i \cong U$ such that $e_1 \otimes t_j$ is mapped to e_1 (in general, an isomorphism $U \otimes T_i \cong U$ maps $e_1 \otimes t_j$ to λe_1).

Recall that $R(\widetilde{G})$ is the $\mathbb{Z}$-algebra generated by the associated characters of the irreducible $(K \subset KG)$-glider representations. From Theorem 4.1.12 we know what the irreducible gliders of essential length 1 look like, namely $V \supset Ka$ where V is a G-representation and $a \in V$ is such that $KGa = V$. To classify these up to isomorphism we use the generalization of Lemma 5.3.1.

Lemma 5.4.4. *Let G be a finite group, U a d-dimensional irreducible G-representation and $m \leq d$. The irreducible $(K \subset KG)$-glider representations $U^{\oplus m} \supset K(u_1 + \ldots + u_m)$ and $U^{\oplus m} \supset K(v_1 + \ldots + v_m)$ are isomorphic if and only if $\langle u_1, \ldots, u_m \rangle$ and $\langle v_1, \ldots, v_m \rangle$ determine the same point in the Grassmanian* $\mathrm{Gr(m, U)}$.

Proof. Extend $\{u_1, \dots, u_m\}$ and $\{v_1, \dots, v_m\}$ to K-bases for U. Then there exists a base change matrix B such that $Bu_i = v_i$ for $1 \leq i \leq m$ if and only if $\langle u_1, \dots, u_m \rangle$ and $\langle v_1, \dots, v_m \rangle$ determine the same point in the Grassmanian $\mathrm{Gr(m, U)}$. □

For an irreducible G-representation U of dimension d we denote $\mathrm{Gr(U)} = \sqcup_{\mathrm{j=1}}^{\mathrm{d}} \mathrm{Gr(j, U)}$ and we denote a point in $\mathrm{Gr(j, U)}$ by $(a_1, \dots, a_j) \in \mathbb{P}^{d-1} \times \dots \times \mathbb{P}^{d-1}$ (all a_k different). For $j = d$, $\mathrm{Gr(d, U)}$ is a singleton which we denote by $\{*_U\}$. We denote by

$$\mathcal{S} = \mathcal{S}_G \tag{5.10}$$

the set of subsets $B \subset \sqcup_{U \in \mathrm{IrrG} \mid \mathrm{dim(U)}>1} \mathrm{Gr}(U)$, such that for all U the intersection $B \cap \mathrm{Gr(j, U)}$ is non-empty for at most one $1 \leq j \leq \dim(U)$ and for this j it is in fact a singleton. We obtain that the irreducible gliders of essential length ≤ 1 are classified by

$$\chi_{(A,B)} \quad A \subset G/G', B \in \mathcal{S},$$

where $G/G' = G^{ab}$ is the Abelianization of the group G and $G' = [G, G]$ is the commutator subgroup of G, i.e. the subgroup generated by all commutators $[g, h] = ghg^{-1}h^{-1}$, $g, h \in G$. Although we work modulo the unique irreducible glider of essential length 0, namely $K \supset 0$, we do reserve a symbol for it, being $\chi_{(\emptyset,\emptyset)}$. By definition, $R(\widetilde{G})$ is the $\mathbb{Z}$-algebra generated by the $\chi_{(A,B)}$. Furthermore, we denote by $M \in \mathcal{S}$ the set $\{*_U \mid U \in \mathrm{Irr(G)}\}$. Finally, when G is Abelian, we use the notation from Section 5.1 and the new notation where $\mathcal{S}_G = \emptyset$ interchangeably, i.e. for $A \subset G, \chi_A = \chi_{(A,\emptyset)}$.

Let $\chi_{(A,B)} \in \mathbb{Q}(\widetilde{H})$ with $A \subset H/H'$ and $B \in \mathcal{S}_H$. If the corresponding glider representation is

$$\bigoplus_{i=1}^{n} W_i^{f_i} \supset Ka,$$

with $a = \sum_{i=1}^{n} a_i^1 + \dots a_i^{f_i}$ and $W = \bigoplus_{i=1}^{n} W_i^{f_i}$, then we define

$$A(\varphi)(\chi_{(A,B)}) = \overline{\varphi}(W) \supset Kb,$$

where

$$b = \iota_{W,\overline{\varphi}(W)}(a) = \sum_{j=1}^{m} \sum_{i=1}^{n} \sum_{k=1}^{f_i} \iota_{W_i,V_j}(a_i^k).$$

Remark 5.4.5. It seems that everything depends on the choice of basis of W. However, we are considering glider representations $W \supset Ka$ of essential length 1, whence working with a different basis of W leads to writing $a \in W$ as another sum of vectors, such that $\iota_{W,\overline{\varphi}(W)}(W \supset Ka)$ is indeed independent of the choice of decomposition.

Proposition 5.4.6. *Let $\chi_{(A,B)}$ be associated with the irreducible $K \subset KH$-glider $\bigoplus_{i=1}^{n} W_i^{f_i} \supset Ka$. Then if $A(\varphi)(\chi_{(A,B)}) = \big(\overline{\varphi}(W) \supset Kb\big)$, we have that $KHb \cong W$ as H-representations.*

Proof. By the irreducibility $\chi_{(A,B)}$ and Theorem 4.1.12 we have that $KH(\sum_{k=1}^{f_i} a_i^k) = W_i^{f_i}$. By construction, we have that

$$KH\iota_{W_i,\overline{\varphi}(W_i)}(a_i^k) \cong W_i$$

as H-representations, whence

$$\begin{aligned} KHb &= KH\sum_{j=1}^{m}\sum_{i=1}^{n}\sum_{k=1}^{f_i} \iota_{W_i,V_j}(a_i^k) \\ &= \sum_{i=1}^{n} KH\sum_{j=1}^{m}\sum_{k=1}^{f_i} \iota_{W_i,V_j}(a_i^k) \\ &= \sum_{i=1}^{n} KH\sum_{k=1}^{f_i}(\iota_{W_i,\overline{\varphi}(W_i)}(a_i^k)) \\ &\cong \sum_{i=1}^{n} KH\sum_{k=1}^{f_i} a_i^k \cong \bigoplus_{i=1}^{n} W_i^{f_i}. \quad \square \end{aligned}$$

By construction, $A(\varphi)(\chi_{(A,B)})$ yields the $(K \subset KG)$-glider $V \supset Kb$ such that V is the "largest" G-representation such that $KHb \cong W$ and $KGb = V$. More precisely,

Proposition 5.4.7. *Let $W = \bigoplus_{i=1}^{n} W_i^{f_i} \supset Ka$ be associated with $\chi_{(A,B)}$. Then*

$$\dim_K(\overline{\varphi}(W)) = \max\{\dim_K(V) \mid \exists \text{ irreducible } KG\text{–glider } V \supset Kv \text{ such that } KHv \cong W \text{ as } H\text{–representations}\}.$$

Proof. Let $V \supset Kv$ be an irreducible $(K \subset KG)$-glider with $KHv \cong W$. The KH-glider $KHv \supset Kv$ is irreducible and $A(\varphi)(KHv \supset Kv) = \overline{\varphi}(KHv) \supset Kv'$ for some v'. Since $KGv = V$, we have that V embeds in $\overline{\varphi}(KHv)$, whence

$$\dim_K(V) \leq \dim_K(\overline{\varphi}(KHv)) = \dim_K(\overline{\varphi}(W)). \quad \square$$

Denote by $N_z(\varphi) = \{W \in \mathrm{Rep(H)} \mid \overline{\varphi}(\mathrm{W}) \neq 0\}$.

Proposition 5.4.8. *The map $A(\varphi) : \mathbb{Q}(\widetilde{H}) \to \mathbb{Q}(\widetilde{G})$ preserves multiplication on the $\mathbb{Z}$-algebra generated by the irreducible gliders $\chi_{A,B} \in \mathbb{Q}(\widetilde{H})$ for which the associated H-representation W is in $N_z(\varphi)$.*

Proof. Let $W \supset Kw$ and $W' \supset Kw'$ be two irreducible $K \subset KH$-glider representations with both $W, W' \in N_z(\varphi)$. By definition, we have that

$$A(\varphi)\Big((W \supset Kw) \otimes (W' \supset Kw')\Big) = KG\iota_{W\otimes W',\overline{\varphi}(W\otimes W')}(w \otimes w') \supset \iota_{W\otimes W',\overline{\varphi}(W\otimes W')}(w \otimes w'),$$

where $KG\iota_{W\otimes W',\overline{\varphi}(W\otimes W')}(w\otimes w')\subset\overline{\varphi}(W\otimes W')$ and

$$A(\varphi)(W\supset Kw)\otimes A(\varphi)(W'\supset Kw')=KG\iota_{W,\overline{\varphi}(W)}(w)\otimes\iota_{W',\overline{\varphi}(W')}(w')\supset\iota_{W,\overline{\varphi}(W)}(w)\otimes\iota_{W',\overline{\varphi}(W')}(w'),$$

where $KG\iota_{W,\overline{\varphi}(W)}(w)\otimes\iota_{W',\overline{\varphi}(W')}(w')\subset\overline{\varphi}(W)\otimes\overline{\varphi}(W')$. The G-linear embedding $\overline{\varphi}(W\otimes W')\hookrightarrow\overline{\varphi}(W)\otimes\overline{\varphi}(W')$ from Lemma 5.4.2 then yields the desired isomorphism as $(K\subset KG)$-glider representations. □

Example 5.4.9. The condition in the previous proposition that $\overline{\varphi}(W)\neq 0$ is necessary. Indeed, consider for example a group G and the projection $\pi: G\twoheadrightarrow G/G'$. Then $\overline{\varphi}(W)=0$ if and only if W decomposes into irreducible G-representations of dimension all bigger than 1. In case $G=Q_8$, we have that the unique two-dimensional irreducible representation U is such that $U\otimes U=T_1\oplus T_2\oplus T_3\oplus T_4$. With notations from [18] we have that

$$A(\varphi)(\chi^2_{(\emptyset,\{[1:0]\})})=A(\varphi)(\chi_{(\{1,a\},\emptyset)})=\chi_{\{1,a\}},$$

which is not equal to

$$A(\varphi)(\chi_{(\emptyset,\{[1:0]\})})^2=0.$$

For a surjective morphism $\varphi: H\twoheadrightarrow G$, we have that

$$N_z(\varphi)=\{W\in\mathrm{Rep(H)}\mid\ker(\varphi)\text{ acts trivial on W}\}.$$

For monomorphisms $\iota: H\hookrightarrow G$ it is clear that $N_z(\iota)=\mathrm{Rep(H)}$.

Suppose now that we have group homomorphisms

$$H\xrightarrow{\varphi}G\xrightarrow{\psi}E$$

and let $\{U_1,\ldots,U_r\}$ be a complete set of irreducible E-representations.

Lemma 5.4.10. *Let* $U\in\mathrm{Irr(E)},\mathrm{W}\in\mathrm{Irr(H)}$, *then*

$$e(U,W)=\sum_{j=1}^{m}e(U,V_j)e(V_j,W).$$

Proof. On the one hand, we have

$$U_{\psi\circ\varphi}\cong\bigoplus_{i=1}^{n}W_i^{e(U,W_i)}.$$

On the other hand, we have

$$U_\psi\cong\bigoplus_{j=1}^{m}V_j^{e(U,V_j)}$$

whence

$$(U_\psi)_\varphi\cong\bigoplus_{j=1}^{m}\bigoplus_{i=1}^{n}W_i^{e(U,V_j)e(V_j,W_i)}\cong\bigoplus_{i=1}^{n}W_i^{\sum_{j=1}^{m}e(U,V_j)e(V_j,W_i)}.$$

Since $U_{\psi\circ\varphi}\cong(U_\psi)_\varphi$, the desired equality follows. □

Proposition 5.4.11. *Let W be an H-representation. Then*

$$\overline{\psi \circ \varphi}(W) = \overline{\psi}(\overline{\varphi}(W))$$

and

$$A(\psi \circ \varphi) = A(\psi) \circ A(\varphi).$$

Proof. Easy corollary of the foregoing lemma. □

Since it is clear that $A(id_G) = id_{\mathbb{Q}(\widetilde{G})}$, we have defined a functor

$$A : Grp_{fin} \to \mathbb{Z}\text{-mod},$$

where Grp_{fin} is the category of finite groups. If we restrict to the subcategory of all finite groups with morphisms being the monomorphisms, then we actually have a functor to commutative rings, where the morphisms do not need to preserve the identity.

5.5 A short exact sequence

Let G be a finite group. From the previous section, we obtain that every subgroup $H \leq G$ yields a monomorphism $\mathbb{Q}(\widetilde{H}) \hookrightarrow \mathbb{Q}(\widetilde{G})$. In this section, we use these monomorphisms to construct a short exact sequence of $\mathbb{Q}(G/G')$-modules. After fixing a group isomorphism $G/G' \cong \widehat{G/G'}$ we denote the one-dimensional G-representations by $\{T_g \mid g \in G/G'\}$. The map $\mathbb{Q}(G/G') \hookrightarrow \mathbb{Q}(\widetilde{G}), e_g \mapsto \chi_{(\{g\},\emptyset)}$ is a ring morphism and this defines a $\mathbb{Q}(G/G')$-module structure on the glider representation ring. It turns out that we can also define a $\mathbb{Q}(G/G')$-module structure on $\mathbb{Q}(\widetilde{H})$. Considered as H-representation through $\varphi : H \hookrightarrow G$, we have $(T_g)_\varphi \cong S = Ks$ for some H-representation and for an irreducible $(K \subset KH)$-glider $W \supset Kw$ we define

$$(T_g \supset Kt_g) \cdot (W \supset Kw) := (S \supset Ks) \otimes (W \supset Kw).$$

We define a subgroup $A_\varphi(H) \leq G/G'$ by

$$A_\varphi(H) = \{g \in G/G' \mid (T_g)_\varphi \cong T_H \text{ as } H\text{-representations}\},$$

where T_H denotes the trivial H-representation.

Lemma 5.5.1. *Let W be an H-representation and T_g, S be such that $(T_g)_\varphi \cong S$ as H-representations. Then*

$$\overline{\varphi}(S \otimes W) \cong T_g \otimes \overline{\varphi}(W).$$

Proof. It suffices to prove this for $W = W_i$ irreducible. The G-representation $\overline{\varphi}(W)$ is determined by decomposing all the $(V_j)_\varphi$ into H-components and checking whether W appears as a component. Let $h \in A_\varphi(H)$, then W_i appears in $(V_j)_\varphi$ if and only if W appears in $T_h \otimes (V_j)_\varphi$. Hence we have two expressions of $\overline{\varphi}(W)$, namely

$$\bigoplus_{j=1}^{m'} V_j \quad \text{and} \quad \bigoplus_{j=1}^{m'} T_h \otimes V_j,$$

for some $m' \leq m$ (up to reordering). It follows that there exists a permutation σ on m' letters such that $V_j \cong T_h \otimes V_{\sigma(j)}$ as G-representations. For every $z \in A_\varphi(H)$ we have that $T_{gz} \otimes V_j$ lies over $S \otimes W$. However, $T_g \otimes V_j \cong T_{gh} \otimes V_{\sigma(j)}$ for all $1 \leq j \leq m'$ and it follows that

$$\bigoplus_{j=1}^{m'} T_g \otimes V_j \cong \bigoplus_{j=1}^{m'} T_{gh} \otimes V_{\sigma(j)}.$$

This shows that

$$\overline{\varphi}(S \otimes W) \cong T_g \otimes \overline{\varphi}(W) \cong T_{gz} \otimes \overline{\varphi}(W) \quad \forall z \in A_\varphi(H). \quad \square$$

Recall that for a group morphism $\varphi : H \to G$ we denoted the associated morphism between glider representation rings by $A(\varphi)$. If φ is a monomorphism, we also denote $A(\varphi)$ by Φ_H^G.

Proposition 5.5.2. *The map* $\Phi_H^G : \mathbb{Q}(\widetilde{H}) \to \mathbb{Q}(\widetilde{G})$ *is a* $\mathbb{Q}(G/G')$*-module morphism.*

Proof. With notations as before we have

$$\begin{aligned} \Phi_H^G\big((T_g \supset Kt_g) \cdot (W \supset Kw)\big) &= \Phi_H^G\big(S \otimes W \supset Ks \otimes w\big) \\ &= \overline{\varphi}(S \otimes W) \supset K\iota_{S\otimes W, \overline{\varphi}(S\otimes W)}(s \otimes w) \\ &\cong T_g \otimes \overline{\varphi}(W) \supset Kt_g \otimes \iota_{W, \overline{\varphi}(W)}(w), \end{aligned}$$

which is isomorphic to $(T_g \supset Kt_g) \otimes \Phi_H^G\big(W \otimes Kw\big)$. $\square$

In order to construct a short exact sequence of $\mathbb{Q}(G/G')$-modules we are forced to introduce three $\mathbb{Q}(G/G')$-modules P, Q and R. In the next section we will on the one hand discuss for which groups one can deduce what these submodules exactly are, which then yields a description of the glider representation ring of G modulo its Jacobson radical $J = J(\mathbb{Q}(\widetilde{G}))$. On the other hand, we explain how these modules are linked with group representation theoretic properties of the group G as mentioned in the introduction of the chapter. The proofs in this section will already make clear some of these connections.

First of all, the glider representation ring of a non-Abelian group is an infinite dimensional vector space. In Section 5.3 we gave a description of the Jacobson radical for $G = Q_8$. We actually defined an ideal I contained in the nilradical $N = N(\mathbb{Q}(\widetilde{Q_8}))$ and showed that the quotient $\mathbb{Q}(\widetilde{Q_8})/I$ was semisimple, from which we could conclude that $I = N = J$. It is not clear that the morphisms Φ_H^G factorize over the Jacobson radical. Since they do factorize over the nilradical, we consider the induced morphisms

$$\overline{\Phi_H^G} : \mathbb{Q}(\widetilde{H})/N \to \mathbb{Q}(\widetilde{G})/N.$$

Here, we denoted the nilradicals of $\mathbb{Q}(\widetilde{H})$ and $\mathbb{G}(\widetilde{H})$ both by N. It will always be clear from the context of which ring R, N is the nilradical.

Let $\chi_{(A,B)} \in \mathbb{Q}(\widetilde{G})$ and consider the cyclic semigroup $\langle \chi_{(A,B)} \rangle$. If this semigroup is finite, then it contains a unique idempotent element $e = \chi_{(C,D)}$. If n is the smallest integer such that $\chi^n_{(A,B)} = e$, then the difference $\chi_{(A,B)} - \chi^{n+1}_{(A,B)}$ is nilpotent. Since the glider representation ring is infinite dimensional for non-Abelian groups, it could be that the cyclic semigroup $\langle \chi_{(A,B)} \rangle \cong \mathbb{N}$. Let P be the $\mathbb{Q}$-vector space with basis the elements $\chi_{(A,B)}$ for which the cyclic semigroup is not finite.

Proposition 5.5.3. *The vector space P is a $\mathbb{Q}(G/G')$-submodule of $\mathbb{Q}(\widetilde{G})$.*

Proof. Let $a = \chi_{(A,B)} \in P$ and $g \in G/G'$. Denote $b = \chi_{(\{g\},\emptyset)}$. If $ba \notin P$, then there exists $n > 0$ such that $b^n a^n = e$ is idempotent. But then $e = b^{|G/G'|n} a^{|G/G'|n} = a^{|G/G'|n}$, contradicting $a \in P$. □

For $\chi_{(A,B)} \notin P$, denote the associated idempotent element by $e(A,B)$. We have the following lemma.

Lemma 5.5.4. *If an element of the form $\chi_{(A,B)} \in \mathbb{Q}(\widetilde{G})$ with $A \neq \emptyset$ is idempotent, then $A \subset G/G'$ is a subgroup.*

Proof. Suppose that $\chi_{(A,B)}$ is idempotent, then $A.A \subset A$. Let $a \in A$, then $e = a^{o(a)} \in A^{o(a)} \subset A$, which entails that $A.A = A$ or that $A \subset G/G'$ is a subgroup. □

Remark 5.5.5. It remains a question whether an element of the form $\chi_{(\emptyset,B)}$ can be idempotent.

The idea is to show that certain idempotent elements $e(A,B)$ are in the image Φ_H^G for some subgroup $H \leq G$. To this extent, we have

Lemma 5.5.6. *Let $C \leq G/G'$ be a subgroup and define $H = \bigcap_{c \in C} \mathrm{Ker}(T_c) \leq G$. Then*

$$A_\iota(H) = \{g \in G/G' \mid (T_g)_{|H} \cong T_H \text{ as } H\text{−representations}\} = C,$$

where $\iota : H \hookrightarrow G$ denotes the embedding

Proof. By definition it follows that $C \subset A_\iota(H)$, from which we obtain

$$\bigcap_{g \in A_\iota(H)} \operatorname{Ker}(T_g) \subset \bigcap_{c \in C} \operatorname{Ker}(T_c) = H.$$

Let $g \in A_\iota(H)$, then $H \subset \operatorname{Ker}(T_g)$. Since this holds for all $g \in A_\iota(H)$, we obtain that

$$\bigcap_{g \in A_\iota(H)} \operatorname{Ker}(T_g) \subset \bigcap_{c \in C} \operatorname{Ker}(T_c) \subset \bigcap_{g \in A_\iota(H)} \operatorname{Ker}(T_g)$$

and it follows that $A_\iota(H) = C$. □

Write $e(A, B) = \chi_{(C,D)}$. If $1 \neq C$ and if there exists an idempotent element $\chi_{(C,D')}$ with $D \subsetneq D'$, then we call $\chi_{(A,B)}$ *non-maximal.* We define Q to be the $\mathbb{Q}$-vector space with basis elements exactly the $\chi_{(A,B)}$ which are non-maximal. By definition of $\mathcal{S}_G = \mathcal{S}$, see (5.10), $D \subsetneq D'$ means that there exists an irreducible G-representation U which appears in the decomposition of $\chi_{(C,D')}$ but not in $\chi_{(C,D)}$. The following example shows that $Q \neq 0$ in general.

Example 5.5.7. Let $G = A_4$. The commutator subgroup equals $A_4 = V_4$ and consider a subgroup $C_2 < V_4$. One shows that under the embedding

$$\Phi_{V_4}^{A_4} : \mathbb{Q}(\widetilde{V_4}) \hookrightarrow \mathbb{Q}(\widetilde{A_4}),$$

the element $\chi_{C_2} = \chi_{(C_2,\emptyset)}$ is sent to $\chi_{(A_4/V_4,\{a\})}$ where $\{a\} \in \mathrm{Gr}(1, \mathrm{U})$, U being the only three dimensional irreducible A_4-representation. Since the former element is idempotent, so is the latter. Since $\chi_{(A_4/V_4,\{*_U\})}$ is also idempotent, $\chi_{(A_4/V_4,\{a\})}$ is indeed non-maximal. We already mention here that

$$\chi_{(A_4/V_4,\{a\})} \in Q \cap \Phi_{V_4}^{A_4}\left(\mathbb{Q}(\widetilde{V_4})\right).$$

Proposition 5.5.8. *The vector space Q is a $\mathbb{Q}(G/G')$-submodule of $\mathbb{Q}(\widetilde{G})$.*

Proof. This follows because $\langle \chi_{(A,B)} \rangle$ and $\langle \chi_{(\{g\},\emptyset)} \chi_{(A,B)} \rangle$ have the same idempotent element (if it exists). □

Lemma 5.5.9. *Let E, H be subgroups of G. If $H \subset E$, then*

$$\Phi_H^G(\mathbb{Q}(\widetilde{H})) \subset \Phi_E^G(\mathbb{Q}(\widetilde{E})).$$

Proof. From the functorial properties of A we have that

$$\Phi_H^G(\mathbb{Q}(\widetilde{H})) = \Phi_E^G(\Phi_H^E(\mathbb{Q}(\widetilde{H}))) \subset \Phi_E^G(\mathbb{Q}(\widetilde{E})). \quad \square$$

Finally, we define the $\mathbb{Q}(G/G')$-module R to be generated by all elements $\chi_{(A,B)}$ for which the associated idempotent element $e(A,B)$ is of the form $\chi_{(\{e\},D)}$ or $\chi_{(\emptyset,D)}$. The reason for including the latter elements comes from Remark 5.5.5. We need one more notion to prove the following theorem: let $V \supset Kv$ be the irreducible glider associated with $\chi_{(A,B)}$, then we call $\alpha = (n_1, \ldots, n_m) \in \mathbb{N}^m$ the *dimension vector* of $\chi_{(A,B)}$ if

$$KGv = V \cong \bigoplus_{j=1}^{m} V^{\oplus n_j}.$$

Theorem 5.5.10. *Let G be a finite group. We have the following short exact sequence of $\mathbb{Q}(G/G')$-modules*

$$0 \longrightarrow \frac{P}{P\cap N} + \frac{Q}{Q\cap N} + \sum_{G'\leq H \triangleleft_{max} G} \overline{\Phi_H^G}(\mathbb{Q}(\widetilde{H})/N) \xrightarrow{\quad\Psi\quad} \mathbb{Q}(\widetilde{G})/N \longrightarrow \frac{R}{R\cap N} \longrightarrow 0$$

The map Ψ denotes the embedding.

Proof. We first prove the statement for G Abelian. It is clear that $P = 0$ since $\mathbb{Q}(\widetilde{G})$ is finite dimensional. Moreover, by definition it is also clear that $Q = 0$ and $R = \mathbb{Q}(G)$. Let $H \triangleleft G$ be a subgroup, then we define the subgroup

$$H'' = \bigcap_{g\in H} \operatorname{Ker}(T_g) \triangleleft G.$$

It holds that $H'' < G$ is a proper subgroup if and only if $H \neq e$. By construction, it follows that $\overline{\Phi_{H''}^G}(1) = \overline{\chi_H} \in \mathbb{Q}(\widetilde{G})/J$. If gH is a left coset, then $(T_g)_{|H''} \cong T_{h'}$ for some $h' \in H''$ and it follows that $\overline{\Phi_{H''}^G}(\chi_{\{h'\}}) = \overline{\chi_{gH}}$. This shows that the generators for the cokernel are exactly the elements of the form $\chi_{\{g\}}$ with $g \in G$.

Now we treat the general case. Let $\overline{\chi_{(A,B)}}$ be a generator of $\mathbb{Q}(\widetilde{G})/N$ with $\chi_{(A,B)} \notin P$. By definition of P, there exists an $n > 0$ such that $\overline{\chi_{(A,B)}}^n = \overline{\chi_{(C,D)}}$ is idempotent. In fact, $\chi_{(A,B)}^n = e(A,B) = \chi_{(C,D)}$ is also idempotent. Lemma 5.5.4 entails that $C \subset G/G'$ is a subgroup. By Lemma 5.5.6, the subgroup

$$G' \leq H = \bigcap_{c\in C} \operatorname{Ker}(T_c) \leq G$$

is such that $A_\iota(H) = C$. This entails that

$$\Phi_H^G(1) = \chi_{(C,\emptyset)},$$

where 1 denotes the unit in $\mathbb{Q}(\widetilde{H})$, i.e. $1 = \chi_{(\{e\},\emptyset)}$. Observe moreover that $G' < H < G$ is proper if and only if $C \neq e$ and $C \neq G/G'$.

Firstly, suppose that $C \neq e$, that is $G' \leq H \lneq G$. If $U \supset Ku$ is the $(K \subset KG)$-glider corresponding to $\chi_{(C,D)}$, then the H-module KHu contains the trivial

H-representation T_H. Moreover, the associated element of $KHu \supset u$ is of the form $\chi_{(C'',D'')}$ for some $1 \in C'' \subset H/H'$ and $D'' \in \mathcal{S}_H$. If $D = \emptyset$, then $D'' = \emptyset$ and $\chi_{(C,\emptyset)} = \Phi_H^G(\chi_{(\{1\},\emptyset)})$. Suppose that $D \neq \emptyset$. By construction, the G-representation U embeds in $V(KHu)$. First of all, if an irreducible H-representation W appears in the decomposition of KHu, then there must be at least one irreducible G-representation V lying over W that appears in U. We also have by definition of H that all one-dimensional components of U are exactly all the one dimensional components that lie over T_H. Therefore, we can write

$$\Phi_H^G(\chi_{(C'',D'')}) = (\bigoplus_{c\in C} T_c \oplus U' \supset Ka),$$

with U' a G-representation. Suppose that the irreducible G-representation V with $\dim(V) > 1$ appears in the decomposition of U; then this implies that there exists an irreducible H-representation W which appears in $\chi_{(C'',D'')}$. The G-representation $V(KHu)$ then contains all other irreducible G-representations V' such that V' lies over this W. If $\dim(V) = \dim(V')$, then by [24, Theorem 5] we know that V and V' differ only in the projective representation of G^{dec}/H, where G^{dec} denotes the decomposition group $H \leq G^{dec} \leq G$. It holds that $G^{dec}/H \leq G/H$ and $G/H \cong (G/G')/(H/G')$, whence G^{dec}/H is Abelian and both representations V and V' only differ in a one-dimensional representation S of G^{dec}/H. By definition of $A_\iota(H) = C$, it follows that there exists $c \in C$ such that $V' \cong V \otimes T_c$ as H-representations. Because the element $\chi_{(C,D)}$ is idempotent, the G-representation $T_c \otimes V$ already appears in U. It follows that

$$\chi_{(C,D)}\Phi_H^G(\chi_{(C'',D'')}) = \Phi_H^G(\chi_{(C'',D'')}) = \chi_{(C,E)},$$

for some $E \in \mathcal{S}$. Hence if $\chi_{(A,B)} \notin Q$, then $E = D$, meaning that $\chi_{(C,D)} = \Phi_H^G(\chi_{(C'',D'')})$.

We return to the element $\chi_{(A,B)}$. It follows that $\chi_{(A,B)}^m$ and $\chi_{(A,B)}^{m'}$ are congruent modulo the nilradical $N(\mathbb{Q}(\widetilde{G}))$ if and only if $m - m' \in n\mathbb{Z}$. Therefore we can replace $\chi_{(A,B)}$ by $\chi_{(A,B)}^{n+1}$ since $\overline{\chi_{(A,B)}} = \overline{\chi_{(A,B)}^{n+1}}$ in $\mathbb{Q}(\widetilde{G})/N$. For $\chi_{(A,B)}^{n+1}$ we have that $A = gC$ for some $g \in G/G'$. Indeed, $\chi_{(A,B)}^n = \chi_{(C,E)}$ for some $E \in \mathcal{S}$ and hence $\chi_{(A,B)}^{n+1} = \chi_{(A'',E'')}$ for some $E'' \in \mathcal{S}$ and where A contains at least one full left coset of C in G/G'. If it would be strictly bigger than one coset, the idempotent element in the cyclic semigroup $\langle\chi_{(A,B)}\rangle$ would be of the form $\chi_{(C'',D'')}$ with $C \lneq C'$, contradiction. In fact, by replacing $\chi_{(A,B)}$ by $\chi_{(A,B)}^{n+1}$ we obtain an equality $\chi_{(A,B)}\chi_{(C,D)} = \chi_{(A,B)}$ inside $\mathbb{Q}(\widetilde{G})$. This shows that the dimension vectors of $\chi_{(A,B)}$ and $\chi_{(C,D)}$ are the same. Since $A = gC$, $\chi_{(A,B)}$ corresponds to the glider

$$\bigoplus_{c\in C} T_{gc} \oplus V \supset K(\sum_{c\in C} t_{cg} + v)$$

for some $v \in V$. The equality $\chi_{(A,B)}\chi_{(C,D)} = \chi_{(A,B)}$ states

$$KG((\sum_{c \in C} t_{cg} \otimes v) \otimes a) \cong \bigoplus_{c \in C} T_{gc} \oplus V.$$

However, since $KG(t_g \otimes a)$ has the same dimension vector as $\chi_{(A,B)}$, this shows that we even have the equality

$$\chi_{(\{g\},\emptyset)}\chi_{(C,D)} = \chi_{(A,B)}.$$

Let S denote the one-dimensional H-representation $(T_g)_{|H}$; then if $\chi_{(A'',B'')}$ denotes the associated element to the $(K \subset KH)$-glider

$$(S \supset Ks) \otimes (KHu \supset Ku),$$

it follows that

$$\Phi_H^G(\chi_{(A'',B'')}) = \Phi_H^G(\chi_{(\{s\},\emptyset)})\Phi_H^G(\chi_{(C'',D'')}) = \chi_{(gC,\emptyset)}\chi_{(C,D)} = \chi_{(\{g\},\emptyset)}\chi_{(C,\emptyset)}\chi_{(C,D)} = \chi_{(\{g\},\emptyset)}\chi_{(C,D)} = \chi_{(A,B)}.$$

Hence we have shown that any generator $\overline{\chi_{(A,B)}}$ of $\mathbb{Q}(\widetilde{G})/N$ with $\chi_{(A,B)} \notin P, \notin Q$ and for which the associated idempotent $e(A,B) = \chi_{(C,D)}$ has $C \neq e$ lies in the image $\overline{\Phi_E^G}(\mathbb{Q}(\widetilde{E})/N)$ where $E = \bigcap_{c \in C} \mathrm{Ker}(T_c) \lhd G$. Since $C \neq e$, $E < G$ is proper and Lemma 5.5.9 entails that $\overline{\chi_{(A,B)}}$ indeed lies in the image of Ψ. By running over all $\overline{\chi_{(A,B)}}$ not in P, not in Q and for which $e(A,B)$ has $C \neq e$, we see that we need all subgroups $G' \leq H \lhd_{max} G$.

The only generators $\overline{\chi_{(A,B)}}$ not in the image of Ψ are the ones with associated idempotent $e(A,B) = \chi_{(\{e\},D)}$ or $\chi_{(\emptyset,D)}$. This shows that the cokernel of Ψ is isomorphic to $R/(R \cap N)$. □

Remark 5.5.11. Observe that the sum on the left is not direct. This was already clear from Example 5.5.7. In the next section we will address this further.

5.6 The structure of P, Q and R

In this section we consider the explicit structure of the three modules P, Q, R for some particular classes of groups. We begin with P. For non-Abelian groups the number of generators $\chi_{(A,B)}$ is infinite, but the number of dimension vectors is not. This allows us to show that under some condition we can ensure that the cyclic semigroup generated by any generator $\langle \chi_{(A,B)} \rangle$ always contains an idempotent, i.e. $P = 0$. Suppose first that $A \neq \emptyset$, then there exists some n such that T appears in $\chi^n_{(A,B)} = \chi_{(C,D)}$. Consider now the sequence

$$\chi_{(C,D)}, \chi^2_{(C,D)}, \chi^4_{(C,D)}, \ldots, \chi^{2^n}_{(C,D)}, \ldots$$

Suppose that $D \cap \mathrm{Gr(j, U)} = \{\mathrm{a}_1, \dots, \mathrm{a_j}\}$, then because T appears we have that at least $\{a_1, \dots, a_j\}$ appears in $\chi^2_{(C,D)}$. This shows that the dimension vector $\alpha(n)$ of $\chi^{2^n}_{(C,D)}$ is an increasing function. However, since there are only a finite number of dimension vectors, this sequence must stabilize, and again using the argument involving the appearance of T, we arrive at an element $\chi_{(E,F)} \in \langle \chi_{(A,B)} \rangle$ which is idempotent.

Proposition 5.6.1. *Suppose that for every $V \in \mathrm{Irr(G)}$, there exists an n such that $V^{\otimes n}$ completely linearizes; then for every generator $\chi_{(A,B)}$ of $\mathbb{Q}(\widetilde{G})$ the cyclic semigroup $\langle \chi_{(A,B)} \rangle$ contains an idempotent.*

Proof. Let $\chi_{(A,B)}$ be a generator of the glider representation ring. If $A \neq \emptyset$, then we follow the argument given above. If $A = \emptyset$, take some irreducible representation V that appears, which has dimension strictly bigger than 1. The condition on G allows the existence of n such that $\chi^n_{(\emptyset,B)} = \chi_{(A,B')}$ with $A \neq \emptyset$ and the statement follows. □

A group G is said to be of nilpotency class smaller or equal to 2 if the commutator subgroup is contained in the center.

Proposition 5.6.2. *The condition in the previous proposition gives an equivalent characterization of groups of nilpotency class 2.*

Proof. Suppose that the nilpotency class of G is larger than 2. Then there exists $g \in G' \setminus Z(G)$. Since $Z(G) = \bigcap_{\chi \in \mathrm{Irr(G)}} Z(\chi)$, there exists an irreducible character χ such that $|\chi(g)| < |\chi(e)|$. Here, $Z(\chi) = \{g \in G \mid |\chi(g)| = \chi(e)\}$ denotes the center of the character χ. If there would exist an $n > 1$ such that χ^n is a positive linear combination of linear characters of G, then on the one hand $\chi^n(g) = \chi^n(e)$, since $g \in G'$ (the commutator subgroup G' is the intersection of the kernels of all linear characters). On the other hand $|\chi^n(g)| < |\chi^n(1)|$, which gives a contradiction. Conversely, suppose that G is of nilpotency class at most 2, i.e. $G' \subset Z(G)$ and let U be an irreducible G-representation. Considered as G'-representation, $U_{|G'} \cong S^{\oplus \dim(U)}$ for some one-dimensional G'-representation S, since $G' \subset Z(G)$. There exists $n \geq 1$ such that $S^{\otimes n}$ is the trivial G'-representation $T_{G'}$. Hence

$$(U_{|G'})^{\otimes n} \cong T_{G'}^{\oplus n \dim(U)}.$$

Hence, $U^{\otimes n}$ decomposes into irreducible G-representations which all lie over the trivial G'-representation, i.e. $U^{\otimes n}$ is a sum of one-dimensional representations. □

Corollary 5.6.3. *If G has nilpotency class at most 2, then $P = 0$.*

To give a measure of P being non-zero, we can look at normal subgroups $N \lhd G$. It is well-known that every normal subgroup can be written as the intersection of kernels of irreducible characters. When

$$N = \bigcap_{i \in I} \mathrm{Ker}(\chi_i)$$

is such that removing one of the $\mathrm{Ker}(\chi_i)$ yields a strictly bigger normal subgroup, we call the presentation *minimal*. We have the following proposition.

Proposition 5.6.4. *Let $N = \bigcap_{i\in I} \mathrm{Ker}(\chi_i)$ be a minimal presentation. If $U \in \mathrm{Irr(G)}$ with associated character χ is such that $N \subset \mathrm{Ker}(\chi)$, then U appears as a component of*

$$(\bigoplus_{i\in I} U_i)^{\otimes n}$$

for some $n \geq 1$ (U_i denotes the irreducible representation associated to χ_i).

Proof. The G-representation $\bigoplus_{i\in I} U_i$ induces a G/N-representation $\overline{V}$ which is faithful. Because $N \subset \mathrm{Ker}(\chi)$, U also induces a G/N-representation $\overline{U}$. Hence there exists $n \geq 1$ such that the inproduct in G/N

$$\langle \overline{U}, \overline{V}^n \rangle_{G/N} \neq 0.$$

The result now follows, because for G-representations W, W' that induce G/N-representations $\overline{W}, \overline{W'}$ we have the equality

$$\langle \overline{W}, \overline{W'} \rangle_{G/N} = \langle W, W' \rangle_G. \quad \square$$

Corollary 5.6.5. *Let G be a finite group and $\chi_{(A,B)} \in \mathbb{Q}(\widetilde{G})$. If there exists $U \in \mathrm{Irr(G)}$ such that $B \cap \mathrm{Gr(U)} = \{*_{\mathrm{U}}\}$, then $\chi_{(A,B)} \notin P$.*

Proof. Because $N = \mathrm{Ker}(\chi_U) \subset \mathrm{Ker}(\chi_T)$, the previous proposition shows there exists n such that T appears in the decomposition of $U^{\otimes n}$. Because we have the liberty of choosing vectors $u_1, \ldots, u_{\dim(U)}$ in $U^{\oplus \dim(U)}$, we can choose them appropriately such that T appears in the decomposition of

$$KG(u_1 + \cdots + u_{\dim(U)})^{\otimes n}.$$

The argument preceding Proposition 5.6.1 then yields the result. $\square$

Last but not least, it follows easily that if $P(G) = 0$, that $P(H) = 0$ for all subgroups $H \leq G$ since we can embed $\mathbb{Q}(\widetilde{H})$ in $\mathbb{Q}(\widetilde{G})$ via Φ_H^G.

SUMMARIZING: starting from an irreducible G-representation U of dimension > 1, there always exists an $n \in \mathbb{N}$ such that the trivial G-representation appears in the decomposition of $U^{\otimes n}$. Working with ($K \subset KG$)-glider representations, however, requires keeping track of a vector $u \in U$ and by definition $\left(U \supset Ku\right)^{\otimes n}$ equals $\left(KG(u \otimes \cdots \otimes u) \supset Ku \otimes \cdots \otimes u\right)$. In general, $KG(u \otimes \cdots \otimes u) \subsetneq U^{\otimes n}$ so we cannot ensure that T indeed appears in the decomposition of $KG(u \otimes \cdots \otimes u)$. It remains a question whether there exist groups with $P \neq 0$. In fact, the same conclusion holds true, when replacing T by another one-dimensional G-representation S. Simple groups only have 1 one-dimensional representation, so maybe they form a counterexample.

Next, let us look at the submodule Q, which keeps track of elements $\chi_{(A,B)}$ that are non-maximal. We recall from Example 5.5.7 that

$$Q \bigcap \sum_{G' \leq H \lhd_{max} G} \overline{\Phi_H^G(\mathbb{Q}(\widetilde{H})/N)} \neq 0.$$

To make a connection with group representation theoretic questions, we alter Q by the quotient

$$\frac{Q}{Q \cap \sum_{G' \leq H \lhd_{max} G} \overline{\Phi_H^G(\mathbb{Q}(\widetilde{H})/N)}},$$

but we still denote this quotient by Q. From the proof of Theorem 5.5.10 we deduce that the only possibility for $\chi_{(A,B)}$ to be non-maximal lies in the existence of irreducible G-representations V, V' of different dimension which lie over the same irreducible H-representation W for some normal subgroup H containing G'.

SUMMARIZING: $Q(G) \neq 0 \Rightarrow \exists V, W \in \mathrm{Irr(G)}$ of different dimension and V, W lie over some irreducible H-representation U for some normal subgroup H containing the commutator subgroup G'.

Let us show a few results related to this question. First of all, in Corollary 4.1.16 it is shown that this situation cannot occur for G nilpotent and $H \lhd G$ a maximal normal subgroup strictly containing the center $Z(G)$. Also, from the obstruction to $Q = 0$, we immediately obtain:

Proposition 5.6.6. *If G has all its irreducible representations of degree ≤ 2 then $Q = 0$.*

The groups satisfying the property stated in the previous proposition are determined in [2]. In [15] the authors show that for these groups we also have $P(G) = 0$.
The equivalent characterization of groups of nilpotency class 2, see Proposition 5.6.2, allows to show that $Q(G) = 0$ for G such a group. To prove this, we first observe the following: Let U, V be two irreducible G-representations. Then the trivial representation T appears as a component of $U \otimes V$ if and only if $V \cong U^*$ and if this is the case its multiplicity is 1. Indeed, we have an isomorphism $(U \otimes V)^G \cong \mathrm{Hom}_G(U, V^*)$, where the left hand side denotes the invariants of $U \otimes V$, the dimension of which corresponds with the number of times T appears as a subrepresentation.

Lemma 5.6.7. *Let $U, V \in \mathrm{Irr(G)}$ such that there exists a one-dimensional representation $S \subset U \otimes V$, then* $\dim(U) = \dim(V)$.

Proof. The condition implies that $T \subset S^* \otimes U \otimes V$, which shows that $V^* \cong S^* \otimes U$. In particular it follows that $\dim(U) = \dim(V)$. □

Lemma 5.6.8. *If there exists an $n > 1$ such that $U^{\otimes n}$ completely linearizes and if*

$$U^{\otimes n-1} = \bigoplus_{i=1}^{k} W_j,$$

with $W_1, \ldots, W_k$ irreducible G-representations, then $\dim(W_1) = \ldots = \dim(W_k) = \dim(U)$.

Proof. This follows from the previous lemma since $U^{\otimes n} \cong \bigoplus_{i=1}^{k} W_i \otimes U \cong \bigoplus_{j=1}^{\dim(U)^n} T_j$. □

Proposition 5.6.9. *Let G be a group of nilpotency 2, then $Q(G) = 0$.*

Proof. Suppose that V, W are irreducible G-representations that lie over a same irreducible H-representation, H some normal subgroup of G which contains G'. Since $G' \subset Z(G)$, $V_{|G'} \cong S_V^{\dim(V)}, W_{|G'} \cong S_W^{\dim(W)}$, for some irreducible G'-representations S_V, S_W. By the assumption on V, W we have $S_V \cong S_W$ as G'-representations. By Proposition 5.6.2 there exists n such that $V^{\otimes n}$ completely linearizes and up to taking a multiple of n, we may assume that T appears as a component of $V^{\otimes n}$. It follows that $S_V^{\otimes n} \cong T_{G'}$, where $T_{G'}$ denotes the trivial G'-representation. This further entails that $V^{\otimes n-1} \otimes W$ contains at least one one-dimensional representation, say S. Decompose $V^{\otimes n-1} \cong U_1 \oplus \cdots \oplus U_k$ into irreducible G-representations, then $\dim(U_i) = \dim(V)$ for $1 \leq i \leq k$ by Lemma 5.6.8. Up to renumbering, it follows that $S \subset U_1 \otimes W$, which shows that $\dim(V) = \dim(U_1) = \dim(W)$ by Lemma 5.6.7. □

Lemma 5.6.10. *Let G be a group with subgroup $G' \leq H \leq G$ containing the commutator subgroup and $V, W \in$ Irr(G) lie over the irreducible H-representation U. Then* $\dim(V) = 1$ *if and only if* $\dim(W) = 1$.

Proof. Suppose that $\dim(V) = 1$, then so is $\dim(U) = 1$. Up to tensoring with a power of V, we can assume that $U = T_H$ is the trivial H-representation. Then W is an irreducible G-representation lying over the trivial G'-representation. Since the only irreducible G-representations lying over $T_{G'}$ are one-dimensional the claim follows. □

Proposition 5.6.11. *Let G be a finite group with maximal subgroup $G' \leq H \leq G$ containing the commutator subgroup G'. If $V, W \in$ Irr(G) *lie over* $U \in$ Irr(H) *then* $\dim(V) = \dim(W)$.

Proof. Suppose $\dim(V) < \dim(W)$. By the previous lemma it follows that $1 < \dim(V) < \dim(W)$. Any subgroup containing the commutator subgroup is normal, since $g^{-1}hg = [g^{-1}, h]h \in H$, whence we can use the results from [24]. In loc. cit. the author shows that V_H either remains irreducible or either decomposes and yields a decomposition group G_V^{dec}. We treat the former case first. In this situation, the H-representation W_H cannot be irreducible (for otherwise $\dim(V) = \dim(W)$) and since $H \triangleleft G$ is maximal G_W^{dec} is either H or G. If $G_W^{dec} = H$, then for some $g_2, \ldots, g_n \in G \setminus H$

$$\begin{aligned} W &= U \oplus g_2 U \oplus \cdots \oplus g_n U \\ &= R_1 \oplus R_2 \oplus \cdots \oplus R_n \end{aligned}$$

with all R_i non-isomorphic H-representations. From [24] we know that $n = [G : G_W^{dec}] = [G : H]$. We also know by [40, Proposition 20.5] that $n^2 \leq [G : H]$. This leads to $[G : H] = 1$, a contradiction. In case $G_W^{dec} = G$, we can write

$$W = R_1 = U \oplus g_2 U \oplus \cdots \oplus g_n U,$$

where the $g_i U$ are isomorphic as H-representations and $U = V$. Hence the same result from [40] now entails the inequality

$$\left(\frac{\dim(W)}{\dim(V)} \right)^2 = n^2 \leq [G : H].$$

Since $H \leq G$ is maximal and normal, G/H has no non-trivial subgroups, whence is cyclic of prime order p. In other words, $[G : H] = p$. Write

$$\begin{gathered} |G| = \dim(W) k_w \\ |H| = \dim(V) k_v \\ \dim(W) = \dim(V) l \end{gathered}$$

Here we used the Frobenius divisibility property, see [28, Theorem 4.16]. It follows that

$$l^2 \leq [G : H] = p = l \frac{k_w}{k_v}.$$

However, $l = p$ contradicts $p^2 \leq p$ and $l = 1$ entails $\dim(V) = \dim(W)$, also a contradiction. This covers the case V_H irreducible. Suppose now that

$$\begin{aligned} V_H &= U \oplus g_2 U \oplus \cdots \oplus g_n U \\ V_{G_V^{dec}} &= R_1 \oplus \cdots \oplus R_m \end{aligned}$$

From [24] we know that the appearing irreducible H-components of W_H are the same ones appearing in V_H, possibly with different multiplicity. We also know that $\dim(R_1) = \ldots = \dim(R_m)$. It follows that

$$W_{G_W^{dec}} = R'_1 \oplus \cdots \oplus R'_m,$$

with $\dim(R'_1) = \ldots = \dim(R'_m)$. Hence $m = [G : G_V^{dec}] = [G : G_W^{dec}]$ and since both decomposition groups contain H, they must be equal; denote this group by G^{dec}. If $G^{dec} = H$, then all $g_i U$ are non-isomorphic H-representations, whence $R_1 = U$. Therefore $R'_1 = U^{\oplus s}$ for some $s \geq 1$, but since R'_1 is an irreducible G^{dec}-representation, s must equal to 1. It then follows that

$\dim(V) = \dim(W)$, a contradiction. If $G^{dec} = G$, then $m = 1$ and $U \cong g_i U$ as H-representation for all $2 \leq i \leq n$. Hence we can write

$$\begin{array}{lcl} V_H = U^{\oplus t} & \Rightarrow & t^2 \leq [G:H] = p \\ W_H = U^{\oplus s} & \Rightarrow & s^2 \leq [G:H] = p \end{array}$$

We can also write $\dim(V) = t\dim(U), \dim(W) = s\dim(U)$ and $|H| = \dim(U)k_u$, whence

$$\begin{aligned} p = [G:H] &= \frac{\dim(V)k_v}{\dim(U)k_u} = t\frac{k_v}{k_u} \\ &= \frac{\dim(W)k_w}{\dim(U)k_u} = s\frac{k_w}{k_u} \end{aligned}$$

Clearly $t < s$, whence $t = 1$ and $s = p$. However, $t = 1$ implies that $V_H = U$ is an irreducible H-rep, a contradiction. □

As was the case for the vanishing of P, we have that the vanishing of Q is preserved for subgroups.

Proposition 5.6.12. *Let G be a finite group such that $Q(G) = 0$, then $Q(H) = 0$ for all subgroups $H \leq G$.*

Proof. Suppose that $e(A,B) = \chi_{(C,D)}$ is a non-maximal idempotent in $\mathbb{Q}(\widetilde{H})$ and let $f = \chi_{(C,D')}$ be an idempotent such that $fe(A,B) = f$. The elements $\Phi_H^G(e(A,B)), \Phi_H^G(f)$ are of the form $\chi_{(C',M)}, \chi_{(C',M')}$ respectively, with $M \subset M'$. Since $Q(G) = 0$, $M = M'$, but because Φ_H^G is injective, $f = e(A,B)$, a contradiction. □

Finally, let us give some comments on the $\mathbb{Q}(G/G')$-module R.

Lemma 5.6.13. *Let G be a finite group of nilpotency class 2. If the element $\chi_{(\{e\},D)}$ is idempotent, then $D = \emptyset$. As a corollary $R = \mathbb{Q}(G/G')$.*

Proof. Let $V \supset Kv$ be associated with $\chi_{(\{e\},D)}$ and suppose that an irreducible G-representation U with $\dim(U) > 1$ appears. By decomposing V, we may find $u \in U$ that appears in the corresponding decomposition of v. There exists an $n > 0$ such that $U^{\otimes n}$ completely linearizes. Because $\chi_{(\{e\},D)}$ is idempotent and by [17, Theorem 3.13] it follows that $KG(u \otimes \cdots \otimes u) \cong T$ which contradicts $KGu = U$. Hence $D = \emptyset$. □

We can come back to Example 5.5.7 to show that R can be strictly bigger than $\mathbb{Q}(G/G')$. Indeed, A_4 has a maximal subgroup $C_3 < A_4$ which is non-normal and the element

$$\Phi_{C_3}^{A_4}(1) = \chi_{(\{1\},\{a\})},$$

where $\{a\} \in \mathrm{Gr}(\mathrm{U},1)$ corresponds to $u \in U$ with the property that $KC_3u = Ku$, is an idempotent element. By definition, this element sits in R. More general, maximal subgroups $H \leq G$ which are not normal, yield idempotent elements of the form $\Phi_H^G(1) = \chi_{(\{1\},D)}$. By construction, we know that

$$\Phi_H^G(1) = \left(T \oplus V_1^{\oplus m_1} \oplus \cdots \oplus V_n^{\oplus m_n} \supset Ka\right),$$

for some irreducible G-representations V_i and $1 \leq m_i \leq \dim(V_i)$. The following proposition shows that when H is maximal but non-normal, then at least one V_i has dimension strictly bigger than one. To prove this we use the Frobenius reciprocity law, which states that for an irreducible G-representation and T_H the trivial H-representations we have the equality

$$\langle V_H, T_H \rangle_H = \langle V, \mathrm{Ind}_{\mathrm{H}}^{\mathrm{G}}(\mathrm{T_H}) \rangle_{\mathrm{G}}.$$

To prove our claim, it suffices to show that $\mathrm{Ind}_{\mathrm{H}}^{\mathrm{G}}(\mathrm{T_H})$ does not completely linearize.

Proposition 5.6.14. *Let G be a group having a maximal subgroup $H \leq G$ which is not normal; then R is strictly bigger than $\mathbb{Q}(G/G')$.*

Proof. Non-normalilty of H implies that $G' \not\subset H$. Furthermore, by using the equalities $g^{-1}hg = [g^{-1}, h]h$ and $gh^{-1}g^{-1} = h^{-1}[h, g]$ one shows that both $G'H$ and HG' are subgroups of G and by maximality of H they are both equal to G. Hence we can write

$$G = H \sqcup g_2' H \sqcup \ldots \sqcup g_k' H$$

with $g_i' \in G' \setminus H, 2 \leq i \leq k$. As observed just before the statement of the proposition, it suffices to show that $\mathrm{Ind}_{\mathrm{H}}^{\mathrm{G}}(\mathrm{T_H})$ does not completely linearize. Hence, suppose it does. This implies that

$$\left(\mathrm{Ind}_{\mathrm{H}}^{\mathrm{G}}(\mathrm{T_H})\right)_{\mathrm{G'}} = \mathrm{T}_{\mathrm{G'}}^{\oplus[\mathrm{G:H}]}.$$

By definition, $\mathrm{Ind}_{\mathrm{H}}^{\mathrm{G}}(\mathrm{T_H}) = \mathrm{T_H} \oplus \mathrm{g}_2' \mathrm{T_H} \oplus \cdots \oplus \mathrm{g}_k' \mathrm{T_H}$ and $g' \cdot g_i' t = g_k' ht = g_k' t$ if $g' g_i' = g_k' h$. However, since $g' \cdot g_i' t = g_i' t$ we have that $g' g_i' \in g_i' H$. In particular, for $g' = g_i'$ we obtain $g_i' g_i' \in g_i' H$, which implies $g_i' \in H$, a contradiction. □

QUESTION: We ask ourselves whether the opposite also holds, that is, if R is strictly bigger than $\mathbb{Q}(G/G')$, that there must exist a maximal subgroup which is not normal.

5.7 Distinguishing some isocategorical groups

In this section we will apply the short exact sequence of Theorem 5.5.10 to obtain a complete structure of the glider representation ring $\mathbb{Q}(1 < G)$ for groups of nilpotency class at most 2. Let G be such a group. We know that $P = 0$ by Corollary 5.6.3, $Q = 0$ by Proposition 5.6.9 and $R(G) = \mathbb{Q}(G/G')$ by Lemma 5.6.13. It follows that the same holds true for all subgroups $H \leq G$. The short exact sequence of G/G'-modules from Theorem 5.5.10 takes the form

$$0 \longrightarrow \sum_{H \lhd_{max} G} \overline{\Phi_H^G}(\mathbb{Q}(\widetilde{H})/N) \longrightarrow \mathbb{Q}(\widetilde{G})/N \longrightarrow \mathbb{Q}(G/G') \longrightarrow 0$$

and is split by the map

$$f : \mathbb{Q}(G/G') \to \mathbb{Q}(\widetilde{G})/N, \ e_{\overline{g}} \mapsto \chi_{(\{\overline{g}\},\emptyset)},$$

from which we deduce the isomorphism as G/G'-modules

$$\mathbb{Q}(\widetilde{G})/N \cong \mathbb{Q}(G^{ab}) \oplus \sum_{G' \leq H \lhd_{max} G} \overline{\Phi_H^G}(\mathbb{Q}(\widetilde{H})/N). \tag{5.11}$$

In fact, we even obtain an isomorphism of rings: let $\{H_i \mid i \in I\}$ be the set of all minimal subgroups $e < H_i < G/G'$. The elements $\chi_{(\{e\},\emptyset)} - \chi_{(H_i,\emptyset)}$ are idempotent elements and look at the monomorphism

$$\alpha : \mathbb{Q}(G^{ab}) \hookrightarrow \mathbb{Q}(\widetilde{G})/N, \ e_{\overline{g}} \mapsto \chi_{(\overline{g},\emptyset)} \underbrace{\Pi_{i\in I}\chi_{(\{e\},\emptyset)} - \chi_{(H_i,\emptyset)}}_{:=\beta}.$$

One checks that

$$\alpha\Big(\mathbb{Q}(G^{ab})\Big) \sum_{G' \leq H \lhd_{max} G} \overline{\Phi_H^G}(\mathbb{Q}(\widetilde{H})/N) = 0,$$

by using that $\overline{\Phi_H^G}(\mathbb{Q}(\widetilde{H})/N) = \overline{\Phi_H^G}(1)\overline{\Phi_H^G}(\mathbb{Q}(\widetilde{H})/N)$ and

$$\overline{\Phi_H^G}(1)\beta = \chi_{(A_\iota(H),\emptyset)}\beta = 0.$$

To make the last term in the decomposition (5.11) into a direct sum, we prove the following.

Lemma 5.7.1. *Let $H, E \lhd G$ be normal subgroups. Then*

$$A_\iota(H)A_\iota(E) = A_\iota(H \cap E).$$

Proof. Write $C = A_\iota(H), D = A_\iota(E)$. Then

$$\begin{aligned} H \cap E &= \bigcap_{c\in C} \mathrm{Ker}(T_c) \cap \bigcap_{d\in D} \mathrm{Ker}(T_d) \\ &\subset \bigcap_{c\in C}\bigcap_{d\in D} \mathrm{Ker}(T_c \otimes T_d) \\ &= \bigcap_{e\in CD} \mathrm{Ker}(T_e). \end{aligned}$$

Since $C, D \subset CD$ we also have

$$\bigcap_{e \in CD} \mathrm{Ker}(T_e) \subset \bigcap_{c \in C} \mathrm{Ker}(T_c) \cap \bigcap_{d \in D} \mathrm{Ker}(T_d) = H \cap E.$$

Lemma 5.5.6 entails that $A_\iota(H \cap E) = CD = A_\iota(H)A_\iota(E)$. □

Proposition 5.7.2. *Let H, E be normal subgroups of G. We have the equality*

$$\overline{\Phi_H^G}(\mathbb{Q}(\widetilde{H})/N) \bigcap \overline{\Phi_E^G}(\mathbb{Q}(\widetilde{E})/N) = \overline{\Phi_{H \cap E}^G}(\mathbb{Q}(\widetilde{E \cap H})/N).$$

Proof. One inclusion follows from Lemma 5.5.9. For the other, let $\overline{\chi_{(A,B)}} \in \overline{\Phi_H^G}(\mathbb{Q}(\widetilde{H})/N) \cap \overline{\Phi_E^G}(\mathbb{Q}(\widetilde{E})/N)$ and we denote $C = A_\varphi(H),\ D = A_\varphi(E)$. Let $\chi_{(V,B')}$ be the unique minimal idempotent in the semigroup $\langle \chi_{(A,B)} \rangle$. It holds that $C \leq V$ and $D \leq V$, whence $A_\varphi(H \cap E) = CD \leq V$ (see Lemma 5.7.1). By the proof of Theorem 5.5.10, we know that $\overline{\chi_{(V,B')}}$ and also $\overline{\chi_{(A,B)}}$ are elements of

$$\overline{\Phi_P^G}(\mathbb{Q}(\widetilde{P})/N),$$

where $P = \bigcap_{v \in V} \mathrm{Ker}(T_v)$. Because we have that

$$P \subset \bigcap_{cd \in CD} \mathrm{Ker}(T_c \otimes T_d) = H \cap E,$$

it follows that $\overline{\chi_{(A,B)}} \in \overline{\Phi_{H \cap E}^G}(\mathbb{Q}(\widetilde{H \cap E})/N)$ by Lemma 5.5.9. □

The next step is to mod out $\sum_{G' \leq H \triangleleft_{max} G} \overline{\Phi_H^G}(\mathbb{Q}(\widetilde{H})/N)$ by $\overline{\Phi_{\bigcap H}^G}(\mathbb{Q}(\widetilde{\cap H})/N)$. In general, this will not be sufficient to arrive at a direct sum over all subgroups $G' \leq H \triangleleft_{max} G$. Nevertheless, we will obtain a partition of these subgroups: suppose that the subset of maximal normal subgroups is indexed by I; then we arrive at a partition

$$I = I_1 \sqcup I_2 \sqcup \cdots \sqcup I_r$$

with $r \geq 2$ and we have the exact sequence of G/G'-modules

$$0 \to \overline{\Phi_{\bigcap H}^G}(\mathbb{Q}(\widetilde{\cap H})/N) \to \textstyle\sum_{H \in I} \overline{\Phi_H^G}(\mathbb{Q}(\widetilde{H})/N) \to \bigoplus_{k=1}^r \frac{\sum_{H \in I_k} \overline{\Phi_H^G}(\mathbb{Q}(\widetilde{H})/N)}{\overline{\Phi_{\bigcap H}^G}(\mathbb{Q}(\widetilde{\cap H})/N)} \to 0$$

Again, we have that the above sequence is split. Indeed, it suffices to send $[\overline{\chi_{(A,B)}}]$ to $\overline{\chi_{(A,B)}}$ and again we arrive at an isomorphism of rings. To see this, let $C = \Pi_H A_\iota(H)$ and consider the idempotent elements

$$\delta = \Pi_D \chi_{(C,\emptyset)} - \chi_{(D,\emptyset)},$$

where the product runs over all subgroups $D \leq G/G'$ minimal over C and

$$\epsilon_H = \Pi_E \chi_{(A_\iota(H),\emptyset)} - \chi_{(E,\emptyset)},$$

where the product runs over all subgroups $E \leq G/G'$ minimal over $A_\iota(H)$. So for any H there exists E and D such that

$$A_\iota(H) \leq E \leq C \leq D,$$

which entails that

$$(\chi_{(C,\emptyset)} - \chi_{(D,\emptyset)})(\chi_{(A_\iota(H),\emptyset)} - \chi_{(E,\emptyset)}) = 0.$$

We have monomorphisms

$$f : \overline{\Phi^G_{\cap H}}(\mathbb{Q}(\widetilde{\cap H})/N) \hookrightarrow \mathbb{Q}(\widetilde{G})/N, \ a \mapsto a\delta$$

and

$$g_H : \overline{\Phi^G_H}(\mathbb{Q}(\widetilde{H})/N) \hookrightarrow \mathbb{Q}(\widetilde{G})/N, \ b \mapsto b\epsilon_H.$$

It follows that

$$f\left(\overline{\Phi^G_{\cap H}}(\mathbb{Q}(\widetilde{\cap H})/N)\right) g_H\left(\overline{\Phi^G_H}(\mathbb{Q}(\widetilde{H})/N)\right) = 0$$

since $f(1) = \chi_{(C,\emptyset)}\delta = \delta$, $g_H(1) = \chi_{(A_\iota(H),\emptyset)}\epsilon_H = \epsilon_H$ and $\delta\epsilon_H = 0$.
Suppose that $|I_1| > 1$; then we can do the same and arrive at the following split exact sequence

$$0 \to \frac{\overline{\Phi^G_{\bigcap_{I_1} Z}}(\mathbb{Q}(\widetilde{\bigcap_{I_1} Z})/N)}{\overline{\Phi^G_{\bigcap H}}(\mathbb{Q}(\widetilde{\cap H})/N)} \to \frac{\sum_{H \in I_1} \overline{\Phi^G_H}(\mathbb{Q}(\widetilde{H})/N)}{\overline{\Phi^G_{\bigcap H}}(\mathbb{Q}(\widetilde{\cap H})/N)} \to \bigoplus_{l=1}^{s} \frac{\sum_{H \in J_s} \overline{\Phi^G_H}(\mathbb{Q}(\widetilde{H})/N)}{\overline{\Phi^G_{\bigcap_{I_1} Z}}(\mathbb{Q}(\widetilde{\bigcap_{I_1} Z})/N)},$$

where

$$I_1 = J_1 \sqcup J_2 \sqcup \cdots \sqcup J_s$$

is a partition. Using the same argument as before, one shows that this again yields an isomorphism of rings. By continuing this procedure we arrive at a partition such that $|I_1| = 1$. Without loss of generalization, we may assume that this was already the case in the first step. By Theorem 5.5.10 we know what remains of

$$\frac{\overline{\Phi^G_H}(\mathbb{Q}(\widetilde{H})/N)}{\overline{\Phi^G_{\cap H}}(\mathbb{Q}(\widetilde{\cap H})/N)},$$

namely

$$\mathbb{Q}(H^{ab}) \oplus \frac{\sum_{H' < E \triangleleft_{max} H} \overline{\Phi^H_E}(\mathbb{Q}(\widetilde{E})/N)}{\overline{\Phi^G_{\cap H}}(\mathbb{Q}(\widetilde{\cap H})/N)}.$$

Hence we have proven the following:

Theorem 5.7.3. *Let G be a finite group of nilpotency class at most 2. Then*

$$\mathbb{Q}(\widetilde{G})/J \cong \bigoplus_H \mathbb{Q}(H^{ab}),$$

where the direct sum runs over all subgroups H of G.

Proof. The equality $J = N$ follows now since J is the smallest ideal such that R/J is semisimple. □

This result allows us to distinguish some iscocategorical groups:

1. There are two non-Abelian groups of prime cube order p^3, which have the same character table. Both groups are of nilpotency class 2 and have $Z(G) = G' = C_p$. It follows that the glider representation rings are non-isomorphic since they have a different number of subgroups. For $p = 2$, we arrive at a distinction between Q_8 and D_8.

2. As another application we recall that two groups G and H are called isocategorical if $\mathrm{Rep}_{\mathbb{C}}(G)$ and $\mathrm{Rep}_{\mathbb{C}}(H)$ are equivalent as tensor category (so without consideration of the symmetry). It was proven by Etingof and Gelaki in [27] that if G and H are isocategorical, then H is 'a twist' of G. As proven by Goyvaerts and Meir in [35], the smallest non-isomorphic, but isocategorical, groups are of order 64. These groups are both of nilpotency class 2 with $Z(G) = G'$. Because both groups have a different amount of subgroups, Theorem 5.7.3 shows that both groups have non-isomorphic glider representation rings.

5.8 Glider characters and Artin's theorem

In this section we will give the definition of the glider trace or glider character of a glider representation. As mentioned in the introduction, the study of glider representation rings actually started with this definition, see [17] and [18], but we termed it generalized trace, generalized character and generalized character ring instead.

Let $\Omega \supset M \supset \ldots \supset M_d \supset 0 \supset \ldots$ be an FKG-glider representation. For $j \leq i$ we have $KG_jM_i \subset M_{i-j}$, which we present in the following upper triangular matrix

$$A(M) = \begin{pmatrix} M_0 & M_1 & \cdots & M_{i-1} & M_i & \cdots & M_{d-1} & M_d \\ & G_1M_1 & \cdots & G_1M_{i-1} & G_1M_i & \cdots & G_1M_{d-1} & G_1M_d \\ & & \ddots & \vdots & \vdots & \ddots & \vdots & \vdots \\ & & & G_{i-1}M_{i-1} & G_{i-1}M_i & \cdots & G_{i-1}M_{d-1} & G_{i-1}M_d \\ & & & & G_iM_i & \cdots & G_iM_{d-1} & G_iM_d \\ & & & & & \ddots & \vdots & \vdots \\ & & & & & & G_{d-1}M_{d-1} & G_{d-1}M_d \\ & & & & & & & G_dM_d \end{pmatrix}$$

The i-th row of $A(M)$ consists of KG_i-modules G_iM_j. We denote the associated characters by ψ_{ij}^M or ψ_{ij}. We propose the following definition

Definition 5.8.1. Let $\Omega \supset M$ be an FKG-glider representation of essential length d and $B(M) = 0$; then the associated glider trace or glider character of $\Omega \supset M$ is the map $\psi_M : G \to K^D$, $D = \frac{(d+1)(d+2)}{2}$ which sends $g \in G_i \setminus G_{i-1}$ to

$$\psi_M(g) = \begin{pmatrix} 0 & \dots & 0 & 0 & 0 & \dots & 0 & 0 \\ & \ddots & \vdots & \vdots & \vdots & \ddots & \vdots & \vdots \\ & & 0 & 0 & 0 & \dots & 0 & 0 \\ & & & \chi_{i,i}(g) & \chi_{i,i+1}(g) & \dots & \chi_{i,d-1}(g) & \chi_{i,d}(g) \\ & & & & \chi_{i+1,i+1}(g) & \dots & \chi_{i+1,d-1}(g) & \chi_{i+1,d}(g) \\ & & & & & \ddots & \vdots & \vdots \\ & & & & & & \chi_{d-1,d-1}(g) & \chi_{d-1,d}(g) \\ & & & & & & & \chi_{d,d}(g) \end{pmatrix}$$

We have written the image $\psi_M(g)$ in matrix form, but it really lives inside K^D. Clearly $\psi_M(g) = \psi_M(g')$ if and only if $g \in c(g') = \{hg'h^{-1} \mid h \in G\}$ and $g, g' \in G_i \setminus G_{i-1}$. By our notation, we have that $\psi_M(g)_{ij} = \psi^M_{i,j}(g)$. The matrix $\psi_M(1)$ lists the dimensions of the G_jM_i and we have the following nice characterization:

Proposition 5.8.2. *Let $\Omega \supset M$ be a glider representation with respect to a finite group algebra filtration. Then M is irreducible of essential length equal to the filtration length and with $B(M) = 0$ if and only if the matrix $\psi(1)$ is symmetric with respect to the anti-diagonal and has a 1 in the upper right corner.*

Proof. For $1 \subset G$, the result is clear since we have irreducibility if and only if $M = GM_1$. We proceed by induction. Let $\Omega \supset M$ be an irreducible FKG-glider with $G_d = G$ such that $B(M) = 0$ and el(M) = d; then Lemma 3.1.1 entails that $M_1 \supset M_2 \supset \dots$ is also irreducible, but since el(M) = d, it is also an irreducible FKG_{d-1}-glider. Suppose that the result holds up to chains of length $d-1$ and let G be some group with a chain of length d. Let $\Omega \supset M$ be an irreducible glider. Then $\Omega \supset M_1 \supset \dots$ is an irreducible FKG_{d-1}-glider and the matrix

$$\chi_M(1) = \begin{pmatrix} \chi_{00}(1) & \dots & \psi_{0i}(1) & \dots & \chi_{0d}(1) \\ & \ddots & \vdots & \ddots & \vdots \\ & & \chi_{ii}(1) & \dots & \chi_{id}(1) \\ & & & \ddots & \vdots \\ & & & & \chi_{dd}(1) \end{pmatrix}$$

is symmetric with respect to the anti-diagonal everywhere above the diagonal. We still have to check whether $\psi_{i,i}(1) = \psi_{d-i,d-i}(1)$ for all $0 \leq i \leq d$. These numbers correspond to the dimension of KG_iM_i and $KG_{d-i}M_{d-i}$, but since M is irreducible we have

$$KG_iM_i = KG_iKG_{d-i}M_d = KG_{\max\{i,d-i\}}M_d = KG_{d-i}KG_iM_d = KG_{d-i}M_{d-i}.$$

□

From the above proof it also follows that the matrix $A(M)$ of an irreducible glider of essential length d and zero body is symmetric with respect to the anti-diagonal and the only entries appearing are exactly the M_i for $0 \leq i \leq d$. Moreover, for $g \in G_i \setminus G_{i-1}$ the matrix $\psi_M(g)$ contains the traces of g working on $M_0, M_1, \ldots, M_{d-i}$.

For a fixed chain $1 \subset G_1 \subset \ldots \subset G_d = G$, we can enlist all the irreducible gliders of essential length d and calculate the 'glider character table' of the chain. For p-groups G with $G_1 = Z(G)$ and G_2 Abelian we can retrieve the classical character table of G from the glider table. Indeed, by looking at $\chi_M(1)$, we find all irreducibles with dimension vector $\dim(M) = (\dim(M_d), \dim(M_{d-1}), \ldots, \dim(M_0)) = (1, 1, \ldots, 1, n)$. One should take care, for this does not imply that M is an irreducible G-representation! Indeed, for $\{1\} \subset \{1, -1\} \subset \mathbb{Z}_4^j \subset Q_8$, we could take $T_1 \oplus T_2 \supset V_3 \supset T \supset T$. By looking at the M with $\dim(M) = (1, \ldots, 1)$ we recover all the one-dimensional representations. Subsequently, look at the M with $\dim(M) = (1, 1, 1, \dim(M_{d-3}), \ldots, \dim(M_1), p^\alpha)$; then we have to look at $\psi_M(z)$ for $z \in G_2$. If $d = 3$, Theorem 4.1.15 ensures that if M is an irreducible G-representation, that the G_2-character of $M_{d-2} = KG_2M_d$ is different from the ones appearing in the irreducible gliders of dimension vector $(1, \ldots, 1)$ already found. In other words, the irreducibility of M as G-representation is being detected by the character of the one-dimensional G_2-rep M_{d-2}.

Example 5.8.3. Consider now $\{1\} \subset \{1, -1\} \subset H \subset Q_8$ and $\{e\} \subset \{e, a^2\} \subset H' \subset D_8$, with H and H' two subgroups of order 4 containing the center. By the above reasoning we can localize the irreducible gliders with M the unique irreducible two-dimensional representation. The Hasse diagram is the same for both chains and its form (4.2) says that there are two irreducible gliders of dimension $(1, 1, 1, 2)$ ending in U. If we then look at $\psi_M(h)$ for $h \in H$, we obtain the characters of the H-representations V_1 and V_2. From this information we can deduce whether H must be isomorphic to C_4 or V_4. In the latter case we have $G = D_8$.

We denote by $\mathcal{A}(\widetilde{G})$ the set of glider class functions, that is, maps $G \to K^D$ that are constant on $c(g) \cap G_i \setminus G_{i-1}$ for $g \in G_i \setminus G_{i-1}$. For any integer $n \in \mathbb{Z}$ we define the glider class function c_n by the map that sends $g \in G_i \setminus G_{i-1}$ to

$$c_n(g) = \begin{pmatrix} 0 & \cdots & 0 & \cdots & 0 \\ & \ddots & \vdots & \ddots & \vdots \\ & & n & \cdots & n \\ & & & \ddots & \vdots \\ & & & & n \end{pmatrix} \longleftarrow i\text{–th row}$$

The set $\mathcal{A}(\widetilde{G})$ carries a $\mathbb{Z}$-module structure

$$\mathbb{Z} \times \mathcal{A}(\widetilde{G}) \to \mathcal{A}(\widetilde{G}),$$

which maps (n, f) to the glider class function $nf = c_n \cdot f$ which is the component-wise multiplication of c_n and f. In fact, we can define c_λ for any $\lambda \in K$ and we write λf for the glider class function $c_\lambda \cdot f$, again obtained by component-wise multiplication of c_λ and f. The idea to generalize the notion of the character ring $R(G)$ of a group via the glider characters does not work however. The problem is that different gliders can yield the same glider traces.

Proposition 5.8.4. *Let $\Omega \supset M, \Omega' \supset M'$ be irreducible glider representations. The glider trace maps ψ_M and $\psi_{M'}$ are equal if and only if the matrices $A(M)$ and $A(M')$ only differ in the top right corner.*

Proof. This follows by the structure of irreducible glider representations over finite algebra filtrations, see Lemma 3.2.3, and by the fact that G_i-representations are determined by their characters. In fact, if $\psi_M = \psi_{M'}$ then the irreducible gliders only differ in the one-dimensional K-vector spaces M_d and M'_d. □

By the previous proposition, we cannot embed $R(\widetilde{G})$ as a subgroup of $\mathcal{A}(\widetilde{G})$. What we can do is write a glider class function as a linear combination of irreducible glider characters, that is, the glider characters associated with irreducible characters. If $f \in \mathcal{A}(\widetilde{G})$ denotes a glider class function, we will write $f \in R(\widetilde{G})$ if there exists irreducible gliders (all of essential length d and zero body) $M_1, \ldots, M_n$ and $\lambda_1, \ldots, \lambda_n \in K$ such that we have an equality

$$f = \lambda_1 \psi_{M_1} + \ldots + \lambda_n \psi_{M_n}$$

as glider class functions.

Our goal is to prove a generalization of Artin's theorem, [65, Theorem 17]. To this extent, let $H \lhd G$ be a normal subgroup and consider the situation from Section 4.3; that is, we start with an FKH-glider N and induce it to an FKG-glider $M = N^G$. To construct the induced glider we choose a set map $\sigma : G/H \to G$ such that $\pi \circ \sigma = id_{G/H}$. By choosing a sequence of transversal sets $T_0 \subset T_1 \subset T_2 \subset \ldots \subset T_d = T$ (T_i transversal for G_i/H_i) one obtains a 2-cocycle $h : G/H \times G/H \to H$ by the equality $\sigma(\overline{g})\sigma(\overline{g'}) = \sigma(\overline{gg'})h(\overline{g}, \overline{g'})$. To define the induced glider we further need that for all $g \in G$, the map $h(-, \overline{g}) : G/H \to H$ is restricting to $G_i/H_i \to H_i$ for all i. We say that the 2-cocycle is filtered. We define $N_i^G = KT \otimes N_i$. Under the condition that all G_i are normal in G we have that the H_i are normal in G, from which it follows that for any $i \leq j$

$$G_i(KT \otimes N_j) = KT \otimes H_i N_j.$$

Suppose now that N is an irreducible FKH-glider of essential length d and zero body. Because $(N^G)_d = M_d = KT \otimes N_d$, the induced glider is not irreducible. However, by choosing T_0 such that $1 \in T_0$, we can consider the subglider

$$KG(1 \otimes n) \supset KG_{d-1}(1 \otimes n) \supset \ldots \supset KG_1(1 \otimes n) \supset K(1 \otimes n),$$

if $N_d = Kn$. This is of course just the irreducible glider determined by the vector $1 \otimes n$ inside $KT \otimes N$. For any $1 \leq i \leq d$ we have that T_i is transversal for G_i/H_i, whence $KG_i(1 \otimes n) = KT_i \otimes N_{d-i}$. The irreducible subglider is thus

$$KT \otimes N \supset KT_{d-1} \otimes N_1 \supset \ldots \supset KT_1 \otimes N_{d-1} \supset K(1 \otimes n).$$

We denote this glider by $S(N^G)$. Observe that in Section 4.3 $T_0 = \{1\}$, but there we assumed that $G_0 = e$. Here we just assume G_0 to be Abelian.

Let us calculate the associated glider trace map of $S(N^G)$. Take $g \in G_i$. From the observations made after Proposition 5.8.2 we have to calculate the trace for g acting on $KT_j \otimes N_{d-j}$ for $i \leq j \leq d$. So let $i \leq j \leq d$. As a vector space, $KT_j \otimes N_{d-j}$ is the direct sum $\bigoplus_{t \in T_j} Kt \otimes H_j N_d$, so to calculate the trace for $g \in G_i$ we must have that $\sigma(\overline{gt}) = \sigma(\overline{t}) = t$. If $g = t_1 h_1$, this is equivalent to

$$t = \sigma(\overline{gt}) = \sigma(\overline{g})\sigma(\overline{t})h(\overline{g}, \overline{t})^{-1} = t_1 t h(\overline{g}, \overline{t})^{-1}$$
$$\Updownarrow$$
$$t^{-1} t_1 t = h(\overline{g}, \overline{t}) \in H_i.$$

So we must sum over $t \in T_j$ such that $t^{-1} t_1 t \in H_i$ or, equivalently, such that

$$t^{-1} g t = t^{-1} t_1 h_1 t \in H_i.$$

Moreover, in this case we have that

$$g \cdot t \otimes n = t \otimes t^{-1} t_1 t t^{-1} h_1 t n = t \otimes t^{-1} g t n.$$

We conclude that

$$\psi_{S(N^G)}(g)_{ij} = \sum_{t \in T_j,\ t^{-1} g t \in H_i} \psi_N(t^{-1} g t)_{ij}.$$

Again, since all G_i are normal in G, we have that

$$t^{-1} g t \in H_i \Leftrightarrow (th)^{-1} g (th) \in H_i \text{ for all } h \in H_j.$$

Hence

$$\psi_{S(N^G)}(g)_{ij} = \frac{1}{|H_j|} \sum_{\substack{g_0 \in G_j, \\ g_0^{-1} g g_0 \in H_i}} \psi_N(g_0^{-1} g g_0)_{ij}. \tag{5.12}$$

We have a morphism of additive groups

$$S_{\widetilde{H}}^{\widetilde{G}} : R(\widetilde{H}) \to R(\widetilde{G}),\ \chi_M \mapsto \chi_{S(M^H)}.$$

Since G_0 is Abelian, so is H_0 and if N is an irreducible FKG-glider of essential length d, then N_d is one-dimensional, and so $KHN_d \supset KH_{d-1}N_d \supset \ldots \supset KH_1 N_d \supset N_d$ defines an element in $R(\widetilde{H})$. Like in Section 4.3, we denote this uniquely determined element in $R(\widetilde{H})$ by χ_{N_H}.

Let M be an irreducible FKH-glider and N an irreducible FKG-glider, both of essential length d and with zero body. Denote $M_d = Km$ and $N_d = Kn$. The product $\chi_{S(M^G)}\chi_N$ in $R(\widetilde{G})$ is the generator of $R(\widetilde{G})$ associated with the irreducible glider determined by the vector $1 \otimes m \otimes n$ inside $KT \otimes M \otimes N$. In $R(\widetilde{H})$ we have that $\chi_M \chi_{N_H}$ corresponds to

$$KH_d(m \otimes n) \supset KH_{d-1}(m \otimes n) \supset \ldots \supset K(m \otimes n).$$

Denote this glider by α. Under the map $S_{\widetilde{H}}^{\widetilde{G}}$, α is sent to the glider

$$KT \otimes KH_d(m \otimes n) \supset KT_{d-1} \otimes KH_{d-1}(m \otimes n) \supset \ldots \supset K(1 \otimes m \otimes n).$$

We conclude that $\chi_{S(M^G)}\chi_N = S_{\widetilde{H}}^{\widetilde{G}}(\alpha)$, which shows that the image $S_{\widetilde{H}}^{\widetilde{G}} R(\widetilde{H}) \lhd R(\widetilde{G})$ is an ideal.

Inspired by the classical case, we introduce the following functions: let $\mathcal{C}$ denote the set of all cyclic subgroups of G and let $H \in \mathcal{C}$. Then we obtain a chain of cyclic groups $H_i = H \cap G_i$ and we define for $h \in H_i \setminus H_{i-1}$

$$\phi_{\widetilde{H}}(h)_{jl} = \begin{cases} \frac{|H_l|}{|G_l| s_{H,i}} & \text{if } i \leq j \leq l \text{ and } \langle h \rangle = H_i, \\ 0 & \text{otherwise.} \end{cases}$$

Here $s_{H,i}$ denotes the number of cyclic subgroups H' of G such that $H' \cap G_i = H_i$. This defines a function $\phi_{\widetilde{H}} : H \to K^D$ for the induced chain on KH and where we used the same indexing as in Definition 5.8.1. Via formula (5.12) we define the induced function $\mathrm{Ind}_{\widetilde{H}}^{\widetilde{G}} \phi_{\widetilde{H}}$. We calculate for $g \in G_i \setminus G_{i-1}$ and $i \leq j \leq l$

$$\begin{aligned} \sum_{H \in C} \mathrm{Ind}_{\widetilde{H}}^{\widetilde{G}} \phi_{\widetilde{H}}(g)_{jl} &= \sum_{H \in C} \frac{1}{|H_l|} \sum_{\substack{g_0 \in G_l, \\ g_0^{-1} g g_0 \in H_j}} \phi_{\widetilde{H}}(g_0^{-1} g g_0)_{jl} \\ &= \sum_{H \in C} \frac{1}{|H_l|} \sum_{\substack{g_0 \in G_l, \\ \langle g_0^{-1} g g_0 \rangle = H_i}} \frac{|H_l|}{|G_l| s_{H,i}}. \end{aligned}$$

Since G_i is normal in G_l, for every $g_0 \in G_l$, the element $g_0^{-1} g g_0$ generates a unique cyclic subgroup $C(g_0)$ in G_i and there are $s_{C(g_0),i}$ cyclic subgroups H of G such that $H_i = C(g_0)$. Hence we have proven

Lemma 5.8.5. *The following equality holds*

$$\sum_{H \in C} \mathrm{Ind}_{\widetilde{H}}^{\widetilde{G}} \phi_{\widetilde{H}} = c_1.$$

The equation of the previous lemma is an equality of functions. The classical result [65, Proposition 28] which states that the function

$$\phi_H : H \to K,\ h \mapsto \begin{cases} |H| & \text{if } \langle h \rangle = H, \\ 0 & \text{otherwise} \end{cases}$$

belongs to $R(H)$ for a cyclic group H remains valid in our situation.

Lemma 5.8.6. *$\phi_{\widetilde{H}} \in R(\widetilde{H})$ for all $H \in \mathcal{C}$.*

Proof. For $H = e$, we have $\phi_{\widetilde{e}} = c_{|\mathcal{C}|} = c_{|\mathcal{C}|}\psi_{T\supset T\supset\ldots\supset T\supset 0\supset\ldots}$ where T denotes the trivial e-representation, so $\phi_{\widetilde{e}} \in R(\widetilde{e})$. Next, use induction on $|H|$, Lemma 5.8.5 and the fact that $c_1 = \psi_T \in R(\widetilde{H})$ where

$$T \supset T \supset \ldots \supset T \supset 0 \supset \ldots$$

is the glider with T the trivial H-representation. □

We finally arrive at the generalization of Artin's theorem, [65, Theorem 17].

Theorem 5.8.7. *If $\Omega \supset M$ is an FKG-glider representation then χ_M is a 'rational' linear combination of characters induced from FKH-glider representations where H runs over the cyclic subgroups of G.*

Proof. Let $H \in \mathcal{C}$. If $\phi_{\widetilde{H}} = \sum_{i=1}^{n} \lambda_i \psi_{M_i}$ for some irreducible FKH-gliders M_i, then we have

$$\mathrm{Ind}_{\widetilde{H}}^{\widetilde{G}} \phi_{\widetilde{H}} = \sum_{i=1}^{n} \lambda_i \psi_{S(M_i^H)}.$$

Together with Lemma 5.8.5 this shows that that $c_1 \in \sum_{H\in\mathcal{C}} S_{\widetilde{H}}^{\widetilde{G}} R(\widetilde{H})$. Since the latter is an ideal, it also contains $c_1\chi_M$, whence the result. □

Classically, the ring of class functions of G carries an inproduct $\langle -, - \rangle$ defined by

$$\langle f, g \rangle = \frac{1}{|G|} \sum_{x\in G} f(x) g(x^{-1})^*.$$

It is well-known that the characters χ_i associated with the irreducible representations V_i - we call them the irreducible characters - form an orthonormal basis for $R(G)$. Again due to Proposition 5.8.4 this is perhaps too much to hope for in the more general glider setting. Nonetheless, we can define an inproduct on the glider representation ring as follows. Fix a chain of groups $1 < G_1 < \ldots < G_d = G$. For $f \in \mathcal{A}(\widetilde{G})$ we denote $f = (f_{ij})_{ij}$ by

$$f(x) = \begin{pmatrix} f_{00}(x) & \cdots & f_{0i}(x) & \cdots & f_{0d}(x) \\ & \ddots & \vdots & \ddots & \vdots \\ & & f_{ii}(x) & \cdots & f_{id}(x) \\ & & & \ddots & \vdots \\ & & & & f_{dd}(x) \end{pmatrix}$$

Definition 5.8.8. We define a K-bilinear map

$$\langle -, - \rangle : \mathcal{A}(\widetilde{G}) \times \mathcal{A}(\widetilde{G}) \to K^D,$$

given by

$$\langle f, g \rangle = \begin{pmatrix} \langle f_{00}, g_{00} \rangle^{G_0} & \cdots & \langle f_{0i}, g_{0i} \rangle^{G_0} & \cdots & \langle f_{0d}, g_{0d} \rangle^{G_0} \\ & \ddots & \vdots & \ddots & \vdots \\ & & \langle f_{ii}, g_{ii} \rangle^{G_i} & \cdots & \langle f_{id}, g_{id} \rangle^{G_i} \\ & & & \ddots & \vdots \\ & & & & \langle f_{dd}, g_{dd} \rangle^{G_d} \end{pmatrix}$$

where the upper index G_i indicates for which group the ordinary inproduct is calculated.

Let us consider the easiest chain $1 < G$, with G Abelian. By Theorem 4.1.12 we observe that non-isomorphic irreducible gliders do yield different irreducible glider characters. Hence in this situation we can embed $R(\widetilde{G})$ in $\mathcal{A}(\widetilde{G})$ and write $\chi_A = \psi_A$ for $A \subset G$, that is, for the irreducible glider representations. We have the following generalization of the orthonormality relations.

Proposition 5.8.9. *Let G be a finite Abelian group. An FKG-glider representation $\Omega \supset M \supset M_1 \supset 0 \supset \ldots$ of essential length 1 and zero body is irreducible if and only if*

$$\langle \psi_M, \psi_M \rangle = \begin{pmatrix} n^2 & 1 \\ 0 & n \end{pmatrix}$$

for some $n \leq |G|$.

Proof. Suppose that $M \supset M_1$ is irreducible. By Theorem 4.1.12 we know that $M_1 = Ka$ is one-dimensional and that the dimension of M equals the number of irreducible components n of $M = KGM_1$, which are moreover all non-isomorphic. We compute

$$\langle \psi_M, \psi_M \rangle = \begin{pmatrix} \dim_K(M)^2 & \dim_K(M_1)^2 \\ 0 & \langle \psi_{GM_1}, \psi_{GM_1} \rangle \end{pmatrix} = \begin{pmatrix} n^2 & 1 \\ 0 & n \end{pmatrix}$$

For the other direction, the 1 in the upper right corner entails that $\dim_K(M_1) = 1$. Decompose $GM_1 = \bigoplus_i V_i^{n_i}$, where all the V_i are one-dimensional (G Abelian!). In fact, since $GM_1 \supset M_1$ is irreducible, all the n_i are 0 or 1. It follows that

$$n = \dim_K(M) \geq \dim_K(GM_1) = \sum_i n_i = \sum_i n_i^2 = n,$$

whence $M = GM_1$ and $M \supset M_1$ is indeed irreducible. □

For $g \in G$, we denote

$$\mathcal{P}^g = \{A \in \mathcal{P} \mid g \in A\}.$$

Recall that we associate K with c_K, i.e. $\lambda \in K$ corresponds to the glider class function c_λ.

Theorem 5.8.10. *The irreducible glider characters* $\{\chi_A \mid A \in \mathcal{P}\}$ *are* K*-linearly independent.*

Proof. Suppose that

$$F = \sum_{A \in \mathcal{P}} a_A \chi_A = 0, \quad \text{for } a_A \in K.$$

Then $F(1)_{12} = \sum_{A \in \mathcal{P} \setminus \{\emptyset\}} a_A = 0$. For $g \in G$ we have that $\langle F, \chi_{\{g\}} \rangle = 0$, so in particular

$$\langle F_{22}, (\chi_{\{g\}})_{22} \rangle^G = \langle F_{22}, \chi_{T_g} \rangle^G = \sum_{A \in \mathcal{P}^g} a_A = 0.$$

Fix an ordering $\{e = g_1, g_2, g_3, \ldots, g_n\}$ of G and choose an ordering $\{A_1, A_2, \ldots, A_k\}$ of $\mathcal{P}$ starting with $\{\{g_1\}, \{g_2\}, \ldots, \{g_n\}, \{g_{n-1}, g_n\}, \ldots\}$. It follows that

$$\underbrace{\left(\begin{array}{cccc} & C & & \\ \hline 1 & 1 & \cdots & 1 \end{array}\right)}_{\widetilde{C}} \begin{pmatrix} a_{A_1} \\ a_{A_2} \\ \vdots \\ a_{A_k} \end{pmatrix} = \begin{pmatrix} 0 \\ 0 \\ \vdots \\ 0 \end{pmatrix},$$

where $C = (c_{ij})_{ij}$ is an $n \times (2^{|G|} - 1)$-matrix defined by

$$c_{ij} = \begin{cases} 1 & \text{if } g_i \in A_j, \\ 0 & \text{otherwise.} \end{cases}$$

Consider the left $(n+1) \times (n+1)$-submatrix of $\widetilde{C}$:

$$A = \left(\begin{array}{cccccc|c} 1 & 0 & \cdots & 0 & 0 & & 0 \\ 0 & 1 & \cdots & 0 & 0 & & 0 \\ \vdots & \vdots & \ddots & \vdots & \vdots & & \vdots \\ 0 & 0 & \cdots & 1 & 0 & & 1 \\ 0 & 0 & \cdots & 0 & 1 & & 1 \\ \hline 1 & 1 & \cdots & 1 & 1 & & 1 \end{array}\right)$$

Then

$$\det(A) = \begin{vmatrix} 1 & 0 & 1 \\ 0 & 1 & 1 \\ 1 & 1 & 1 \end{vmatrix} = -1.$$

Hence $\mathrm{rk}(\widetilde{\mathrm{C}}) = \mathrm{n} + 1$ and it follows that all $a_A = 0$ for $A \neq \emptyset$. □

To end this section, we consider a slightly more elaborate chain, namely a chain $1 < H < G$ of length 2 with G Abelian. For these chains, the irreducible gliders of essential length 2 and zero body look like

$$\bigoplus_{g \in A} V_g \supset KHv \supset Kv$$

for some $A \subset \mathcal{P}(G)$. In general, if a glider $M \supset M_1 \supset \ldots \supset M_i \supset \ldots$ is irreducible, then so is the glider $M_m \supset M_{m+1} \supset \ldots$ for any $m \geq 0$, see Lemma 3.1.1. In casu, for the glider $\mathbb{C}H \supset \mathbb{C}v$, the group G does not play any role, hence it is also irreducible as FKH-glider and $KHv = \oplus_{h \in B} Z_h$ for some subset $B \subset H$. By general glider theory, it is clear that A determines B. Concretely, choose an isomorphism $\varphi : G \to \widehat{G}$ between G and its character group, and do the same for H, i.e. $\psi : H \xrightarrow{\cong} \widehat{H}$. The composition morphism

$$G \xrightarrow{\phi} \widehat{G} \xrightarrow{res} \widehat{H} \xrightarrow{\psi^{-1}} H$$

is surjective and we denote it by $\pi : G \twoheadrightarrow H$. Let us consider the example $1 \subset C_n \subset C_{nm}$, where $C_{nm} = \langle a \rangle$ and $C_n = \langle a^m \rangle$. For a cyclic group $C_N = \langle b \rangle$ we have an isomorphism

$$\phi : C_N \to \widehat{C}_N, \; b^i \mapsto [b \mapsto \omega^i],$$

where ω is an N-th primitive root of unity. By composing with the isomorphism $\widehat{C_N} \to \mathbb{Z}_N, \; \phi(b^i) \mapsto \bar{i}$ we obtain that

$$\pi : C_{nm} \twoheadrightarrow C_n, \; a^i \mapsto a^{\bar{i}},$$

corresponds to the canonical projection map

$$\pi : \mathbb{Z}_{nm} \twoheadrightarrow \mathbb{Z}_n.$$

The subsets of C_N correspond to subsets of $\{0, 1, \ldots, N-1\}$ so we will work over subsets over the integers, rather than over subsets of the cyclic group. For $A \in \mathcal{P}(\{0, 1, \ldots, nm-1\})$ we denote $\pi(A)$ by $\overline{A} = \{\pi(\bar{i}) \mid i \in A\}$. Hence

$$KC_n v = \oplus_{i \in \overline{A}} Z_i,$$

where the $Z_i, 0 \leq i < n$ denote the irreducible C_n-representations. We introduce the following set

$$C_{n,nm} = \{(|A|, |\pi(A)|) \in \mathbb{N}^2 \mid A \in \mathcal{P}(\{0, 1, \ldots, nm-1\})\}.$$

Proposition 5.8.9 extends to the following:

Proposition 5.8.11. *An FKC_{nm}-glider representation $\Omega \supset M \supset M_1 \supset M_2 \supset 0 \supset \ldots$ of essential length 2 and zero body is irreducible if and only if*

$$\langle \psi_M, \psi_M \rangle = \begin{pmatrix} k^2 & l^2 & 1 \\ 0 & l & l \\ 0 & 0 & k \end{pmatrix}$$

for some $(k, l) \in C_{n,nm}$.

As another example, consider the chain $1 \subset C_n \subset C_n \times C_m$. In this case we have a 1-to-1 correspondence between the irreducible gliders and $\mathcal{P}(\{0, 1 \ldots, n-1\}) \times \mathcal{P}(\{0, 1, \ldots, m-1\})$. By picking the same isomorphisms as before, the map $\pi : C_n \times C_m \to C_n$ is also just the canonical projection map. Analogously one then proves the following

Proposition 5.8.12. *An FKC_nC_m-glider representation $\Omega \supset M \supset M_1 \supset M_2 \supset 0 \supset \ldots$ of essential length 2 and zero body is irreducible if and only if*

$$\langle \psi_M, \psi_M \rangle = \begin{pmatrix} (kl)^2 & l^2 & 1 \\ 0 & l & l \\ 0 & 0 & kl \end{pmatrix}$$

for some $k \leq m, l \leq n$.

Remark 5.8.13. For $(n, m) = 1$, the group C_{nm} sits in both cases. Let $(k, l) \in C_{n,nm}$, then $l = \pi_{C_n}(k)$. Let $l' = \pi_{C_m}(k)$. Because n and m are coprime it follows that $k = ll'$ and we see that the results from both propositions are the same.

5.9 Another multiplication for chains of length 2

In determining in Section 5.1 a well-defined multiplication on the additive group $R(\widetilde{G})$ generated by all irreducible FKG-gliders $R(\widetilde{G})$ we deduced that G_0 has to be Abelian, but also, that we have to work modulo all irreducible gliders of essential length strictly smaller than the length of the chain of subgroups of G. However, in the special case of chain $1 < H < G$ of length 2, where H is Abelian, we can define another multiplication on $R(\widetilde{G})$. And this time, we do not have to work modulo irreducible gliders of essential length < 2. For explicit computations, we also assume here that G is Abelian, but this is not necessary.

To avoid confusion we will denote the additive group generated by all irreducible $(1 < H < G)$-gliders by $R(1 < H < G)$. From the discussions at the end of Section 5.8 we know that the irreducible gliders of essential length 2 are determined by $\{(A, \pi(A)) \in \mathcal{P}(G) \times \mathcal{P}(H)\}$. Recall that $\pi : G \twoheadrightarrow H$ is the surjective group morphism obtained by choosing group isomorphisms between G, H and their character groups. Denote by $K = \mathrm{Ker}(\pi)$.

Example 5.9.1. Let $V_4 = \langle a, b \mid a^2 = 1 = b^2 \rangle$ and denote $c = ab$. Let $H = \{1, c\}$. If we denote by T_g ($g = a, b, c$) the irreducible representation which sends g and 1 to 1 and the other two elements to -1, then $\pi : V_4 \twoheadrightarrow \{1, c\}$ is given by

$$1 \mapsto 1, \; a \mapsto c, \; b \mapsto c, \; c \mapsto 1,$$

whence $K = \{1, c\}$. If we would have chosen the isomorphism $V_4 \cong \widehat{V_4}, a \mapsto T_c, b \mapsto T_a, c \mapsto T_b$, then K would be $\{1, a\}$.

For $A \in \mathcal{P}(G)$ we denote the associated element in $R(1 < H < G)$ by $\chi_{(A,\pi(A))}$.

The irreducible gliders of essential length 1 are of the form $\Omega \supset M \supset Km \supset 0 \supset \ldots$, with $M = KHm$. It follows that $M \supset M \supset Km$ is a $(1 < H)$-glider. Of course, given Ω, we can associate with it the irreducible FKG-glider

$$\Omega \supset KGm \supset M \supset Km \supset 0 \supset \ldots$$

and we see that different $\Omega's$ yield different gliders. In any case, we see that the irreducible gliders of essential length 1 correspond to the irreducible gliders for $1 < H$, which we know are labeled by the subsets of H. For $B \in \mathcal{P}(H)$ we write $\chi_{(\emptyset,B)}$ for the element in $R(1 < H < G)$. Finally, the irreducible glider $K \supset 0 \supset \ldots$ of essential length 0 is denoted by $\chi_{(\emptyset,\emptyset)}$.

Let us determine the multiplication rules in $R(1 < H < G)$. First, let $M \supset M_1 \supset Km \supset 0 \supset \ldots$ and $N \supset N_1 \supset Kn \supset 0 \supset \ldots$ be two irreducible gliders of essential length 2 corresponding to A and B in $\mathcal{P}(G)$. In the fragment decomposition of $M \otimes N$ there appears only one irreducible glider of essential length 2, namely

$$\bigoplus_{c \in AB} T_c \supset \bigoplus_{d \in \pi(AB)} S_d \supset Km \otimes n,$$

where T_a denotes an irreducible representation of G, and S_b one of H. As H-representations, we have

$$M_1 \otimes N_1 \cong (\bigoplus_{h \in \pi(A)} S_h) \otimes (\bigoplus_{h' \in \pi(B)} S_{h'}) \cong \bigoplus_{c \in \pi(AB)} S_c^{n_c},$$

for some $n_c \geq 1$. In fact, whenever $n_c > 1$, we have $n_c - 1$ irreducible gliders of essential length 1 splitting off. Indeed, let $c \in \pi(AB)$ be such that $n_c > 1$; then take any additional copy of S_c appearing in the decomposition as H-representations of $M_1 \otimes N_1$. Since $HS_c = H$, the irreducible $(1 < H)$-glider or $(1 < H < G)$-glider $S_c \supset S_c \supset 0 \supset \ldots$ of essential length 1 splits off. We observe that these gliders are isomorphic to $\chi_{(\emptyset,\{c\})}$. Finally, in the decomposition of $M \otimes N$ there can appear irreducible gliders of essential length 0, from which it follows that

$$\chi_{(A,\pi(A))}\chi_{(B,\pi(B))} = \chi_{(AB,\pi(AB))} + \sum_{c \in \pi(AB)} (n_c - 1)\chi_{(\emptyset,\{c\})} + d\chi_{(\emptyset,\emptyset)},$$

for some $d \geq 0$. We observe that the uniqueness of the fragment decomposition in this case is ensured by the length of the chain being 2.

Next, we consider the product of $\chi_{(A,\pi(A))}$ and $\chi_{(\emptyset,B)}$ for $A \subset G$, $B \subset H$. Their tensor product becomes

$$M \otimes N \supset M_1 \otimes Kn \supset 0 \supset \dots$$

It follows that

$$\chi_{(A,\pi(A))}\chi_{(\emptyset,B)} = \sum_{h\in\pi(A)} \chi_{(\emptyset,Bh)} + d\chi_{(\emptyset,\emptyset)},$$

for some $d \geq 0$. To ensure uniqueness of the decomposition as a fragment direct sum, we use here that H is Abelian. Indeed, by Theorem 4.1.12

$$M_1 = \oplus_{h\in\pi(A)} S_h,$$

and $KH(S_h \otimes Kn) = S_h \otimes N$, and this does not depend on the choice of decomposition of M_1.

Finally, it is straightforward that

$$\chi_{(\emptyset,B)}\chi_{(\emptyset,C)} = \chi_{(\emptyset,BC)} + d\chi_{(\emptyset,\emptyset)},$$

for $B, C \in \mathcal{P}(H)$ and some $d \geq 0$.
Combining everything we see that $R(1 < H < G)$ is the $\mathbb{Z}$-algebra generated by the elements of

$$S = \{\chi_{(A,\pi(A))} \mid (A,\pi(A)) \in \mathcal{P}(G) \times \mathcal{P}(H)\} \cup \{\chi_{(\emptyset,B)} \mid (\emptyset, B), B \in \mathcal{P}(H)\} \cup \{\chi_{(\emptyset,\emptyset)}\}.$$

The element $\theta = \chi_{(\emptyset,\emptyset)}$ is a 0-element; hence we can work modulo the two-sided ideal $\mathbb{Z}\chi_{(\emptyset,\emptyset)}$ and we obtain the following relations

$$\begin{aligned}
\chi_{(A,\pi(A))}\chi_{(B,\pi(B))} &= \chi_{(AB,\pi(AB))} + \sum_{c\in\pi(AB)} (n_c - 1)\chi_{(\emptyset,\{c\})}, \quad (5.13)\\
\chi_{A,\pi(A))}\chi_{(\emptyset,B)} &= \sum_{h\in\pi(A)} \chi_{(\emptyset,Bh)},\\
\chi_{(\emptyset,B)}\chi_{(\emptyset,C)} &= \chi_{(\emptyset,BC)}.
\end{aligned}$$

We wonder whether we can find nilpotent and idempotent elements in

$$\mathbb{Q}(1 < H < G) := \mathbb{Q} \otimes_{\mathbb{Z}} R(1 < H < G)$$

(in fact we work modulo the two-sided ideal $\mathbb{Q}\chi_{(\emptyset,\emptyset)}$). In Section 5.2 we calculated $\mathbb{Q}(1 < H) = \mathbb{Q} \otimes_{\mathbb{Z}} R(1 < H)$, with the multiplication (5.1) on $R(1 < H)$. One checks, however, that for chains of the form $G_0 < G_1$ with G_0 Abelian, one can introduce a similar product as we just did above. In fact, when working modulo the 0-element $\theta = \chi_{(K\supset 0\supset\dots)}$ both products are the same. From the multiplication rules above, we see that the element $\chi_{(\emptyset,\{1\})}$ is a central idempotent and one verifies that we obtain a ring isomorphism

$$\mathbb{Q}(1 < H < G) \xrightarrow{\cong} \mathbb{Q}(1 < H < G)(1 - \chi_{(\emptyset,\{1\})}) \times \mathbb{Q}(1 < H), \quad (5.14)$$

which is defined by

$$\chi_{(A,\pi(A))} \mapsto (\chi_{(A,\pi(A))} - \sum_{h\in\pi(A)} \chi_{(\emptyset,\{h\})}, \sum_{h\in\pi(A)} \chi_{(\emptyset,\{h\})})$$

and

$$\chi_{(\emptyset,B)} \mapsto (0, \chi_{(\emptyset,B)}).$$

We already know the structure of the Jacobson radical of $\mathbb{Q}(1 < H)$ by Theorem 5.2.8 and the observation just stated above, so we focus on $T := \mathbb{Q}(1 < H < G)(1 - \chi_{(\emptyset,\{1\})})$. Denote by $f : \mathbb{Q}(H < G) \to T$ the ring morphism obtained by composing the above isomorphism with the projection on T. Because

$$f(\chi_{(A,\pi(A))})f(\chi_{(B,\pi(B))}) = f(\chi_{(AB,\pi(A)\pi(B))}),$$

the calculations from Section 5.2 that lead to the nilpotent elements $\chi_A - \chi_{gn(A)}$ if $A \subset gn(A)$ go through here. Explicitly, for $A \subset G$ there exist $g \in A$ and a subgroup $N \leq G$ such that $A \subset gN$ and then $f(\chi_{(A,\pi(A))} - \chi_{(gN,\pi(g)\pi(N))})$ is nilpotent in T. If we denote by I the ideal of T generated by these elements we arrive at the following:

Proposition 5.9.2. *We have an isomorphism of rings*

$$\bigoplus_{N\leq G} \mathbb{Q}(G/N) \to \frac{\mathbb{Q}(1 < H < G)(1 - \chi_{(\emptyset,\{1\})})}{I},$$

where the unit element e_N of $\mathbb{Q}(G/N)$ is sent to $\Pi_{\overline{N'}\in\mathcal{M}(G/N)}(f(\chi_{(N,\pi(N))}) - f(\chi_{(N',\pi(N'))}))$.

Proof. By the discussion above, the proof is analogous to the proof of Theorem 5.2.8. □

As mentioned in the beginning of this section, it is not necessary that G is Abelian. We wonder whether for more general chains of higher length there can also be defined other multiplications.

Chapter 6

Glider representations of a chain of semisimple Lie algebras

In the third chapter of Part II we enter the world of Lie algebras. Concretely, we consider chains of semisimple Lie algebras $\mathfrak{g}_1 \subset \mathfrak{g}_2 \subset \ldots \subset \mathfrak{g}_n$, which yield finite algebra filtrations on the universal enveloping algebra $U(\mathfrak{g}_n)$ by putting $F_iU(\mathfrak{g}_n) = U(\mathfrak{g}_{i+1})$ for $i = 0, \ldots, n-2$, $F_mU(\mathfrak{g}_n) = U(\mathfrak{g}_n)$ for $m \geq n-1$. When dealing with Lie algebras one uses the beautiful geometry of the root systems, which appear by considering some Cartan subalgebra $\mathfrak{h}$. For a chain of Lie algebras one can fix a chain of such Cartan subalgebras $\mathfrak{h}_1 \subset \mathfrak{h}_2 \subset \ldots \subset \mathfrak{h}_n$. Elements of the root system live in the dual space $\mathfrak{h}^*$ and the above chain of Cartan subalgebras yields a sequence of projections

$$\mathfrak{h}_n^* \twoheadrightarrow \mathfrak{h}_{n-1}^* \twoheadrightarrow \ldots \twoheadrightarrow \mathfrak{h}_1^*.$$

The traditional starting point for representation theory of semisimple Lie algebra is to study all finite dimensional $U(\mathfrak{g})$-modules. More generally, Bernstein, Gelfand and Gelfand introduced in [11] the BGG category $\mathcal{O}$, which encompasses the finite dimensional representations. A famous and interesting class of infinite dimensional $U(\mathfrak{g})$-modules is given by the so called Verma modules, introduced and studied in depth by Verma in his Ph.D. thesis, [73]. It is our aim to generalize the notion of a Verma module to a Verma glider. After fixing some Cartan subalgebra $\mathfrak{h}$, Verma modules $M(\lambda)$ are indexed by functionals $\lambda \in \mathfrak{h}^*$, so when fixing a chain of Cartan subalgebras $\mathfrak{h}_1 \subset \ldots \subset \mathfrak{h}_n$ as before, we would like to make a connection between the different labeling sets $\mathfrak{h}_i^*$. To do this, we derive a condition on the embeddings $\mathfrak{g}_i \subset \mathfrak{g}_{i+1}$ appearing in the chain of semisimple Lie algebras, such that the inclusion $\iota : U(\mathfrak{g}_i) \hookrightarrow U(\mathfrak{g}_{i+1})$ of the universal enveloping algebras behaves nicely. Explicitly, we mean the following: let $\pi : \mathfrak{h}_{i+1}^* \twoheadrightarrow \mathfrak{h}_i^*$ denote the canonical projection. We want that

$$\forall \alpha \in \Phi_i : \ \#\{\beta \in \Phi_{i+1} \mid \pi(\beta) = \alpha\} = 1. \tag{6.1}$$

We will argue why this behavior makes it possible to relate Verma modules for the different Lie algebras $\mathfrak{g}_i$. We devote a first section to determining the right condition and we provide both examples and counterexamples.

In the next section we introduce Verma gliders for chains of semisimple Lie algebras satisfying condition (6.1). They will be special glider representations

$\Omega \supset M \supset \ldots \supset M_i \supset \ldots$ with regard to the positive algebra filtration on the universal enveloping algebra $U(\mathfrak{g}_n)$ given by the chain. Starting from functionals $\lambda_i \in \mathfrak{h}_i^*$ for $i = 1, \ldots, n$ we explain how to construct such Verma gliders. The idea is to embed a $\mathfrak{g}_i$-Verma module inside a $\mathfrak{g}_{i+1}$-Verma module, the inclusion being an embedding of $U(\mathfrak{g}_i)$-modules. This leads to the quest for elements $z \in U(\mathfrak{n}_{i+1}^-)$ such that $U(\mathfrak{n}_i) \cdot zv^+ = 0$, where v^+ is the highest weight vector of a $\mathfrak{g}_{i+1}$-Verma module. If the element z is an eigenvector in $\mathfrak{g}_{i+1}$, we call such an element an embedding element for $\mathfrak{g}_i$ in $\mathfrak{g}_{i+1}$. Imposing a mild additional condition on the inclusions of Lie algebras, we derive that the embedding elements all lie in the centralizer $C_{\mathfrak{n}_{i+1}^-}(\mathfrak{n}_i) = \{z \in \mathfrak{n}_{i+1}^- \mid [\mathfrak{n}_i, z] = 0\}$ of $\mathfrak{n}_i$ inside $\mathfrak{n}_{i+1}$. Of course, there is no harm in choosing $z \in U(\mathfrak{n}_{i+1}^-)$ and it follows that any element in the centralizer $C_{U(\mathfrak{n}_{i+1}^-)}(\mathfrak{n}_i)$ satisfies $U(\mathfrak{n}_i) \cdot zv^+ = 0$. We prove that a PBW-monomial $z = y_{\alpha_1}^{r_1} \ldots y_{\alpha_m}^{r_m}$ lies in the centralizer $C_{U(\mathfrak{n}_{i+1}^-)}(\mathfrak{n}_i)$ if and only if all appearing y_{α_i} are embedding elements (the notation is explained in Section 6.1). However, it appears not to be true, that every element in the centralizer $C_{U(\mathfrak{n}_{i+1}^-)}(\mathfrak{n}_i)$ is generated by the embedding elements, see Example 6.2.11.

In Section 6.3 we study the irreducibility of Verma gliders, in particular for chains of simple Lie algebras of the same type A, B or D. It is known that ordinary Verma modules $M(\lambda)$ are irreducible exactly when $\lambda \in \Lambda^+$ is dominant integral. We are able to extend this result for Verma gliders, see Theorem 6.3.9. Finally, in Section 6.4 we answer a question that arose by looking at the glider theory for chains of Lie algebras. In constructing Verma gliders, we introduced the so called embedding elements for an inclusion $\mathfrak{g}_1 \subset \mathfrak{g}_2$ of Lie algebras. These particular eigenvectors are nilpotent elements of $\mathfrak{g}_2$, hence lie in some nilpotent orbit. We ask ourselves which nilpotent orbits we reach by only considering the embedding elements and all their linear combinations. We give an answer to this question for Lie algebras $\mathfrak{g}_1 \subset \mathfrak{g}_2$ of resp. rank $n < m$ and of the same type A, B, C or D. In fact, this question was inspired by email correspondence with prof. J. Alev.

The content of this chapter is based on [14]. For generalities on Lie algebra theory, the reader may consult [31], [37]. For a nice overview on the BGG category $\mathcal{O}$ and properties of Verma modules, we refer to [38]. For the readers' convenience, we have tried to introduce all the necessary Lie algebra related tools and concepts. We chose to use the notations from [37], except when we deal with the classical Lie algebras of type A, B, C, D. For these, we took the notations from [31], since this reference neatly discusses the root systems and representation theory of the simple Lie algebras.

6.1 Inclusions of semisimple Lie algebras

Throughout we work with complex finite dimensional semisimple Lie algebras $\mathfrak{g}$ with Lie bracket $[-,-]$. A Lie subalgebra $\mathfrak{h} \subset \mathfrak{g}$ is a subvector space such that $[\mathfrak{h},\mathfrak{h}] \subset \mathfrak{h}$. If such a subalgebra $\mathfrak{h}$ consists entirely of semisimple elements, that is, elements for which $\mathrm{ad_x} \in \mathrm{End}(\mathfrak{g})$ is diagonalizable , we call $\mathfrak{h}$ toral. A toral subalgebra is Abelian and if $\mathfrak{h}$ is a maximal toral subalgebra, then the centralizer $C_{\mathfrak{g}}(\mathfrak{h})$ equals $\mathfrak{h}$. A maximal toral subalgebra is also referred to as a Cartan subalgebra.

Fix some Cartan subalgebra $\mathfrak{h} \subset \mathfrak{g}$. Since $\mathfrak{h}$ is Abelian, $\{\mathrm{ad_h}, \mathrm{h} \in \mathfrak{h}\}$ is a commuting family of semisimple endomorphisms of $\mathfrak{g}$. Hence we can simultaneously diagonalize this family and obtain a decomposition

$$\mathfrak{g} = \mathfrak{h} \oplus \bigoplus_{\alpha\in\Phi} \mathfrak{g}_\alpha,$$

where $\mathfrak{g}_\alpha = \{x \in \mathfrak{g} \mid \forall h \in \mathfrak{h} : [h,x] = \alpha(h)x\}$ and $\Phi \subset \mathfrak{h}^*$. Of course, $\mathfrak{h} = \mathfrak{g}_0$ is the eigenspace of $\mathfrak{g}$ with eigenvector 0. We call the elements of Φ the roots of α, Φ is called the root system of $\mathfrak{g}$ and the above decomposition is termed a root space decomposition. Although the Cartan subalgebra $\mathfrak{h}$ is not unique, the root system is uniquely determined by the Lie algebra. In fact, more is true; the root system can be divided into two subsets $\Phi = \Phi^+ \cup \Phi^-$, where Φ^+ denotes the set of positive roots, Φ^- the set of negative roots and such that $\Phi^- = \{-\alpha \mid \alpha \in \Phi^+\}$. For any positive root $\alpha \in \Phi^+$, there exist elements $x_\alpha \in \mathfrak{g}_\alpha, y_\alpha \in \mathfrak{g}_{-\alpha}, h_\alpha \in \mathfrak{h}$ such that the vector space generated by $\{h_\alpha, x_\alpha, y_\alpha\}$ is a Lie algebra isomorphic to $\mathfrak{sl}_2$. We denote this Lie algebra by $\mathfrak{sl}_\alpha$.

The geometry of the root system is determined by a symmetric bilinear form κ, defined by $\kappa(x,y) = \mathrm{Tr}(\mathrm{ad_x} \circ \mathrm{ad_y})$. This κ is called the Killing form of $\mathfrak{g}$ and it is nondegenerate if and only if $\mathfrak{g}$ is semisimple. In this case, the Killing form restricted to $\mathfrak{h}$ is also nondegenerate and by transferring the inner product, $\mathfrak{h}^*$ becomes Euclidean. We denote the inner product on $\mathfrak{h}^*$ by $\langle -,- \rangle$. We refer to [37, Chapter 8] for details about the geometric properties of the root space Φ. A subset $\Delta \subset \Phi$ is called a base if Δ is a base for $E = \mathfrak{h}^*$ and if each root $\alpha \in \Phi$ can be written as a linear combination of the elements of Δ with integral coefficients, which moreover are all nonnegative or nonpositive, thus leading to the decomposition $\Phi = \Phi^+ \cup \Phi^-$. The elements of such a base are called simple roots. Denote by P_α the hyperplane in E perpendicular to $\alpha \in E$. Let $\gamma \in E \setminus \bigcup_{\alpha\in\Phi} P_\alpha$; then the set $\Phi^+(\gamma) = \{\alpha \in \Phi,\ \langle\gamma,\alpha\rangle > 0\}$ consists of all roots lying on the same side of the hyperplane orthogonal to γ. It can be shown that the subset $\Delta(\gamma) \subset \Phi^+(\gamma)$ of all indecomposable vectors in $\Phi^+(\gamma)$ forms a base for Φ and that every base Δ for Φ is of the form $\Delta(\gamma)$

for some $\gamma \in E \setminus \bigcup_{\alpha\in\Phi} P_\alpha$. In fact, this shows that the number of bases is in one-to-one correspondence with the number of Weyl chambers of E. Elements $\gamma \in E \setminus \bigcup_{\alpha\in\Phi} P_\alpha$ are called regular.

If one fixes a base $\Delta = \Delta(\gamma) = \{\alpha_1, \dots, \alpha_n\}$ for some suitable $\gamma \in E = \mathbb{R}^n$, where $n = \dim(\mathfrak{h}) = \mathrm{rk}(\mathfrak{g})$, then we can write $\mathfrak{g} = \mathfrak{n}^- \oplus \mathfrak{h} \oplus \mathfrak{n}$, with $\mathfrak{n} = \bigoplus_{\alpha\in\Phi^+} \mathfrak{g}_\alpha$. The Poincaré-Birkhoff-Witt theorem states that there is an isomorphism $U(\mathfrak{g}) \cong U(\mathfrak{n}^-) \otimes U(\mathfrak{b})$, where $\mathfrak{b} = \mathfrak{h} \oplus \mathfrak{n}$. To calculate in the universal enveloping algebra, we fix an ordering of the positive roots $\alpha_1, \dots, \alpha_m$ and pick elements $h_i = h_{\alpha_i}, x_i = x_{\alpha_i}, y_i = y_{\alpha_i}$ generating the $\mathfrak{sl}_{\alpha_i}$. Then the elements of the form

$$y_1^{r_1} \dots y_m^{r_m} h_1^{s_1} \dots h_n^{s_n} x_1^{t_1} \dots x_m^{t_m}, \quad r_i, s_i, t_i \in \mathbb{N}$$

form a base for $U(\mathfrak{g})$. We call them PBW-monomials. Such an element belongs to the weight space $U(\mathfrak{g})_\omega$, where $\omega = \sum_{i=1}^m (t_i - r_i)\alpha_i \in \mathfrak{h}^*$ and where $U(\mathfrak{g})_\omega = \{z \in U(\mathfrak{g}) \mid h \cdot z = \omega(h)z \text{ for all } h \in \mathfrak{h}\}$. To any $\lambda \in \mathfrak{h}^*$ we can associate a left $U(\mathfrak{g})$-module $M(\lambda)$ as follows: since $\mathfrak{h} \cong \mathfrak{b}/\mathfrak{n}$ as Lie algebras, λ yields a one-dimensional $\mathfrak{b}$-module $\mathbb{C}_\lambda$ with trivial $\mathfrak{n}$-action. By the PBW theorem, [37, Chapter 17], $U(\mathfrak{g})$ has a canonical $(U(\mathfrak{g}), U(\mathfrak{b}))$-bimodule structure. Define $M(\lambda) := U(\mathfrak{g}) \otimes_{U(\mathfrak{b})} \mathbb{C}_\lambda$, which we call the Verma module associated with λ. Observe that $M(\lambda) \cong U(\mathfrak{n}^-) \otimes \mathbb{C}_\lambda$, which is a free $U(\mathfrak{n}^-)$-module of rank 1. It is easy to see that $M(\lambda)$ is a highest weight module with maximal vector $v_\lambda^+ = 1 \otimes 1$ of weight λ. Moreover, the set of weights is $\lambda - \Gamma$, where Γ is the set of all $\mathbb{Z}^+$-linear combinations of simple roots. In other words, the action highly depends on the structure of the root system.

As stated in the introduction of the chapter, we would like to study glider representations for a chain of universal enveloping algebras $U(\mathfrak{g}_1) \subset U(\mathfrak{g}_2) \subset \dots \subset U(\mathfrak{g}_n)$ associated with a chain of semisimple Lie algebras $\mathfrak{g}_1 \subset \mathfrak{g}_2 \subset \dots \subset \mathfrak{g}_n$. More specifically, we want to study so called Verma gliders, the definition of which is postponed to Section 6.2. Nevertheless, we already mention here that our study of Verma gliders naturally leads to embeddings of $\mathfrak{g}_i$-Verma modules as $U(\mathfrak{g}_i)$-modules inside $\mathfrak{g}_{i+1}$-Verma modules. By the structure of $U(\mathfrak{g})$ we know that the elements $y_1^{r_1} \dots y_m^{r_m}$ form a base of $U(\mathfrak{n}^-)$, hence we must make some logical connection between roots of $\mathfrak{g}_i$ and roots of $\mathfrak{g}_{i+1}$. In doing so, we are forced to put some condition on the inclusion of Lie algebras, as we explain now.

Consider an embedding $\iota : \mathfrak{g}_1 \hookrightarrow \mathfrak{g}_2$ of semisimple Lie algebras and choose some Cartan subalgebra $\mathfrak{h}_1$ in $\mathfrak{g}_1$. One can extend $\mathfrak{h}_1$ to a Cartan subalgebra $\mathfrak{h}_2$ of $\mathfrak{g}_2$, so we obtain the following root space decompositions

$$\mathfrak{g}_1 = \mathfrak{h}_1 \oplus \bigoplus_{\alpha\in\Phi_1} \mathfrak{g}_{1,\alpha}, \quad \mathfrak{g}_2 = \mathfrak{h}_2 \oplus \bigoplus_{\beta\in\Phi_2} \mathfrak{g}_{2,\beta}.$$

The root systems Φ_1 and Φ_2 are subsets of $\mathfrak{h}_1^*, \mathfrak{h}_2^*$ respectively, and the inclusion $\mathfrak{h}_1 \subset \mathfrak{h}_2$ gives rise to a projection $\pi : \mathfrak{h}_2^* \twoheadrightarrow \mathfrak{h}_1^*$ given by restriction. We want the following condition to hold

$$\forall \alpha \in \Phi_1 : \#\{\beta \in \Phi_2 \mid \pi(\beta) = \alpha\} = 1. \tag{6.2}$$

Suppose that this condition holds and let $\alpha \in \Phi_1$. If β is the unique root in Φ_2 such that $\pi(\beta) = \alpha$, then $x_\alpha^1 \in \mathfrak{g}_{2,\beta}$, which entails that $\iota(x_\alpha^1) = c_\beta x_\beta^2$, with $c_\beta \in K$ (the upper index i refers to the Lie algebra $\mathfrak{g}_i$ we are working in). In fact, there is some liberty in choosing the element x_β^2, so one may assume that $c_\beta = 1$. Here we did not fix a base yet, so we did not speak about positive or negative roots. If bases are fixed, then you have to adjust the notations. E.g. if α is negative and β is positive, then this means that $\iota(y_{-\alpha}^1) = x_\beta^2$. The condition allows us to appoint a single root of the bigger Lie algebra to any root of the smaller one. We denote by Φ_1^* the set of roots obtained in this way. We have an equivalent characterization in terms of eigenvectors.

Proposition 6.1.1. *us Let $\Delta(i)$ be a base of Φ_i for $i = 1, 2$. Condition (6.2) is equivalent to saying that for all $\alpha \in \Phi_1^+, x_\alpha^1$ is an eigenvector in $\mathfrak{g}_2$ with regard to $\mathfrak{h}_2$ such that if $x_\alpha^1 = z_\beta^2$ ($z_\beta^2 = x_\beta^2$ or y_β^2, $\beta \in \Phi_2^+$) then $h_\beta^2 \in \mathfrak{h}_1$.*

Proof. We already showed that if (6.2) holds, that the x_α^1 are eigenvectors. If $x_\alpha^1 = \lambda x_\beta^2$, then $y_\alpha^1 = \mu y_\beta^2$, from which it follows that $h_\beta^2 = [x_\beta^2, y_\beta^2] = \frac{1}{\lambda\mu}[x_\alpha^1, y_\alpha^1] = \frac{1}{\lambda\mu} h_\alpha^1 \in \mathfrak{h}_1$. The case $x_\alpha^1 = \lambda y_\beta^2$ is analogous. Conversely, suppose that $\beta, \gamma \in \Phi_2$ are such that $\pi(\beta) = \pi(\gamma) = \alpha \in \Phi_1$. Without loss of generality, we may assume that $x_\alpha^1 \in \mathfrak{g}_{2,\beta}$, which implies that $h_\beta^2 \in \mathfrak{h}_1$. It follows that $\alpha(h_\beta^2) = \beta(h_\beta^2) = \gamma(h_\beta^2) = 2$. Hence

$$2 = \gamma(h_\beta^2) = \langle \gamma, \beta^\vee \rangle = 2\cos(\theta_{\gamma,\theta}) \frac{\|\gamma\|}{\|\beta\|},$$

where $\beta^\vee = 2\beta/\langle \beta, \beta \rangle$ is the coroot of β. By the geometry of root systems, we must have that $\frac{\|\gamma\|}{\|\beta\|} = 1$ and $\cos(\theta_{\gamma,\beta}) = 1$, hence $\gamma = \beta$. □

To show that this condition is independent of the choice of Cartan subalgebras $\mathfrak{h}_1 \subset \mathfrak{h}_2$ we have to recall some facts concerning automorphisms of Lie algebras. There is a normal subgroup $E(\mathfrak{g})$ of the automorphism group $\mathrm{Aut}(\mathfrak{g})$ generated by all $\exp(\mathrm{ad}_\mathrm{x})$, x a strongly ad-nilpotent element. For semisimple Lie algebras it holds that $E(\mathfrak{g}) = \mathrm{Int}(\mathfrak{g})$, where the latter group is the subgroup of $\mathrm{Aut}(\mathfrak{g})$ generated by all $\exp(\mathrm{ad}_\mathrm{x})$, x a nilpotent element. Apparently, any two Cartan subalgebras $\mathfrak{h}, \mathfrak{h}'$ of a Lie algebra $\mathfrak{g}$ are conjugated by an element $\sigma \in E(\mathfrak{g})$, i.e. $\sigma(\mathfrak{h}) = \mathfrak{h}'$ and with regard to Lie subalgebras there is a nice functorial behavior. If $\mathfrak{g}' \subset \mathfrak{g}$ is a Lie subalgebra, we can look at the subgroup $E(\mathfrak{g}; \mathfrak{g}')$ of $E(\mathfrak{g})$ generated by all $\exp(\mathrm{ad}_\mathrm{x}^\mathfrak{g})$, $\mathrm{x} \in \mathfrak{g}'$ strongly ad-nilpotent. By restricting the automorphisms of $E(\mathfrak{g}; \mathfrak{g}')$ one obtains the group $E(\mathfrak{g}')$. We refer the reader to [37, Chapter 8] for a detailed overview of these facts.

Let $\mathfrak{g}$ be a Lie algebra with Cartan subalgebras $\mathfrak{h}, \mathfrak{h}'$. If $\sigma \in E(\mathfrak{g})$ is such that $\sigma(\mathfrak{h}) = \mathfrak{h}'$, then the roots with regard to $\mathfrak{h}'$ are exactly the functionals $\beta\sigma^{-1} : \mathfrak{h}' \to K$ where $\beta : \mathfrak{h} \to K$ is a root with regard to $\mathfrak{h}$. Moreover, it holds that $\sigma(x_\beta) = x_{\beta\sigma^{-1}}$.

Proposition 6.1.2. *Condition* (6.2) *is independent under the orbit* $\mathcal{O}_{\mathfrak{h}_1 \subset \mathfrak{h}_2}$ *for the action of* $E(\mathfrak{g}_2; \mathfrak{g}_1)$.

Proof. Let $\widetilde{\sigma} \in E(\mathfrak{g}_2; \mathfrak{g}_1)$ and denote its restriction to $\mathfrak{g}_1$ be σ. We define $\mathfrak{h}_1' = \sigma(\mathfrak{h}_1)$ and $\mathfrak{h}_2' = \widetilde{\sigma}(\mathfrak{h}_2)$. We have the following commutative diagram

$$\begin{array}{ccc} \mathfrak{h}_2^* & \xrightarrow{\ \pi\ } & \mathfrak{h}_1^* \\ {\scriptstyle \widetilde{\sigma}^*}\big\downarrow & & \big\downarrow{\scriptstyle \sigma^*} \\ \widetilde{\sigma}(h_2)^* & \xrightarrow{\ \pi\ } & \mathfrak{h}_1'^* \end{array}$$

If $\beta, \gamma \in \Phi_2$ are such that $\pi(\beta) = \pi(\gamma) = \alpha \in \Phi_1$, then $\sigma^*(\alpha) = (\pi \circ \widetilde{\sigma}^*)(\beta) = (\pi \circ \widetilde{\sigma}^*)(\gamma)$, which shows that the roots $\beta\sigma^{-1}, \gamma\widetilde{\sigma}^{-1} \in \Phi_2'$ both restrict to the root $\alpha\sigma^{-1} \in \Phi_1'$. □

To show that (6.2) is independent of the choice of inclusions $\mathfrak{h}_1 \subset \mathfrak{h}_2$ of Cartan subalgebras, it would suffice to show that the condition is independent for inclusions $\mathfrak{h}_1 \subset \mathfrak{h}_2$ and $\mathfrak{h}_1 \subset \mathfrak{h}_2'$. Imposing a condition on the ranks of the $\mathfrak{g}_i$, this follows from the following:

Proposition 6.1.3. *Assume that* $\mathrm{rk}(\mathfrak{g}_2) = \mathrm{rk}(\mathfrak{g}_1) + 1$. *Then* (6.2) *is independent of the choice of Cartain subalgebras* $\mathfrak{h}_1 \subset \mathfrak{h}_2$.

Proof. From the remark above, it suffices to show independence for inclusions $\mathfrak{h}_1 \subset \mathfrak{h}_2$ and $\mathfrak{h}_1 \subset \mathfrak{h}_2'$. Suppose that (6.2) holds for $\mathfrak{h}_1 \subset \mathfrak{h}_2$ and let $\alpha \in \Phi_1$. We know there exists a unique $\beta \in \Phi_2$ such that $x_\alpha^1 = \lambda x_\beta^2$ (by adjusting bases we may assume that both α and β are positive within their respective root system and by rescaling we may assume that $\lambda = 1$). Take some $\tau \in E(\mathfrak{g}_2)$ sending $\mathfrak{h}_2$ to $\mathfrak{h}_2'$. For $h \in \mathfrak{h}_2 \cap \mathfrak{h}_2'$, we have that $h = \tau(h')$ for some $h' \in \mathfrak{h}_2 \cap \mathfrak{h}_2'$. This shows that τ restricts to an automorphism of $\mathfrak{h}_2 \cap \mathfrak{h}_2'$ which equals $\mathfrak{h}_1$ by the hypotheses on the ranks of $\mathfrak{g}_1$ and $\mathfrak{g}_2$. It follows that $h^2_{\beta\tau^{-1}} = \tau(h^2_\beta) = \tau(h^1_\alpha) \in \mathfrak{h}_1$ and also that the root $\alpha\tau^{-1}_{|\mathfrak{h}_1}$ with regard to $\tau^{-1}(\mathfrak{h}_1) = \mathfrak{h}_1$ has associated element $x^1_{\alpha\tau^{-1}_{|\mathfrak{h}_1}} = \tau(x^1_\alpha) = \tau(x^2_\beta) = x^2_{\beta\tau^{-1}}$. We have shown that for all roots $\alpha' = \alpha\tau^{-1}_{|\mathfrak{h}_1} \in \Phi_1$, the associated element $x^1_{\alpha'} = x^2_{\beta\tau^{-1}}$ is an eigenvector such that $h^2_{\beta\tau^{-1}} \in \mathfrak{h}_1$. □

Remark 6.1.4. Without imposing the condition on the ranks of the Lie algebras, we are not able to prove this, neither can we come up with a counterexample.

Not all embeddings $\mathfrak{g}_1 \subset \mathfrak{g}_2$ satisfy condition (6.2).

Example 6.1.5. Consider $\mathfrak{sl}_2 \subset \mathfrak{so}_4 \cong \mathfrak{sl}_2 \times \mathfrak{sl}_2$, where the $\mathfrak{sl}_2$ is embedded diagonally. Choose the Cartan subalgebra $\mathfrak{h}_2$ to be the subalgebra generated by the diagonal matrices $H_1 = E_{11} - E_{33}$ and $H_2 = E_{22} - E_{44}$. If $\{L_1, L_2\}$ denotes a dual base for $\{H_1, H_2\}$ in $\mathfrak{h}_2^*$, then the positive roots of $\mathfrak{so}_4$ are $L_1 - L_2$ and $L_1 + L_2$. The copies of $\mathfrak{sl}_2$ for $L_1 - L_2$, resp. $L_1 + L_2$ are generated by

$$\begin{aligned} h_{L_1-L_2} &= H_1 - H_2, & & h_{L_1+L_2} = H_1 + H_2, \\ x_{L_1-L_2} &= E_{12} - E_{43}, & \text{resp.} \quad & x_{L_1+L_2} = E_{14} - E_{23}, \\ y_{L_1-L_2} &= E_{21} - E_{34}, & & y_{L_1+L_2} = E_{32} - E_{41}. \end{aligned}$$

We see that the nilpositive element of $\mathfrak{g}_1 = \mathfrak{sl}_2$ is $x_\alpha = x_{L_1-L_2} + x_{L_1+L_2}$. The vector space generated by $2H_1 = (H_1 - H_2) + (H_1 + H_2)$ is a Cartan subalgebra $\mathfrak{h}_1$ of $\mathfrak{g}_1$. With regard to $\mathfrak{h}_1$, the root α is $2L_1$ and we have that $\pi(L_1 - L_2) = \pi(L_1 + L_2)$, where $\pi : \mathfrak{h}_2^* \twoheadrightarrow \mathfrak{h}_1^*$ denotes the projection.

An element $x \in \mathfrak{g}$ is semisimple, resp. nilpotent if $\mathrm{ad}_{\mathrm{x}} \in \mathrm{End}(\mathfrak{g})$ is semisimple, resp. nilpotent. We recall the Jacobson-Morozov theorem, [25, Theorem 3.3.1], which states that for any nonzero nilpotent element x, there exists a standard triple $\{h, x, y\}$. By this we mean that the subspace generated by these three elements is isomorphic to $\mathfrak{sl}_2$ with $\mathfrak{h} = Kh$ a Cartan subalgebra, $x = x_\alpha$ and $y = y_\alpha$.

Lemma 6.1.6. *Let E be an n-dimensional Euclidean space with inner product $\langle -, - \rangle$ and let $\alpha_1, \ldots, \alpha_n$ be linearly independent vectors in E. Then there exists a hyperplane $P_\gamma, \gamma \in E$, such that all vectors $\alpha_i, 1 \leq i \leq n$ lie on the same side of the hyperplane.*

Proof. We proceed by induction, the case $n = 2$ being trivial. So assume the result holds in $n-1$-dimensional Euclidean spaces and take n linearly independent vectors $\alpha_1, \ldots, \alpha_n$ in an n-dimensional space E. Consider $E' = \langle \alpha_1, \ldots, \alpha_{n-1} \rangle$ and let $\delta \in E$ be perpendicular to all $\alpha_i, 1 \leq i \leq n-1$. By induction, there exists $\gamma' \in E'$ such that $\langle \gamma', \alpha_i \rangle > 0$ for all $1 \leq i \leq n-1$. If $\langle \alpha_n, \gamma' \rangle > 0$ then $\gamma = \gamma'$ satisfies. If not, then choose $\lambda \in K$ such that $\langle \alpha_n, \gamma' + \lambda\delta \rangle > 0$ and then $\gamma = \gamma' + \lambda\delta$ satisfies. □

Proposition 6.1.7. *Let $\mathfrak{g}$ be a semisimple Lie algebra of rank > 2 with root space decomposition $\mathfrak{g} = \mathfrak{n}^- \oplus \mathfrak{h} \oplus \mathfrak{n}$ with regard to some Cartan subalgebra $\mathfrak{h}$. If $\alpha, \beta \in \Phi^+$, then the elements $x_\alpha + x_\beta$ and $x_\alpha + y_\beta$ are nilpotent.*

Proof. The endomorphism $\mathrm{ad}_{\mathrm{x}_\alpha + \mathrm{x}_\beta}$ maps a weight vector to a linear combination of weight vectors with higher weight, so $x_\alpha + x_\beta$ is clearly nilpotent. By Lemma 6.1.6 we know there exists some $\gamma \in E = \mathfrak{h}^*$ such that $\langle \gamma, \alpha \rangle > 0$ and $\langle \gamma, -\beta \rangle > 0$. We may assume that γ is regular. With regard to the base $\Delta(\gamma)$ both α and $-\beta$ are positive roots, whence $x_\alpha + y_\beta = x_\alpha^\gamma + x_{-\beta}^\gamma$ is nilpotent, where x_δ^γ denotes the associated weight vector of a positive root δ with regard to $\Delta(\gamma)$. □

From the previous proposition we see that the inclusion of Example 6.1.5 corresponds to the standard triple of the nilpotent element $x_\alpha + x_\beta$ for positive roots α, β. In general, the inclusion of the associated triple $\{h, x, y\}$ of a nilpotent element of the form $x = x_{\alpha_1} + \ldots + x_{\alpha_m}$, where $\alpha_1, \ldots, \alpha_m$ are positive roots in a root system Φ of a semisimple Lie algebra $\mathfrak{g}$ does not satisfy condition (6.2).
Embeddings that do satisfy the condition are for example Levi subalgebras. For background on these subalgebras we refer to [25, Section 3.8]. We use the notations from loc. cit.

Proposition 6.1.8. *Let $\mathfrak{g}$ be a Lie algebra with fixed Cartan subalgebra $\mathfrak{h}$. Let $\mathfrak{l}_I \subset \mathfrak{g}$ be the Levi subalgebra associated with a subset $I \subset \Delta$ and $\mathfrak{g}_I = [\mathfrak{l}_I, \mathfrak{l}_I]$ the derived subalgebra. The embedding $\mathfrak{g}_I \subset \mathfrak{g}$ satisfies condition* (6.2).

Proof. As Cartan subalgebra of $\mathfrak{g}_I$ we take $\mathfrak{h}_I = \bigoplus_{\alpha \in I} K h_\alpha$. Then $\mathfrak{g}_I = \mathfrak{n}_I^- \oplus \mathfrak{h}_I \oplus \mathfrak{n}_I$, with $\mathfrak{n}_I = \bigoplus_{\alpha \in \Phi_I^+} \mathfrak{g}_\alpha$. The result easily follows from Proposition 6.1.1. □

Example 6.1.9. Embeddings $\mathfrak{g}_1 \subset \mathfrak{g}_2$ of simple Lie algebras of the same type A, B, C or D corresponding to a connected subdiagram of the associated Dynkin diagram all satisfy condition (6.2). To clarify what we mean, we give the example of the inclusion $\mathfrak{sp}_{2n} \subset \mathfrak{sp}_{2m}$ given by

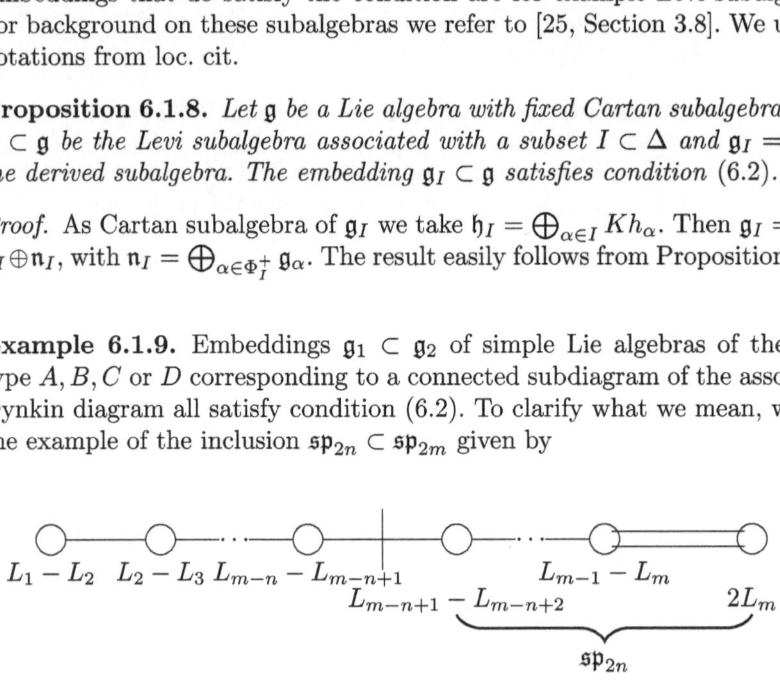

(6.3)

We use the notations from [31]. In other words, they just correspond to the derived Lie subalgebra of a suitable Levi subalgebra, and the claim follows from the previous proposition. A final word here on embeddings of type A: we only consider embeddings $\mathfrak{sl}_n \subset \mathfrak{sl}_m$ corresponding to a connected subdiagram of the A_{m-1}-diagram containing one of the end points.

Let $\mathfrak{g}_1 \subset \mathfrak{g}_2$ be an inclusion satisfying condition (6.2) (with regard to $\mathfrak{h}_1 \subset \mathfrak{h}_2$) and with $\mathrm{rk}(\mathfrak{g}_1) = \mathrm{n} < \mathrm{rk}(\mathfrak{g}_2) = \mathrm{m}$. Pick a base $\Delta(1) = \{\alpha_1, \ldots, \alpha_n\}$ for $\mathfrak{g}_1$ and denote by $\alpha_i^*, (i = 1, \ldots, n)$ the unique root in Φ_2 restricting to α_i. Lemma 6.1.6 allows us to pick a base $\Delta(2)$ such that all α_i^* are positive roots. This entails that the embedding $\iota : U(\mathfrak{g}_1) \hookrightarrow U(\mathfrak{g}_2)$ maps $U(\mathfrak{n}_1)$ inside $U(\mathfrak{n}_2)$. Indeed, if $\alpha_1 + \alpha_2$ is a positive root, then $x^1_{\alpha_1+\alpha_2}$ is an eigenvector in $\mathfrak{g}_2$. Since

$$x^1_{\alpha_1+\alpha_2} = \lambda[x^1_{\alpha_1}, x^1_{\alpha_2}] = \lambda'[x^2_{\alpha_1^*}, x^2_{\alpha_2^*}] \subset \mathfrak{g}_{2,\alpha_1^*+\alpha_2^*},$$

we see that $\alpha_1^* + \alpha_2^* \in \Phi_2$ is the unique root restricting to $\alpha_1 + \alpha_2$. We end this section with the following:

Proposition 6.1.10. *If $\alpha, \beta \in \Phi_1^+$ are such that $\alpha^* + \beta^* \in \Phi_2^+$, then $\alpha + \beta \in \Phi_1$ and $(\alpha + \beta)^* = \alpha^* + \beta^*$.*

Proof. This is straightforward, since $[x_\alpha^1, x_\beta^1] = [x_{\alpha^*}^2, x_{\beta^*}^2] = \lambda x_{\alpha^*+\beta^*}^2 \in \mathfrak{g}_1$. □

6.2 Verma gliders

Consider now a chain $\mathfrak{g}_1 \subset \ldots \subset \mathfrak{g}_n$ of semisimple Lie algebras and an associated chain of maximal toral subalgebras such that all inclusions $\mathfrak{g}_i \subset \mathfrak{g}_{i+1}, (i = 1, \ldots, n-1)$ satisfy condition (6.2). This fixed chain of Lie algebras determines a positive algebra filtration on the universal enveloping algebra $U(\mathfrak{g}_n)$ given by $F_iU(\mathfrak{g}_n) = U(\mathfrak{g}_{i+1})$ for $i = 0, \ldots, n-2$, $F_mU(\mathfrak{g}_n) = U(\mathfrak{g}_n)$ for $m \geq n-1$. By the remark at the end of the previous section, we can pick bases $\Delta(i)$ of $\mathfrak{g}_i$ such that $\iota(U(\mathfrak{n}_i)) \subset U(\mathfrak{n}_{i+1})$ for $i = 1, \ldots, n-1$. In fact, $\iota(\mathfrak{n}_1) = \mathfrak{n}_1^* = \bigoplus_{\alpha^* \in \Phi_1^* \cap \Phi_2^+} g_{2,\alpha^*}$, but we will just write $\mathfrak{n}_1$ instead of $\mathfrak{n}_1^*$. Also, we denote the intersection $\Phi_1^* \cap \Phi_2^+$ by $(\Phi_1^*)^+$.

Definition 6.2.1. Let $\mathfrak{g}_1 \subset \ldots \subset \mathfrak{g}_n$ be a chain of semisimple Lie algebras as above. A glider representation, of essential length $n-1$,

$$\Omega \supset M \supset M_1 \supset \ldots \supset M_{n-1} \supset 0 \supset \ldots$$

with regard to the finite algebra filtration of length $n-1$ $F_iU(\mathfrak{g}_n) = U(\mathfrak{g}_{i+1})$ on $U(\mathfrak{g}_n)$ is called a Verma glider if it satisfies the following conditions

1. $\Omega = M(\lambda_n), \lambda_n \in \mathfrak{h}_n^*$ is a $\mathfrak{g}_n$-Verma module;
2. $M_{n-1} = M(\lambda_1), \lambda_1 \in \mathfrak{h}_1^*$ is a $\mathfrak{g}_1$-Verma module;
3. all $M_{n-1-i}, i = 1, \ldots, n-2$ are contained in some $\mathfrak{g}_{i+1}$-Verma module $M(\lambda_{i+1}), \lambda_{i+1} \in \mathfrak{h}_{i+1}^*$.

One of the purposes of glider theory is to provide information between representations of the various $U(\mathfrak{g}_i)$ appearing in the chain. There is a nice way to construct Verma gliders by starting from a set of Verma modules:

$$\begin{array}{ll} M(\lambda_1) & \lambda_1 : \mathfrak{h}_1 \to \mathbb{C}, \\ M(\lambda_2) & \lambda_2 : \mathfrak{h}_2 \to \mathbb{C}, \\ \vdots & \vdots \\ M(\lambda_n) & \lambda_n : \mathfrak{h}_n \to \mathbb{C}. \end{array}$$

The Verma module $M(\lambda_n)$ will play the role of Ω, whereas the M_{n-i} will be embedded in $M(\lambda_i)$ for $i = 1, \ldots, n$. To establish this we must embed a 'smaller' $\mathfrak{g}_1$-Verma module in a 'bigger' $\mathfrak{g}_n$-Verma module. Geometrically, one can think of a Verma module in terms of its weight space $\lambda - \Gamma$; hence such an embedding comes down to viewing $\lambda_1 - \Gamma_1$ as a subset of $\lambda_n - \Gamma_n$, in some sense at least. In fact, we want to do this step by step; that is, we would like to embed $M(\lambda_i)$ inside $M(\lambda_{i+1})$ as a $U(\mathfrak{g}_i)$-module for $i = 1, \ldots, n-1$. If we succeed in doing so, $\Omega = M = M(\lambda_n) \supset M(\lambda_{n-1}) \supset \ldots \supset M(\lambda_1)$ becomes a Verma glider. Admittedly, $\Omega = M$ in this particular example, but this need not always be the case, which we will see below. Nonetheless, in Section 6.3, we deal with irreducible gliders, for which it holds that $M = U(\mathfrak{g}_n)M_{n-1}$ is a $U(\mathfrak{g}_n)$-module. For example, when λ_n is antidominant (we recall its definition later on), $\Omega = M(\lambda_n)$ is simple, thus equals M if the glider is irreducible. But we are running ahead of things here.

Let $\lambda_1 \in \mathfrak{h}_1^*, \lambda_2 \in \mathfrak{h}_2^*$. Since $M(\lambda_1)$ is a highest weight module with highest weight vector, say $v_{\lambda_1}^+$, an embedding $M(\lambda_1) \subset M(\lambda_2)$ as $U(\mathfrak{g}_1)$-modules is given by an element $z \in U(\mathfrak{n}_2^-)$; that is, it is given by sending $v_{\lambda_1}^+$ to zv^+, with v^+ a highest weight vector of $M(\lambda_2)$. Order the positive roots $\alpha_1, \ldots, \alpha_m$ of $\mathfrak{g}_2$. If $z = \sum_i' \mu_i y_1^{r_1^i} \ldots y_m^{r_m^i} \in U(\mathfrak{n}_2^-)$ (notations from the previous section), we denote by $\widetilde{z}$ the functional

$$\widetilde{z} : \mathfrak{h}_2 \to \mathbb{C},\ h \mapsto -\sum_i{}' \sum_{j=1}^{m} \mu_i r_j^i \alpha_j(h).$$

The element z must satisfy two conditions, namely

$$\lambda_1 = \pi(\lambda_2 - \widetilde{z}), \tag{6.4}$$

$$U(\mathfrak{n}_1) \cdot zv^+ = 0. \tag{6.5}$$

The first condition depends on the choice of the λ_i $(i = 1, 2)$, but the second one solely depends on the structure of the root systems and, more importantly, on how both Φ_1 and Φ_2 are related! In fact, the second condition will determine for which pairs (λ_1, λ_2) we obtain Verma gliders. How can we determine the elements z satisfying condition (6.5)? A starting point would be to determine which of the elements $y_i = y_{\alpha_i}$ for positive roots α_i satisfy. The crucial point will be the following:

Lemma 6.2.2. *[37, Lemma 10.2.A]*
If α is a positive root but not simple, then $\alpha - \beta$ is a (necessarily positive) root for some $\beta \in \Delta$.

Let $\Delta(1)^*$ denote the set of roots extending the simple roots of $\Delta(1)$ and then pick a suitable base $\Delta(2)$ such that $\Delta(1)^* \subset \Phi_2^+$. If $\alpha \in \Phi_2^+$ is not simple, then by the lemma there exists a simple root $\beta \in \Delta(2)$ such that $\alpha - \beta$ is a root, hence $[x_\beta, y_\alpha] \neq 0$ in $U(\mathfrak{n}_2^-)$. If $\beta \in \Delta(1)^*$, then $z = y_\alpha$ does not satisfy the second condition. In general, if some root $\gamma \in \Phi_1^*$ is such that $\alpha - \gamma$ is a (positive) root, then y_α does not apply. We will see below that often it holds that $\Delta(1)^* \subset \Delta(2)$, and in this case the previous reasoning immediately shows that x_γ with $\gamma \in \Delta(2) \setminus \Delta(1)^*$ an additional simple root satisfies condition (6.5). Observe that we made a choice of base $\Delta(1)$, from which we determined a suitable base $\Delta(2)$. Fortunately, we have the following

Proposition 6.2.3. *The number of roots $\gamma \in \Phi_2$ such that $z = y_\gamma$ satisfies condition* (6.5) *above, is independent of our choice of bases $\Delta(1), \Delta(2)$, i.e. such that $\Delta(1)^* \subset \Phi_2^+$.*

Proof. A base change from $\Delta(1)$ to $\Delta(1)'$ is given by an element σ of the Weyl group W_1 of $\mathfrak{g}_1$. The Weyl group is generated by the $s_\alpha, \alpha \in \Phi_1$, where s_α is the reflection in $\mathfrak{h}_1^*$ with regard to the hyperplane orthogonal to α. Algebraically, $s_\alpha(\beta) = \beta - \beta(h_\alpha)\alpha$, from which it follows that $\pi(s_{\alpha^*}(\beta^*)) = s_\alpha(\beta)$. In other words, the element s_{α^*} in the Weyl group W_2 of $\mathfrak{g}_2$ maps Φ_1^* to itself. This shows that to σ there is an associated element $\sigma^* \in W_2$, which maps Φ_1^* to itself. It follows that $\sigma^*(\Delta(2))$ is a suitable base and the result follows easily because $\sigma^*(\beta - \gamma) = \sigma^*(\beta) - \sigma^*(\gamma)$ for $\beta, \gamma \in \Phi_2$. It remains to show that if, given any $\Delta(1)$, two bases $\Delta(2)$, $\Delta(2)'$ suffice, that the number of roots is still the same. Let $\tau \in W_2$ be the (unique) element sending $\Delta(2)$ to $\Delta(2)'$. In other words, if $\Delta(2) = \Delta(\gamma)$ for a regular $\gamma \in E_2$, then $\Delta(2)' = \Delta(\tau(\gamma))$. Since $\pi(\gamma)$ and $\pi(\tau(\gamma))$ must lie on the same side of each hyperplane $P_\alpha, \alpha \in \Phi_1$ in E_1, τ must be a composition of reflections with regard to a hyperplane containing the space generated by Φ_1^*. This shows that τ sends Φ_1^* to itself. Hence if $\beta \in \Phi_2^+$ is such that $\beta - \alpha^*$ is a positive root for $\alpha \in \Phi_1^+$, then $\tau(\beta) - \tau(\alpha^*) = \tau(\beta - \alpha^*)$ is a positive root in Φ_2 with regard to $\Delta(2)'$, and we are done. □

Definition 6.2.4. An element $z = y_\gamma$ for some positive root $\gamma \in \Phi_2$ satisfying condition (6.5) is called an embedding element.

Example 6.2.5. Consider the embedding $\mathfrak{sl}_n \subset \mathfrak{sl}_m$ $(n < m)$ corresponding to the first $n - 1$ nodes of the A_{m-1}-diagram. Then

$$\Delta(1)^* = \{L_1 - L_2, \ldots, L_{n-1} - L_n\}.$$

As observed above, the $m-n$ additional basis elements are possible candidates, and for obvious reasons, so are the roots which are positive linear combinations of these. Of course, they form an A_{m-n}-diagram and so we already obtain $\frac{(m-n)^2+(m-n)}{2}$ candidates. The $\frac{n^2+n}{2}$ roots from $\mathfrak{sl}_n$ obviously do not apply. The only roots that are left are the ones that go through the last node of A_{n-1} and the first node of A_{m-n}. There are $(m - n)n$ roots of this form, that is, with $i \leq n < j$. These roots also do not apply. In total we checked all of them, since

$$\frac{n(n+1)}{2} + \frac{(m-n)(m-n+1)}{2} + n(m-n) = \frac{m(m+1)}{2}.$$

So we arrive at the amount of $\frac{(m-n)^2+(m-n)}{2}$.

Of course, if $\gamma \in \Phi_2$ is such that $y_\gamma \in C_{\mathfrak{n}_2^-}(\mathfrak{n}_1) = \{z \in \mathfrak{n}_2^- \mid [\mathfrak{n}_1, z] = 0\}$, the centralizer of $\mathfrak{n}_1$ inside $\mathfrak{n}_2^-$, then y_γ automatically satisfies (6.5). However, for some inclusions $\mathfrak{g}_1 \subset \mathfrak{g}_2$ it could be that $\alpha^* = \beta + \gamma$ with $\beta, \gamma \in \Delta(2) \setminus \Phi_1^*$. Since β, γ are additional roots, y_β, y_γ are both embedding elements, but they are not in the centralizer $C_{\mathfrak{n}_2^-}(\mathfrak{n}_1)$. Indeed,

$$[x_{\alpha^*}, y_\beta]v^+ = \lambda x_\gamma v^+ = 0, \quad \lambda \in K.$$

Also, we do not have that $y_\beta y_\gamma$ also satisfies (6.5). Indeed

$$\begin{aligned}
[x_{\alpha^*}, y_\beta y_\gamma]v^+ &= [x_{\alpha^*}, y_\beta]y_\gamma v^+ + y_\beta[x_{\alpha^*}, y_\gamma]v^+ \\
&= \lambda x_\gamma y_\gamma v^+ + \mu y_\beta x_\beta v^+ \\
&= 2\lambda v^+ \neq 0.
\end{aligned}$$

We intend to apply the glider theory first to chains of simple Lie algebras of the same type A,B,C,D with inclusions as in Example 6.1.9. For these inclusions we can choose $\Delta(1), \Delta(2)$ such that $\Delta(1)^* \subset \Delta(2)$, hence we do not have the peculiar behavior from above. Therefore we assume from now on that the inclusion $\mathfrak{g}_1 \subset \mathfrak{g}_2$ satisfies condition (6.2) but that the situation above does not occur. We deduced that for these inclusions we have that y_γ is an embedding element if and only if $y_\gamma \in C_{\mathfrak{n}_2^-}(\mathfrak{n}_1)$.

Lemma 6.2.6. *Let $\beta, \gamma \in \Phi_2^+$ be such that $\beta + \gamma$ is also a root. If two out of three elements of the set $\{y_\beta, y_\gamma, y_{\beta+\gamma}\}$ are embedding elements, then so is the third one.*

Proof. This follows directly from the Jacobi identity as $y_{\beta+\gamma} = \lambda[y_\beta, y_\gamma]$ for some $\lambda \in K$. □

In particular, if β, γ are roots such that their associated elements y_β, y_γ are embedding elements, then if $\beta + \gamma$ is a root, then $y_{\beta+\gamma}$ is also an embedding element.

Lemma 6.2.7. *Let β and γ be positive roots such that y_β and y_γ are embedding elements, then for all $n, m \geq 0$, $y_\beta^n y_\gamma^m$ also satisfies condition* (6.5).

Proof. This is straightforward, as $y_\beta^n y_\gamma^m \in C_{U(\mathfrak{n}_2^-)}(\mathfrak{n}_1)$. □

Lemma 6.2.8. *Let $\beta \in \Phi_2$ be a positive root. The following are equivalent*

1. *y_β is an embedding element;*
2. *$\forall n > 0$, $y_\beta^n \in C_{U(\mathfrak{n}_2^-)}(\mathfrak{n}_1)$;*
3. *$\exists n > 0$, $y_\beta^n \in C_{U(\mathfrak{n}_2^-)}(\mathfrak{n}_1)$.*

Proof. (1) $\Rightarrow$ (2) is Lemma 6.2.7 and (2) $\Rightarrow$ (3) is trivial. Assume that $y_\beta^n \in C_{U(\mathfrak{n}_2^-)}(\mathfrak{n}_1)$ for some $n > 1$. From [37, Lemma 21.4] we know that for any $\alpha^* \in (\Phi_1^*)^+$ we have the equality

$$\begin{aligned} 0 &= [x_{\alpha^*}, y_\beta^n] \\ &= n[x_{\alpha^*}, y_\beta]y_\beta^{n-1} + \binom{n}{2}[[x_{\alpha^*}, y_\beta], y_\beta]y_\beta^{n-2} + \binom{n}{3}[[[x_{\alpha^*}, y_\beta]y_\beta]y_\beta]y_\beta^{n-3}. \end{aligned}$$

If $[x_{\alpha^*}, y_\beta] = \lambda y_{\beta-\alpha^*}$ for some $\lambda \in K$, then we would have a linear dependence relation of PBW-polynomials. Indeed, either $[[x_{\alpha^*}, y_\beta], y_\beta] = 0$ or equals y_γ for some root $\gamma \in \Phi_2^+$. The same holds for$[[[x_{\alpha^*}, y_\beta]y_\beta]y_\beta]$, but with a different root γ'. This is, however, impossible. Observe moreover that we wrote $y_{\beta-\alpha^*}$, because $\alpha^* - \beta$ cannot be positive by the assumption on the inclusion $\mathfrak{g}_1 \subset \mathfrak{g}_2$. □

Lemma 6.2.9. *Let γ, β be positive roots. Then $y_\beta y_\gamma \in C_{U(\mathfrak{n}_2^-)}(\mathfrak{n}_1)$ if and only if y_β, y_γ are embedding elements.*

Proof. We only need to show the 'only if' direction, so suppose that $y_\beta \notin C_{U(\mathfrak{n}_2^-)}(\mathfrak{n}_1)$ and let $\alpha^* \in (\Phi_1^*)^+$ be such that $0 \neq [x_{\alpha^*}, y_\beta] \in \mathfrak{g}_{2,\alpha^*-\beta}$. By the assumption on the inclusion $\mathfrak{g}_1 \subset \mathfrak{g}_2$, $\alpha^* - \beta$ must be a negative root, whence we can write $[x_{\alpha^*}, y_\beta] = \lambda y_{\beta-\alpha^*}$. We obtain

$$\begin{aligned} 0 = [x_{\alpha^*}, y_\beta y_\gamma] &= x_{\alpha^*} y_\beta y_\gamma - y_\beta y_\gamma x_{\alpha^*} \\ &= y_\beta x_{\alpha^*} y_\gamma + \lambda y_{\beta-\alpha^*} y_\gamma - y_\beta y_\gamma x_{\alpha^*} \\ &= y_\beta [x_{\alpha^*}, y_\gamma] + \lambda y_{\beta-\alpha^*} y_\gamma. \end{aligned}$$

It follows that $[x_{\alpha^*}, y_\gamma] \neq 0$, which means that it equals $\mu y_{\gamma-\alpha^*}$ for some $\mu \in K$. Hence we obtain the equality

$$\lambda y_{\beta-\alpha^*} y_\gamma = -\mu y_\beta y_{\gamma-\alpha^*}$$

in $U(\mathfrak{n}_2^-)$. If the monomials on both the left and right hand side are PBW-monomials, then $\beta - \alpha^* = \beta$, which is absurd. If, say, the right hand side is not written in PBW form, then

$$\lambda y_{\beta-\alpha^*} y_\gamma = -\mu y_{\gamma-\alpha^*} y_\beta + \rho [y_\beta, y_{\gamma-\alpha^*}].$$

If the bracket $[y_\beta, y_{\gamma-\alpha^*}] = 0$, then $\beta = \gamma$. But then it follows from Lemma 6.2.8 that y_β is an embedding element, contradicting our assumption. If the bracket is not zero, we obtain a linear dependence relation between three PBW-monomials which is impossible. Hence, our contradiction is wrong, i.e. y_β is an embedding element. It follows then that the same is true for y_γ. □

Proposition 6.2.10. *Let $z = y_{\alpha_1}^{r_1} \dots y_{\alpha_m}^{r_m}$ be a PBW-monomial in $U(\mathfrak{n}_2^-)$. Then $z \in C_{U(\mathfrak{n}_2^-)}(\mathfrak{n}_1)$ if and only if all y_{α_i} are embedding elements ($i = 1, \dots, m$).*

Proof. We proceed by induction on the number n of positive roots α_i appearing in the PBW-monomial z. The case $n = 1$ is just a restatement of Lemma 6.2.8. Assume now that the result holds for $n-1$ and let $z = y_\alpha^m z'$, where z' has $n-1$ roots $\alpha_i \neq \alpha$ appearing. Suppose that y_α is not an embedding element and let $\beta^* \in (\Phi_1^*)^+$ be such that $[x_{\beta^*}, y_\alpha^m] = \sum_k' Y_k$ is a non-zero sum of PBW-monomials. We have the equality

$$0 = [x_{\beta^*}, z] = y_\alpha^m [x_{\beta^*}, z'] + \sum_k{}' Y_k z'.$$

As in the proof of Lemma 6.2.9, we conclude that $[x_{\beta^*}, z'] = \sum_l' Z_l$ is a non-zero sum of PBW-monomials and we obtain the equality

$$-y_\alpha^m \sum_l{}' Z_l = \sum_k{}' Y_k z'.$$

By the proof of Lemma 6.2.8 we know that no y_α^m appears in the monomials Y_k and it also does not appear in z', so we arrive at a contradiction. We conclude that y_α is an embedding element, whence $z' \in C_{U(\mathfrak{n}_2^-)}(\mathfrak{n}_1)$. The result now follows via induction. □

The previous results hint at a possibility for the centralizer $C_{U(\mathfrak{n}_2^-)}(\mathfrak{n}_1)$ to be generated by the embedding elements y_γ. The only thing we still need to check is that when the sum $z_1 + z_2$ of two PBW-monomials is in $C_{U(\mathfrak{n}_2^-)}(\mathfrak{n}_1)$, then so are both elements z_1, z_2. If the weight of z_1 is different from the weight of z_2, this is trivial. Also, even if z_1, z_2 have the same weight, but different degree, then it follows again that both z_1, z_2 are in the centralizer. By the degree of a PBW-monomial $z = y_1^{r_1} \dots y_m^{r_m} h_1^{s_1} \dots h_n^{s_n} x_1^{t_1} \dots x_m^{t_m}$ we mean the sum $\sum_{i=1}^m (t_i - r_i)$. To see this, suppose that $z_1 = y_{\alpha_1}^{r_1} \cdots y_{\alpha_m}^{r_m}, z_2 = y_{\alpha_1}^{s_1} \cdots y_{\alpha_m}^{s_m}$, with $-\sum_{i=1}^m r_i < -\sum_{i=1}^m s_i$ and such that $z_1 + z_2 \in C_{U(\mathfrak{n}_2^-)}(\mathfrak{n}_1)$ but both $z_1, z_2 \notin C_{U(\mathfrak{n}_2^-)}(\mathfrak{n}_1)$. Then there exists an $\alpha^* \in (\Phi_1^*)^+$ such that

$$0 \neq [x_{\alpha^*}, z_1] = -[x_{\alpha^*}, z_2].$$

When expressing the left hand side of the above equality as a linear combination of PBW-monomials there appears exactly one monomial of degree $1 - \sum_{i=1}^m r_i$, namely $z_1 x_{\alpha^*}$ and all the other PBW-monomials in this combination have higher degree. Similarly, in the expression of the right hand side, the lowest degree appearing is $1 - \sum_{i=1}^m s_i > 1 - \sum_{i=1}^m r_i$, a contradiction. The only ingredient missing for proving that the embedding elements are a generating set is when z_1, z_2 have the same degree. Unfortunately, we have a counterexample.

Example 6.2.11. Consider the inclusion $\mathfrak{sl}_2 \subset \mathfrak{sl}_4$ embedded in the top left hand corner. We know from Example 6.2.5 that $y_{L_1-L_4}$ and $y_{L_1-L_3}$ are not embedding elements. Hence Lemma 6.2.9 entails that $z_1 = E_{41} \otimes E_{32} = y_{L_1-L_4}y_{L_2-L_3}$ and $z_2 = E_{42} \otimes E_{31} = y_{L_2-L_4}y_{L_1-L_3}$ are not in the centralizer. For $\alpha^* = L_1 - L_2$ we have

$$[E_{12}, E_{41} \otimes E_{32} - E_{42} \otimes E_{31}] = -E_{42} \otimes E_{32} + E_{42} \otimes E_{32} = 0,$$

which shows that $z_1 - z_2 \in C_{U(\mathfrak{n}_2^-)}(\mathfrak{n}_1)$.

Now, let us finally go back to the construction of Verma gliders. For chains of length two, that is, just an inclusion $\mathfrak{g}_1 \subset \mathfrak{g}_2$, we must embed a $\mathfrak{g}_1$-Verma module $M(\lambda_1)$ in a $\mathfrak{g}_2$-Verma module $M(\lambda_2)$ by means of an element z in $U(\mathfrak{n}_2^-)$ that satisfies (6.5). For chains of bigger length we add a small remark.

Remark 6.2.12. Our rather long digression on the centralizer $C_{U(\mathfrak{n}_2^-)}(\mathfrak{n}_1)$ showed that 'many' elements $z \in U(\mathfrak{n}_2^-)$ satisfy (6.5). Also, if $z \in U(\mathfrak{n}_2^-)$ satisfies, then so does kz for any $k \in K$. However, there is a good reason why we always want $k = 1$. Indeed, if we look at chains of Lie algebras of bigger length, say 3, then we consider functionals

$$\lambda_i : \mathfrak{h}_i \to \mathbb{C}, \quad i = 1, 2, 3.$$

The idea remains the same: we want to embed $M(\lambda_1)$ inside $M(\lambda_2)$ as $U(\mathfrak{g}_1)$-modules and $M(\lambda_2)$ inside $M(\lambda_3)$ as $U(\mathfrak{g}_2)$-modules, such that the composition embeds $M(\lambda_1)$ as an $U(\mathfrak{g}_1)$-module as well. Suppose that $s, r \in \mathbb{C}$, $y \in U(\mathfrak{n}_2^-)$ and $z \in U(\mathfrak{n}_3^-)$ are such that

$$v_{\lambda_1}^+ \mapsto syv_{\lambda_2}^+, \quad v_{\lambda_2}^+ \mapsto rzv_{\lambda_3}^+$$

give the right embeddings. Composition is given by $v_{\lambda_1}^+ \mapsto rsyzv_{\lambda_3}^+$ but for $h_1 \in \mathfrak{h}_1$ we have that

$$\begin{aligned} \lambda_1(h) &= \lambda_2(h) - s\widetilde{y}(h) \\ &= \lambda_3(h) - r\widetilde{z}(h) - s\widetilde{y}(h) \\ &= \lambda_3(h) - rs(\widetilde{y}(h) + \widetilde{z}(h)). \end{aligned}$$

Hence we must have that $r = s = rs$ or that $r = s = 1$.

Example 6.2.13. Consider the tower $\mathfrak{sl}_2 \subset \mathfrak{sl}_3 \subset \mathfrak{sl}_4$, with embeddings graphically depicted as

$$\left(\begin{array}{cc|c|c} *_1 & *_1 & *_2 & 0 \\ *_1 & *_1 & *_2 & 0 \\ \hline *_2 & *_2 & *_2 & 0 \\ \hline 0 & 0 & 0 & 0 \end{array}\right)$$

With notations as before, let $\Delta(1)^* = \{L_1 - L_2\}, \Delta(2)^* = \{L_1 - L_2, L_2 - L_3\}$ and $\Delta(3) = \{L_1 - L_2, L_2 - L_3, L_3 - L_4\}$ and the dual basis of $\{L_i\}$ for $\mathfrak{h}$ is denoted by $\{H_1, \ldots, H_4\}$. For the first embedding, the only candidate is $\alpha = L_2 - L_3$ and for the second we have $\beta = L_3 - L_4$. Hence

$$(\lambda_2 - \lambda_1)(H_1 - H_2) = \alpha(H_1 - H_2) = -1$$

and

$$(\lambda_3 - \lambda_2)(H_1 - H_2,\ H_2 - H_3) = \beta(H_1 - H_2,\ H_2 - H_3) = (0,\ -1).$$

For example, take $\lambda_1 = 3$, $\lambda_2 = (2\ 1)$ and $\lambda_3 = (2\ 0\ 0)$ where the functionals are represented with regard to the basis $\{H_1 - H_2, H_2 - H_3, H_3 - H_4\}$. So we have $M_2 = U(\mathfrak{sl}_2)y_{L_2-L_3}y_{L_3-L_4}v^+$. The embedding of $M(\lambda_2)$ into Ω uses the simple root $L_3 - L_4$ and for simple roots α we can easily check whether $s_\alpha \cdot \lambda_3 < \lambda_3$, i.e. $\lambda_3 - s_\alpha \cdot \lambda_3 \in \Gamma_3$. Recall that for an element σ of the Weyl group and $\lambda \in \mathfrak{h}^*$, the dot action $\sigma \cdot \lambda = \sigma(\lambda + \rho) - \rho$ with ρ half the sum of the positive roots. Since α is simple, $s_\alpha(\rho) = \rho - \alpha$, whence

$$s_\alpha \cdot \lambda - \lambda = s_\alpha(\lambda + \rho) - \rho - \lambda = -(\lambda(h_\alpha) + 1)\alpha.$$

So it suffices that

$$\langle \lambda_3, (L_3 - L_4)^\vee \rangle = \lambda_3(H_3 - H_4) = 0 \in \mathbb{Z}^+.$$

There are two conditions on M_1, namely

$$M_2 = U(\mathfrak{sl}_2)y_{L_2-L_3}y_{L_3-L_4}v^+ \subset U(\mathfrak{sl}_3)y_{L_2-L_3}y_{L_3-L_4}v^+ \subset M_1,$$

and

$$M_1 \subset U(\mathfrak{sl}_3)y_{L_3-L_4}v^+ \subset M(s_{L_3-L_4} \cdot \lambda_3) \subset M \subset \Omega = M(\lambda_3).$$

The second condition comes from the fact that we want M_1 to be contained in the $\mathfrak{sl}_3$-Verma module $M(\lambda_2)$. For M_1 we can choose any $U(\mathfrak{sl}_2)$-module satisfying both conditions. For example, $M_1 = M(s_{L_3-L_4} \cdot \lambda_3)$ satisfies. For M we can add the $y_{L_1-L_2}$-string starting from $y_\alpha v^+$ with α any positive root in $\Phi_3 \setminus \Phi_2^*$. For example

$$\Omega = M(\lambda_3) \supset M = M(s_{L_3-L_4} \cdot \lambda_3) + U(\mathfrak{sl}_2)y_{L_1-L_4}v^+ \supset M(s_{L_3-L_4} \cdot \lambda_3) \supset U(\mathfrak{sl}_2)y_{L_2-L_3}y_{L_3-L_4}v^+.$$

Another Verma glider would be

$$\Omega = M(\lambda_3) \supset M(s_{L_3-L_4} \cdot \lambda_3) + U(\mathfrak{sl}_2)y_{L_1-L_4}v^+ \supset U(\mathfrak{sl}_3)y_{L_3-L_4}v^+ \supset U(\mathfrak{sl}_2)y_{L_2-L_3}y_{L_3-L_4}v^+.$$

Example 6.2.14. In the special case that $(\lambda_i)_{|\mathfrak{h}_{i-1}} = \lambda_{i-1}$ for $i = 2, \ldots, n$, we can take the element $z = 1$ to define the embedding at every stage. We obtain the glider representation

$$M(\lambda_n) = U(\mathfrak{u}_n^-) \otimes \mathbb{C}_{\lambda_n} \supset U(\mathfrak{u}_{n-1}^-) \otimes \mathbb{C}_{\lambda_n} \supset \cdots \supset U(\mathfrak{u}_1^-) \otimes \mathbb{C}_{\lambda_n}.$$

6.3 Irreducible gliders

To investigate when Verma gliders are irreducible, we recall the notion of antidominant weights. A functional $\lambda \in \mathfrak{h}^*$ is called antidominant if

$\langle \lambda + \rho, \alpha^\vee \rangle \notin \mathbb{Z}^{>0}$, where ρ is half the sum of the positive roots (or the sum of the fundamental weights $\overline{\omega}_i$, which are obtained by base change via the Cartan matrix). Antidominant weights play an important role in the study of Verma modules since $M(\lambda)$ is simple if and only if λ is antidominant. We also recall the following important theorem due to Verma.

Theorem 6.3.1. *Let $\lambda \in \mathfrak{h}^*$. Suppose that $\mu := s_\alpha \cdot \lambda \leq \lambda$ for some $\alpha \in \Phi^+$. Then there exists an embedding $M(\mu) \subset M(\lambda)$.*

Moreover, since $\dim \mathrm{Hom}(M(\mu), M(\lambda)) \leq 1$ for all μ, λ, the above embedding is unique up to some scalar. In the particular situation of Example 6.2.14, the answer as to whether a Verma glider is irreducible follows from the classical representation theory of Verma modules.

Proposition 6.3.2. *With assumptions and notations of Example 6.2.14, we have that $M(\lambda_n)$, as a fragment, is an irreducible fragment if and only if $\lambda_1 \in \mathfrak{h}_1^*$ is antidominant.*

Proof. If $M(\lambda_n)$ is irreducible, then $U(\mathfrak{u}_1^-) \otimes \mathbb{C}_{\lambda_n}$ must be a simple $U(\mathfrak{g}_1)$-module. But this module is just the ordinary Verma module $M(\lambda_1)$ and so λ_1 is antidominant. The converse follows easily since by definition $U(\mathfrak{g}_i)M^{\mathfrak{g}_1}(\lambda_1) = U(\mathfrak{u}_i^-) \otimes \mathbb{C}_{\lambda_n}$, for all $i = 1, \ldots, n$ (the upper index in $M^{\mathfrak{g}_1}(\lambda_1)$ means that we consider the $\mathfrak{g}_1$-Verma module). □

Example 6.3.3. Consider $\mathfrak{sl}_2 \subset \mathfrak{sl}_3$ with the embedding as in Example 6.2.13. The root vectors $L_3 - L_1 = -\rho$ and $L_2 - L_3$ both restrict to $\frac{-1}{2}(L_1 - L_2)$ on $\mathfrak{h}_1$, which is antidominant (every antidominant weight λ is minimal in its linkage class $W \cdot \lambda$ and in $\mathfrak{sl}_2$ only λ and $-\lambda - 2$ are linked). By the proposition, both Verma gliders

$$\Omega = M = M(-\rho) \supset M^{\mathfrak{sl}_2}(-\tfrac{1}{2}(L_1 - L_2)),$$
$$\Omega = M = M(L_2 - L_3) \supset M^{\mathfrak{sl}_2}(-\tfrac{1}{2}(L_1 - L_2))$$

are irreducible. However, since $\langle L_2 - L_3 + \rho, (L_2 - L_3)^\vee \rangle = 3$, $L_2 - L_3$ is not antidominant. Of course, $-\rho$ is antidominant.

When there appear non-trivial embedding elements, Proposition 6.3.2 is no longer true. Of course, it is a necessary condition for a fragment of essential length $n - 1$ to be irreducible that $M_{n-1}/B(M)$ is a simple F_0R-module. By definition, this $M_{n-1} = M(\lambda_1)$, so irreducibility of the glider indeed implies that λ_1 is antidominant.

Example 6.3.4. Consider $\mathfrak{sl}_2 \subset \mathfrak{sl}_3$ embedded in the same way as in the previous example. If $\lambda_1 = -\frac{1}{2}(L_1 - L_2)$ and $\lambda_2 = 3(L_2 - L_3)$, then $z = y_{L_2 - L_3}^2$ is an embedding element that satisfies. Hence we have the Verma glider

$$\Omega = M(3(L_2 - L_3)) = M \supset U(\mathfrak{sl}_2) y_{L_2 - L_3}^2 v_{\lambda_1}^+.$$

Since $\langle 3(L_2 - L_3), (L_2 - L_3)^\vee \rangle = 6$, we know that

$$M(s_{L_2-L_3} \cdot \lambda_2) = U(\mathfrak{sl}_3)y^7_{L_2-L_3} \subset M(\lambda_2).$$

Hence $M(s_{L_2-L_3} \cdot \lambda_2) \supset U(\mathfrak{sl}_2)y^2_{L_2-L_3}v^+_{\lambda_1}$ is a non-trivial subfragment.

This example shows that the study of irreducible gliders, even for chains of length 2, is already of a much higher complexity. In some cases, however, we can say something more. To state the result we recall that each Verma module $M(\lambda)$ has a unique maximal submodule $N(\lambda)$ and unique simple quotient $L(\lambda) = M(\lambda)/N(\lambda)$. It is a natural question to ask for which λ the simple quotients are finite dimensional. To this extent, we recall the notion of dominant integral weights.

The root system Φ of $\mathfrak{g}$ determines a root lattice Λ_r, which is just the $\mathbb{Z}$-span of Φ. There is also a natural dual lattice, called the integral weight lattice Λ defined by

$$\Lambda = \{\lambda \in \mathfrak{h}^* \mid \langle \lambda, \alpha^\vee \rangle \in \mathbb{Z} \text{ for all } \alpha \in \Phi\}.$$

Clearly, $\Lambda_r \subset \Lambda$ and their quotient Λ/Λ_r is a finite group, the fundamental group of the Lie algebra $\mathfrak{g}$. The subset Λ^+ denotes the set of elements of Λ for which the inproduct is nonnegative for all $\alpha \in \Phi$. We call Λ^+ the set of dominant integral weights. Their importance is given by

Theorem 6.3.5. *[38, Theorem 1.6]*
The simple module $L(\lambda)$ is finite dimensional if and only if $\lambda \in \Lambda^+$.

Now, from Proposition 6.4.1 we know that if $\mathfrak{g}_1 \subset \mathfrak{g}_2$ is an embedding of simple Lie algebras of type A,B or D and such that $\mathrm{rk}(\mathfrak{g}_2) = \mathrm{rk}(\mathfrak{g}_1) + 1$, there is only one embedding element y_α with α the additional simple root $\alpha \in \Delta(2)\backslash\Delta(1)^*$. So if z is a PBW-monomial in $U(\mathfrak{n}_2^-)$ that determines the embedding, then $z = y^n_\alpha$ by Proposition 6.2.10. We have

Proposition 6.3.6. *Let $\lambda_2 \in \Lambda_2^+, \lambda \in \mathfrak{h}_1^*$ and $z = y^n_\alpha$ be such that $\Omega = M(\lambda_2) \supset M \supset U(\mathfrak{g}_1)y^n_\alpha v^+$ is a Verma glider. Then the Verma glider is irreducible if and only if λ_1 is antidominant and $n < m = \langle \lambda_2, \alpha^\vee \rangle$.*

Proof. Suppose that λ_1 is antidominant and $n < m$. If $M \supset M_1$ is not irreducible then $U(\mathfrak{g}_2)y^n_\alpha v^+ \subsetneq M$. In particular, it follows that

$$U(\mathfrak{g}_2)y^n_\alpha v^+ \subset N(\lambda_2) = \sum_{\alpha_i \in \Delta(2)} M(s_{\alpha_i} \cdot \lambda_2).$$

Because $n < m$, we have that $y^n_\alpha v^+ \notin M(s_\alpha \cdot \lambda_2)$, hence $y^n_\alpha v^+ \notin N(\lambda_2)$, a contradiction. Conversely, if $M \supset M_1$ is irreducible, then λ_1 must be antidominant. Also, since $U(\mathfrak{g}_2)y^m_\alpha v^+ = M(s_\alpha \cdot \lambda_2) \subsetneq M = U(\mathfrak{g}_2)y^n_\alpha v^+$, we have that $m > n$. □

For general embeddings $\mathfrak{g}_1 \subset \mathfrak{g}_2$ we have the following.

Proposition 6.3.7. *Let $\mathfrak{g}_1 \subset \mathfrak{g}_2$ be a chain of semisimple Lie algebras and suppose that $\lambda_i \in \mathfrak{h}_i^*$, $i = 1, 2$ are such that a Verma glider $\Omega = M(\lambda_2) \supset M \supset M(\lambda_1)$ exists. If λ_2 is dominant integral and λ_1 is not, then $M(\lambda_1) = M_1 \subset N(\lambda_2)$ and $U(\mathfrak{g}_2)M(\lambda_1) \subset M \cap N(\lambda_2)$.*

Proof. Since $N(\lambda_2) \cap M_1$ is an $U(\mathfrak{g}_1)$-submodule of $M_1 = M(\lambda_1)$, we have that $N(\lambda_2) \cap M_1 \subset N(\lambda_1)$ or that $N(\lambda_2) \cap M_1 = M_1$. Since λ_2 is dominant integral we have that $L(\lambda_2)$ is finite dimensional. If the first case holds, we have an isomorphism of vector spaces

$$\frac{M_1}{N(\lambda_1)} \cong \frac{\frac{M_1}{N(\lambda_2)\cap M_1}}{\frac{N(\lambda_1)}{N(\lambda_2)\cap M_1}}.$$

Since $M_1/(N(\lambda_2) \cap M_1)$ embeds in $M(\lambda_2)/N(\lambda_2)$ it follows that λ_1 is dominant integral, a contradiction. So we have that $M_1 \subset N(\lambda_2)$. The last statement then automatically follows by the definition of a glider representation. □

Corollary 6.3.8. *In the situation of the previous proposition, a Verma glider $\Omega = M(\lambda_2) \supset M \supset M(\lambda_1)$ with $\Omega = U(\mathfrak{g}_2)M$ and λ_2 dominant integral, is never irreducible.*

Proof. By the previous proposition and by the hypothesis $\Omega = U(\mathfrak{g}_2)M$, the glider representation $N(\lambda_2) \cap M \supset M(\lambda_1)$ is a non-trivial subfragment. □

For chains $\mathfrak{g}_1 \subset \ldots \subset \mathfrak{g}_n$ we restrict to Verma gliders of the form

$$\Omega = M = M(\lambda_n) \supset M(\lambda_{n-1}) \supset \ldots \supset M(\lambda_1),$$

for functionals $\lambda_i \in \mathfrak{h}_i^*$. If $z_i \in U(\mathfrak{n}_{i+1}^-)$ is the element that determines the embedding $M(\lambda_i)$ inside $M(\lambda_{i+1})$, then we can rewrite such a glider as

$$M = U(\mathfrak{g}_n)v^+ \supset U(\mathfrak{g}_{n-1})z_{n-1}v^+ \supset \ldots \supset U(\mathfrak{g}_2)z_2z_3\ldots z_{n-1}v^+ \supset U(\mathfrak{g}_1)z_1\ldots z_{n-1}v^+. \quad (6.6)$$

Again, if all Lie algebras $\mathfrak{g}_i$ are of the same type A,B or D and $\mathrm{rk}(\mathfrak{g}_{i+1}) = \mathrm{rk}(\mathfrak{g}_i) + 1$ and the z_i are PBW-monomials, then $z_i = y_{\alpha_i}^{k_i}$ with α_i in $\Delta(i+1) \setminus \Delta(i)^*$ the additional simple root. We have the generalization of Proposition 6.3.6.

Theorem 6.3.9. *Consider a Verma glider of the form (6.6) with $\lambda_i \in \Lambda_i^+$ for $i = 2, \ldots, n$. The Verma glider is irreducible if and only if λ_1 is antidominant and $k_i < m_i = \langle \lambda_{i+1}, \alpha_i^\vee \rangle$ for all $i = 1, \ldots, n-1$.*

Proof. This is analogous to the proof of Proposition 6.3.6. □

6.4 Nilpotent orbits

We recall the construction of nilpotent orbits from [25]. To any complex Lie algebra $\mathfrak{g}$ one can associate its adjoint group G_{ad}, which is a connected complex Lie group. In case $\mathfrak{g}$ is semisimple, $G_{ad} = \mathrm{Aut}(\mathfrak{g})^\circ$ is the identity component of the automorphism group

$$\mathrm{Aut}(\mathfrak{g}) = \{\varphi \in \mathrm{End}(\mathfrak{g}) \mid [\varphi(\mathrm{X}), \varphi(\mathrm{Y})] = \varphi[\mathrm{X}, \mathrm{Y}], \text{ for all } \mathrm{X}, \mathrm{Y} \in \mathfrak{g}\}.$$

For any $X \in \mathfrak{g}$ we then obtain the adjoint orbit

$$\mathcal{O}_X := G_{ad} \cdot X = \{\varphi(X) \mid \varphi \in G_{ad}\}.$$

Such an adjoint orbit $\mathcal{O}_X$ is called semisimple, resp. nilpotent when X is semisimple, resp. nilpotent. Since for any $\varphi \in \mathrm{Aut}(\mathfrak{g})$ and $X \in \mathfrak{g}$ it holds that

$$\varphi \cdot \mathrm{ad}_{\mathrm{X}} \cdot \varphi^{-1} = \mathrm{ad}_{\varphi(\mathrm{X})},$$

a semisimple, resp. nilpotent orbit consists entirely of semisimple, resp. nilpotent elements. Apparently, the number of semisimple orbits can be infinite, but the number of nilpotent orbits is always finite.

The construction of Verma gliders leads to the existence of embedding elements. Obviously, these embedding elements are nilpotent elements, hence belong to some nilpotent orbit. For a complex semisimple Lie algebra $\mathfrak{g}$ these nilpotent orbits are classified by the Dynkin-Kostant classification, see e.g. [25, Chapter 3] for a nice overview. For $\mathfrak{g}_1 \subset \mathfrak{g}_2$ two complex semisimple Lie algebras, we ask ourselves which nilpotent orbits we reach by just looking at the embedding elements. We restrict here to Lie algebras of the same type A, B, C and D and embeddings as in Example 6.1.9. In Example 6.2.5 we determined the embedding elements for type A. One can perform similar reasonings for the other types to obtain the following:

Proposition 6.4.1. *Let $\mathfrak{g}_1 \subset \mathfrak{g}_2$ be a canonical embedding of simple Lie algebras of the same type A, B, C or D of rank $n < m$. The number of embedding elements for each type is given by*

$$\begin{array}{ll} \text{type } A: & \frac{(m-n)^2+(m-n)}{2} \\ \text{type } B: & (m-n)^2 \\ \text{type } C: & (m-n)^2+(m-n) \\ \text{type } D: & (m-n)^2 \end{array}$$

Let us start by looking at type A, i.e. at $\mathfrak{sl}_n \subset \mathfrak{sl}_m$. In this case, the classification is given by partitions of m. We introduce some notation (following [25]).

A partition of m is a tuple $[d_1^{i_1}, d_2^{i_2}, \ldots, d_k^{i_k}]$ with d_j and i_j positive integers such that

$$d_1 \geq d_2 \geq \ldots \geq d_k > 0 \text{ and } i_1 d_1 + i_2 d_2 + \ldots + i_k d_k = m.$$

For a positive integer i, we denote the elementary Jordan block of type i by

$$J_i = \begin{pmatrix} 0 & 1 & 0 & \ldots & 0 & 0 \\ 0 & 0 & 1 & \ldots & 0 & 0 \\ \vdots & \vdots & \vdots & \ddots & \vdots & \vdots \\ 0 & 0 & 0 & \ldots & 0 & 1 \\ 0 & 0 & 0 & \ldots & 0 & 0 \end{pmatrix} \in M_i(\mathbb{C}).$$

For a partition $[d_1^{i_1}, d_2^{i_2}, \ldots, d_k^{i_k}]$ of m we form the diagonal sum of elementary Jordan blocks

$$X_{[d_1^{i_1}, d_2^{i_2}, \ldots, d_k^{i_k}]} = \begin{pmatrix} J_{d_1} & 0 & 0 & \ldots & 0 \\ 0 & J_{d_1} & 0 & \ldots & 0 \\ \vdots & \vdots & \vdots & \ddots & \vdots \\ 0 & 0 & 0 & \ldots & J_{d_k} \end{pmatrix}$$

where there are i_1 blocks J_{d_1}, i_2 blocks J_{d_2}, etc. The matrix $X_{[d_1^{i_1}, d_2^{i_2}, \ldots, d_k^{i_k}]}$ in $\mathfrak{sl}_m$ is nilpotent and generates the nilpotent orbit

$$\mathcal{O}_{[d_1^{i_1}, d_2^{i_2}, \ldots, d_k^{i_k}]} = PSL_m \cdot X_{[d_1^{i_1}, d_2^{i_2}, \ldots, d_k^{i_k}]}.$$

There exists a partial order on the set of partitions m. Namely, let $\mathbf{d} = [d_1^{i_1}, d_2^{i_2}, \ldots, d_k^{i_k}], \mathbf{f} = [f_1^{j_1}, f_2^{j_2}, \ldots, f_l^{j_l}]$ be two partitions of m. Write $\mathbf{d} = [D_1, D_2, \ldots, D_m] = [d_1, d_1, \ldots, d_1, d_2, \ldots, d_k, 0, \ldots, 0]$, with i_1 d_1's, i_2 d_2's, etc. and $m - \sum_{a=1}^{k} i_a$ zeros. Do the same for $\mathbf{f}$. Then $\mathbf{d} \geq \mathbf{f}$ if and only if

$$\sum_{1 \leq a \leq b} D_a \geq \sum_{1 \leq a \leq b} F_a \text{ for } 1 \leq b \leq m.$$

Gerstenhaber and Hesselink showed that this partial ordering corresponds to the partial ordering on the set of nilpotent orbits determined via the Zariski closure, see [25, Theorem 6.2.5]. We denote by $H_{\mathfrak{sl}_n}$ the Hasse diagram of the nilpotent orbits of $\mathfrak{sl}_n$. E.g. $H_{\mathfrak{sl}_3}$ is given by

$$\begin{array}{c} [3] \\ | \\ [2,1] \\ | \\ [1^3] \end{array}$$

If $\iota : \mathfrak{g}_1 \hookrightarrow \mathfrak{g}_2$ is an inclusion of complex semisimple Lie algebras then we denote by $\iota(H_{\mathfrak{g}_1})$ the Hasse subdiagram of $H_{\mathfrak{g}_2}$ containing those orbits $\mathcal{O}$ that have an element $\iota(X)$, $X \in \mathfrak{g}_1$ nilpotent.

Lemma 6.4.2. *Let $A \in M_m(\mathbb{C})$ have Jordan normal form $J(A) = J$; then the Jordan normal form of*

$$J(\left(\begin{array}{c|c} A & 0 \\ \hline 0 & 0 \end{array}\right)) = \left(\begin{array}{c|c} J & 0 \\ \hline 0 & 0 \end{array}\right).$$

Proof. If $J = S^{-1}AS$, then

$$\left(\begin{array}{c|c} J & 0 \\ \hline 0 & 0 \end{array}\right) = \left(\begin{array}{c|c} S^{-1} & 0 \\ \hline 0 & I \end{array}\right)\left(\begin{array}{c|c} A & 0 \\ \hline 0 & 0 \end{array}\right)\left(\begin{array}{c|c} S & 0 \\ \hline 0 & I \end{array}\right). \qquad \square$$

Remark 6.4.3. In the lemma, the 0 can also denote any matrix of size $n \times m$ with all entries zero.

Proposition 6.4.4. *Let $\mathfrak{sl}_n \subset \mathfrak{sl}_m$; then $\iota(H_{\mathfrak{sl}_n})$ is classified by the partitions $[d_1^{i_1}, d_2^{i_2}, \ldots, d_k^{i_k}]$ of m with $d_k = 1$ and $i_k \geq m - n$.*

Proof. Let $X = X_{[d_1^{i_1}, d_2^{i_2}, \ldots, d_k^{i_k}]} \in \mathfrak{sl}_n$ be the nilpotent element orbit associated to the partition $[d_1^{i_1}, d_2^{i_2}, \ldots, d_k^{i_k}]$ of n. Under the inclusion, X is sent to $X_{[d_1^{i_1}, d_2^{i_2}, \ldots, d_k^{i_k}, 1^{m-n}]}$. Conversely, if $[d_1^{i_1}, d_2^{i_2}, \ldots, d_{k-1}^{i_{k-1}}, 1^k]$ is a partition of m with $k \geq m - n$, then the first upper diagonal of $X_{[d_1^{i_1}, d_2^{i_2}, \ldots, d_{k-1}^{i_{k-1}}, 1^k]}$ has zeros on the last $m - n$-entries, hence it belongs to $\iota(\mathfrak{sl}_n)$. $\square$

By Proposition 6.4.1 we know that there are $((m-n)^2 + (m-n))/2$ embedding elements. The embedding elements are situated in the following positions:

$$\left(\begin{array}{ccc|cccccc} 0 & & & & & & & & \\ & \ddots & & & & & & & \\ & & 0 & & & & & & \\ \hline & & & 0 & * & * & \cdots & * \\ & & & & 0 & * & \cdots & * \\ & & & & & \ddots & \ddots & \vdots \\ & & & & & & 0 & * \\ & & & & & & & 0 \end{array}\right) \tag{6.7}$$

If we only make linear combinations of the $m - n$ elements on the first upper diagonal then we obtain the partitions of $m - n + 1$. Proposition 6.4.4 shows that these form $\iota(H_{\mathfrak{sl}_{m-n+1}})$.

Theorem 6.4.5. *Let $\mathfrak{sl}_n \subset \mathfrak{sl}_m$; then the nilpotent orbits generated by the embedding elements correspond to the inclusion of the Hasse subdiagram $\iota(H_{\mathfrak{sl}_{m-n+1}})$.*

Proof. We already observed that we reach $\iota(H_{\mathfrak{sl}_{m-n+1}})$. If X is a nilpotent element with only non-zero coefficients on the $*$-positions of (6.7), then Lemma 6.4.2 shows that X has Jordan normal form with (possibly) only 1's on the starred positions. This shows that $X \in \mathcal{O}_{[d_1^{i_1}, d_2^{i_2}, \ldots, d_k^{i_k}]}$, with $d_k = 1$ and $i_k \geq n-1$. Thus $\mathcal{O}_X \in \iota(H_{\mathfrak{sl}_{m-n+1}})$ by Proposition 6.4.4. $\square$

Next, we consider Lie algebras of type C. Recall from [25, Theorem 5.1.3] that the nilpotent orbits of $\mathfrak{sp}_{2m}$ are in one-to-one correspondence with the set of partitions of $2m$ in which odd parts occur with even multiplicity. The root system of $\mathfrak{sp}_{2m}$ is $\{\pm L_i \pm L_j, \pm 2L_i \mid 1 \leq i, j \leq m,\ i \neq j\}$ and we make the standard choice $\{L_i \pm L_j, 2L_k \mid 1 \leq i < j \leq m, 1 \leq k \leq m\}$ of positive roots. With respect to this choice of basis, the authors give in [25] a recipe for constructing a standard triple $\{X, H, Y\}$ associated with a partition $\mathbf{d}$ of $2m$. We quickly recall this. Given $\mathbf{d} \in \mathcal{P}(2m)$, break it up into chunks of the following two types: pairs $\{2r+1, 2r+1\}$ of equal odd parts, and single even parts $\{2q\}$. We attach sets of positive (but not necessarily simple) roots to each chunk $\mathcal{C}$ as follows. If $\mathcal{C} = \{2q\}$, choose a block $\{j+1, \ldots, j+q\}$ of consecutive indices and let $\mathcal{C}^+ = \mathcal{C}^+(2q) = \{L_{j+1} - L_{j+2}, L_{j+2} - L_{j+3}, \ldots, L_{j+q-1} - L_{j+q}, 2L_{j+q}\}$. If $\mathcal{C} = \{2r+1, 2r+1\}$, choose a block $\{l+1, \ldots, l+2r+1\}$ of consecutive indices and let $\mathcal{C}^+ = \mathcal{C}^+(2r+1, 2r+1) = \{L_{l+1} - L_{l+2}, \ldots, L_{l+2r} - L_{l+2r+1}\}$. (Note that $\mathcal{C}^+$ is empty if $\mathcal{C} = \{1, 1\}$). We define X to be the sum of the X_α, where α appears in some of the $\mathcal{C}^+$.

The inclusion $\mathfrak{sp}_{2n} \subset \mathfrak{sp}_{2m}$ sends a matrix $X \in \mathfrak{sp}_{2n}$ to $\widetilde{X}$ by adding some nonzero rows and columns. Let X be the nilpotent element associated with a partition $\mathbf{d} = [d_1^{n_1}, \ldots, d_k^{n_k}] \in \mathcal{P}(2n)$ given by the above procedure. Since $\mathcal{C}^+(1,1) = \emptyset$, a chunk of the form $\{1^{2k}\}$ of $\mathbf{d}$ does not contribute to X. Hence we see that $\widetilde{X}$ is the associated element of the partition $[d_1^{n_1}, \ldots, d_k^{n_k}, 1^{2(m-n)}]$. This gives the analogue of Proposition 6.4.4 for type C. By Proposition 6.4.1 we know that there are $(m-n)^2 + (m-n)$ embedding elements, amongst which we have the $m-n$ simple roots $L_i - L_{i+1}$, $1 \leq i \leq m-n$. In determining the embedding elements, one deduces that the roots $2L_i$ for $1 \leq i \leq m-n$ also satisfy. In fact, we have all the positive roots of an $\mathfrak{sp}_{2(m-n)}$. We see that we already reach the nilpotent orbits $\iota(H_{\mathfrak{sp}_{2(m-n)}})$. However, if $m-n+1$ is odd, we can form the set of simple roots $\{L_1 - L_2, \ldots, L_{m-n} - L_{m-n+1}\}$ and this corresponds to a chunk $\{2\frac{m-n}{2}+1, 2\frac{m-n}{2}+1\}$, leading to an additional partition $[(m-n+1)^2, 1^{2(n-1)}]$.

Lemma 6.4.6. *Let* $\begin{pmatrix} A & B \\ C & D \end{pmatrix}$ *have Jordan normal form J; then the Jordan normal form of*

$$J\left(\left(\begin{array}{cc|cc} A & 0 & B & 0 \\ 0 & 0 & 0 & 0 \\ \hline C & 0 & D & 0 \\ 0 & 0 & 0 & 0 \end{array}\right)\right) = \left(\begin{array}{c|c} 0 & 0 \\ \hline 0 & J \end{array}\right),$$

and

$$J(\left(\begin{array}{c|cc} 0 & 0 & 0 \\ \hline 0 & A & B \\ 0 & C & D \end{array}\right)) = \left(\begin{array}{c|c} 0 & 0 \\ \hline 0 & J \end{array}\right),$$

where we have the same behavior of 0 as in Remark 6.4.3

Theorem 6.4.7. *Let $\mathfrak{sp}_{2n} \subset \mathfrak{sp}_{2m}$, then the nilpotent orbits generated by the embedding elements are*

$$\begin{array}{rl} H_{\mathfrak{sp}_{2m-2n}} & \text{if } m-n \text{ is odd,} \\ H_{\mathfrak{sp}_{2m-2n}} \cup \mathcal{O}_{[(m-n+1)^2, 1^{2(n-1)}]} & \text{if } m-n \text{ is even.} \end{array}$$

Proof. This follows by the above discussion and Lemma 6.4.6. □

For type B the result is less straightforward. We know by [25, Theorem 5.1.2] that the nilpotent orbits of $\mathfrak{so}_{2m+1}$ are in one-to-one correspondence with the set of partitions of $2m+1$ in which even parts occur with even multiplicity. The root system equals the root system of $\mathfrak{sp}_{2m}$ with the $2L_i$ replaced by L_i. Hence, an inclusion $\mathfrak{so}_{2n+1} \subset \mathfrak{so}_{2m+1}$ can be depicted by (6.3) with the $2L_m$ replaced by L_m. The difference with the symplectic case, however, is that amongst the embedding elements we do not have the L_i for $1 \leq i \leq m-n$. Hence we certainly do not reach all nilpotent orbits of an $\mathfrak{so}_{2(m-n)+1}$. An element of $\mathfrak{so}_{2m+1}$ has the form

$$\begin{pmatrix} 0 & u & v \\ -v^t & Z_1 & Z_2 \\ -u^t & Z_3 & -Z_1^t \end{pmatrix}, \quad u, v \in \mathbb{C}^m, Z_i \in M_m(\mathbb{C}), Z_2, Z_3 \text{ skew}-\text{symmetric},$$

and the positive roots α for which the root vector X_α has non-zero u or v are exactly the L_i, $1 \leq i \leq m$. Since we do not have access to these guys, we get restrictions on the orbits we reach. Anyway, an orbit of $\mathfrak{so}_{2n+1}$ given by a partition $[d_1^{n_1}, \ldots, d_k^{n_k}]$ corresponds to the orbit $[d_1^{n_1}, \ldots, d_k^{n_k}, 1^{2(m-n)}]$ in $\mathfrak{so}_{2m+1}$. In [25] a recipe is given to construct the nilpotent element $X_{[d_1^{n_1}, \ldots, d_k^{n_k}]}$. One has to break up the partition into chunks of three types: pairs $\{r, r\}$ of equal parts, pairs $\{2s+1, 2t+1\}$ of unequal parts and one single odd part $\{2u+1\}$. One then associates positive roots to all three types of chunks and we observe that we only need some L_i for a chunk of the last type $\{2u+1\}$. Moreover, if $u = 0$ then we do not need such an L_i! Hence, we get access to embedded orbits of an $\mathfrak{so}_{2(m-n)+1}$ represented by a partition $\mathbf{d}$ of $2(m-n)+1$ for which the unique chunk $\{2u+1\}$ has $u = 0$. A moment's thought leads to the observation that a partition $\mathbf{d}$ of $2m+1$ having at least one 1, can be broken up into chunks such that $\{2u+1\} = \{1\}$. So we arrive at

Theorem 6.4.8. *Let $\mathfrak{so}_{2n+1} \subset \mathfrak{so}_{2m+1}$; then the nilpotent orbits generated by the embedding elements correspond to the partitions of $2(m-n)+1$ having at least one 1.*

Proof. The above discussion shows that we can construct the $X_{\mathbf{d}}$ for $\mathbf{d}$ a partition of $2m+1$ corresponding to a partition of $2(m-n)+1$ having at least one 1. Let X be a nilpotent element constructed out of the embedding elements; then it is of the form

$$X = \left(\begin{array}{c|cc|cc} 0 & 0 & 0 & 0 & 0 \\ \hline 0 & A & 0 & B & 0 \\ 0 & 0 & 0 & 0 & 0 \\ \hline 0 & C & 0 & D & 0 \\ 0 & 0 & 0 & 0 & 0 \end{array}\right)$$

and Lemma 6.4.6 entails that its Jordan form equals

$$\left(\begin{array}{c|c} 0 & 0 \\ \hline 0 & J(\begin{pmatrix} A & B \\ C & D \end{pmatrix}) \end{array}\right),$$

so the associated partition has at least one 1. □

For example, for $\mathfrak{so}_5 \subset \mathfrak{so}_{11}$, we have $2(m-n)+1 = 7$ and so we do not reach the orbits $[3, 2^2, 1^4]$ and $[7, 1^4]$.

Finally, we discuss type D.

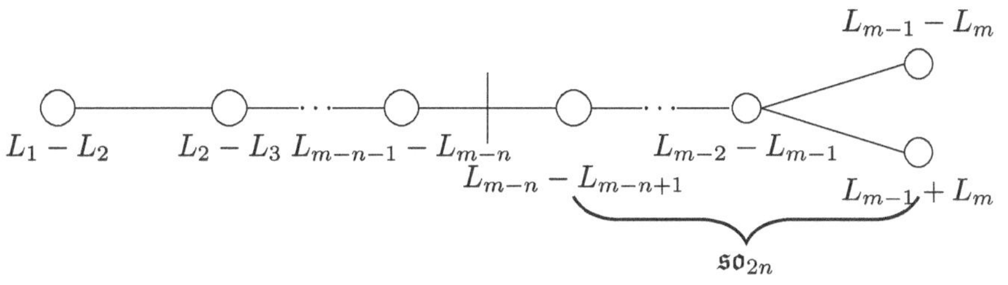

(6.8)

With respect to this embedding, one deduces that amongst the embedding elements we have the roots $L_i \pm L_j$, $1 \leq i < j \leq m-n$. These form the root system of an $\mathfrak{so}_{2(m-n)}$, which has $(m-n)(m-n-1)$ positive roots. Proposition 6.4.1 says that there are $m-n$ more embedding elements out there. Explicitly, these are the $L_i + L_{m-n+1}$ for $1 \leq i \leq m-n$. Springer and Steinberg showed that the nilpotent orbits in $\mathfrak{so}_{2m}$ are parametrized by partitions on $2m$ in which even parts occur with even multiplicity, except that very even partitions (those with only even parts, each having even multiplicity) correspond to two orbits. In [25] a recipe is given to construct the nilpotent elements $X_{\mathbf{d}}$ corresponding to some partition $\mathbf{d}$ and we see that only for these very even partitions the roots $L_i + L_{i+1}$ are needed. Again, we automatically obtain the nilpotent orbits of an $\mathfrak{so}_{2(m-n)}$. In some cases however, we also reach one of the two orbits associated with the very even partitions of $2m$! For such a very even partition $\mathbf{d}$ of $2m$ the recipe shows that we always need access to the root $L_{m-1} + L_m$, so we must have that $m - n + 1 = m$ or $n = 1$. Of course, for $2m$ to have a very even partition in the first place, m must be even. Hence we are in the situation $\mathfrak{sl}_2 \subset \mathfrak{so}_{4m'}$. We conclude

Theorem 6.4.9. *Let $\mathfrak{so}_{2n} \subset \mathfrak{so}_{2m}$; then the nilpotent orbits generated by the embedding elements correspond to the inclusion of the Hasse subdiagram $\iota(H_{\mathfrak{so}_{2(m-n)}})$. If in addition, $n = 1$ and m is even, i.e. $\mathfrak{sl}_2 \subset \mathfrak{so}_{4m'}$ then we also reach one of the two orbits associated with a very even partition of $2m$.*

Part III

Unbounded and standard filtrations

Chapter 7

Sheaves of glider representations

In the third part of the book we study glider representation theory of unbounded filtrations FR and we begin by considering standard filtrations. These filtrations are induced on R via an epimorphism $K[\underline{X}] \to R$ or $K\langle \underline{X} \rangle \to R$, where $K[\underline{X}]$ is the ring of polynomials on a set of variables $\underline{X}$ and $K\langle \underline{X} \rangle$ is the free K-algebra on $\underline{X}$. In each case, the filtration on $K[\underline{X}]$ or $K\langle \underline{X} \rangle$ is obtained by letting $\oplus_\alpha K X_\alpha$ be the part of filtration degree 1 and $F_n = F_1^n$ for $n \geq 0$. Well-known examples of these positive standard filtrations are the Weyl algebras $A_n(K)$, rings of differential operators on a smooth variety, and also $K[V]$ the coordinate rings of algebraic varieties over a field K. For a variety W and a morphism $V \to W$ such that $K[W] \subset K[V]$ we may define a standard filtration, $fK[V]$, with $f_0 K[V] = K[W]$ by selecting arbitrary $K[W]$-ring generators for $K[V]$.

In this chapter we develop the glider theory in both commutative and non-commutative algebraic geometry. Maybe a starting point could be Example d) from Section 3.3: to every (positively) filtered ring FR there is associated a Rees ring $\widetilde{R} = \oplus_{n\in\mathbb{Z}} F_n R$ and associated graded ring $G(R) = \oplus_{n\in\mathbb{Z}} F_n R / F_{n-1} R$. The Rees ring or blow-up ring $\widetilde{R}$ may be viewed as a graded subring $\sum_n F_n R X^n \subset R[X]$, where X is a regular, central homogeneous element of degree 1. The three rings $R, \widetilde{R}$ and $G(R)$ are related in the following way

$$\begin{array}{ccccc} \widetilde{R} & \longrightarrow & \widetilde{R}/X\widetilde{R} = G(R) & \qquad & \widetilde{R}_X \cong R[X, X^{-1}] \\ & \searrow & & & \\ & & \widetilde{R}/(1-X)\widetilde{R} = R & \qquad & (\widetilde{R}_X)_0 = R. \end{array}$$

It appears that every $\widetilde{R}$-module is realized as an FR-fragment! In case $R = K[V] = K[\underline{X}]/I(V)$ is the coordinate ring of an affine variety $V \subset \mathbb{A}^n$ with standard filtration induced by putting $\deg(X_i) = 1$, we can take the projective closure $\overline{V} \subset \mathbb{P}^n$. This is done by identifying $\mathbb{A}^n$ with the open subset $U_0 = \{X_0 \neq 0\}$ of $\mathbb{P}^n$ by the map $\varphi_0 : (x_0, x_1, \ldots, x_n) \mapsto (\frac{x_1}{x_0}, \ldots, \frac{x_n}{x_0})$ and taking the closure $\overline{V}$ of $\varphi_0^{-1}(V)$ inside $\mathbb{P}^n$. One then shows that the ideal $I(\overline{V})$ is generated by $\beta(I(V))$, where β is the homogenization map. Since $\mathbb{P}^n = \mathrm{Proj}(\mathrm{K}[\mathrm{X}_0, \mathrm{X}_1, \ldots, \mathrm{X}_n])$ with $\deg(X_i) = 1$ we see that $\overline{V}$ corresponds to the Rees ring of $K[X_0, X_1, \ldots, X_n]/\beta(I(V))$ with standard filtration, i.e.

$\overline{V}$ = Proj K[$X_0, X_1, \ldots, X_n$]/β(I(V)). On the open subset $U_0 = \{X_0 \neq 0\}$ we obtain the Rees ring $\widetilde{R}$ with regard to the standard filtration on $R = K[V]$. On the Zariski closed set $\{X_0 = 0\}$, that is, the part at infinity, we obtain Proj(G(R)). Indeed, if f is a generator of $I(V)$, then $\beta(f) = f^l$ in $\{X_0 = 0\}$, where f^l is the homogeneous part of highest degree of f. In the associated graded $G(K[V])$, $\overline{f} = \overline{f^l}$, from which the statement follows. In total we have that $\overline{V} \supset V$ as an open and $V_\infty \subset \overline{V}$ as its closed complement. In general, for commutative R, we interpret Proj($\widetilde{\mathrm{R}}$) as the projective closure of Spec(R) by "gluing" Proj(G(R)) to it as the part at infinity. This is in fact a meta-property that remains valid in the non-commutative situation, after suitably defining all geometric concepts, cf. [69], [72]. For the commutative case we refer to [36]. The Rees ring with regard to the standard positive filtration is also called the global blow-up ring. One can also consider the negative m-adic filtration where $m \in \max(K[V])$ is a point in Spec(K[V]). The Rees ring (considered as positively graded) corresponds in this case to the blow-up at the point m. Geometrically, the associated graded $G(K[V])$ corresponds to the tangent cone at m, see [50].

For some reason, the only situation in algebraic geometry where the standard filtration on the coordinate ring is used is exactly in constructing the Rees ring and its part at infinity. For example, one neglects the additional filtration information when developing scheme theory. One of our purposes is to extend the theory of schemes to schemes of glider representations. Because the degree zero part M_0 of a glider is not (necessarily) an R-module, this scheme structure is not obvious! For modules there is a theory of structure schemes over a base scheme, the structure sheaf of the ring; in fact this is classical for a commutative ring in algebraic geometry, but it can also be done for a non-commutative ring in terms of Spec(R), cf. [51], R-tors the lattice of torsion theories, cf. [33], a non-commutative topology, cf. [70] or a non-commutative Grothendieck topology, cf. [69]. So the problem we face here is to construct a structure sheaf of a glider representation over the topological space associated with the ring R. The glider contains extra information connected to the structure of the chain used to construct R from S by an iterated construction step (in the filtration). One very interesting example we have in mind is almost commutative rings. Roughly said an almost commutative ring is one where $G(R)$ is commutative, e.g. a PBW-deformation of a commutative ring (see [13]).

Scheme theory classically depends on localization of modules at torsion theories (cf. [33], [34], [60]) on R-mod. Hence, the first step is to study a quotient filtration on filtered rings and modules, and the second step is to extend this to glider representations by modifying the localization technique. There are some technical problems if you want the localized filtration of a separated filtration on the ring R to remain separated. We will need a notion of κ-separatedness for a torsion theory κ and this leads to the definition of a so called strong characteristic variety. Classically, for a ring of differential operators R , see [12], (or more general for an almost commutative ring) and a filtered R-module M, the characteristic variety $\chi(M)$ is defined by the variety $V(\text{annG(M)})$ where annG(M) is the annihilator in $G(R)$ of the graded module $G(M)$. We are interested in a subvariety $\xi(M) \subset \chi(M)$ which contains those prime ideals P such that $G(M)$ is $(G(R) \setminus P)$-torsion free. For such a prime P, the localization at $G(R) \setminus P$ is the stalk at P in the structure sheaf of $G(M)$. We will generalize this to filtered rings with commutative associated graded $G(R)$ and define a strong characteristic variety as a closed subset in $G(R)$-tors. We will prove that the separatedness of the quotient filtration at some localization at a torsion theory on R-mod, say κ, follows from torsion freeness of $G(R)$ (also $G(M)$ for a module) at $G\kappa$ some left exact radical associated to κ on $G(R)$-gr. Hence we obtain in some sense a non-commutative version of the characteristic variety!

In fact we present two related but different approaches, one on the level of filtered R-modules, R-filt, and one on the level of graded modules $\widetilde{R}$-gr. In the first approach we start from a localization on R-mod applied to filtered modules, leading to a scheme theory for gliders $\Omega \supset M$ over the lattice R-tors with separability related to the strong characteristic variety $\xi(\Omega)$ as indicated above. The second approach starts from a graded localization at $\widetilde{\kappa}$ on $\widetilde{R}$-gr, cf. [55], and links it to a localization κ on R-mod plus a graded localization $G\kappa$ on $G(R)$-gr for those $\widetilde{\kappa}$ in some affine part of the localizations of $\widetilde{R}$. By this affine part we mean the graded kernel functors $\widetilde{\kappa}$ that correspond well to some κ on R-mod via dehomogenization. The complement of this affine part reduces to the correspondence of $\widetilde{\kappa}$ and some $G(\kappa)$-gr. So in the second approach the scheme theory fits in the philosophy of the projective closure via $\widetilde{R}$-gr (of the affine part that may be viewed as R-tors) and the geometry "at infinity" via $G(R)$-tors.

The localization of a glider $\Omega \supset M$ may then be obtained as the part of degree zero of the quotient filtration on a localization of Ω, much in the spirit of the construction of a projective scheme. In fact, such and other projective aspects are already discussed in [53]. A very interesting class of examples is given by localization at Ore sets (geometrically corresponding to affine open subsets!). For example, we show that they allow a scheme theory over a non-commutative Grothendieck topology (cf. [69], [72]). This requires some facts about consecutive multiple localizations, but since composition of non-commutative localization functors is not a localization functor (see [68]), this forces the consideration of "localization" at left exact preradicals, adding some extra, though unavoidable technicality. The gain is that a global section theorem a la Serre, [63], then holds, Theorem 7.4.19.

We start with an introductory section. We introduce standard fragments and investigate the possibility to localize such a fragment at an Ore set S. We deduce some conditions on S and arrive at Proposition 7.1.17. The reason for including this section is twofold. On the one hand, many geometrical examples are standard filtered rings with graded associated being a commutative domain (e.g. rings of differential operators with the Σ-filtration). The conditions we put on the Ore sets S are then automatically satisfied. On the other hand, dealing with Ore sets gives a lot of inspiration to deal with the more general case of localizing at kernel functors. Also, if the reader is not familiar with general torsion theory and localization, he or she can always think in terms of Ore sets. Before laying out the first approach in Section 7.2 we recall the definition of localization at a kernel functor. For a detailed overview of localization theory, the reader may consult [33], [34], [60], [66] or [71]. In the second approach, we start from a kernel functor on the category of graded modules $\widetilde{R}$-gr, so we also include a review on this. The section further contains a short digression on the strong characteristic variety and we also explain how to deal with the composition of kernel functors. The latter will be important to arrive at a scheme theory for the non-commutative site. The second approach is carried out in Section 7.3, which unravels the existence of affine, pseudo-affine and non-affine kernel functors $\widetilde{\kappa}$ on $\widetilde{R}$-mod or $\widetilde{R}$-gr. In fact, this again shows how we can view the space $\widetilde{R}$-grtors as the gluing of $G(R)$-grtors to R-tors. Section 7.4 deals with the construction of schemes over the various topological spaces Spec(R), R-tors and the non-commutative site. In Section 7.5 we include some applications to commutative algebraic geometry and the final Section 7.6 deals with some projective aspects of glider representations.

We are aware that there is a lot of technical baggage to handle to read this chapter. We have tried to provide additional information where needed to make the reading more accessible. The content of this chapter is based on [20].

7.1 Localization at Ore sets

A positive filtration FA with $F_0A = B$ is said to be a standard filtration if FA is finite in the sense that every F_nA is a finitely generated left B-module and $F_nA = (F_1A)^n$ for every $n \in \mathbb{N}$, or equivalently: $F_nAF_mA = F_{n+m}A$ for every $n, m \in \mathbb{N}$.

Examples 7.1.1. 1. Let A be a K-algebra (K some field) with a positive filtration FA such that $F_0A = K$ and every F_nA a finite dimensional K-vector space, $F_nA = (F_1A)^n$ for every $n \in \mathbb{N}$;

2. $A = K\langle X_1, \ldots, X_n\rangle/I = K\langle a_1, \ldots, a_n\rangle$ an affine K-algebra, $F_1A = Ka_1 \oplus \cdots \oplus Ka_n$, $F_nA = (F_1A)^n$. Generic example for $I = 0$.
Every A, FA as in 1. is obtained as an example in 2. because if $F_1A = Ka_1 \oplus \cdots \oplus Ka_n$ then $A = K\langle a_1, \ldots, a_n\rangle$;

3. Let fA be a strong filtration, i.e. $f_nAf_mA = f_{n+m}A$ for all $n, m \in \mathbb{Z}$. Then fA is finite over f_0A. The positive filtration FA given as f^+A, $F_nA = f_nA$ for $n \geq 0$ is a standard filtration.

Definition 7.1.2. Let A have standard filtration FA. An FA-fragment M is said to be a standard fragment if and only if $F_nAM_m = M_{m-n}$ for all $n \leq m$, equivalently: for all $n \in \mathbb{N}, F_1AM_n = M_{n-1}$.

Proposition 7.1.3. *If $f : M \to N$ is a fragment morphism then* Im(f) *is a fragment. In case M is standard, so is* Im(f).

Proof. $\mathrm{Im}(f)_{n-1} = f(M_{n-1}) = f(F_1AM_n) = F_1Af(M_n) = F_1A\mathrm{Im}(f)_n$. □

Corollary 7.1.4. *If N is a strict subfragment of a standard fragment M then M/N is standard.*

Proof. Direct from $(M/N)_n = M_n/N_n$ for every $n \geq 0$. □

Let FA be a standard filtration with $F_0A = B$. The associated graded ring $G(A)$ is given by $\oplus_n F_nA/F_{n-1}A$ and for $a \in F_nA \setminus F_{n-1}A$, we denote by $\sigma(a)$ the image of a in $G(A)_n$. A left Ore set $S \subset A$ is a multiplicatively closed subset such that $1_A \in S$ and such that the following two Ore conditions are satisfied

1. If $as = bs$ for $a, b \in A$ and $s \in S$, then $\exists s' \in S$ such that $s'a = s'b$;

2. For any $a \in A, s \in S$ there $\exists a' \in A, s' \in S$ such that $s'a = a's$.

One analogously defines right Ore sets. In the sequel however, if we refer to the Ore conditions, we always mean the left Ore conditions. If $S = S_{sat} = \{s \in A \mid \sigma(s) \in \sigma(S)\}$ then we call S a saturated Ore set. If S is saturated Ore then $\sigma(S)$ is Ore in $G(A)$. Indeed, any lift s of $\overline{s} \in \sigma(S)$ sits in $S_{sat} = S$.

Consider an Ore set S in A such that $\sigma(S)$ is an Ore set in $G(A)$ and $S \not\subset B$. If $G(A)$ is $\sigma(S)$-torsion free then A is S-torsion free; indeed, if for $s \in S$, $sa = 0$, then $\sigma(s)\sigma(a) = 0$ by definition of the product in $G(A)$, hence $\sigma(a)$ in $t_{\sigma(S)}(G(A))$ and $\sigma(a) \neq 0$. Here, $t_S(A) = \{a \in A \mid \exists s \in S \text{ such that } sa = 0\}$ denotes the (left) S-torsion elements of A.

Lemma 7.1.5. *In the situation as above: $S^{-1}A$ is strongly filtered, hence $F^+S^{-1}A$ is a standard filtration.*

Proof. The quotient filtration $FS^{-1}A$ is defined by putting:

$$F_nS^{-1}A = \{x \in S^{-1}A \text{ such that } sx \in A \text{ for } s \in A, \text{ say } \sigma(s) \in G(A)_p, \text{ entails } sx \in F_{n+p}A\}.$$

To check whether this is well-defined, look at an $x \in S^{-1}A$ such that $sx \in A$ and $sx \in F_{n+p}A$. Assume there is a $t \in S$, $t \in \dot{F}_qA := F_qA \setminus F_{q-1}A, tx \in A$ but assume $tx \notin F_{n+q}A$. By exhaustivity of FA, $tx \in \dot{F}_{n+q+\delta}A$ for some

$\delta > 0$. Since S is an Ore set, there are $s' \in S, t' \in A$ such that $s't = st'$. Since $G(A)$ is $\sigma(S)$-torsion free, we have for every $s \in S$ that $s\dot{F}_nA \subset \dot{F}_{n+p}A$ where $p = \deg\sigma(s)$. From $tx \in \dot{F}_{n+q+\delta}A$ we get $s'tx \in \dot{F}_{n+q+\delta+p'}A$ where $p' = \deg\sigma(s')$. Put $q' = \deg\sigma(t')$. Again by torsion freeness, $p'+q = \deg(s't) = \deg(st') = p+q'$, so δ must equal zero, a contradiction. We leave it to the reader to check that this indeed gives a separated filtration on $S^{-1}A$. Exhaustivity follows from exhaustivity of FA. Clearly, if $s \in F_pA \cap S$, then $s^{-1} \in F_{-p}S^{-1}A$ and $F_nS^{-1}A \cap A = F_nA$ for all $n \in \mathbb{N}$.
Consider $x \neq 0$ in $F_nS^{-1}A$, then $sx \in F_{n+p}A$ for some $s \in S, p = \deg\sigma(s)$. Since $F_{n+p}A = (F_1A)^{n+p}$ we may write $sx = \sum_i' a_1^i \cdots a_{n+p}^i$ for $a_j^i \in F_1A$. Thus we have that $x = \sum_i' s^{-1}(a_1^i \cdots a_{n+p-1}^i)a_{n+p}^i$ with $s^{-1}(a_1^i \cdots a_{n+p-1}^i)$ in $F_{n-1}S^{-1}A$ and $a_{n+p}^i \in F_1A \subset F_1S^{-1}A$. Hence $F_nS^{-1}A = F_{n-1}S^{-1}AF_1S^{-1}A$. For $n = 0$ this yields the strongly filtered condition. □

Now consider an Ore set S in A with $S \cap B$ Ore in A (and thus also in B) and such that $\sigma(S)$ is Ore in $G(A)$; take for example a saturated Ore set in A over an Ore set intersection with B. Assume $G(A)$ is $\sigma(S)$-torsion free and $S \not\subset B$. We look at $S \setminus B$. By torsion freeness, $S \setminus B$ is multiplicatively closed.

Lemma 7.1.6. *$S \setminus B$ is an Ore set in A.*

Proof. Consider $a \in A, s \in S \setminus B$, then since S is Ore in A, $\exists s' \in S, a'$ in A such that $s'a = sa'$. If $s' \in S \setminus B$ then we are done. Assume $s' \in S \cap B$, then (ss')a = (sa')s with $ss' \in S \setminus B$ ($\deg\sigma(ss') > 0$) and $sa' \in A$. □

Remark 7.1.7. Since $G(A)$ is $\sigma(S)$-torsion free, A is S-torsion free, hence if $as = 0$ for $a \in A, s \in S \setminus B$ then $s'a = 0$ for some $s' \in S$ but then $a = 0$ since $t_S(A) = 0$. So the first Ore condition is trivially satisfied.

Proposition 7.1.8. *$S \setminus B$ is an Ore set in $(S \cap B)^{-1}A$.*

Proof. Take $s \in S\setminus B, b \in (S\cap B)^{-1}A$, thus for some $s_0 \in S\cap B, s_0b \in A$ (note $A \hookrightarrow (S \cap B)^{-1}A$ since A is $S \cap B$-torsion free). By the previous lemma, $\exists t \in S \setminus B, a \in A$ such that $ts_0b = as$. Since $ts_0 \in S \setminus B$ and $a \in A \subset (S \cap B)^{-1}A$, the Ore condition for $S \setminus B$ in $(S \cap B)^{-1}A$ follows. □

Corollary 7.1.9. $Q_{S\setminus B}(Q_{S\cap B}(A)) = Q_S(A) = S^{-1}A$.

Symmetrically, we have

Proposition 7.1.10. *$S \cap B$ is an Ore set in $(S \setminus B)^{-1}A$.*

Proof. Take $b \in (S\setminus B)^{-1}A$, then $\exists t \in S\setminus B, tb \in A$. For given s in $S\cap B, \exists s' \in S \cap B$ such that $s'tb = a's$ with $a' \in A$ ($S \cap B$ Ore in A). $s't \in S \setminus B$ since $\deg(s't) = \deg(t) > 0$, hence $(s't)^{-1}b \in (S \setminus B)^{-1}A$ and thus $1.b = (s't)^{-1}a's$ with $1 \in S \cap B$ and $(s't)^{-1}a' \in (S \setminus B)^{-1}A$.

□

Corollary 7.1.11. $Q_{S\cap B}Q_{S\setminus B}(A) = Q_S(A) = Q_{S\setminus B}Q_{S\cap B}(A)$.

Corollary 7.1.12. *The localizations at $S \cap B$ and $S \setminus B$ are compatible (see [68] for the notion of compatibility).*

Let FA be the positive part of a strong filtration on A (e.g. when A is $S^{-1}A'$ for some standard filtered ring A'). Consider a standard fragment M over FA.

Proposition 7.1.13. *If $N \subset M$ is a strict subfragment, then N is standard.*

Proof. $N_m = M_m \cap N$ thus $F_1AN_m \subset N_{m-1} = M_{m-1} \cap N = F_1AM_m \cap N$. If $n \in F_1AM_m \cap N$ then $F_{-1}An \subset F_{-1}AF_1AM_m = M_m$ and $F_{-1}An \subset N \cap M_m = N_m$ or $n \in F_1AN_m$. Thus $F_1AN_m = F_1AM_m \cap N = M_{m-1} \cap N = N_{m-1}$, i.e. N is standard. □

Proposition 7.1.14. *If FA is the positive part of a strong filtration on A and M is a natural FA-fragment, then M is standard.*

Proof. $M_n = \{m \in M, F_nRm \subset M\}$, so if $m \in M_{n-1}$ then $F_{-1}Am \subset M_n$ since $F_nAF_{-1}Am = F_{n-1}Am \subset M$. Thus $m \in F_1AM_n \subset M_{n-1}$, i.e. M is standard. □

Sometimes it suffices that $S \cap B$ is Ore in B. For example, if A is an extension of B, then $S \cap B$ Ore in B implies that $S \cap B$ is Ore in A. Recall that A is an extension of B if $A = BZ_B(A)$, where

$$Z_B(A) = \{a \in A,\ ba = ab \text{ for all } b \in B\}.$$

In this case $S \cap B$ Ore in B implies $S \cap B$ Ore in A. Indeed, take $a \in A$, $s \in S \cap B$, say $a = \sum b_i z_i$ with $b_i \in B$, $z_i \in Z_B(A)$. Take $s_i \in S \cap B$ such that $s_i b_i = b_i' s$ for all i (finite sum) and take $s'' \in \bigcap_i Bs_i$, $s'' \in S \cap B$ ($S \cap B$ Ore in B and $1 \in S \cap B$). Then we obtain that $s''b_i = b_i''s$ with $b_i'' \in B$, for all i. Hence

$$s''a = \sum s''b_i z_i = \sum b_i'' s z_i = \sum b_i'' z_i s = (\sum b_i'' z_i)s,$$

with $\sum b_i'' z_i \in A$, whence $S \cap B$ is Ore in A.

By the above, we can assume that S is Ore in A such that $S \cap B = 1$ and such that $\sigma(S)$ is an Ore set of $G(A)$, the latter being $\sigma(S)$-torsion free. Now, let M be a standard glider representation, $\Omega = AM \supset M$. We consider on Ω the filtration $f\Omega, f_{-n}\Omega = M_n, n \geq 0$ and $f_m\Omega = F_mAM$ for $m \geq 0$. The associated graded $G_f(\Omega)$ has for the negative part $g(M) = \oplus_{i \geq 0} M_i/M_{i+1}$ and in Section 1.9 we showed that this is itself a fragment with respect to the positive part of the gradation filtration of $G(A)$. We assume $G_f(\Omega)$ is $\sigma(S)$-torsion free; thus in particular, elements of $g(M)$ cannot be annihilated by elements of $\sigma(S)$. We also assume (however not necessary) that the body $B(M) = \bigcap_n M_n = 0$. We define for $d \in \mathbb{Z}$

$$Q_S(\Omega)_d = \{z \in Q_S(\Omega), sz \in F_{p+d}\Omega, s \in S \cap \dot{F}_pA\}.$$

Just as in the proof of Lemma 7.1.5 one shows that the d is depending only on z and not on the specific s.

Proposition 7.1.15. *The left $S^{-1}A$-module $Q_S(\Omega)$ with chain defined above is filtered with regard to the localization filtration on $S^{-1}A$.*

Proof. Let $x \in S^{-1}A_m, z \in Q_S(\Omega)_d$ and let $s \in S \cap \dot{F}_pA, t \in S \cap \dot{F}_qA$ be such that $sx = a \in F_{m+p}A, tz = \omega \in F_{d+q}\Omega$. By the Ore condition there exist $t' \in S, a' \in A$ such that $t'a = a't$. Since $G(A)$ is $\sigma(S)$-torsion free, we have that $\deg(\sigma(a')) = \deg(\sigma(a)) + \deg(\sigma(t')) - \deg(\sigma(t))$. We have

$$t'sxz = t'az = a'tz = a'\omega \in F_{\deg(\sigma(a'))+d+q}\Omega.$$

Because $\deg(\sigma(a')) + d + q = \deg(\sigma(a)) + \deg(\sigma(t')) - \deg(\sigma(t)) + d + q = \deg(\sigma(t')) + p + m + d$ it follows that $xz \in Q_S(\Omega)_{m+d}$. □

We have that $Q_S(\Omega) = \bigcup_d Q_S(\Omega)_d$ and we call this the quotient filtration on $S^{-1}\Omega$.

Lemma 7.1.16. *Let $X = FX$ be a filtered FA-module, with FA strongly filtered. Then X is strongly filtered.*

Proof. Look at $F_1AF_{n-1}X \subset F_nX$. Since $F_{-1}AF_nX \subset F_{n-1}X$ we have that $F_1AF_{n-1}X \supset F_1AF_{-1}AF_nX = F_nX$ or $F_1AF_{n-1}X = F_nX$ for every n. From this, one easily deduces that FX is strongly filtered. □

By the proof of Lemma 7.1.5 we obtain that $S^{-1}\Omega$ is a strongly filtered and standard filtration. Observe that we need the assumption that $G_f(\Omega)$ is $\sigma(S)$-torsion free. We define the localization $Q_S(M)$ of M at S to be $F_0S^{-1}\Omega$. We have the following:

Proposition 7.1.17. *$Q_S(M)$ with $Q_S(M)_d = F_{-d}S^{-1}\Omega$ is a standard fragment with respect to $F^+S^{-1}A$. Moreover, $Q_S(M) \cap \Omega = M$ and $Q_S(M)_d \cap \Omega = M_d, d \geq 0$.*

Proof. The first statement follows from Example 1.3.5. If $\omega \in Q_S(M) \cap \Omega$ then $s\omega \in F_p\Omega$ for some s with $\deg \sigma(s) = p$. Since multiplication by s cannot lower the degree of an element of Ω ($G(\Omega)$ is $\sigma(S)$-torsion free!) it follows that $\omega \in F_0\Omega = M$. Similarly, if $\omega \in Q_S(M)_d \cap \Omega$, $s\omega \in F_{p-d}\Omega$, hence ω in $F_{-d}\Omega, d \geq 0$, i.e. $w \in M_d$. □

Observation 7.1.18. One may easily verify that $Q_S(M) = F_0Q_S(A)M \subset Q_S(\Omega)$. Observe moreover that $Q_S(M)$ is natural, since $S^{-1}A$ is strongly filtered. Indeed, Lemma 7.1.16 entails that $FQ_S(\Omega)$ is strongly filtered. Hence for any z in $Q_S(M)$, $F_dS^{-1}Az \subset F_0Q_S(\Omega)$ implies that $z \in F_{-d}S^{-1}AQ_S(\Omega) = Q_S(M)_d$.

We call $Q_S(M)$ the quotient glider of M over $S^{-1}A$. In the next section we will discuss how we can localize glider representations at more general kernel functors. We chose to treat this particular case independently, since many geometrical examples are standard filtered rings with graded associated being a commutative domain (e.g. rings of differential operators with the Σ-filtration).

7.2 Quotient filtrations - first approach

We begin by introducing the quotient filtration on the localization of a filtered ring at a kernel functor κ on R-mod; this is inherent in [61] but there it is applied for saturated localizations because one is mainly interested in lifting a localization from the associated graded ring to arrive at microlocalizations at multiplicatively closed sets S of a filtered ring FR. In general, however, it is not possible to lift the kernel functor κ to kernel functors on the Rees or associated graded level. Let us recall the necessary notions from localization theory.

Let R be a ring and denote by R-mod the category of left R-modules. A preradical ρ on R-mod is a subfunctor of the identity functor. The class of preradicals on R-mod is denoted by $\mathcal{Q}(R)$. A preradical ρ such that $\rho\rho = \rho$ is said to be idempotent. A preradical ρ such that $\rho(M/\rho(M)) = 0$ for all $M \in R$-mod, is said to be radical. To a preradical ρ, we associate two classes $(\mathcal{T}_\rho, \mathcal{F}_\rho)$ of R-modules given by

$$\begin{aligned} \mathcal{T}_\rho &= \{M \in R\text{-mod} \mid \rho(M) = M\} \quad \text{(torsion class)} \\ \mathcal{F}_\rho &= \{M \in R\text{-mod} \mid \rho(M) = 0\} \quad \text{(torsion free class).} \end{aligned}$$

These classes determine a torsion theory in the Abelian category R-mod. Such a torsion theory is a couple (T, F) of additive subcategories satisfying the following conditions:

1. $\text{Hom}(T, F) = 0$;
2. $\text{Hom}(T, Y) = 0 \Rightarrow Y \in \text{Ob}(F)$;
3. $\text{Hom}(X, F) = 0 \Rightarrow X \in \text{Ob}(T)$;
4. for all $X \in R{-}\text{mod}$, there exists $Y \subset X, Y \in \text{Ob}(T)$ and $X/Y \in \text{Ob}(F)$.

If the additive subcategory T is closed under submodules, we call the torsion theory hereditary.

A left exact idempotent radical is called a kernel functor. In fact, it suffices for a preradical ρ to be left exact and radical in order to be a kernel functor by

Proposition 7.2.1. *[70, Proposition 2.11]*
For $\rho \in \mathcal{Q}(R)$, the following are equivalent:

1. *ρ is left exact;*
2. *For every submodule N of M, $\rho(N) = \rho(M) \cap N$;*
3. *ρ is idempotent and $\mathcal{T}_\rho$ is closed under subobjects.*

The pair $(\mathcal{T}_\kappa, \mathcal{F}_\kappa)$ of the torsion class and the torsion free class for a kernel functor κ determines a hereditary torsion theory, and there is in fact a bijective correspondence between the class of hereditary torsion theories and kernel functors. Therefore, we denote by R-tors the class of all hereditary torsion theories or kernel functors. To any left exact preradical κ one associates a filter $\mathcal{L}(\kappa)$ of left ideals:

$$\mathcal{L}(\kappa) = \{L \text{ left ideal of } R \mid \kappa(R/L) = R/L\},$$

which satisfies the following four properties from [34]:

1. If $I \in \mathcal{L}(\kappa)$ and $I \subset J$ for some left ideal J, then $J \in \mathcal{L}(\kappa)$;
2. If $I,\ J \in \mathcal{L}(\kappa)$, then $I \cap J \in \mathcal{L}(\kappa)$;
3. If $I \in \mathcal{L}(\kappa), r \in R$, then $(I : r) = \{a \in R, ar \in I\} \in \mathcal{L}(\kappa)$;
4. If M is an R-module, then $m \in \kappa(M)$ if and only if $\exists I \in \mathcal{L}(\kappa)$ such that $Im = 0$.

We denote these properties by $G1, G2, G3$ and $G4$. One can show that the radical property $\kappa(M/\kappa(M)) = 0$ is equivalent to

5. If $I \in \mathcal{L}(\kappa)$ and $J \subset I$ is such that for every $x \in I, (J : x) \in \mathcal{L}(\kappa)$ then $J \in \mathcal{L}(\kappa)$.

If R is moreover Noetherian, then this condition is equivalent to

5′. If $I \in \mathcal{L}(\kappa)$ and $J \subset I$ such that I/J is $\kappa-$torsion , then $J \in \mathcal{L}(\kappa)$.

As a corollary one obtains that the filter $\mathcal{L}(\kappa)$ is multiplicatively closed. Indeed, for $I, J \in \mathcal{L}(\kappa)$ we have that $I \subset (IJ : x)$ for all $x \in J$. If the filter $\mathcal{L}(\kappa)$ has a cofinal set of two-sided ideals, then we call κ a symmetric kernel functor. In the sequel, to prove that some preradical κ is actually a kernel functor, we will show that the associated filter $\mathcal{L}(\kappa)$ satisfies the properties 1,2,3 and 5. In this case, we call $\mathcal{L}(\kappa)$ the associated Gabriel filter of κ. A kernel functor κ determines a localization functor $Q_\kappa : R\text{-mod} \to R\text{-mod}$ which associates to an R-module M its localization

$$Q_\kappa(M) = \varinjlim_{I \in \mathcal{L}(\kappa)} \mathrm{Hom}_R(I, M/\kappa(M)).$$

The reader should be aware that idempotent in [34] or [67] means radical in [70]. In fact, there is a one-to-one correspondence between filters of left ideals $\mathcal{L}$ satisfying the five properties above and left exact idempotent radicals, i.e. kernel functors! If $\mathcal{L}$ only satisfies the first four, then one loses the radical property and the localization theory becomes different. This observation is crucial when we deal with composition of kernel functors to handle the non-commutative site, see below.

In the following section we will start from a kernel functor κ on $\widetilde{R}$-gr, the category of (left) graded $\widetilde{R}$-modules. In fact, the notions recalled above can more generally be defined on any Grothendieck category $\mathcal{C}$ so there is no harm in speaking of a kernel functor on $\widetilde{R}$-gr. One associates with such a κ the set of graded left ideals

$$\mathcal{L}^g(\kappa) = \{L \text{ graded left ideal of } \widetilde{R} \,|\, \kappa(\widetilde{R}/L) = \widetilde{R}/L\}$$

and one shows that the following properties hold

1. If $I \in \mathcal{L}^g(\kappa)$ and $I \subset J$ for some graded left ideal J, then $J \in \mathcal{L}^g(\kappa)$;
2. If $I, J \in \mathcal{L}^g(\kappa)$, then $I \cap J \in \mathcal{L}^g(\kappa)$;
3. If $I \in \mathcal{L}^g(\kappa), \widetilde{r} \in \widetilde{R}$, then $(I : \widetilde{r}) = \{\widetilde{a} \in \widetilde{R}, \widetilde{a}\widetilde{r} \in I\} \in \mathcal{L}^g(\kappa)$;
4. If $I \in \mathcal{L}^g(\kappa)$ and $J \subset I$ for some graded left ideal J is such that $(J : x) \in \mathcal{L}^g(\kappa)$ for all $x \in I$ then $J \in \mathcal{L}^g(\kappa)$.

Any non-empty set of graded left ideals of $\widetilde{R}$ satisfying these four properties is called a graded filter in $\widetilde{R}$. One can show that there is a 1-to-1 correspondence between hereditary rigid torsion theories in $\widetilde{R}$-gr, rigid kernel functors on $\widetilde{R}$-gr and graded filters in $\widetilde{R}$. Rigid means that the corresponding torsion and torsion free classes are closed under shift of grading. Some kernel functors on $\widetilde{R}$-gr can actually be induced by a kernel functor on $\widetilde{R}$-mod. To state the result, we introduce graded kernel functors on $\widetilde{R}$-mod. These are kernel functors $\underline{\kappa}$ such that the associated Gabriel filter $\mathcal{L}(\underline{\kappa})$ of left ideals of $\widetilde{R}$ possesses a cofinal set of graded left ideals. One then shows that there is a bijective correspondence between graded kernel functors on $\widetilde{R}$-mod and rigid kernel functors on $\widetilde{R}$-gr. If $\underline{\kappa}$ is a graded kernel functor on $\widetilde{R}$-mod which corresponds uniquely to κ on $\widetilde{R}$-gr then $\mathcal{L}^g(\kappa) = L_g(\widetilde{R}) \cap \mathcal{L}(\underline{\kappa})$, where $L_g(\widetilde{R})$ denotes the set of graded left ideals of $\widetilde{R}$. We refer the reader to [55, Section II.9] for proofs of the statements.

A filtered ring FR is said to be *Zariskian* if the Rees ring $\widetilde{R}$ associated with FR is both left and right Noetherian and if $F_{-1}R$ is contained in the Jacobson radical $J(F_0R)$ of F_0R, see [46, Chapter 2, Section 2] for other characterizations. Consider a filtered ring R with Zariskian filtration FR. A filtration FM on an R-module M will always be assumed exhaustive, i.e. $\bigcup_n F_nM = M$; FR will always be assumed to be separated, i.e. $\bigcap_n F_nR = 0$. The associated graded $G(R)$-module of FM will be denoted by $G(M)$, the Rees module by $\widetilde{M}$. Since FR is separated, there exists for homogenous $\widetilde{r} \in \widetilde{R}$ a unique integer n such that $\widetilde{r} = rX^{n(r)}$, where $r \in \dot{F}_nR = F_nR \setminus F_{n-1}R$. We denote this integer by $n = \deg(r)$. Recall: R-filt $\to \mathcal{F}_X, M \mapsto \widetilde{M}$ is an equivalence of categories, where $\mathcal{F}_X$ denotes the full subcategory of X-torsion free $\widetilde{R}$-modules. Here, X stands for the central element of degree 1 such that $\widetilde{R} = \sum_n F_nRX^n \subset R[X, X^{-1}]$.

Let R be a Zariski filtered ring and let κ be a localization functor (i.e. a kernel functor) on R-mod given by its Gabriel filter $\mathcal{L}(\kappa)$. One may define

$$\mathcal{L}(G(\kappa)) = \{J \text{ left ideal of } G(R) \mid J \supset G(L) \text{ for some } L \in \mathcal{L}(\kappa)\}$$

and show that this gives rise to a left exact preradical $G(\kappa)$ on $G(R)$-gr in general and to a kernel functor if $G(R)$ is a commutative domain.

Lemma 7.2.2. *$\mathcal{L}(G(\kappa))$ defines a topology on $G(R)$, i.e. the associated functor $G(\kappa)$ is a left exact preradical. If $G(R)$ is a commutative domain, then $\mathcal{L}(G(\kappa))$ is a Gabriel filter, i.e. $G(\kappa)$ is a kernel functor.*

Proof. a) If $J \in \mathcal{L}(G(\kappa))$ and $M \supset J$ then obviously $M \in \mathcal{L}(G(\kappa))$.
b) If $I, J \in \mathcal{L}(G(\kappa))$ then $I \cap J \supset G(H_1) \cap G(H_2) \supset G(H_1 \cap H_2)$ for some $H_1, H_2 \in \mathcal{L}(\kappa)$ with $G(H_1) \subset I$ and $G(H_2) \subset J$. Thus $I \cap J \in \mathcal{L}(G(\kappa))$.
c) If $L \in \mathcal{L}(G(\kappa))$ and $\overline{y} \in G(R)$ then we have to find $H \in \mathcal{L}(G(\kappa))$ such that $H\overline{y} \subset L$ and it is sufficient to do this for homogeneous $\overline{y}$ since $G(R)$ is graded, say $y \in \dot{F}_{\deg(y)}R$. Since $L \supset G(I)$ for some $I \in \mathcal{L}(\kappa)$, there exists an $H \in \mathcal{L}(\kappa)$ such that $Hy \subset I$. For homogeneous $\overline{h} \in G(H)$, say h in $\dot{F}_{\deg(h)}R$ is a representative for $\overline{h}$, it holds that $\overline{h}\overline{y} = \overline{hy}\delta_{\deg(hy),\deg(h)+\deg(y)}$ in $G(R)_{\deg(h)+\deg(y)}$ ($hy \in \dot{F}_{\deg(hy)}R$, with $\deg(yh) \leq \deg(y) + \deg(h)$). Thus $G(H)\overline{y} \subset G(Hy) \subset G(I) \subset L$ with $G(H) \in \mathcal{L}(G(\kappa))$ as desired.
Assume now that $G(R)$ is a commutative domain. Because FR is Zariskian, $G(R)$ is Noetherian (see [46]) so we can check condition 5'. Let $L \in \mathcal{L}(G(\kappa))$ and $H \subset L$ such that L/H is $G(\kappa)$-torsion. We have to establish that $H \in \mathcal{L}(G(\kappa))$. If $I \in \mathcal{L}(\kappa)$ is such that $G(I) \subset L$ then for every $\overline{i} \in G(I)$ there is a $G(J_i) \in \mathcal{L}(G(\kappa))$ such that $G(J_i)\overline{i} \subset H$. The ideal $G(I)$ is finitely generated, say by $\overline{i_1}, \ldots, \overline{i_m}$. Then $J = \cap_{k=1}^m J_{i_k} \in \mathcal{L}(\kappa)$ satisfies $G(J)G(I) = G(JI) \subset H$ ($G(R)$ commutative domain!) We conclude that $H \in \mathcal{L}(G(\kappa))$ since the Gabriel filter $\mathcal{L}(\kappa)$ is closed under multiplication. □

Remark that if $G(R)$ is a domain and $\mathcal{L}(\kappa)$ is symmetric, then the same conclusion holds. Fortunately, we only need the notion of $G(\kappa)$- or $\overline{\kappa}$-torsion, so in fact we can forget about the above lemma and simply define as follows:

Definition 7.2.3. Let M be a graded $G(R)$-module. An element $m \in M$ is called $\overline{\kappa}$-torsion if there exists $I \in \mathcal{L}(\kappa)$ such that $G(I)m = 0$. $\overline{\kappa}(M)$ is the set of $\overline{\kappa}$-torsion elements of M. If $\overline{\kappa}(M) = 0$, we say that M is $\overline{\kappa}$-torsion free.

Lemma 7.2.4. *Let M be a filtered R-module. Then $G(\kappa(M)) \subset \overline{\kappa}(G(M))$. This induces an epimorphism*

$$G(M/\kappa(M)) \twoheadrightarrow G(M)/\overline{\kappa}G(M).$$

Proof. Let $\overline{n} \in G(\kappa(M))_{\deg(n)}$ and take a representative n in $\dot{F}_{\deg(n)}\kappa(M)$. There exists $I \in \mathcal{L}(\kappa)$ such that $In = 0$, hence $G(I)\overline{n} = 0$ in $G(M)$. □

Observe that we do not know whether $G(M)/\overline{\kappa}G(M)$ is $\overline{\kappa}$-torsion free, since we do not know in general whether $\overline{\kappa}$ is radical! In order to define a filtration on the localization, we need the notion of κ-separatedness.

Definition 7.2.5. Let κ be a kernel functor with associated filter $\mathcal{L}(\kappa)$ and let M be a filtered R-module. If for some $I \in \mathcal{L}(\kappa)$ such that if $m \in \dot{F}_n M$ with $F_\gamma I m \subset F_{\gamma+n-1}M$ for all γ implies that $m \in \kappa(M)$, then we say that M is κ-separated.

In fact, we do not need that κ is radical; that is, the definition can also be applied to left exact preradicals.

Lemma 7.2.6. *If $M/\kappa(M)$ is κ-separated, then $G(\kappa(M)) = \overline{\kappa}(G(M))$ and $G(M/\kappa(M)) = G(M)/\overline{\kappa}G(M)$.*

Proof. Let $\overline{z} \in \overline{\kappa}(G(M))_n$ and let $I \in \mathcal{L}(\kappa)$ such that $G(I)\overline{z} = 0$. Let $z \in \dot{M}_n$ be a representative for $\overline{z}$. Since FR is Zariskian, the induced filtration on I is good, so there exist $i_1, \ldots, i_k \in I$ and $n_1, \ldots, n_k$ such that $F_n I = \sum_{j=1}^{k} F_{n-n_j} R i_j$ for all n. Moreover, $\deg(i_j) \leq n_j$ (we even have equality if this is a minimal generating set, since FR is Zariskian). For any j, $\overline{i_j}\overline{z} = 0$ in $G(M)_{n+\deg(i_j)}$, i.e. $i_j z \in F_{n+\deg(i_j)-1}M$. Hence

$$F_\gamma I \overline{z} = \textstyle\sum_{j=1}^{k} F_{\gamma-n_j} R i_j \overline{z} \subset \sum_{j=1}^{k} F_{\gamma-n_j} R F_{n+\deg(i_j)-1} M/\kappa(M) \subset F_{\gamma+n-1}M/\kappa(M),$$

where $\overline{z}$ denotes the class of z in $M/\kappa(M)$. Since $M/\kappa(M)$ is κ-separated, $z \in \kappa(M)$, or $\overline{z} \in G(\kappa(M))$. □

Remark 7.2.7. We consider $M/\kappa(M)$ because of the definition of localization. The proof also works for M κ-separated.

Example 7.2.8. In general, we do not have that $G(\kappa(M)) = \overline{\kappa}(G(M))$. Consider for example the ring $R = \mathbb{C}[X,Y]/(XY-1)$ with standard filtration. Then the associated graded ring is $G(R) = \mathbb{C}[x,y]/(xy)$ with $x = \overline{X}, y = \overline{Y}$. Let $S = S_X$ be the multiplicative set $\{1, X, X^2, \ldots\}$. Then $\overline{\kappa_S}$ is given by the multiplicative set $\overline{S} = S_x = \{1, x, x^2, \ldots\}$ in $G(R)$. Since R is a domain, we have that $G(\kappa_S(R)) = G(0) = 0$. But $\kappa_{\overline{S}}(G(R)) = (y)$.

Lemma 7.2.9. *If $G(M)$ is $\overline{\kappa}$-torsion free then M is κ-separated.*

Proof. Let $m \in \dot{F}_n M, \overline{m} = \sigma(m) \in G(M)_n$ and assume that for $I \in \mathcal{L}(\kappa)$ and all $\gamma, F_\gamma I m \subset F_{\gamma+n-1}M$. Then clearly: $G(I)\overline{m} = 0$. Since $G(M)$ is $\overline{\kappa}$-torsion free, $\overline{m} = 0$, a contradiction, unless $m = 0 \in \kappa(M)$. In fact, $\kappa(M) = 0$ because if $Jm = 0$ for some $J \in \mathcal{L}(\kappa)$ then $J_\gamma m = 0 \subset F_{\gamma+n-1}M$ for $m \in \dot{F}_n M$ $(m \neq 0)$. As above, this leads to a contradiction. □

Corollary 7.2.10. *If $G(R)$ is a commutative domain then R is κ-separated for every kernel functor (or more generally, for every preradical) κ on R-mod.*

Before we state and prove the theorem on localizations in our first approach, we deduce the existence of a strong characteristic variety, which will turn out to be a closed set in R-tors such that the localizations remain separated.

For rings of differential operators R, see [12], [46, p. 196] and a filtered R-module M, the characteristic variety of M is defined over the commutative ring $G(R)$ by taking $\mathrm{ann}_{\mathrm{G(R)}}\mathrm{G(M)} = \mathrm{I}$ and letting the characteristic variety $\chi(M)$ be given by $V(I)$ in Spec(G(R)). For a prime $P \in \chi(M)$, we have that $(G(R) \setminus P) \cap I = \emptyset$, but this is a more general statement than to say that $G(M)$ is $\overline{\kappa}_P$-torsion free, the statement we want by Lemma 7.2.9. So we have to look at a smaller set of prime ideals. Now in the generality considered here, i.e. both R and $G(R)$ being non-commutative, where we may assume R to be positively filtered and $G(R)$ Noetherian (hence FR is a Zariskian filtration), we introduce the strong characteristic variety for an arbitrary separated filtered R-module, M say, as a subset $V = \xi(M)$ in $G(R)$-tors. The following theorem will yield a separated quotient filtration on the localization of M at κ with which $\overline{\kappa} \in \xi(M)$ in $G(R)$-tors is associated.

Define $\mathcal{A}\mathrm{nn}(M)$ with script $\mathcal{A}$ as the set

$$\mathcal{A}\mathrm{nn}(M) = \{\overline{L} \text{ left ideal of } G(R) \,|\, \overline{L}\overline{m} = 0 \text{ for some } \overline{m} \in G(M)\}.$$

Define $\xi(M) = V \subset G(R)$-tors by putting

$$V = \{\overline{\kappa} \,|\, \mathcal{L}(\overline{\kappa}) \cap \mathcal{A}\mathrm{nn}(M) = \emptyset\}.$$

Thus we have for $\overline{\kappa} \in V$ that $G(M)$ is $\overline{\kappa}$-torsion free, hence M is κ-separated by Lemma 7.2.9. Observe that $\kappa(M) = \overline{\kappa}(G(M))$ and $G(M/\kappa M) = G(M)/\overline{\kappa}G(M)$ by Lemma 7.2.6. The subset V is closed in the gen-topology of $G(R)$-tors (see Section 4 for the definition of the gen-topology). Indeed, for $\overline{\tau} \in G(R)\text{-tors} \setminus \mathrm{V}$ we have $\mathcal{L}(\tau) \cap \mathcal{A}\mathrm{nn}(M) \neq \emptyset$ or $\overline{\tau}(G(M)) \neq 0$. If $\overline{\gamma} \geq \overline{\tau}$ then $\overline{\gamma}(G(M)) \neq 0$ too or $\overline{\gamma} \in G(R)\text{-tors} \setminus \mathrm{V}$, proving that $\mathrm{gen}(\overline{\tau}) \subset \mathrm{G(R)\text{-}tors} \setminus \mathrm{V}$, i.e. that $G(R)\text{-tors} \setminus \mathrm{V}$ is gen-open, and thus V is gen-closed. If $\kappa \in R$-tors is such that $\overline{\kappa} \in V$ (observe that the whole construction works for $\overline{\kappa}$ a left exact preradical, then V is constructed in the non-commutative topology $G(R)$-pretors) then $\kappa \in V_R$ (i.e. those $\tau \in R$-tors with $\overline{\tau} \in V \subset G(R)$-pretors). Hence by the following theorem, $Q_\kappa(M)$ has a separated quotient filtration. Moreover, $G(\kappa(M)) = \overline{\kappa}(G(M)) = 0$ yields that $\kappa(M) = 0$ as $\kappa(M)$ is separated filtered by $FM \cap \kappa(M)$.

Going back to the classical setting of rings of differential operators, we recall that the associated Gabriel filter of an Ore set S is given by

$$\mathcal{L}(S) = \mathcal{L}(\kappa_S) = \{L \leq R \text{ left ideal} \,|\, \exists s \in S \cap L\}. \tag{7.1}$$

We denote the corresponding kernel functor by κ_S and for $P \in \mathrm{Spec(R)}$ we write $\kappa_P := \kappa_{S_P} = \kappa_{R \setminus P}$. Suppose that there exists a prime ideal P such that $\kappa_P \in \chi(M) \setminus \xi(M)$. We still obtain a filtered localization $S_P^{-1}M$, but the filtration is no longer separated. Indeed, lifts of $\overline{S_P}$-torsion elements are in the core of the filtration (see proof of theorem). However, $P \in \chi(M)$, so the core of the filtration is not the whole of the localized module! Hence, even in the classical setting of almost commutative rings there are questions of how both characteristic varieties are related. This is work in progress, but we include an example.

Example 7.2.11. Let $A_1 = \mathbb{C}\langle X, \partial_X\rangle/(X\partial_X - \partial_X X - 1)$ be the first Weyl algebra and consider the holonomic module $M = A_1/A_1P$, where $P = X\partial_X$. The sigma-filtration on A_1 is good and Zariskian, so we have that the induced filtration on M is good and $G(M) = \mathbb{C}[X, \overline{\partial_X}]/(X\overline{\partial_X})$. Hence the characteristic variety is $\chi(M) = V((X\overline{\partial_X}))$ and in particular $(X) \in \chi(M)$. However, $(X) \notin \xi(M)$, because $G(A_1)\overline{\partial_X} \in \mathcal{L}(\overline{\kappa_(X)}) \cap \mathcal{A}\mathrm{nn}(M)$. So in general, the characteristic variety is strictly bigger than the strong characteristic variety.

Theorem 7.2.12. *Let FM be a filtered module over the Zariskian filtered ring FR such that $M/\kappa(M)$ is κ-separated. The localized module $Q_\kappa(M)$ has a separated filtration $FQ_\kappa(M)$ making the localization morphism $j_\kappa : M \to Q_\kappa(M)$ into a strict filtered morphism. Similar for FR such that $R/\kappa(R)$ is κ-separated, $j_\kappa : R \to Q_\kappa(R)$ is a strict filtered morphism and a filtered ring morphism. Moreover, $Q_\kappa(M)$ with the quotient filtration defined here is a filtered $Q_\kappa(R)$-module with respect to the quotient filtration on $Q_\kappa(R)$.*

Proof. We have a strict exact sequence $0 \to \kappa M \to M \to M/\kappa M \to 0$; hence by exactness of G on strict sequences we obtain an exact sequence

$$0 \to G(\kappa M) \to G(M) \to G(M/\kappa M) \to 0$$

of graded $G(R)$-modules. Since $M/\kappa(M)$ is κ-separated, $G(\kappa M) = \overline{\kappa}G(M)$ and $G(M/\kappa M)$ equals $G(M)/\overline{\kappa}G(M)$. This allows us to reduce the situation to the case where M is κ-torsion free and the canonical map $j_\kappa : M \to Q_\kappa(M)$ is injective.

For $x \neq 0$ in $Q_\kappa(M)$ there is an $I \in \mathcal{L}(\kappa)$ such that $Ix \subset M$. Since FR is Zariskian, the filtration induced by FR on I is a good filtration; hence for all $n \in \mathbb{Z}, F_nI = \sum_{i=1}^{s} F_{n-d_i}R\zeta_i$, for $\zeta_1, \ldots, \zeta_s \in I$ with $\deg \sigma(\zeta_i) \leq d_i \in \mathbb{Z}$. Then $\zeta_i x \in F_{d_i+\gamma_i}M$ for some $\gamma_i \in \mathbb{Z}$. Put $\gamma = \max\{\gamma_i, i = 1, \ldots, s\}$, then $Ix \subset M$ and for all $n \in \mathbb{Z}$

$$F_nIx \subset F_{n+\gamma}M, \tag{7.2}$$

because $F_nIx \subset \sum F_{n-d_i}R\zeta_i x \subset F_{n-d_i+d_i+\gamma}M = F_{n+\gamma}M$.

Since FM is separated (the original FM was separated, so $FM/\kappa M$ is separated as well) and $Ix \neq 0$ as $Q_\kappa(M)$ is κ-torsion free, there is a minimal γ such that for all $n, F_nIx \subset F_{n+\gamma}M$. For this γ we have that there is an $n \in \mathbb{Z}$ such that $F_nIx \not\subset F_{n+\gamma-1}M$. Let us check that γ depends on x but not on the $J \in \mathcal{L}(\kappa)$ chosen. Pick $\zeta \in F_nI$, n as just mentioned, such that

$\zeta x \in F_{n+\gamma}M \setminus F_{n+\gamma-1}M$ and now look at $J \in \mathcal{L}(\kappa)$ such that $Jx \subset M$, with $F_\alpha Jx \subset F_{\alpha+\gamma-1}M$ for all $\alpha \in \mathbb{Z}$. Then $(J:\zeta) \in \mathcal{L}(\kappa)$ and for all $m \in \mathbb{Z}$,

$$F_m(J:\zeta)\zeta x \subset F_{m+n}Jx \subset F_{m+n+\gamma-1}M.$$

Since M is κ-separated, $\zeta x \in \kappa(M) = 0$, a contradiction. So we define

$$\begin{aligned} F_\gamma Q_\kappa(M) &= \{q \in Q_\kappa(M) \mid \exists I \in \mathcal{L}(\kappa) \text{ such that } F_n Iq \subset F_{n+\gamma}M \text{ for all } n \in \mathbb{Z}\} \\ &= \{q \in Q_\kappa(M) \mid v(q) \le \gamma\}, \end{aligned}$$

where the *filtration degree* v is given by $v(q) = \gamma$, γ as constructed above. It is obvious that $F_\gamma Q_\kappa(M)$ defines an ascending chain of additive subgroups of $Q_\kappa(M)$. Now first look at $M = R$ and $x, y \in Q_\kappa(R)$ such that $I_1 x \subset R, I_2 y \subset R$ and both satisfying (7.2). Put $J = (I_2 : x)$; then for all $n \in \mathbb{Z}$: $F_n(J \cap I_1)x \subset F_{n+v(x)}R \cap I_2 = F_{n+v(x)}I_2$, and also

$$F_n(J \cap I_1)xy \subset F_{n+v(x)}I_2 y \subset F_{n+v(x)+v(y)}R.$$

Hence $FQ_\kappa(R)$ makes $Q_\kappa(R)$ into a filtered ring. That $F_nR \subset F_nQ_\kappa(R)$ is obvious (we reduced to the torsion free case), on the other hand $E_n = F_nQ_\kappa(R) \cap R$ is the κ-closure of F_nR in R, say $y \in E_n, Iy \subset R$ with $F_\alpha Iy \subset F_{\alpha+n}R$ for all $\alpha \in \mathbb{Z}$. If $y \notin F_nR$, say $y \in \dot{F}_\delta R$ with $\delta > n$, then $F_\alpha Iy \subset F_{\alpha+\delta-1}R$ as $n \le \delta - 1$. By κ-separatedness, $y = 0$, a contradiction. Hence $F_nQ_\kappa(R) \cap R = F_nR$ follows. This establishes that $j_\kappa : R \to Q_\kappa(R)$ is a strict filtered map of rings (note that $R \to R/\kappa R$ is already strict so the restriction to the κ-torsion free case did not harm this).
Now for $x \in Q_\kappa(R), y \in Q_\kappa(M)$ with $I_1 x \subset R, I_2 y \subset M$ both satisfying (7.2), put $J = (I_2 : x)$ and as in the preceding argument it follows that $Q_\kappa(M)$ is a filtered $Q_\kappa(R)$-module. Also the proof that $F_nQ_\kappa(M) \cap M = F_nM$ goes through in the same way as for R. Finally observe that $FQ_\kappa(M)$ (resp. $FQ_\kappa(R)$) is separated. Indeed, put $E = \bigcap F_nQ_\kappa(M)$, then $E \cap M = \bigcap F_nM = 0$, but for $z \in E, Iz \subset M$ for some $I \in \mathcal{L}(\kappa)$, thus $Iz \subset E \cap M = 0$ contradicting that $Q_\kappa(M)$ is κ-torsion free, unless $E = 0$. □

Example 7.2.13. Suppose we are in the 'almost commutative' case, that is $G(R)$ is a commutative domain (then $Q_\kappa(R)$ has a filtration for every kernel functor κ by Corollary 7.2.10). Let M be a glider representation, $\Omega = RM \supset M$. We consider on Ω the filtration $f\Omega$ given by

$$\begin{aligned} F_n\Omega &= \textstyle\sum_{i-j=n} F_iRM_j, \\ F_{-n}\Omega &= M_n, \quad n \ge 0. \end{aligned} \tag{7.3}$$

The associated graded $G_f(\Omega)$ has for the negative part $g(M) = \oplus_{i\ge0} M_i/M_{i+1}$ and this is itself a fragment with respect to the positive part of the gradation filtration of $G(A)$ (see Section 1.9). If $\Omega/\kappa(\Omega)$ is κ-separated, e.g. when $G_f(\Omega/\kappa(\Omega))$ is $\overline{\kappa}$-torsion free, or when $G_f(\Omega)$ is a faithful $G(R)$-module, then the degree 0 part $F_0Q_\kappa(\Omega)$ becomes a fragment w.r.t the positive part of the

filtration $F^+Q_\kappa(R)$ by Example 1.3.5. We call $Q_\kappa(M) := F_0Q_\kappa(\Omega)$ the localized fragment of M w.r.t the kernel functor κ.
As a particular case, if S is an Ore set in a standard filtered ring FR, we obtain the quotient fragment $Q_S(M)$ from the previous section. In fact, the standard assumption is no longer needed in our more general setting.

Example 7.2.14.
(*i*) Let V be an affine variety with coordinate ring $\Gamma(V)$. Then one considers the ring of differential operators $D(V)$ with the Σ-filtration. This is a standard filtered ring with $F_0D(V) = \Gamma(V)$ and with associated graded being a commutative domain. In forthcoming work, we will exploit this setting more, investigating for example the link between Ore sets in $D(V)$ and $\Gamma(V)$ (as the previous section uncovered) or the characteristic variety of a glider fragment. For details on rings of differential operators we refer to [12].
(*ii*) Let V and W be varieties embedded in resp. $\mathbb{A}_n(K), \mathbb{A}_m(K)$ such that $\Gamma(V) \subset \Gamma(W)$, $\Gamma(V) = K[a_1, \ldots, a_n], \Gamma(W) = K[a_1, \ldots, a_m][b_1, \ldots, b_{n-m}]$. We view the V-filtration on $\Gamma(W)$ given by

$$F_0\Gamma(W) = \Gamma(V),$$

$$F_1\Gamma(W) = \Gamma(V)b_1 + \cdots + \Gamma(V)b_{n-m} = \Gamma(V)[B^1],$$

where $B^1 = Kb_1 + \cdots + Kb_{n-m}$, $B = \{b_1, \ldots, b_{n-m}\}, \Gamma(W) = \Gamma(V)[B]$. For $n \geq 1$ we define

$$F_n\Gamma(W) = \Gamma(V)[B^n],$$

where $B^n = \sum_{i_1,\ldots,i_n} Kb_{i_1} \ldots b_{i_n}$, $b_{i_j} \in B, j = 1, \ldots, n$.
W-glider representations over V are then by definition $\Omega \supset M$, Ω a $\Gamma(W)$-module, M an $F\Gamma(W)$-fragment structure induced by the $\Gamma(W)$-module Ω.
There are other interesting geometric filtrations on $\Gamma(W)$, e.g. the ring filtration $\Gamma(V) \subset \Gamma(V)[b_1] \subset \ldots \subset \Gamma(V)[b_1, \ldots, b_d] = \Gamma(W)$, where each $\Gamma(V)[b_1, \ldots, b_i]$ corresponds to a W_i with $\Gamma(V) \subset \Gamma(W_i) \subset \Gamma(W)$ and W_i embedded in $\mathbb{A}_{n+i}(K)$ say, $W = W_d \to W_{d-1} \to \ldots \to W_1 \to V$. This commutative theory of glider representations has to be connected to the geometric properties of V and W.

Let M be a filtered R-module with filtration FM. We recall the definition of a *good filtration*, see [46, p. 46].

Definition 7.2.15. Let M be a filtered FR-module with filtration FM. If there exist $m_1, \ldots, m_s \in M$, $k_1, \ldots, k_s \in \mathbb{Z}$ such that for all $n \in \mathbb{Z}$

$$F_nM = \sum_{i=1}^{s} F_{n-k_i}Rm_i,$$

then FM is called a *good filtration* on M.

Let M be a filtered R-module. A submodule N of M is said to be closed if $\bigcap_\gamma N + F_\gamma M = N$. If FM is good and if FR is Zariskian then for every $N \subset M$ we have that N is closed (see [46, p. 83]). In particular if FM is good then κM is closed and therefore $M/\kappa M$ with the induced filtration is separated. Indeed if $y \in M$ maps to $F_\gamma(M/\kappa M)$ for all γ, then $y \in \kappa M + F_\gamma M$ for all γ, hence $y \in \bigcap_\gamma \kappa M + F_\gamma M = \kappa M$ and y maps to zero. Without the assumption that FM is good we have

Proposition 7.2.16. *If M is κ-separated then κM is closed and $M/\kappa M$ is separated.*

Proof. Suppose $y \in \bigcap_\gamma \kappa M + F_\gamma M$ and $y \notin \kappa M$. Then $F_\gamma M \neq 0$ and $F_\gamma M \not\subset \kappa M$ for all γ. Since FM is separated, $y \in \dot{F}_n M$ for some $n \in \mathbb{Z}$. Take $\gamma \leq n-1, y \in \kappa M + F_\gamma M$, say $y = t_\gamma + f_\gamma$ where $t_\gamma \in \kappa M, f_\gamma \neq 0$ in $F_\gamma M$. Choose I such that $It_\gamma = 0$ and look at $F_\tau I y = F_\tau I f_\gamma \subset F_{\tau+\gamma} M \subset F_{\tau+n-1} M$. This holds for all $\tau \in \mathbb{Z}$, so since M is κ-separated, $y \in \kappa M$ follows. Thus κM is closed and $M/\kappa M$ is separated for the induced filtration from FM. □

Corollary 7.2.17. *If M is κ-separated, then $FQ_\kappa(M)$ is a separated filtration.*

Proof. If $\bigcap_\gamma F_\gamma Q_\kappa(M) = E$, then $E \cap M/\kappa M = \bigcap_\gamma F_\gamma(M/\kappa M)$. By the previous proposition, $E \cap M/\kappa M = 0$. Now if $E \neq 0$ then, say $x \neq 0, x$ in $E, Ix \subset M/\kappa M$ for some $I \in \mathcal{L}(\kappa)$. Thus $Ix \subset E \cap M/\kappa M = 0$ but then $x \in \kappa(Q_\kappa(M)) = 0$. □

The situation in the above corollary occurs for example when $G(M)$ is $\overline{\kappa}$-torsion free, i.e. when $\overline{\kappa}G(M) = 0$.

Suppose now that we have two kernel functors κ and τ. We can consider the composition $\tau\kappa$, which in general is only a preradical. In other words, $\tau\kappa$ is not a localization functor, that is, the associated torsion class $\mathcal{T}_{\tau\kappa}$ is not hereditary. The latter torsion class is defined by the R-modules M such that there exists a submodule $N \subset M$ such that N is τ-torsion and M/N is κ-torsion, or equivalently $M/\tau(M)$ is κ-torsion. The associated filter of left ideals is denoted by $\mathcal{L}(\tau\kappa)$ and consists of left ideals containing a left ideal $\sum'_\alpha I_\tau x_\kappa^\alpha$ where $I_\kappa = \sum'_\alpha R x_\kappa^\alpha \in \mathcal{L}(\kappa)$ and $I_\tau \in \mathcal{L}(\tau)$. We denote this by $I_\tau \cdot I_\kappa$ (FR is Zariskian so R is Noetherian and the existence of a finite set of generators for any left ideal I is guaranteed).

Lemma 7.2.18. *The filter $\mathcal{L}(\tau\kappa)$ satisfies the first three properties $G1, G2$ and $G3$.*

Proof. We only prove the third property $G3$. Let $I \cdot J \in \mathcal{L}(\tau\kappa)$, with $J = \sum'_\alpha R x_\kappa^\alpha$ and take $r \in R$. Then there exists $y_\kappa^\beta \in R$ with

$(J:r) = \sum'_\beta R y_\kappa^\beta \in \mathcal{L}(\kappa)$. Let $y_\kappa^\beta r = \sum r_{\beta,\alpha} x_\kappa^\alpha \in J$. We find an $I' \in \mathcal{L}(\tau)$ such that $I' r_{\beta,\alpha} \subset I$ for all α, β. Then

$$I' \cdot (J:r)r = \sum_\beta I' y_\kappa^\beta r = \sum_{\alpha,\beta} I' r_{\beta,\alpha} x_\kappa^\alpha \subset I \cdot J. \quad \square$$

Although we do not know in general whether $\tau\kappa$ is radical, we can still consider the canonical morphism $j_{\tau\kappa} : M \to Q_\kappa Q_\tau(M)$ for any R-module M. This is just the composition

$$M \xrightarrow{j_\tau} Q_\tau(M) \xrightarrow{j'_\kappa} Q_\kappa Q_\tau(M),$$

in which the latter is the κ-localization morphism of $Q_\tau(M)$. We have

$$j_{\tau\kappa}(M) = \frac{M/\tau(M)}{M/\tau(M) \cap \kappa Q_\tau(M)} = \frac{M/\tau(M)}{\kappa(M/\tau(M))},$$

whence $j_{\tau\kappa}(M) = 0$ if and only if $M \in \mathcal{T}_{\tau\kappa}$ and this implies $Q_\kappa Q_\tau(M) \in \mathcal{T}_{\tau\kappa}$ (indeed, $M/\tau(M) = \kappa(M/\tau(M))$ implies that $M/\tau(M) \subset \kappa Q_\tau(M)$, whence $Q_\tau(M)/\kappa Q_\tau(M)$ is τ-torsion as it is a quotient of the τ-torsion module

$$Q_\tau(M)/(M/\tau(M)).$$

Therefore, $Q_\tau(M)/\kappa Q_\tau(M) \subset Q_\kappa Q_\tau(M)$ with $Q_\kappa Q_\tau(M)/(Q_\tau(M)/\kappa Q_\tau(M))$ being κ-torsion shows that $Q_\kappa Q_\tau(M) \in \mathcal{T}_{\tau\kappa}$. To show that

$$Q_\kappa Q_\tau(M)/j_{\tau\kappa}(M) \in \mathcal{T}_{\tau\kappa}$$

for any R-module M, one considers the submodule

$$\frac{Q_\tau(M)/\kappa Q_\tau(M)}{j_{\tau\kappa}(M)},$$

which is τ-torsion. To see this, one uses the isomorphism

$$\frac{Q_\tau(M)/\kappa Q_\tau(M)}{(M/\tau(M))/\kappa(M/\tau(M))} \cong \frac{Q_\tau(M)}{M/\tau(M)}.$$

Proposition 7.2.19. *[70, Proposition 3.10]*
For $M \in R$-mod, the following statements are equivalent

1. $j_{\tau\kappa}(M) = 0$;
2. $M \in \mathcal{T}_{\tau\kappa}$;
3. *For every $m \in M$ there is an $L \in \mathcal{L}(\tau\kappa)$ such that $Lm = 0$.*

The above proposition shows that for any $x \neq 0$ in the double localization $Q_\kappa Q_\tau(M)$ there exists an $I \cdot J \in \mathcal{L}(\tau\kappa)$ such that $I \cdot Jx \subset j_{\tau\kappa}(M)$. One also verifies that the kernel $\mathrm{Ker}(j_{\tau\kappa})$ equals $\tau\kappa(M)$, the elements of M annihilated by an ideal of $\mathcal{L}(\tau\kappa)$.

Lemma 7.2.20. *Let M be a filtered R-module. Then $m \in \tau\kappa(M)$ if and only if $\overline{m} \in \kappa(M/\tau M)$.*

Proof. If $\overline{m} \in \kappa(M/\tau M)$ with original $m \in M$, then for some $I = \sum'_\alpha Rx^\alpha_\kappa \in \mathcal{L}(\kappa)$ we have $Im \subset \tau M$. Let $J \in \mathcal{L}(\tau)$ such that $Jx^\alpha_\kappa m = 0$ for all α, then $J \cdot Im = 0$, or $m \in \tau\kappa(M)$. The converse is similar. □

The previous lemma shows that $\kappa(M/\tau M) = \tau\kappa(M)/\tau(M)$, whence

$$j_{\tau\kappa}(M) = \frac{M/\tau M}{\kappa(M/\tau M)} = \frac{M/\tau M}{\tau\kappa M/\tau M} \cong M/\tau\kappa M.$$

Corollary 7.2.21. *Let M be a filtered R-module. Then $M/\tau\kappa M$ is $\tau\kappa$-separated if and only if $(M/\tau M)/\kappa(M/\tau M)$ is κ-separated.*

Proposition 7.2.22. *Let τ, κ be kernel functors and M a filtered R-module. If $M/\tau\kappa M$ is $\tau\kappa$-separated, then the double localization module $Q_\kappa Q_\tau(M)$ has a filtration $FQ_\kappa Q_\tau(M)$ making the localization morphism $j_{\tau\kappa}(M)$ into a strict filtered morphism. Similar for $R/\tau\kappa R$ being $\tau\kappa$-separated, $j_{\tau\kappa} : R \to Q_\kappa Q_\tau(R)$ is a strict filtered ring morphism.*

Proof. One follows the proof of Theorem 7.2.12 without reducing to the torsion free case ($\tau\kappa$ no longer radical!) and one checks that only the first three properties for filters of ideals are used. Explicitly

$$F_\gamma Q_\kappa Q_\tau(M) = \{q \in Q_\kappa Q_\tau(M), \exists I \cdot J \in \mathcal{L}(\tau\kappa) \text{ such that } F_n(I \cdot J)q \subset F_{n+q} j_{\tau\kappa}(M) \text{ for all } n \in \mathbb{Z}\}.$$

□

If the filtered module $M/\tau M$ is τ-separated, we obtain a filtered module $FQ_\tau(M)$. If $Q_\tau(M)/\kappa(Q_\tau(M))$ is κ-separated, we get a filtration $FQ_\kappa Q_\tau(M)$. If however, $M/\tau\kappa(M)$ is also $\tau\kappa$-separated, then we obtain a second filtration $F'Q_\kappa Q_\tau(M)$. We clearly have an inclusion $F' \subset F$ of filtered modules. In case κ is coming from a $\widetilde{\kappa}$ (see second approach below), then we obtain equality, for in this case we have for $I = \sum'_\alpha Rx^\alpha_\kappa \in \mathcal{L}(\kappa)$ and $J \in \mathcal{L}(\tau)$ that $F_\gamma(J \cdot I) = \sum_\alpha F_{\gamma - \deg(x^\alpha_\kappa)} Jx^\alpha_\kappa$ for any γ, since $(\widetilde{JI})_\gamma = \sum_{\sigma+\tau=\gamma} \widetilde{J}_\sigma \widetilde{I}_\tau$. We do not know whether we have equality in general. If one works with Ore sets in the 'almost commutative' case everything holds on the other hand.

Lemma 7.2.23. *Let S, T be Ore sets in R with $G(R)$ being a commutative domain. Then under the above assumptions, $F_n Q_S Q_T(M) \subset F'_n Q_S Q_T(M)$ for all $n \in \mathbb{Z}$.*

Proof. For $s \in S, t \in T$ we have that $Rt \cdot Rs = Rts$ and $F_\gamma(Rts) = F_{\gamma - \deg(st)} Rst$, so the conclusion easily follows. □

For two Ore sets S, T we have the following diagram for an R-filtered module M.

$$\begin{array}{ccc} Q_T(M) & \longrightarrow & Q_T(Q_S(M)) \\ & \nearrow\!\!\!\!\!\!\searrow & \\ Q_S(M) & \longrightarrow & Q_S(Q_T(M)) \end{array} \tag{7.4}$$

Suppose that M is such that the corresponding quotients with induced filtrations are S-, T-, ST-, resp. TS-separated, that $Q_T(M)/SQ_T(M)$ is S-separated and that $Q_S(M)/TQ_S(M)$ is T-separated. Both diagonal morphisms are the canonical localization morphisms, hence filtered. The upper horizontal arrow is the localized morphism $Q_T(j_S)$, which is also filtered by the following:

Lemma 7.2.24. *Let $f : M \to N$ be a filtered morphism between filtered FR-modules M, N and κ an idempotent kernel functor. If both M and N are κ-separated, then $Q_\kappa(f) : Q_\kappa(M) \to Q_\kappa(N)$ is a filtered morphism with respect to the localization filtrations.*

Proof. This follows from the fact that $Q_\kappa(f)$ is (left) R-linear and from the definition of the localization filtration. □

In Section 7.4 we discuss the lattice structure on R-tors. We already mention here that for kernel functors κ, τ, $\kappa \leq \tau$ means that $\mathcal{L}(\kappa) \subset \mathcal{L}(\tau)$ or that $\kappa(M) \subset \tau(M)$ for all R-modules M.

Proposition 7.2.25. *If $\kappa \leq \tau$ and M is κ-separated and τ-separated then there is a canonical filtered morphism $\rho_\tau^\kappa : Q_\kappa(M) \to Q_\tau(M)$ yielding a commutative diagram*

$$\begin{array}{ccc} Q_\kappa(M) & \xrightarrow{\rho_\tau^\kappa} & Q_\tau(M) \\ \big\uparrow{\scriptstyle j_\kappa} & & \big\uparrow{\scriptstyle j_\tau} \\ M/\kappa(M) & \xrightarrow{\pi_\tau^\kappa} & M/\tau(M) \end{array}$$

where the vertical morphisms are strict filtered injections and π_τ^κ is the canonical filtered epimorphism.

Proof. The statements about j_κ, j_τ and π_τ^κ are obvious. We know that π_τ^κ extends to an R-linear ρ_τ^κ by general localization theory. Let $y \in F_n Q_\kappa(M)$, that is, there is an $I \in \mathcal{L}(\kappa)$ such that for all $\gamma \in \mathbb{Z}, F_\gamma I y \subset F_{\gamma+n} M/\kappa M$. Applying ρ_τ^κ yields $F_\gamma I \rho_\tau^\kappa(y) = \rho_\tau^\kappa(F_\gamma I y) = \pi_\tau^\kappa(F_\gamma I y) \subset F_{\gamma+n}(M/\tau M)$ since π_τ^κ is filtered. Now, $F_\gamma I \rho_\tau^\kappa(y) \subset F_{\gamma+n}(M/\tau M)$ for all γ means that $\rho_\tau^\kappa(y)$ in $F_n Q_\tau(M)$ and thus ρ_τ^κ is a filtered morphism. □

With an eye toward the non-commutative site, we add an additional kernel functor σ. Since $\frac{M}{\tau\kappa(M)} = \frac{M/\tau(M)}{\kappa(M/\tau(M))}$, we have that $\mathcal{T}_{(\tau\kappa)\sigma} = \mathcal{T}_{\tau(\kappa\sigma)}$. We denote this torsion class by $\mathcal{T}_{\tau\kappa\sigma}$ and the associated filter $\mathcal{L}(\tau\kappa\sigma)$ consists of left ideals containing some $I_\tau \cdot I_\kappa \cdot I_\sigma = \sum'_{\alpha,\beta} I_\tau y_\kappa^\beta x_\sigma^\alpha$ with

$I_\tau \in \mathcal{L}(\tau), I_\kappa = \sum'_\beta R y_\kappa^\beta \in \mathcal{L}(\kappa), I_\sigma = \sum'_\alpha R x_\sigma^\alpha \in \mathcal{L}(\sigma)$. This filter also satisfies the first three properties $G1, G2$ and $G3$ and the canonical localization morphism

$$j_{\tau\kappa\sigma} : M \xrightarrow{j_\tau} Q_\tau(M) \xrightarrow{j'_\kappa} Q_\kappa Q_\tau(M) \xrightarrow{j'_\sigma} Q_\sigma Q_\kappa Q_\tau(M)$$

has image

$$\frac{M/\tau(M)}{\kappa(M/\tau(M))} / \sigma \left(\frac{M/\tau(M)}{\kappa(M/\tau(M))} \right),$$

which equals $M/\tau\kappa\sigma(M)$ by Lemma 7.2.20. Hence, we again have that $Q_\sigma Q_\kappa Q_\tau(M)/j_{\tau\kappa\sigma}(M) \in \mathcal{T}_{\tau\kappa\sigma}$ and if $M/\tau\kappa\sigma(M)$ is $\tau\kappa\sigma$-separated, the localization module $Q_\sigma Q_\kappa Q_\tau(M)$ has a filtration $FQ_\sigma Q_\kappa Q_\tau(M)$ such that $j_{\tau\kappa\sigma}$ is strict filtered. Observe that in Proposition 7.2.25 it actually follows that the filtered morphism ρ_τ^κ is strict. Indeed, $M/\tau(M) = \frac{M/\kappa(M)}{\tau(M)/\kappa(M)}$ with induced filtration of $M/\kappa(M)$ on $\tau(M)/\kappa(M)$ shows that π_ρ^κ is strict, which shows strictness of ρ_τ^κ. The previous observations imply that the $\rho_W^{W'} : Q_{W'}(M) \to Q_W(M)$ for words in kernel functors $W' \hookrightarrow W$ are also strict filtered maps. Moreover, one checks that the results following Proposition 7.2.22 remain valid (up to restriction to Ore sets) and that we can extend diagram (7.4). In fact, one can consider words of finite length of kernel functors. We will restrict to Ore sets when dealing with the non-commutative site, but we chose the more general approach using kernel functors as the results most likely will be extendable to the so called non-commutative affine site.

7.3 Quotient filtrations - second approach

Now, we start with a graded kernel functor $\widetilde{\kappa}$ on $\widetilde{R}$-mod. Recall that $\widetilde{\kappa}$ corresponds to a unique rigid kernel functor on $\widetilde{R}$-gr and that the associated Gabriel functor $\mathcal{L}(\widetilde{\kappa})$ has a cofinal set of graded left ideals. We define

$$\mathcal{L}(G\kappa) = \{L \subset G(R) \mid L \supset \pi(I) \text{ for some } I \in \mathcal{L}(\widetilde{\kappa})\}.$$

where $\pi : \widetilde{R} \to G(R), \widetilde{r} \mapsto \widetilde{r} \mod \widetilde{R}X$.

Proposition 7.3.1. *$G\kappa$ is a graded kernel functor on $G(R)$-mod.*

Proof. a) If $L' \supset L$ with $L \in \mathcal{L}(G\kappa)$, then $L' \in \mathcal{L}(G\kappa)$.
b) If $L_1, L_2 \in \mathcal{L}(G\kappa)$ then $L_1 \cap L_2 \supset \pi(I_1 \cap I_2), I_1 \cap I_2 \in \mathcal{L}(\widetilde{\kappa})$.
c) Look at $\overline{r} \in G(R)$ and $\widetilde{r} \in \widetilde{R}$ such that $\pi(\widetilde{r}) = \overline{r}$. Take $L \in \mathcal{L}(G\kappa)$ and $(L : \overline{r}) \subset G(R)$. Pick $I \in \mathcal{L}(\widetilde{\kappa})$ such that $L \supset \pi(I)$ then $(I : \widetilde{r}) \in \mathcal{L}(\widetilde{\kappa})$ since the latter is a Gabriel filter. From $(I : \widetilde{r})\widetilde{r} \subset I$ it follows that $\pi(I : \widetilde{r})\overline{r} \subset \pi(I) \subset L$ with $\pi(I : \widetilde{r}) \in \mathcal{L}(G\kappa)$, whence $(L : \overline{r}) \in \mathcal{L}(G\kappa)$.

d) Consider $L \in \mathcal{L}^g(G\kappa)$ and a graded $J \subset L$ such that for all $a \in L$ there is an $I_a \in \mathcal{L}(G\kappa)$ such that $I_a a \subset J$. Since $L \supset \pi(I)$ for some $I \in \mathcal{L}(\widetilde{\kappa})$ we may put $K = \pi^{-1}(J) \cap I$, thus $\pi(K) \subset J \cap \pi(I)$. If $\widetilde{i} \in I$, say $a = \pi(\widetilde{i}) \in L$, then there is an $I_a \in \mathcal{L}(G\kappa)$ such that $I_a a \subset J$. Thus $J_a \widetilde{i} \subset \pi^{-1}(J) \cap I = K$, with $J_a \in \mathcal{L}(\widetilde{\kappa})$ such that $\pi(J_a) \subset I_a$. This holds for every $\widetilde{i} \in I$, thus $K \in \mathcal{L}(\widetilde{\kappa})$ and $\pi(K) \in \mathcal{L}(G\kappa)$. This implies that $J \cap \pi(I) \in \mathcal{L}(G\kappa)$ and $J \in \mathcal{L}(G\kappa)$. Since $\mathcal{L}(\widetilde{\kappa})$ is graded, so is $\mathcal{L}(G\kappa)$. Therefore, if $L \in \mathcal{L}(G\kappa)$ and $J \subset L$ are not graded, then $L \supset \pi(I)$ still holds with $\pi(I)$ a graded ideal and $J \cap \pi(I) \subset \pi(I)$. Let $J^g \subset J$ be the largest graded $G(R)$-submodule of J. Since $(J \cap \pi(I))/(J^g \cap \pi(I))$ and $\pi(I)/(J \cap \pi(I))$ are $G\kappa$-torsion, so is $\pi(I)/J^g \cap \pi(I)$. By the first part of this proof $J^g \cap \pi(I)$ and thus $J^g \in \mathcal{L}(G\kappa)$. It follows that $J \in \mathcal{L}(G\kappa)$. □

We call $G\kappa$ the graded kernel functor on $G(R)$-mod induced by $\widetilde{\kappa}$ on $\widetilde{R}$-mod. We also write $G\kappa = \overline{\kappa}$. Sometimes $\overline{\kappa}$ is trivial, e.g. when $\widetilde{R}X \in \mathcal{L}(\widetilde{\kappa})$ then $0 \in \mathcal{L}(\overline{\kappa})$ and $\overline{\kappa}$ is indeed trivial. If $\mathcal{L}^g(\widetilde{\kappa})$ does not contain $\widetilde{R}X$ then $\overline{\kappa}$ is non-trivial. Observe moreover that the $\overline{\kappa}$ defined here from $\widetilde{\kappa}$ is not the same as the $\overline{\kappa}^1$ defined from κ as in the first approach. Recall that $\overline{\kappa}^1$ has a filter $\mathcal{L}(\overline{\kappa}^1) = \{L \subset G(R), L \supset \pi(\widetilde{I})$ for some $I \in \mathcal{L}(\kappa)\}$, while $\mathcal{L}(\overline{\kappa}) = \{L \subset G(R), L \supset \pi(I)$ for some $I \in \mathcal{L}(\widetilde{\kappa})\}$. Both filters coincide exactly when $\mathcal{L}(\widetilde{\kappa})$ has a filterbasis consisting of $\widetilde{I}$ with $I \in \mathcal{L}(\kappa)$, which is the filter in R-mod defined by

$$\mathcal{L}(\kappa) = \{J \subset R \,|\, \widetilde{J} \in \mathcal{L}(\widetilde{\kappa})\},$$

The $\widetilde{J}$ are always calculated with respect to the FJ induced by FR on J. Graded kernel functors $\widetilde{\kappa}$ on $\widetilde{R}$-mod satisfying the above property are called pseudo-affine. For now, we forget about this property. Our definition of $\mathcal{L}(\kappa)$ is satisfactory since

Proposition 7.3.2. *$\mathcal{L}(\kappa)$ is a Gabriel filter, that is, the associated preradical κ is a kernel functor.*

Proof. a) Consider $H \supset J$ with $J \in \mathcal{L}(\kappa)$, then $\widetilde{H} \supset \widetilde{J}$ yields $\widetilde{H} \in \mathcal{L}(\widetilde{\kappa})$.
b) Take $J_1, J_2 \in \mathcal{L}(\kappa)$ then $(J_1 \cap J_2)^\sim = \widetilde{J_1} \cap \widetilde{J_2}$. Indeed $(J_1 \cap J_2)^\sim \subset \widetilde{J_1} \cap \widetilde{J_2}$ is obvious. If $\widetilde{j} \in \widetilde{J_1} \cap \widetilde{J_2}$ then $\widetilde{j} \in \widetilde{J_1}$ implies that $\widetilde{j} = j_1 X^n$ where $j_1 \in J_1$, $n = \deg(j_1)$, $\widetilde{j} \in \widetilde{J_2}$ yields $\widetilde{j} = j_2 X^n$ where $n = \deg \widetilde{j}$ and $\deg(j_2) = n$. From $(j_1 - j_2)X^n = 0$ and $\widetilde{R}$ being X-torsion free it follows that $j_1 = j_2$, hence $\widetilde{j} = jX^n$ with $j \in J_1 \cap J_2$, or $\widetilde{J_1} \cap \widetilde{J_2} \subset (J_1 \cap J_2)^\sim$.
c) Take $r \in R, J \in \mathcal{L}(\kappa)$, then from $\widetilde{r} \in \widetilde{R}$ and $\widetilde{J} \in \mathcal{L}(\widetilde{\kappa})$ it follows that $(\widetilde{J} : \widetilde{r}) \in \mathcal{L}(\widetilde{\kappa})$. Let $y \in (J : r)$, i.e. $\widetilde{y} \in (J : r)^\sim$. If $\deg(yr) = \gamma$ with $\gamma \leq \deg(y) + \deg(r) = m + n$, then $\widetilde{yr} = yrX^\gamma, \widetilde{y}\widetilde{r} = \widetilde{yr}X^{m+n-\gamma}$. Thus $(yr)^\sim \subset \widetilde{J}$ yields $\widetilde{y}\widetilde{r} \in X^{n+m-\gamma}\widetilde{J} \subset \widetilde{J}$, i.e. $\widetilde{y} \in (\widetilde{J} : \widetilde{r})$. Conversely, from $\widetilde{z}\widetilde{r} \in \widetilde{J}$ we have $\widetilde{z} = zX^q$ for some $z \in R$, then $zrX^{q+n} \in \widetilde{J}$. $zr \in \dot{F}_\gamma R$ for some $\gamma \leq q + n$, whence $zrX^{q+n} = zrX^\gamma X^{q+n-\gamma} \in \widetilde{J}$. Since $J \subset R$ has the induced filtration, $\widetilde{R}/\widetilde{J}$ is X-torsion free. Therefore $zrX^\gamma = \widetilde{zr} \in \widetilde{J}$, or $zr \in J$ and $z \in (J : r)$. So we have established $(J : r)^\sim = (\widetilde{J} : \widetilde{r})$ and $\widetilde{J} \in \mathcal{L}(\widetilde{\kappa})$ then yields $(J : r) \in \mathcal{L}(\kappa)$.

d) Take $J \in \mathcal{L}(\kappa), H \subset J$ such that $\forall j \in J, \exists I_j \in \mathcal{L}(\kappa)$ such that $I_j j \subset H$. So we have $\widetilde{J} \in \mathcal{L}(\widetilde{\kappa}), \widetilde{H} \subset \widetilde{J}$ (FH induced by FR like FJ). Suppose first that $\widetilde{J}$ is graded. Let us check $\widetilde{I_j}\widetilde{j} \subset \widetilde{H}$ for all homogeneous $\widetilde{j} \in \widetilde{J}$; then idempotency of $\widetilde{\kappa}$ yields $\widetilde{H} \in \mathcal{L}(\widetilde{\kappa})$ or $H \in \mathcal{L}(\kappa)$. For $\widetilde{i_j} \in \widetilde{I_j}$, $\widetilde{i_j}\widetilde{j} = (i_j j)^{\sim} X^{n_i+n_j-n'}, n_i = \deg \widetilde{i_j}, n_j = \deg \widetilde{j}$ and $n' = \deg i_j j$. Thus $\widetilde{i_j}\widetilde{j} \in X^{n_i+n_j-n'}\widetilde{H} \subset \widetilde{H}$, $n_i + n_j - n' \geq 0$. This holds for all homogeneous $\widetilde{i_j} \in \widetilde{I_j}$, hence $\widetilde{I_j}\widetilde{j} \subset \widetilde{H}$. If $\widetilde{J}$ is not graded, then since $\widetilde{\kappa}$ is a graded kernel functor, there exists a graded $L \in \mathcal{L}(\widetilde{\kappa})$ with $L \subset \widetilde{J}$. Then $\pi^{-1}(L) \subset J$, and by the first part, we know that $\pi^{-1}(L) \cap H$, whence $H \in \mathcal{L}(\kappa)$. □

Lemma 7.3.3. *With assumptions and notations as before: For a separated filtered FR-module M, $\widetilde{\kappa}(\widetilde{M}) = \kappa(M)^{\sim}$ and $\widetilde{M}/\widetilde{\kappa}(\widetilde{M}) = (M/\kappa M)^{\sim}$.*

Proof. Obviously, $\widetilde{\kappa}(\widetilde{M})$ is X-torsion free, thus $\widetilde{\kappa}(\widetilde{M}) = \widetilde{N}$ for some filtered R-submodule N in M. If $\widetilde{m} \in \widetilde{M}$ is X-torsion modulo $\widetilde{\kappa}(\widetilde{M})$, i.e. $X^d\widetilde{m} \in \widetilde{\kappa}(\widetilde{M})$, then $\widetilde{I}X^d\widetilde{m} = 0$ for some $\widetilde{I} \in \mathcal{L}(\widetilde{\kappa})$, hence $\widetilde{I}\widetilde{m} = 0$ or $\widetilde{m} \in \widetilde{\kappa}(\widetilde{M})$. Consequently $\widetilde{\kappa}(\widetilde{M})$ is X-closed in $\widetilde{M}$, which means that N is a strict filtered submodule of M, i.e. FN is induced by FM. If $n \in N$ then $\widetilde{n} \in \widetilde{N}$ is $\widetilde{\kappa}$-torsion so $L\widetilde{n} = 0$ for some $L \in \mathcal{L}(\widetilde{\kappa})$. We have $\mathrm{cl_X(L)} \in \mathcal{L}(\widetilde{\kappa})$ and since FR is Zariskian, $\mathrm{cl_X(L)}$ is finitely generated, hence there is an $m \in \mathbb{N}$ such that $X^m\mathrm{cl_X(L)} \subset \mathrm{L}$, as X is central in $\widetilde{R}$. Then $X^m\mathrm{cl_X(L)}\tilde{\mathrm{n}} = 0$ yields $\mathrm{cl_X(L)}\tilde{\mathrm{n}} = 0$ since $\widetilde{N}$ is X-torsion free. Now $\mathrm{cl_X(L)}$ is X-closed in $\widetilde{R}$ hence $\mathrm{cl_X(L)} = \tilde{\mathrm{I}}$ for some I and $I \in \mathcal{L}(\kappa)$ since $\mathrm{cl_X(L)} \in \mathcal{L}(\widetilde{\kappa})$. Now from $\widetilde{i}\widetilde{n} = \widetilde{in}X^{\delta}$ where $\delta = \deg_F i + \deg_F n - \deg_F in$, $\deg_F$ being defined since FM is separated, we obtain $in = 0$ and this holds for all $i \in I$ with $I \in \mathcal{L}(\kappa)$, thus $n \in \kappa(M)$ and $\widetilde{n} \in (\kappa(M))^{\sim}$. Consequently $(\kappa(M))^{\sim} \supset \widetilde{\kappa}(\widetilde{M})$. Conversely, if $\widetilde{m}$ is homogeneous in $(\kappa(M))^{\sim}$ then $\widetilde{m} = mX^d$ for some $m \in \dot{F}_d(\kappa(M))$, thus $Lm = 0$ for some $L \in \mathcal{L}(\kappa)$. Then from $(Lm)^{\sim} \supset \widetilde{L}\widetilde{m}$ it follows that $\widetilde{m} \in \widetilde{\kappa}(\widetilde{M})$, therefore $\widetilde{\kappa}(\widetilde{M}) = (\kappa(M))^{\sim}$.
We checked that $\widetilde{M}/\widetilde{\kappa}(\widetilde{M})$ is X-torsion free above. Moreover $\widetilde{M}/\widetilde{\kappa}(\widetilde{M}) = \widetilde{M}/\kappa(M)^{\sim}$. We have an exact sequence of strict filtered morphisms:

$$0 \to \kappa(M) \to M \to M/\kappa(M) \to 0$$

Therefore, applying $\sim$ to this sequence, we get an exact sequence in $\widetilde{R}$-gr:

$$0 \to \kappa(M)^{\sim} \to \widetilde{M} \to (M/\kappa(M))^{\sim} \to 0$$

Thus $\widetilde{M}/\widetilde{\kappa}(\widetilde{M}) = \widetilde{M}/(\kappa(M))^{\sim} = (M/\kappa(M))^{\sim}$. □

We call a graded kernel functor $\widetilde{\kappa}$ on $\widetilde{R}$-mod or a kernel functor $\widetilde{\kappa}$ on $\widetilde{R}$-gr "affine" if $\widetilde{R}X \in \mathcal{L}(\widetilde{\kappa})$. In this case we write $\widetilde{\kappa} \in \mathrm{Afftors}(\widetilde{R})$.

Observation 7.3.4.

1. If $\widetilde{\kappa}$ is affine, then $0 \in \mathcal{L}(\overline{\kappa})$, and localization on the associated graded level is trivial;

2. $\widetilde{\kappa}$ is affine if and only if: $I \in \mathcal{L}(\widetilde{\kappa})$ if and only if $X^n I \in \mathcal{L}(\widetilde{\kappa})$ for all $n \in \mathbb{N}$.
   Proof. If $\widetilde{R}X \in \mathcal{L}(\widetilde{\kappa})$ and $I \in \mathcal{L}(\widetilde{\kappa})$, then $(\widetilde{R}X)I \subset XI$ shows that $XI \in \mathcal{L}(\widetilde{\kappa})$. By induction, it follows that $X^n I \in \mathcal{L}(\widetilde{\kappa})$ for all $n \in \mathbb{N}$. Conversely, $X^n I \in \mathcal{L}(\widetilde{\kappa})$ yields for $I = \widetilde{R} \in \mathcal{L}(\widetilde{\kappa})$ that $\widetilde{R}X^n$ and $\widetilde{R}X$ are in $\mathcal{L}(\widetilde{\kappa})$;

3. $\widetilde{\kappa}$ is affine if $I \in \mathcal{L}(\widetilde{\kappa})$ if and only if $\mathrm{cl_X(I)} \in \mathcal{L}(\widetilde{\kappa})$.
   Proof. Suppose $\mathrm{cl_X(I)} \in \mathcal{L}(\widetilde{\kappa})$, since $\mathrm{cl_X(I)}$ is finitely generated and X is central in $\widetilde{R}$, there is an $X^n, n \in \mathbb{N}$ such that $X^n \mathrm{cl_X(I)} \subset \mathrm{I}$, thus $\widetilde{R}X^n \mathrm{cl_X(I)} \subset \mathrm{I}$ and hence $I \in \mathcal{L}(\widetilde{\kappa})$. The other direction follows from the equality $\mathrm{cl_X(\widetilde{R}X)} = \widetilde{\mathrm{R}}$;

4. If $\widetilde{\kappa}$ is affine then $\mathcal{L}(\widetilde{\kappa})$ has a filter basis consisting of $X^n \widetilde{I}$ with $n \in \mathbb{N}$ and $\widetilde{I}$ such that $\widetilde{R}/\widetilde{I}$ is X-torsion free or $\mathrm{cl_X(\widetilde{I})} = \widetilde{\mathrm{I}}$, in other words $\widetilde{I}$ is coming from $I \in \mathcal{L}(\kappa)$.
   Proof. From observation 3, $I \in \mathcal{L}(\widetilde{\kappa})$ iff $\mathrm{cl_X(I)} \in \mathcal{L}(\widetilde{\kappa})$ hence $\mathrm{cl_X(I)} = \widetilde{\mathrm{J}}$ for some $J \in \mathcal{L}(\kappa)$ and $X^n \widetilde{J} \subset I$ for some $n \in \mathbb{N}$ as in observation 3, thus I contains some $X^n \widetilde{J}$ as claimed.

The complement of affine graded localizations on $\widetilde{R}$-mod consists of the $\widetilde{\kappa}$ such that $\widetilde{R}X \notin \mathcal{L}(\widetilde{\kappa})$. Then it is obvious that $0 \notin \mathcal{L}(\overline{\kappa})$. The non-commutative geometry of $\widetilde{R}$-mod has an "open set" of affine torsion theories; this can be identified with $\widetilde{R}_X$-grtors where $\widetilde{R}_X \cong R[X, X^{-1}]$ via the torsion theories such that $\widetilde{\kappa} \geq \widetilde{\kappa}_X$[1], where $\widetilde{\kappa}_X$ is the kernel functor associated with

$$\mathcal{L}(\widetilde{\kappa}_X) = \{\widetilde{R}X^n, n \in \mathbb{N}\}$$

and it is known that $\widetilde{R}_X$-grtors $\cong$ gen$(\widetilde{\kappa}_X)$ because $\widetilde{\kappa}_X$ is central and perfect. Since $R[X, X^{-1}]$ is strongly graded, we have an equivalence of categories $\widetilde{R}_X$-gr $\cong$ R-mod (see [26]), justifying the statement in the introduction. For non-affine $\widetilde{\kappa}$ we do not necessarily have κ being trivial but in any case $\widetilde{\kappa}$ is not obtained as an obvious $\sim$-construction from κ. Indeed, starting from a kernel functor κ one could attempt to define $\mathcal{L}(\widetilde{\kappa})$ as the filter of left ideals containing some $\widetilde{I}$ for some $I \in \mathcal{L}(\kappa)$. However, in proving the radical property one needs the assumption that $X^n J \in \mathcal{L}(\widetilde{\kappa})$ if $J \in \mathcal{L}(\widetilde{\kappa})$, but this is just saying that $\widetilde{\kappa}$ is affine. Indeed, suppose that $\overline{m} \in \widetilde{\kappa}(M/\widetilde{\kappa}(M))$ for some $\widetilde{R}$-module M. Then we find $I, J \in \mathcal{L}(\kappa)$ such that $\widetilde{J}\widetilde{I}m = 0$. Assuming that FR is Zariskian, we have that $\widetilde{J}\widetilde{I} \subset (JI)^{\sim}$ and we find some $n \in \mathbb{N}$ such that $\widetilde{J}\widetilde{I} = (JI)^{\sim} X^n$ (see

[1]For the lattice structure on R-tors we refer to Section 7.4.

the proof of Proposition 7.3.2). In order to conclude that $\widetilde{JI} \in \mathcal{L}(\widetilde{\kappa})$ we must know that $(JI)^\sim X^n \in \mathcal{L}(\widetilde{\kappa})$. In the set of non-affine $\widetilde{\kappa}$ there is nonetheless a subset which relates well to both κ and $\overline{\kappa}$; these are the pseudo-affine kernel functors.

Definition 7.3.5. $\widetilde{\kappa}$ is said to be pseudo-affine if $\mathcal{L}(\widetilde{\kappa})$ has a filter basis consisting of $\widetilde{I}$ with $I \in \mathcal{L}(\kappa)$.

In situations where both κ and $\overline{\kappa}$ are used or linked we have to restrict to pseudo-affine kernel functors. The most important example is given by Zariskian filtered rings with a domain for the associated graded ring and localizations deriving from Ore sets (see below).

Starting from a κ on R-mod we may define $\mathcal{L}(\widetilde{\kappa})$ on $\widetilde{R}$-gr by putting $\mathcal{L}(\widetilde{\kappa}) = \{J \subset_g \widetilde{R}, J \supset \widetilde{I}$ for some $I \in \mathcal{L}(\kappa)\}$. One easily verifies that the $\widetilde{\kappa}$ defines a left exact preradical, but it is a priori not radical, as we observed above. Even if κ is defined from a kernel functor $\widetilde{\kappa}$ then the $\mathcal{L}(\widetilde{\kappa})$ is not obtained from $\mathcal{L}(\kappa)$ by taking the $\widetilde{I}, I \in \mathcal{L}(\kappa)$ for a filterbasis! Of course, for a pseudo-affine $\widetilde{\kappa}$ the κ defined as before does yield the $\widetilde{\kappa}$ in the procedure sketched above. Another way of trying to solve this is to define $\mathcal{L}(\kappa)$ as the filter generated by the filtered left ideals L of R such that $\widetilde{L} \in \mathcal{L}(\kappa)$, i.e. not viewing only filtered left ideals I such that $0 \to I \to R$ is a strict filtered exact chain! In fact, in this way we consider the category R-filt which has been neglected. This may be possible but the whole torsion theory has to be adapted, so we do not go deeper into this.

For pseudo-affine kernel functors we obviously have

$$\mathcal{L}(\overline{\kappa}) = \{L \subset G(R) \,|\, L \supset \pi(\widetilde{I}) \text{ for some } I \in \mathcal{L}(\kappa)\}.$$

As far as sheaves over the lattice R-tors will go, we will work only with localizations and quotient filtrations defined in the first approach, say for R with $G(R)$ a domain. As soon as one desires good relations on the sheaf level between the R-, the $\widetilde{R}$- and the $G(R)$-levels, one restricts attention to pseudo-affine graded kernel functors $\widetilde{\kappa}$ on $\widetilde{R}$-mod and the associated κ on R-mod and $\overline{\kappa}$ on $G(R)$-mod. In the next section we will discuss for example the non-commutative site in the 'almost commutative' case, where words of Ore sets are considered. To this extent, we have

Lemma 7.3.6. *Let $\widetilde{S}$ be an Ore set in $\widetilde{R}$ such that $\widetilde{S} \cap \widetilde{R}X = \emptyset$. Define $S \subset R, S = \{s, \widetilde{s} \in \widetilde{S}\}$. Then S is an Ore set in R.*

Proof. Take $s, t \in S$, then $\widetilde{s}\widetilde{t} = (st)^\sim X^{n(s)+n(t)-n(st)}$ where $n(s) = \deg \widetilde{s}, n(t) = \deg \widetilde{t}$ and $n(st) = \deg (st)^\sim$. If $n(st) < n(s) + n(t)$ then $\widetilde{s}\widetilde{t} \in \widetilde{S}$ and $\widetilde{s}\widetilde{t} \in \widetilde{R}X$, a contradiction. Hence $n(st) = n(s) + n(t)$ and $st \in S$. Observe moreover that S is an Ore set. Indeed take $s \in S, r \in R$ and look at

$\widetilde{s} \in \widetilde{S}, \widetilde{r} \in \widetilde{R}$ then $\widetilde{s'}\widetilde{r} = \widetilde{r'}\widetilde{s}$ for some $\widetilde{s'} \in \widetilde{S}, \widetilde{r'} \in \widetilde{R}$ because $\widetilde{S}$ is an Ore set. Put $\deg_F s'r = \gamma, \deg_F r's = \tau$ then

$$(s'r)^{\sim} = s'rX^{\gamma}, \widetilde{s'}\widetilde{r} = (s'rX^{\gamma})X^{\deg(s')+\deg(r)-\gamma}$$
$$(r's)^{\sim} = r'sX^{\tau}, \widetilde{r'}\widetilde{s} = (r'sX^{\tau})X^{\deg(r')+\deg(s)-\tau},$$

where $\deg(s') + \deg(r) = \deg(r') + \deg(s) = \epsilon$ because $\widetilde{s'}\widetilde{r} = \widetilde{r'}\widetilde{s}$. We conclude that $s'r = r's$. Furthermore, if $rs = 0$ for some $r \in R, s \in S$, then $\widetilde{r}\widetilde{s} = (rs)^{\sim}X^{\epsilon}$ where $\epsilon = \deg(\widetilde{r}) + \deg(\widetilde{s}) - \deg(rs)^{\sim}, \epsilon \geq 0$. Thus $\widetilde{r}\widetilde{s} = 0$ and since $\widetilde{S}$ is an Ore set in $\widetilde{R}$, there is an $\widetilde{s'}$ in $\widetilde{S}$ such that $\widetilde{s'}\widetilde{r} = 0$, hence $s'r = 0$. □

To an Ore set as in the previous lemma, we associate $\overline{S} = \pi(\widetilde{S})$, where

$$\pi : \widetilde{R} \twoheadrightarrow G(R)$$

is the canonical epimorphism. Since $\widetilde{S} \cap \widetilde{R}X = \emptyset$, $0 \notin \overline{S}$ and obviously $\overline{S}$ is multiplicatively closed.

Lemma 7.3.7. *In the situation of the previous lemma, $\overline{S}$ satisfies the second Ore condition. In the case when $G(R)$ is commutative or a domain then $\overline{S}$ is an Ore set.*

Proof. Given $\overline{r} \in G(R), \overline{s} \in \overline{S}$ and $\widetilde{r}$, resp. $\widetilde{s}$ representatives in $\widetilde{R}$, resp. $\widetilde{S}$. Then $\exists \widetilde{s'} \in \widetilde{S}, \widetilde{r'} \in \widetilde{R}$ such that $\widetilde{s'}\widetilde{r} = \widetilde{r'}\widetilde{s}$, consequently $\pi(\widetilde{s'})\pi(\widetilde{r}) = \pi(\widetilde{r'})\pi(\widetilde{s})$ or $\overline{s'}\overline{r} = \overline{r'}\overline{s}$ for some $\overline{s'} \in \overline{S}, \overline{r'} \in G(R)$. If $G(R)$ is commutative and $\overline{r}\overline{s} = 0$ then $\overline{s}\overline{r} = 0$ and the first Ore condition holds with $\overline{s'} = \overline{s}$. If $G(R)$ is a domain, then $\overline{r}\overline{s} = 0$ with $\overline{s} \neq 0$ yields $\overline{r} = 0$ and of course $\overline{s}\overline{r} = 0$. □

If $G(R)$ is $\overline{S}$-torsion free on the right then in the situation of the previous two lemmas, $\overline{S}$ is an Ore set. The foregoing results also work for right Ore sets and for left and right Ore sets. So summarizing all of this yields:

Proposition 7.3.8. *If $G(R)$ is $\pi(\widetilde{S})$-torsion free left and right and $\widetilde{S}$ is as in the lemmas, then S and $\overline{S}$ are Ore sets (left if $\widetilde{S}$ is left, left and right if $\widetilde{S}$ is left and right). Moreover, $\widetilde{\kappa}_{\widetilde{S}}$ defines κ_S and $\overline{\kappa}_{\overline{S}}$ because $\widetilde{\kappa}_{\widetilde{S}}$ is pseudo-affine.*

Proof. Observe that $\widetilde{\kappa}_{\widetilde{S}}$ is indeed pseudo-affine because by definition its Gabriel filter is given by

$$\mathcal{L}(\widetilde{\kappa}_{\widetilde{S}}) = \{L \leq \widetilde{R} \text{ left ideal} \,\big|\, \exists \widetilde{s} \in \widetilde{S} \cap L\},$$

and the ideals $\widetilde{R}\widetilde{s} = (Rs)^{\sim}$ (the equality holds since $G(R)$ is $\overline{S}$-torsion free!) for $s \in S$ yield a filterbasis. □

In the 'almost commutative' case, we have for an Ore set S in R that $\widetilde{S} = \{\widetilde{s}, s \in S\}$ is an Ore set in $\widetilde{R}$ not intersecting $\widetilde{R}X$. For a word in Ore sets, say $S_1S_2 \ldots S_n$, we have that $(Rs_1 \ldots s_n)^{\sim} = \widetilde{R}\widetilde{s_1} \ldots \widetilde{s_n}$ ($G(R)$ domain!), so the preradical $\kappa_{S_1}\kappa_{S_2} \ldots \kappa_{S_n}$ with associated filter $\mathcal{L}(\kappa_{S_1}\kappa_{S_2} \ldots \kappa_{S_n})$ is indeed pseudo-affine.

Let us return to our aim in defining a quotient filtration. The following two lemmas and corollary correspond to Lemma 7.2.4, Lemma 7.2.6 and Lemma 7.2.9 from the results starting from a kernel functor κ on R-mod. Since our approach is different, we do include the proofs in this setting, although they are similar.

Lemma 7.3.9. *$G(\kappa(M)) \subset \overline{\kappa}G(M)$ and there is an epimorphism*

$$G(M/\kappa(M)) \twoheadrightarrow G(M)/\overline{\kappa}G(M).$$

Proof. Note that $G(\kappa(M)) = \pi_{\widetilde{M}}(\kappa(M)^{\sim}) = \pi_{\widetilde{M}}(\widetilde{\kappa}(\widetilde{M}))$, where the last equality follows from Lemma 7.3.3 and where

$$\pi_{\widetilde{M}} : \widetilde{M} \twoheadrightarrow G(M) = \widetilde{M}/X\widetilde{M}$$

is the canonical epimorphism. Hence $G(\kappa(M))$ is $\overline{\kappa}$-torsion since $\widetilde{I}\widetilde{m} = 0$ for $\widetilde{m} \in \kappa(M)^{\sim}$ and $\widetilde{I} \in \mathcal{L}(\widetilde{\kappa})$ yields $\widetilde{I}\pi_{\widetilde{M}}(\widetilde{m}) = \pi(\widetilde{I})\pi_{\widetilde{M}}(\widetilde{m}) = 0$ with $\pi(\widetilde{I}) \in \mathcal{L}(\overline{\kappa})$. Thus $G(\kappa(M)) \subset \overline{\kappa}(G(M))$. □

Lemma 7.3.10. *If M is κ-separated and κ is pseudo-affine or else if $M/\kappa(M)$ is κ-separated then $G(\kappa(M)) = \overline{\kappa}(G(M))$ and $G(M/\kappa(M)) = G(M)/\overline{\kappa}G(M)$.*

Proof. Take $\overline{z} \in \overline{\kappa}(G(M))_n$; we want to establish that $\overline{z} \in G(\kappa(M))$. The relation between $\kappa, \overline{\kappa}$ and $\widetilde{\kappa}$ entails that for $\overline{J} \in \mathcal{L}(\overline{\kappa})$ there is a $J \in \mathcal{L}(\kappa)$ such that $\overline{J} \supset G(J)$ and for every $J \in \mathcal{L}(\kappa)$ we have $G(J) \in \mathcal{L}(\overline{\kappa})$ (because κ is pseudo-affine). If $z \in \dot{F}_nM$ is such that $\sigma(z) = \overline{z}$ then $\overline{J}\overline{z} = 0$ for some $\overline{J} \in \mathcal{L}(\overline{\kappa})$ yields $G(J)\overline{z} = 0$ for some $J \in \mathcal{L}(\kappa)$. Then $\widetilde{I}\widetilde{z} \subset \mathrm{Ker}(\pi_{\widetilde{M}})$ for some $\widetilde{I} \in \mathcal{L}(\widetilde{\kappa}), \widetilde{I}\widetilde{z} \subset X\widetilde{M}$. If $i \in \dot{F}_\gamma I, I \in \mathcal{L}(\kappa)$, then $\widetilde{i}\widetilde{z} = izX^{\deg i + \deg z} = izX^{\gamma+n}$. Hence $izX^{\gamma+n} = X\widetilde{m} = mX^{\deg_F(m)+1}$ for some $m \in \dot{M}_{\deg_F(m)}$ entails that $\deg_F(m) = \gamma + n - 1$, i.e. $iz \in F_{\gamma+n-1}M$. The latter holds for all γ and for all $i \in F_\gamma I$, thus we have $F_\gamma Iz \subset F_{\gamma+n-1}M$. The κ-separatedness of M then yields that $z \in \kappa(M)$ and finally $\overline{z} \in G(\kappa(M))$.
The exact sequence of strict filtered morphisms:

$$0 \to \kappa(M) \to M \to M/\kappa(M)$$

yields the exact sequence

$$0 \to G(\kappa(M)) \to G(M) \to G(M/\kappa(M)) \to 0$$

and from $G(\kappa(M)) = \overline{\kappa}(G(M))$ it follows that $G(M/\kappa(M)) = \frac{G(M)}{\overline{\kappa}G(M)}$. □

Corollary 7.3.11. *Let κ be pseudo-affine. We have*

1. *If M is κ-separated, then $G(M/\kappa(M))$ is $\overline{\kappa}$-torsion free;*

2. *If $G(M)$ is $\overline{\kappa}$-torsion free, then M is κ-separated.*

Proof. 1. is obvious from the foregoing lemma. We prove 2: suppose $m \in \dot{F}_n M$ is such that $F_\gamma I m \subset F_{n+\gamma-1}M$ for all γ and some $I \in \mathcal{L}(\kappa)$. Put $\overline{m} = \sigma(m) \in G(M)_n$. Then clearly $\widetilde{I}\widetilde{m} \subset X\widetilde{M}$ with $\widetilde{I} \in \mathcal{L}(\widetilde{\kappa})$ and it follows that $\pi_{\widetilde{M}}(\widetilde{I}\widetilde{m}) = 0$ or $\pi(\widetilde{I})\overline{m} = 0$. Since $\overline{\kappa}G(M) = 0$ and $\pi(\widetilde{I}) \in \mathcal{L}(\overline{\kappa})$, we must have $\overline{m} = 0$, contradicting $m \in \dot{F}_n M$. □

Corollary 7.3.12. *With assumptions as before, if M is κ-separated then $M/\kappa(M)$ is κ-separated.*

Proof. $G(M/\kappa(M)) = G(M)/\overline{\kappa}G(M)$ is $\overline{\kappa}$-torsion free; thus the filtered R-module $M/\kappa(M)$ (with induced filtration by FM) is κ-separated. □

Theorem 7.3.13. *1. Let FR be Zariskian, $M \in R$-filt is separated and $\widetilde{\kappa}$ is pseudo-affine. If M is κ-separated (or $M/\kappa(M)$ is κ-separated in the affine case) then the localized $Q_\kappa(M)$ has a filtration $FQ_\kappa(M)$ making the localization morphism $j_{\kappa,M} : M \to Q_\kappa(M)$ into a strict filtered morphism (similar for FR when R is κ-separated, then $j_\kappa : R \to Q_\kappa(R)$ is strict filtered and a ring morphism). We call $FQ_\kappa(M)$ (resp. $FQ_\kappa(R)$) the quotient filtration of FM (resp. FR). Moreover $Q_\kappa(M)$ with $FQ_\kappa(M)$ is a filtered $Q_\kappa(R)$-module with respect to $FQ_\kappa(R)$;*

2. *In the situation of 1. with $\widetilde{\kappa}$ pseudo-affine, $Q_\kappa(M)^\sim = Q^g_{\widetilde{\kappa}}(\widetilde{M})$, where $Q^g_{\widetilde{\kappa}}$ is the graded localization functor associated to $\widetilde{\kappa}$ on $\widetilde{R}$-gr;*

3. *In the situation of 2., we also have $G(Q_\kappa(M)) \subset Q^g_{\overline{\kappa}}(G(M))$, where $Q^g_{\overline{\kappa}}$ is the graded localization functor w.r.t. $\overline{\kappa}$ on $G(R)$-gr.*

Proof. The proof of 1. is completely analogous to the proof of Theorem 7.2.12. 2) First we check that $Q^g_{\widetilde{\kappa}}(\widetilde{M})$ is X-torsion free. It follows from Lemma 7.3.3 that $\widetilde{M}/\widetilde{\kappa}(\widetilde{M}) = (M/\kappa(M))^\sim$, hence the reduction to M κ-torsion free also leads to $\widetilde{M}$ being $\widetilde{\kappa}$-torsion free. Now $\widetilde{M}/\widetilde{\kappa}(\widetilde{M})$ is X-torsion free (in fact it is $\widetilde{M}$ now). If q is homogeneous in $Q^g_{\widetilde{\kappa}}$ such that $Xq = 0$ then there is a $\widetilde{J} \in \mathcal{L}(\widetilde{\kappa})$ such that $\widetilde{J}q \subset \widetilde{M}$ and also $X\widetilde{J}q = \widetilde{J}Xq = 0$ hence $\widetilde{J}q = 0$ since $\widetilde{M}$ is X-torsion free. Then $\widetilde{J}q = 0$ contradicts the fact that $Q^g_{\widetilde{\kappa}}(\widetilde{M})$ is $\widetilde{\kappa}$-torsion free. By 1. the exact sequence

$$0 \to M \to Q_\kappa(M) \to Q_\kappa(M)/M = T \to 0$$

is strict exact, so we arrive at an exact sequence in $\mathcal{F}_X$:

$$0 \to \widetilde{M} \to Q_\kappa(M)^\sim \to \widetilde{T} \to 0$$

with $\widetilde{T}$ being $\widetilde{\kappa}$-torsion ($t \in T$, then $It = 0$ for some $I \in \mathcal{L}(\kappa), \forall i \in I, it = 0$ yields $\widetilde{it} = itX^{\deg it}X^{\deg i + \deg t - \deg it} = 0$). Thus $Q_\kappa(M)^\sim \subset Q^g_{\widetilde{\kappa}}(\widetilde{M})$ (recall that if N/M is κ-torsion, then $N \subset Q_\kappa(M)$). The latter is X-torsion free. Let us check that $Q^g_{\widetilde{\kappa}}(\widetilde{M})/Q_\kappa(M)^\sim$ is X-torsion free. Suppose that $\widetilde{q} \in Q^g_{\widetilde{\kappa}}(\widetilde{M})_n$ is such that $X\widetilde{q} \in Q_\kappa(M)^\sim$. Since $\widetilde{\kappa}$ is pseudo-affine we have for some $I \in \mathcal{L}(\kappa)$ that $\widetilde{I}\widetilde{q} \subset \widetilde{M}$ and $\widetilde{I}X\widetilde{q} \subset X\widetilde{M}$. Since $X\widetilde{q}$ is in $(Q_\kappa(M)^\sim_{n+1})$ we may write $X\widetilde{q} = mX^{n+1}$ with $m \in F_{n+1}Q_\kappa(M)$. If $m \in F_nQ_\kappa(M)$ then $\widetilde{q} = mX^n \in Q_\kappa(M)^\sim$ as desired. So look at the case $m \notin F_nQ_\kappa(M)$. From $\widetilde{i}\widetilde{q} \in \widetilde{M}_{\gamma+n}$ if $\widetilde{i} \in \widetilde{I}_\gamma$, it follows that $\widetilde{i}X\widetilde{q} \in \widetilde{M}_{\gamma+n+1}$ and $\widetilde{i}X\widetilde{q} = \widetilde{i}(mX^{n+1}) = imX^{\gamma+n+1}$, thus $\widetilde{i}\widetilde{q} = imX^{\gamma+n}$ since $Q^g_{\widetilde{\kappa}}(\widetilde{M})$ is X-torsion free. From $\widetilde{i}\widetilde{q} \in \widetilde{M}_{\gamma+n}$ we then must have $im \in F_{\gamma+n}M$ (note that $\widetilde{I}$ is actually the tilde of an I and not just an ideal in $\mathcal{L}(\widetilde{\kappa})$). Thus for all γ, $F_\gamma Im \subset F_{\gamma+n}M$, and $m \in \dot{F}_{n+1}Q_\kappa(M)$. Then $\sigma(m) \in G(M)_{n+1}$ has $G(I)\sigma(m) = 0$, thus $\sigma(m) \in \overline{\kappa}(G(M))$. Since we reduced to the κ-torsion free case $\overline{\kappa}(G(M)) = G(\kappa(M))(M$ κ-separated) hence $\overline{\kappa}G(M) = 0$, thus $\sigma(m) = 0$, a contradiction. Therefore $Q^g_{\widetilde{\kappa}}(\widetilde{M})/Q_\kappa(M)^\sim$ is an X-torsion free $\widetilde{R}$-module and thus $Q^g_{\widetilde{\kappa}}(\widetilde{M}) = \widetilde{N}$ for a filtered R-module N such that $0 \to Q_\kappa(M) \to N$ is a strict filtered exact sequence. Since $Q^g_{\widetilde{\kappa}}(\widetilde{M})/Q_\kappa(M)^\sim$ is $\widetilde{\kappa}$-torsion, $N/Q_\kappa(M)$ is κ-torsion, because $(N/Q_\kappa(M))^\sim = Q^g_{\widetilde{\kappa}}(\widetilde{M})/Q_\kappa(M)^\sim(\in \mathcal{F}_X \subset \widetilde{R}\text{-gr})$. Thus if $n \in N/Q_\kappa(M)$ then $\widetilde{I}\widetilde{n} = 0$ for some $\widetilde{I} \in \mathcal{L}(\widetilde{\kappa})$ and $\widetilde{i}\widetilde{n} = (in)^\sim X^{\deg i + \deg n - \deg in} = 0$. Hence $(in)^\sim = 0$ because $(N/Q_\kappa(M))^\sim$ is X-torsion free. It follows that $n \in \kappa(N/Q_\kappa(M))$. In other words, $N/Q_\kappa(M)$ is κ-torsion and this implies that N/M is κ-torsion. Hence $N \subset Q_\kappa(M)$ and therefore we arrive at $Q^g_{\widetilde{\kappa}}(\widetilde{M}) = Q_\kappa(M)^\sim$.

3) Since $G(\kappa(M)) = \overline{\kappa}G(M)$ follows from the κ-separatedness of M we may also restrict in this part the problem to the κ-torsion free case, i.e. suppose $\kappa(M) = 0$ and $\overline{\kappa}G(M) = 0$. Consider $\overline{q} \in G(Q_\kappa(M))_n$, i.e. $\overline{q} = \sigma(q)$ for q in $\dot{F}_nQ_\kappa(M)$. There is an $I \in \mathcal{L}(\kappa)$ such that $F_\gamma Iq \subset F_{n+q}M$ for all $\gamma \in \mathbb{Z}$. For $i \in \dot{F}_\gamma I$ we have that $\sigma(i)\sigma(q)$ is either 0, or else $\sigma(i)\sigma(q) = \sigma(iq) \in G(M)$, thus always $\sigma(i)\sigma(q) \in G(M)$ and thus $G(I)\sigma(q) \subset G(M)$ or $\sigma(q) = \overline{q}$ is in $Q_{\overline{\kappa}}(G(M))$. Consequently $G(Q_\kappa(M)) \subset Q_{\overline{\kappa}}(G(M))$. □

If for every left ideal $L \in \mathcal{L}(\kappa)$ it holds that $Q_\kappa(R)j_\kappa(L) = Q_\kappa(R)$, we say that the localization at κ is perfect. An equivalent condition is that the localization functor Q_κ is right exact and commutes with direct sums. A kernel functor κ is said to be of finite type if its associated Gabriel filter $\mathcal{L}(\kappa)$ has a filterbasis of finitely generated (left) ideals. Suppose that the graded kernel functor $\widetilde{\kappa}$ on $\widetilde{R}$-mod is of finite type; then the corresponding kernel functor $\widetilde{\kappa}^g$ on $\widetilde{R}$-gr has a filter basis of finitely generated graded (left) ideals for the filter $\mathcal{L}^g(\widetilde{\kappa}^g)$ (we actually often omit the supindex ^{g}). For finitely generated graded $\widetilde{R}$-modules M, N it holds that $\mathrm{HOM}_{\widetilde{R}}(M,N) = \mathrm{Hom}_{\widetilde{R}}(M,N)$, where $\mathrm{HOM}_{\widetilde{R}}$ denotes graded Hom (see [56, Section 2.4]); hence by definition of (graded) localization (direct limit of graded Hom's) it follows that $Q^g_{\widetilde{\kappa}}(M) = Q_{\widetilde{\kappa}}(M)$ for $M \in \widetilde{R}$-gr.

Proposition 7.3.14. *In the situation of the theorem, if $\widetilde{\kappa}$ is perfect, then $\overline{\kappa}$ is perfect and $G(Q_\kappa(R)) = Q_{\overline{\kappa}}(G(R)), G(Q_\kappa(M)) = Q_{\overline{\kappa}}(G(M))$.*

Proof. Since R and $\widetilde{R}$ are Noetherian, $\widetilde{\kappa}$ has finite type and as a graded kernel functor we then have $Q_{\widetilde{\kappa}}(\widetilde{M}) = Q^g_{\widetilde{\kappa}}(\widetilde{M})$. Hence $Q_{\widetilde{\kappa}}(\widetilde{M}) = Q_\kappa(M)^\sim$ and $G(Q_\kappa(M)) = Q_{\widetilde{\kappa}}(\widetilde{M})/XQ_{\widetilde{\kappa}}(\widetilde{M})$. By perfectness of $\widetilde{\kappa}$: $Q_{\widetilde{\kappa}}(\widetilde{M})/XQ_{\widetilde{\kappa}}(\widetilde{M}) = Q_{\widetilde{\kappa}}(\widetilde{M}/X\widetilde{M}) = Q_{\widetilde{\kappa}}(G(M))$. As $\widetilde{\kappa}$ induces $\overline{\kappa}$ on $G(R)$-mod (also on $G(R)$-gr), we have $Q_{\widetilde{\kappa}}(G(M)) = Q_{\overline{\kappa}}(G(M))$. It follows that $G(Q_\kappa(M)) = Q_{\overline{\kappa}}(G(M))$. Concerning the perfectness of $\overline{\kappa}$, take $L \in \mathcal{L}(\overline{\kappa})$. Then $L \supset G(I)$ for some $I \in \mathcal{L}(\kappa)$. We have that

$$\pi(Q_{\widetilde{\kappa}}(\widetilde{R})\widetilde{I}) = G(Q_\kappa(R))G(I) = Q_{\overline{\kappa}}(G(R))G(I).$$

Now by the perfectness of $\widetilde{\kappa}$, $Q_{\widetilde{\kappa}}(\widetilde{R})\widetilde{I} = Q_{\widetilde{\kappa}}(\widetilde{R})$, hence

$$Q_{\overline{\kappa}}(G(R))G(I) = Q_{\overline{\kappa}}(G(R))$$

holds for all $I \in \mathcal{L}(\kappa)$ and then for all $L \in \mathcal{L}(\overline{\kappa})$ we have that $Q_{\overline{\kappa}}(G(R))L = Q_{\overline{\kappa}}(G(R))$. This means $\overline{\kappa}$ is perfect. □

Remark 7.3.15. If S is an Ore set of R then even if $\sigma(S)$ is an Ore set of $G(R)$ we may not have that $\overline{\kappa}_S = \kappa^g_{\sigma(S)}$; indeed $G(Rs) \supsetneq G(R)\sigma(s)$ is possible (we always have $\overline{\kappa}_S \leq \kappa^g_{\sigma(S)}$). However, if $G(R)$ is $\sigma(S)$-torsion free then $\sigma(r)\sigma(s) = \sigma(rs)$ for all $r \in R$ and $G(Rs) = G(R)\sigma(s)$ and $\overline{\kappa}_S = \kappa^g_{\sigma(S)}$ follow. So when R is a Zariskian ring with $G(R)$ being a domain then R is κ-separated for all κ, and localization at Ore sets behaves very nicely (e.g. $\widetilde{S}$ is an Ore set in $\widetilde{R}$ then too).

Proposition 7.3.16. *In the situation of the previous proposition, κ is perfect too.*

Proof. Consider $I \in \mathcal{L}(\kappa)$ and $Q_\kappa(R)I$ (we may reduce to the case where R and M are κ-torsion free as before). Since $\widetilde{\kappa}$ is perfect $Q_\kappa(R)^\sim = Q^g_{\widetilde{\kappa}}(\widetilde{R}) = Q^g_{\widetilde{\kappa}}(\widetilde{R})\widetilde{I}$, hence $1 = \sum'_{j=1} \widetilde{q}_j\widetilde{i}_j$ with $\widetilde{q}_j \in Q^g_{\widetilde{\kappa}}(\widetilde{R}), \widetilde{i}_j \in \widetilde{I}$ homogeneous and we have $\widetilde{q}_j\widetilde{i}_j \in Q^g_{\widetilde{\kappa}}(\widetilde{R})_0$, i.e. $\widetilde{q}_j \in Q^g_{\widetilde{\kappa}}(\widetilde{R})_{-n}, \widetilde{i}_j \in Q^g_{\widetilde{\kappa}}(\widetilde{R})_n$ for some $n \in \mathbb{Z}$. Then $\widetilde{q}_j = q_jX^{-n}$ with $q_j \in F_{-n}Q_\kappa(R), \widetilde{i}_j = i_jX^n$ with $i_j \in F_nI$ and $1 = \sum'_{j=1} q_ji_jX^0 = \sum'_{j=1} q_ji_j \in Q_\kappa(R)I$. From $Q_\kappa(R)I = Q_\kappa(R)$ for all $I \in \mathcal{L}(\kappa)$ we have that κ is perfect. □

To end this section we continue our discussion on Ore sets from the previous section, Lemma 7.2.23, but now we focus more on the relations between the affine, Rees and associated graded levels. We suppose that $G(R)$ is a commutative domain and let S, T be Ore sets in R. Then $G(R)$ is $\sigma(S)$- and $\sigma(T)$-torsion free, whence R is S- and T-torsion free. By Remark 7.3.15 we know that $\overline{\kappa}_U = \kappa^g_{\sigma(U)}$, for $U = S, T$. Corollary 7.3.11 entails that R is κ_S and κ_T-separated, so we can localize at κ_T and we obtain a filtered ring $Q_T(R)$, extending FR.

Lemma 7.3.17. *$Q_T(R)$ is κ_S-separated.*

Proof. Let $s \in S$ and $z \in \dot{F}_n Q_T(R)$ be such that for all $\gamma \in \mathbb{Z}$:

$$F_\gamma(Rs)z \subset F_{\gamma+n-1}Q_T(R).$$

Denote by $m(s) = \deg(\sigma(s))$ the degree of $\sigma(s)$ in $G(R)$. By the localization filtration, we know that there exists $t \in T$ such that $tz \in F_{m(t)+n}R$ ($\kappa_T(R) = 0$). Since S is an Ore set, there exists $s' \in S, a \in R$ such that $as = s't$ and because $G(R)$ is a domain, we have that $m(a)+m(s) = m(s')+m(t)$. Therefore

$$\begin{aligned} asz = s'tz &\in F_{m(a)+m(s)+n-1}Q_T R \cap F_{m(s')+m(t)+n}R \\ &= F_{m(s')+m(t)+n-1}R, \end{aligned}$$

and for all $\gamma \in \mathbb{Z}$ we obtain

$$F_\gamma(Rs')(tz) \subset F_{\gamma+m(t)+n-1}R.$$

Since R is κ_S-separated, this shows that $tz \in \kappa_S(R) = 0$, whence $z = 0$ in $Q_T(R)$, or $z \in \kappa_S(Q_T(R))$. □

This shows that we can localize the filtered R-module $Q_T(R)$ at κ_S and obtain $Q_S(Q_T(R))$. By our construction, this is a filtered $Q_S(R)$-module. We do not have to worry about killing the torsion, since one easily shows that $Q_T(R)$ is κ_S-torsion free. The filtration is given by

$$F_n Q_S Q_T(R) = \{z \in Q_S Q_T(R) \mid \exists s \in S, t \in T \text{ such that } tsz \in F_{m(t)+m(s)+n}R\}.$$

The localization morphism

$$j_S : Q_T(R) \to Q_S Q_T(R)$$

is strict filtered and we have that

$$\widetilde{Q_S Q_T(R)} = Q^g_{\widetilde{\kappa}_S}(Q^g_{\widetilde{\kappa}_T}(R)).$$

Because localizing at Ore sets is perfect, we can apply Proposition 7.3.14. We arrive at $G(Q_S Q_T(R)) = Q_{\widetilde{\kappa}_S} Q_{\widetilde{\kappa}_T}(G(R))$. Because $G(R)$ is commutative, both $\sigma(S)$ and $\sigma(T)$ are Ore sets and because in the commutative case the composition of kernel functors commute, we have

$$\begin{aligned} G(Q_S Q_T(R)) &= Q^g_{\sigma(S)} Q^g_{\sigma(T)}(G(R)) \\ &= Q^g_{\sigma(T)} Q^g_{\sigma(S)}(G(R)) \\ &= G(Q_T Q_S(R)). \end{aligned}$$

Hence, although $Q_T Q_S(R) \neq Q_S Q_T(R)$ are not equal in general (they are equal if T and S are compatible), they become equal on the associated graded level!

7.4 Sheaves of glider representations

In this section we establish some sheaf theories of glider representations. Simultaneously, we obtain some sheaves of filtered modules, by which we mean an ordinary sheaf of modules, such that on every open set, the sections give a filtered module and the restriction morphisms are also filtered. Concerning the gluing axiom, the gluing element x of a compatible set of sections x_i must have a degree not exceeding the highest appearing degree of the x_i. Before we get to this however, we need some additional results. We assume either that all kernel functors κ occurring below are coming from some $\widetilde{\kappa}$ (second approach), or else we assume that $G(R)$ is a domain and $G(M)$ is a faithful $G(R)$-module (first approach). The crucial property we will need is namely that $G(\kappa(\Omega)) = \overline{\kappa}(G(\Omega))$, see Lemma 7.2.6 and Lemma 7.3.10. First we observe that the filtered morphism ρ_τ^κ for $\kappa \leq \tau$ from Proposition 7.2.25 restricts to $F_0Q_\kappa(\Omega) \to F_0Q_\tau(\Omega)$, hence to $Q_\kappa(M) \to Q_\tau(M)$ for a glider $\Omega \supset M$, with filtration $f\Omega$ as defined in Example 7.2.13. Additionally, all this remains valid for left exact preradicals.

Remark 7.4.1. By defining $Q_\kappa(M)$ via $Q_\kappa(\Omega)$ we obtain a notion of localization of a glider representation $\Omega \supset M$ which depends on Ω, so correctly we should adopt the notation $Q_\kappa(M, \Omega)$. However, we will not do this for reason of simplicity. There remains a question therefore: how does $Q_\kappa(M, \Omega)$ depend on Ω? Since $RM \subset \Omega$, and RM in Ω are not easily related at first sight, this remains non-trivial. We postpone this aspect to forthcoming work. So we write $Q_\kappa(M)$ for $Q_\kappa(M, \Omega)$ when Ω is fixed.

In Subsection 7.4.2 we discuss sheaves on the lattice R-tors, with the gen-topology. We recall that $\tau \leq \tau'$ if $\mathcal{L}(\tau) \subset \mathcal{L}(\tau')$, where τ, τ' are kernel functors.

Lemma 7.4.2. *Let Ω be a filtered R-module and let κ_i be kernel functors such that Ω is κ_i-separated. Then Ω is $\wedge\kappa_i$-separated.*

Proof. Take $x \in \dot{F}_n\Omega$ and suppose there is an $I \in \mathcal{L}(\wedge\kappa_I)$ such that $F_\gamma Ix \subset F_{\gamma+n-1}\Omega$ for all γ. Since $I \in \mathcal{L}(\kappa_i)$ for all i, we get that $x \in \kappa_i\Omega$ for all i. Hence $x \in \wedge\kappa_i(\Omega)$. □

Lemma 7.4.3. *Let $\tau \geq \kappa$ and $\Omega \in R$-filt such that Ω is τ, κ-separated. Then $\Omega/\kappa\Omega$ is τ-separated.*

Proof. From $(\Omega/\kappa(\Omega))/(\tau(\Omega)/\kappa(\Omega)) \cong \Omega/\tau(\Omega)$ we have

$$\frac{G(\Omega/\kappa(\Omega))}{G(\tau(\Omega)/\kappa(\Omega))} \cong G(\Omega/\tau(\Omega)) = G(\Omega)/\overline{\tau}G(\Omega)$$

because $G(\tau(\Omega)) = \overline{\tau}G(\Omega)$ since Ω is τ-separated and because $\tau(\Omega)$ has by definition the induced filtration from Ω. Thus $G((\Omega/\kappa(\Omega))/\tau(\Omega/\kappa(\Omega)))$ is $\overline{\tau}$-torsion free, hence $\Omega/\kappa(\Omega)$ is τ-separated. □

Remark 7.4.4. There is a direct proof via the $\widetilde{\kappa}, \widetilde{\tau}, \kappa \leq \tau$. Although the proof is longer, it highlights the property that the gradation yields $(\widetilde{JI})_{\tau+\gamma} = \sum_{\tau'+\gamma'=\tau+\gamma} \widetilde{J}_{\tau'}\widetilde{I}_{\gamma'}$. We omit the proof here.
Let $\Omega \supset M$ be a glider representation and let κ_i be a finite number of kernel functors such that Ω is $\kappa_i \wedge \kappa_j$-separated for all i, j. Since $\wedge\kappa_i \leq \kappa_i$,

$$Q_{\wedge\kappa_i}(\Omega) \xrightarrow{\rho_i} Q_{\kappa_i}(\Omega)$$

is filtered, so we can restrict to the degree 0 part

$$Q_{\wedge\kappa_i}(M) \xrightarrow{\rho_i} Q_{\kappa_i}(M)$$

Proposition 7.4.5. *With the above assumptions we have that if x_i in $F_0Q_{\kappa_i}(\Omega) = Q_{\kappa_i}(M)$ are such that $\rho^i_{ij}(x_i) = \rho^j_{ij}(x_j)$ and if there exists an $x \in Q_{\wedge\kappa_i}(\Omega)$ such that $\rho_i(x) = x_i$ for all i, then $x \in F_0Q_{\wedge\kappa_i}(\Omega) = Q_{\wedge\kappa_i}(M)$.*

Proof. Suppose $x \in \dot{F}_nQ_{\wedge\kappa_i}(\Omega)$ for some $n > 0$. There exists $I \in \mathcal{L}(\wedge\kappa_i)$ such that for all γ, $F_\gamma Ix \subset F_{n+\gamma}(\Omega/\wedge\kappa_i\Omega)$ and for some γ, $F_\gamma Ix \not\subset F_{n+\gamma-1}(\Omega/\wedge\kappa_i\Omega)$. Then $\rho_i(F_\gamma Ix) = F_\gamma Ix_i \subset F_{n+\gamma}(\Omega/\kappa_i\Omega)$. Since $I \in \mathcal{L}(\kappa_i)$ for every i and $F_\gamma Ix_i \subset F_{n+\gamma}(\Omega/\kappa_i\Omega)$, the relation $F_\gamma Ix_i \not\subset F_{n+\gamma-1}(\Omega/\kappa_i\Omega)$ would contradict that $x_i \in F_0Q_{\kappa_i}(\Omega)$ ($\Omega/\kappa_i\Omega$ is κ_i-separated so $\deg x_i$ does not depend on the ideal in $\mathcal{L}(\kappa_i)$ used to define it!). Consequently, $F_\gamma Ix_i \subset F_\gamma(\Omega/\kappa_i\Omega)$ for all γ, i.e.

$$F_\gamma Ix \subset F_\gamma(\Omega/\wedge\kappa_j\Omega) + \kappa_i(\Omega/\wedge\kappa_j\Omega).$$

Since $F_\gamma Ix \subset F_{n+\gamma}(\Omega/\wedge\kappa_j\Omega)$ with $n > 0$ we obtain

$$\begin{aligned} F_\gamma Ix &\subset (F_\gamma(\Omega/\wedge\kappa_j\Omega) + \kappa_i(\Omega/\wedge\kappa_j\Omega)) \cap F_{n+\gamma}(\Omega/\wedge\kappa_j\Omega) \\ &\subset F_\gamma(\Omega/\wedge\kappa_j\Omega) + F_{n+\gamma}\kappa_i(\Omega/\wedge\kappa_j\Omega), \end{aligned}$$

hence for $i_\gamma \in F_\gamma I$ with $i_\gamma x \notin F_{n+\gamma-1}(\Omega/\wedge\kappa_j\Omega)$:

$$i_\gamma x \mod F_{n+\gamma-1}(\Omega/\wedge\kappa_j\Omega) \in \frac{F_{n+\gamma}\kappa_i(\Omega/\wedge\kappa_j\Omega)}{F_{n+\gamma-1}\kappa_i(\Omega/\wedge\kappa_j\Omega)}.$$

The latter means that $\overline{i_\gamma x} \in G_{n+\gamma}(\kappa_i(\Omega/\wedge\kappa_j\Omega)) = \overline{\kappa_i}G_{n+\gamma}(\Omega/\wedge\kappa_j\Omega)$, where the last equality follows from Lemma 7.3.10. This holds for all i, hence $\overline{i_\gamma x} \in \overline{\wedge\kappa_i}G(\Omega/\wedge\kappa_i\Omega)$ but as Ω is $\wedge\kappa_i$ separated by Lemma 7.4.2 we have that $\overline{\wedge\kappa_i}G(\Omega/\wedge\kappa_i\Omega) = G(\wedge\kappa_i(\Omega/\wedge\kappa_i\Omega)) = 0$. But then $i_\gamma x \in F_{n+\gamma-1}(\Omega/\wedge\kappa_i\Omega)$, a contradiction. Hence $x \in F_0Q_{\wedge\kappa_i}(\Omega) = Q_{\wedge\kappa_i}(M)$. □

Remark 7.4.6. In fact, one analogously proves a slightly more general statement, saying that if $x_i \in F_mQ_{\kappa_i}(\Omega)$ for some $m \in \mathbb{Z}$ are compatible on the intersections, then if a gluing element x exists, then it must be in $F_mQ_{\wedge\kappa_i}(\Omega)$. This proves for instance that if an R-module Ω yields a sheaf for some topology (e.g. the classical $\widetilde{\Omega}$ if R is commutative), then a filtration such that $G(\Omega)$ is faithful as $G(R)$-module, yields a sheaf of filtered modules. In particular, if FR is 'almost commutative', then we obtain a filtered structure sheaf $\mathcal{O}_X$ for X some suitable topological space (e.g. the Zariski topology). In this case, we denote the filtered sheaf by $\mathcal{F}\mathcal{O}_X$.

The previous proposition allows us to define sheaves of glider representations for various topological spaces.

Definition 7.4.7. Let X be some topological space and FR a filtered ring such that the structure sheaf $\mathcal{F}\mathcal{O}_X$ is filtered. A sheaf $\mathcal{F}$ of glider representations is a subpresheaf of a presheaf $\mathcal{G}$ of filtered $\mathcal{F}\mathcal{O}_X$-modules such that for every open set $U \subset X$, $\mathcal{F}(U) = F_0\mathcal{G}(U)$ and such that $\mathcal{F}$ satisfies the separability and gluing axioms.

Assume from now on that FR is an 'almost commutative' Zariskian ring. In particular, we know that R is κ-separated for every kernel functor (or preradical) on R-mod.

7.4.1 Spec(R) for R a Noetherian prime ring

First we consider the Zariski topology Spec(R) for a Noetherian prime ring (see [51] for its construction). There is a basis of the Zariski topology consisting of open sets $X(I)$ where $I \lhd R$ is an ideal. The kernel functor κ_I associated with an ideal I is determined by the filter

$$\mathcal{L}(\kappa_I) = \{\text{left ideals } L \text{ of } R \text{ such that } L \supset I^n \text{ for some positive integer } n\},$$

and κ_I is symmetric. Let M be a left R-module, which is κ_I-torsion free for all ideals I. Assigning $Q_{\kappa_I}(M) = Q_I(M)$ to the open set $X(I)$ defines a sheaf $\widetilde{M}$ of R-modules on Spec(R) (cf. [51, Theorem 14]). Since $X(I) \subset X(J)$ if and only if $\kappa_J \leq \kappa_I$, we have by Proposition 7.2.25 that the restriction map

$$Q_J(M) \to Q_I(M)$$

is filtered. Moreover, if $X(I) = \cup X(I_j)$ is a finite cover, then it is shown in [51, Theorem 13] that $\cap \kappa_{I_j}(N) = \kappa_I(N)$ for all $N \in R$-mod. Hence $\mathcal{T}_{\kappa_I} = \cap \mathcal{T}_{\kappa_{I_j}} = \mathcal{T}_{\wedge \kappa_{I_j}}$ and $\kappa_I = \wedge \kappa_{I_j}$ follows.

If $\Omega \supset M$ is a glider representation, we denote by $\widetilde{M}$ the presheaf obtained by restricting $\widetilde{\Omega}$ to the degree zero part, that is, $\Gamma(X(I), \widetilde{M}) = F_0Q_I(\Omega) = Q_I(M)$. We denote by $\mathcal{F}^+\widetilde{R}$ the presheaf given by $\Gamma(X(I), \mathcal{F}^+\widetilde{R}) = F^+Q_I(R)$.

Proposition 7.4.8. *Let R be a Zariskian prime ring (hence Noetherian) and M a left R-module such that $G(M)$ is a faithful $G(R)$-module. Then $\widetilde{M}$ becomes a sheaf of filtered $\mathcal{F}\widetilde{R}$-modules. If $\Omega \supset M$ is glider such that $G(\Omega)$ is a faithful $G(R)$-module, then $\widetilde{M}$ is a sheaf of glider $\mathcal{F}^+\widetilde{R}$-representations.*

Proof. For the separated presheaf condition, we refer to Proposition 7.4.10 where it is shown in a more general setting. Observe that we need that $\kappa_I = \wedge \kappa_{I_j}$ if $X(I) = \cup X(I_j)$ is a cover. The gluing axiom follows directly from Proposition 7.4.5. □

Remark 7.4.9. Since $G(R)$ is commutative, we have $Y = \text{Proj(G(R))}$, the classical Zariski topology. In [69, Section 2.4.A] it is shown that

$$\begin{cases} \kappa_{<P>} = \vee\{\kappa_{<f>}, \overline{f} \notin P\} = \vee\{\kappa_{<I>}, \overline{I} \not\subset P\}, \\ \kappa_{<f>} = \wedge\{\kappa_{<P>}, \overline{f} \notin P\}, \\ \kappa_{<I>} = \wedge\{\kappa_{<P>}, \overline{I} \not\subset P\} = \wedge_{i=1}^{d}\kappa_{<f_i>}, \text{ if } \overline{I} = (\overline{f}_1, \ldots, \overline{f}_n), \end{cases} \tag{7.5}$$

with notations as in loc. cit. By deriving sheaf-conditions on the level of torsion theories, one obtains sheaves $\underline{M}_Y$ for arbitrary R-modules M (with no torsion freeness assumptions!) (cf. Theorem 2.4.5 and 2.5.6 in loc. cit.). In order to let everything work in our filtered setting, it suffices to have a basis of the Zariski topology for which the filtered module M is separated. The filtration on a general open set can then be defined by an inverse limit, cf. [46, Observation before Remark 2.6, p.8].

7.4.2 R-tors

Next, we consider R-tors, the set of all kernel functors on R-mod, which has a (distributive) lattice structure (see [33, Chapter II] for a detailed overview). We define a topology on R-tors by defining a basis consisting of the open sets $\text{gen}(\tau) = \{\tau' \in \text{R-tors} \mid \tau \leq \tau'\}$. We call this the gen-topology of R-tors. Recall that $\tau \leq \tau'$ if $\mathcal{L}(\tau) \subset \mathcal{L}(\tau')$. We have

$$\text{gen}(\kappa) \cap \text{gen}(\tau) = \text{gen}(\kappa \vee \tau),$$

$$\text{gen}(\kappa) \cup \text{gen}(\tau) = \text{gen}(\kappa \wedge \tau).$$

The distributivity of the lattice implies that

$$\text{gen}(\kappa) \cup (\text{gen}(\sigma) \cap \text{gen}(\tau)) = (\text{gen}(\kappa) \cup \text{gen}(\sigma)) \cap (\text{gen}(\kappa) \cup \text{gen}(\tau)).$$

To a basic open set $U = \text{gen}(\tau)$ we associate the localized ring $Q_{\wedge U}(R) = Q_\tau(R)$ which is filtered by Theorem 7.2.12. Since $\text{gen}(\tau) \subset \text{gen}(\kappa)$ if and only if $\kappa \leq \tau$, the restriction morphism $Q_\kappa(R) \to Q_\tau(R)$ is also filtered. Hence we obtain a presheaf $\mathcal{F}\overline{Q}(-, R)$ of filtered rings. The same result holds for a filtered R-module N such that $G(N)$ is faithful as a $G(R)$-module, i.e. then $\overline{Q}(-, N)$ is a presheaf of filtered $\mathcal{F}\overline{Q}(-, R)$-modules. For $\Omega \supset M$ a glider representation, we can restrict the sections $\overline{Q}(\text{gen}(\tau), \Omega) = \text{Q}_\tau(\Omega)$ to the degree zero part $F_0\overline{Q}(\text{gen}(\tau), \Omega)$. We denote the latter by $\overline{Q}(\text{gen}(\tau), \text{M})$ and we obtain in this way a presheaf of $\mathcal{F}^+\overline{Q}(-, R)$ glider representations. We also use the notations $\mathcal{F}\mathcal{O}_X, \mathcal{O}_\Omega, \mathcal{O}_M$ for the respective presheaves.

Proposition 7.4.10. *Let $\Omega \supset M$ be a glider representation such that $G(\Omega)$ is faithful as a $G(R)$-module. If $\overline{Q}(-, \Omega)$ is a sheaf of filtered $\mathcal{F}\overline{Q}(-, R)$-modules, we obtain a sheaf $\overline{Q}(-, M)$ of glider $\mathcal{F}^+\overline{Q}(-, R)$-representations.*

Proof. We first have to check the separated presheaf condition, so consider $x \in Q_{\wedge\kappa_i}(M)$ such that $\rho_i(x) = 0$ for every i in the covering of $\wedge\kappa_i$ by the κ_i.

Take $I \in \mathcal{L}(\wedge\kappa_i)$ such that $F_\gamma Ix \subset F_{\gamma+n}M/\wedge\kappa_i M$. Then $\rho_i(F_\gamma Ix) = 0$ means $F_\gamma Ix \subset \kappa_i(M/\wedge\kappa_i M)$. This holds for all i, so $F_\gamma Ix \subset \wedge\kappa_i M/\wedge\kappa_i M = 0$. This holds for all γ, i.e. $Ix = 0$, but then $x \in (\wedge\kappa_i)(Q_{\wedge\kappa_i}(M)) = 0$.
Concerning the gluing condition, let $\text{gen}(\kappa) = \cup_i \text{gen}(\kappa_i)$ be a finite cover. By [33, Proposition 8.4] we have that $\kappa = \wedge\kappa_i$ and the result follows by applying Proposition 7.4.5. □

As was the case for Spec(R), we do not need to restrict to faithful modules. Indeed, it suffices again to have a basis $\text{gen}(\kappa)$ for which M is κ-separated and use inverse limits to define filtrations.

Recall that on $\widetilde{R}$-grtors we have an open set given by $\text{gen}(\widetilde{\kappa}_X)$ of the affine torsion theories. Let $\widetilde{\kappa} \geq \widetilde{\kappa}_X$ be an affine kernel functor with associated κ on R-mod. Let M be an R-module. The sections over the open set $\text{gen}(\kappa)$ are given by the R-module $Q_\kappa(M)$ with quotient filtration. Under the equivalence FR-filt $\cong \mathcal{T}_X$, the filtered module $Q_\kappa(M)$ corresponds to $\widetilde{Q_\kappa(M)}$, which is a graded X-torsion free $\widetilde{R}$-module. Under the equivalence of categories $\widetilde{R}_X$-gr $\cong R$-mod, the graded $\widetilde{R}$-module $\widetilde{Q_\kappa(M)}$, after localizing at S_X, corresponds to the R-module $Q_\kappa(M)$. Since localizing at X is perfect and since $\mathcal{L}(\widetilde{\kappa})$ contains $\widetilde{R}X^n$ for all $n \in \mathbb{N}$, we obtain that

$$Q^g_{\widetilde{\kappa}}(\widetilde{M}) = Q^g_{\widetilde{\kappa}} Q^g_{\widetilde{\kappa}_X}(\widetilde{M}) = Q^g_{\widetilde{\kappa}}(\widetilde{M}_X) = Q^g_{\widetilde{\kappa}}(\widetilde{M})_X = (Q_\kappa(M)^\sim)_X.$$

Hence we have shown that the sheaf $\mathcal{O}^g_{\widetilde{M}}$ of graded modules associated with $\widetilde{M}$ restricted to the open set $\text{gen}(\widetilde{\kappa}_X) \subset \widetilde{\text{R}}$-grtors is isomorphic to the sheaf $\mathcal{O}_M$ of R-modules(!) on R-tors.

7.4.3 The non-commutative site

Finally, we discuss the non-commutative site from [69, Section 1.3] in the filtered case. We give a concise survey of how this site is built. We refer the reader to loc. cit. for a more detailed treatment.

Definition 7.4.11. A non-commutative ring R is said to be affine schematic if there exists a finite set of non-trivial Ore sets of R, say $S_1, \ldots, S_n$ such that for every choice of $s_i \in S_i, i = 1, \ldots, n$, we have that $\sum_{i=1}^n Rs_i = R$ or equivalently: $\cap_i \mathcal{L}(S_i) = \{R\}$, in which $\mathcal{L}(S)$ is the Gabriel filter associated to the Ore set S (7.1).

In torsion theoretic language, the latter means that the infimum of the kernel functors κ_{S_i} associated with S_i equals the trivial kernel functor, i.e. $\wedge_{i=1}^n \kappa_{S_i} = \kappa_1$ (κ_1 associated with the trivial Ore set $\{1\}$). One considers the free monoid $\mathcal{W}(R)$ on $\theta(R)$, the set of left Ore sets S of R and one introduces the category $\underline{\mathcal{W}}(R)$ with objects the elements of $\mathcal{W}(R)$ and concerning morphisms: if $W = S_1 \ldots S_n, W' = T_1 \ldots T_m$, then $\text{Hom}(W', W) = \{W' \to W\}$ or $\emptyset$ depending upon whether there exists a strictly increasing map $\alpha : \{1,\ldots,n\} \to \{1,\ldots,m\}$ for which $S_i = T_{\alpha(i)}$. Let $\{W_i,\ i \in I\}$ be a finite subset of objects of $\underline{\mathcal{W}}$ such that $\cap_{i\in I}\mathcal{L}(W_i) = \mathcal{L}(\kappa_1)$, in which $\mathcal{L}(W) = \{L,\ w \in L \text{ for some } w \in W\}$ is the associated filter of left ideals. We call such a subset a global cover of $Y = \underline{\mathcal{W}}(R)$. Observe that the schematic property exactly states that there exists such a cover. For $W \in \underline{\mathcal{W}}$ let Cov(W) be $\{W_iW \to W,\ i \in I\}$. Then one shows that we obtain a (non-commutative) Grothendieck topology on this category.

Let $W = S_1 \ldots S_n$. At the end of Section 7.2 we associated the filter of left ideals $\mathcal{L}(\kappa_{S_1}\kappa_{S_2}\ldots\kappa_{S_n})$ with W. By choosing s_j as generators for Rs_j we have that $Rs_1 \cdot Rs_2 \cdot \ldots \cdot Rs_n = Rs_1s_2\ldots s_n$, but one may take different generators, yielding different left ideals in the filter. In [69, p. 45], however, one considers

$$\mathcal{L}(W) = \{L \supset Rs_1 \ldots s_n, \text{ with } s_i \in S_i, i = 1,\ldots,n\}.$$

Under some torsion freeness assumptions, we have that both filters are the same.

Lemma 7.4.12. *Assume that $G(R)$ is a domain. We have $\mathcal{L}(\kappa_{S_1}\kappa_{S_2}\ldots\kappa_{S_n}) = \mathcal{L}(W)$.*

Proof. Let $L \in \mathcal{L}(W)$, i.e. L contains some $Rs_1 \ldots s_n$. By choosing s_j as generators for Rs_j we have that $Rs_1 \cdot Rs_2 \cdot \ldots \cdot Rs_n = Rs_1s_2\ldots s_n$ and we see that $L \in \mathcal{L}(\kappa_{S_1}\kappa_{S_2}\ldots\kappa_{S_n})$. For the other inclusion, we begin by choosing for $Rs_i \in \mathcal{L}(\kappa_{S_i})$ another generator, x say, $Rx = Rs_i$, thus $s_i = u_ix$ and $x = v_is_i$. This yields $s_i = u_iv_is_i$ and by S-torsion freeness of R, $1 = u_iv_i$. Since $G(R)$ is a domain, R is a domain and then u_i is invertible in R. The left ideals $Rs_1u_1s_2\ldots u_{n-1}s_n$ where the u_j are invertible elements in R are part of a filterbasis for $\mathcal{L}(\kappa_{S_1}\kappa_{S_2}\ldots\kappa_{S_n})$. By the Ore condition for s_1 we have $s_1'u_1^{-1} = r_1's_1$ for some $s_1' \in S_1, r_1' \in R$. Thus $s_1' = r_1's_1u_1$ and $Rs_1's_2u_2\ldots u_{n-1}s_n = Rr_1's_1u_1s_2u_2\ldots u_{n-1}s_n$, or $Rs_1's_2u_2\ldots u_{n-1}s_n \subset Rs_1u_1s_2u_2\ldots u_{n-1}s_n$. Now R satisfies the second Ore condition with respect to S_1S_2 so $s_1''s_2''u_2^{-1} = r_2's_1's_2$ for some $s_1'' \in S_1, s_2'' \in S_2, r_2' \in R$. Hence $s_1''s_2'' = r_2's_1's_2u_2$ and therefore:

$$Rs_1''s_2''s_3u_3\ldots u_{n-1}s_n \subset Rs_1's_2u_2s_3\ldots u_{n-1}s_n \subset Rs_1u_1s_2u_2\ldots u_{n-1}s_n.$$

We repeat this procedure until we obtain $s_i^{(n)} \in S_i$ such that

$$Rs_1^{(n)}\ldots s_n^{(n)} \subset Rs_1u_1s_2\ldots u_{n-1}s_n,$$

proving that $Rs_1u_1s_2\ldots u_{n-1}s_n \in \mathcal{L}(W)$ as desired.
Now let us look at Rt, Rs for s, t in Ore sets S and T resp. and we look at $Rt\cdot Rs$ corresponding to writing $Rs = Rus + Rvs$; that is, now we consider more than one generator. So we have $Ru + Rv = R$, say $r_1u + r_2v = 1$. By the Ore conditions there are $t', t'' \in T$ and $x, y \in R$ such that $t'u = xt$ and $t''v = yt$. Take $t_1 \in Rt' \cap Rt'' \cap Rt$ then $t_1u = x't$ and $t_1v = y't$ for some $x', y' \in R$. Again by the Ore conditions, $t_2r_1 = r_1't_1, t_3r_2 = r_2't_1$ and we may take again

$t''' \in Rt_2 \cap Rt_3$ and obtain $t'''r_1 = r_1''t_1$ and $t'''r_2 = r_2''t_1$ for some $r_1'', r_2'' \in R$. Now from $r_1u + r_2v = 1$ we obtain that $t''' = t'''r_1u + t'''r_2v = r_1''t_1u + r_2''t_1v$. Thus

$$t'''s = r_1''t_1us + r_2''t_1vs.$$

Since we took $t_1 \in Rt$ we have $t'''s = r_1'''tus + r_2'''tvs$ for some $r_1''', r_2''' \in R$, i.e. $t'''s \in Rt \cdot Rs$ with respect to the chosen generators us, uv for Rs. The extension to more Ore sets in the word considered can easily be obtained by induction since the Ore property holds with respect to every word $S_1 \dots S_n$. For example, add another Ore set U and consider $Ru \cdot Rt \cdot Rs$ corresponding to writing $Rs = Ras + Rbs$ and $Rt = Rct + Rdt$. By induction there exist $t''' \in T, u''' \in U$ such that

$$t'''s = r_1tas + r_2tbs, \quad u'''t = r_3uct + r_4udt, \quad r_i \in R.$$

By the Ore condition we find $u_1, u_2 \in U, r_1', r_2' \in R$ such that $u_1r_1 = r_1'u'''$ and $u_2r_2 = r_2'u'''$. Take some $u_3 \in Ru_1 \cap Ru_2$, then $u_3t'''s \in Ru \cdot Rt \cdot Rs$ with respect to the chosen generators. The extension to more generators is easy but technical to write down; essentially it follows from taking some Rs_0 in a finite intersection $Rs_1 \cap \dots \cap Rs_n$; we leave the details to the reader. So we have proved that $\mathcal{L}(W) = \mathcal{L}(\kappa_{S_1}\kappa_{S_2}\dots;\kappa_{S_n})$, the localization of R at the Ore set generated by the Ore sets $S_1, \dots, S_n$.

□

In [69, Section 1.3], one constructed the structure sheaf $\mathcal{O}_R$ as the functor $\underline{\mathcal{W}}(R)^{op} \to R$-mod sending $W = S_1 \dots S_n$ to $Q_{S_n} \dots Q_{S_1}(R)$. The $Q_{S_i}(R)$ are rings but the $Q_W(R)$ are in general not rings. Nevertheless, we have a multiplication defined by

$$Q_{S_n} \dots Q_{S_1}(R) \times Q_{T_m} \dots Q_{T_1}(R) \to Q_{S_n} \dots Q_{S_1}Q_{T_m} \dots Q_{T_1}(R)$$

Recall that the $Q_{S_n} \dots Q_{S_1}(R)$ may be seen as the n-fold tensor product over R: $Q_{S_n}(R) \otimes_R \dots \otimes_R Q_{S_1}(R)$ for every $W = S_1 \dots S_n$. In fact, $Q_W(R)$ is an R-submodule of $Q_{S_n \vee \dots \vee S_1}(R)$.

In [69], one also defined $\mathcal{O}_M$ for an R-module M and where we called $\mathcal{O}_R$ a structure sheaf of rings by "abus de language" we also call $\mathcal{O}_M$ a structure sheaf of modules over the non-commutative Grothendieck topology considered. Again $\mathcal{O}_M$ is defined as the functor $\underline{\mathcal{W}}(R)^{op} \to R$-mod sending the object W to $Q_W(M) = Q_{S_n} \dots Q_{S_1}(M)$. So $Q_W(M)$ is not a $Q_W(R)$-module but there is for every $Q_T(R)$ a scalar multiplication $Q_T(R) \times Q_W(M) \to Q_{WT}(M)$, hence also $Q_W(R) \times Q_W(M) \to Q_{WW}(M)$. In particular, $Q_W(M)$ is a $Q_{S_n}(R)$-module!

Definition 7.4.13. We will call a presheaf of $\mathcal{O}_R$-modules, say $\mathcal{M}$, an $\mathcal{O}_R$-module if there are operations $\varphi(T, W) : Q_T(R) \times \mathcal{M}(W) \to \mathcal{M}(WT)$ which respect the additions and act associatively i.e. $\varphi(ST, W) = \varphi(S, WT)$.

In the case where we consider a filtration on R, we will usually restrict to a Zariskian filtration, in fact even a positive filtration on a Noetherian ring. Then when we assume that $G(R)$ is a domain, R is a Noetherian domain and it has a classical total quotient ring $Q(R)$ which is a simple Artinian ring. The multiplication considered in $\mathcal{O}_R$ is just obtained from the multiplication in $Q(R)$, i.e. $Q_T(R)$ and $Q_W(R)$ can be multiplied in $Q(R)$ and in particular in $Q_{TW}(R) \subset Q(R)$. Similarly, a filtered R-module M such that $G(M)$ is $G(R)$-faithful is then a submodule of $Q(R) \otimes_R M$ and the scalar multiplications $Q_W(R) \times Q_T(M) \to Q_{TW}(M)$ may be viewed as carried out in $Q(R) \otimes_R M$.

Definition 7.4.14. A filtered $\mathcal{O}_R$-module is a presheaf of filtered $\mathcal{O}_R$-modules such that the restriction morphisms in R-filt are strict filtered morphisms. In particular, for a filtered $\mathcal{O}_R$-module $\mathcal{M}$ each $\mathcal{M}(S_1 \dots S_n)$ is an $FQ_{S_n}(R)$-filtered module.
An $\mathcal{O}_R$-glider $\mathcal{F}$ is a subpresheaf of a filtered $\mathcal{O}_R$-module $\mathcal{G}$ obtained by taking for every W, $\mathcal{F}(W) = F_0\mathcal{G}(W)$ (compare with Definition 7.4.7).
In case $\mathcal{G}$ is a sheaf such that the gluing axiom holds in $F_n\mathcal{G}$ for every n, then we call $\mathcal{G}$ a filtered $\mathcal{O}_R$-module sheaf. Similarly, when $\mathcal{F}$ is a sheaf with the gluing axiom holding with respect to $\mathcal{F}_n$ for every n, then $\mathcal{F}$ is said to be an $\mathcal{F}^+\mathcal{O}_R$-glider sheaf.

In the remainder of this section we restrict to a non-commutative geometry situation: R is Noetherian and positively filtered by FR such that $F_0R = K$ is a field, and R is affine schematic (cf. Definition 7.4.11). Since we aim to work with separated filtrations on the localizations of R we assume moreover that $G(R)$ is a domain. In the book [69] there are given many examples of such schematic algebras and these algebras have moreover standard filtrations. E.g. Weyl algebras, enveloping algebras of Lie algebras, Sklyanin algebras, Witten gauge algebras, rings of differential operators on regular varieties, quantum-sl(2), many quantum groups obtained as iterated Ore extensions, etc.
In this case we consider $\underline{\mathcal{W}}(R)$, the set of words in Ore sets in non-trivial Ore sets, that is $S \in \underline{\mathcal{W}}(R)$ if $S \cap K = \{1\}$.

Proposition 7.4.15. *In the situation considered above, we consider an $\mathcal{F}^+\mathcal{O}_R$-glider $\mathcal{F}$ and its filtered $\mathcal{O}_R$-module $\mathcal{G}$, such that $\mathcal{F}(W) = F_0\mathcal{G}(W)$ for all $W \in \underline{\mathcal{W}}(R)$. Then $\mathcal{F}$ is an $\mathcal{F}^+\mathcal{O}_R$-glider sheaf if and only if $\mathcal{G}$ is a filtered $\mathcal{O}_R$-module sheaf.*

Proof. We only have to check the separation and gluing axiom.
1. Assuming that $\mathcal{G}$ is a filtered $\mathcal{O}_R$-module sheaf, both the separation and gluing axiom are easily verified for $\mathcal{F}$ (using that the gluing axiom on $\mathcal{G}$ holds in fact in $F_n\mathcal{G}$ for every n).
2. Assume that $\mathcal{F}$ is an $\mathcal{F}^+\mathcal{O}_R$-glider sheaf. Look at a cover $\{W \to W_iW\}$ and suppose $x \in \mathcal{G}(W)$ maps to 0 in each $\mathcal{G}(W_iW)$. If $W = S_1 \dots S_n$ then $\mathcal{G}(W)$ is a $Q_{S_n}(R)$-module and for every $s_n \in S_n$ we have $s_n^{-1} \in Q_{S_n}(R)$. Since $G(R)$ is a domain $\deg(s_n^{-1}) = -\deg(s_n) < 0$ since $S \cap K = \{1\}$. So we can choose $\deg(s_n)$ high enough so that $s_n^{-1}x \in F_0\mathcal{G}(W)$ and $s_n^{-1}x$ still maps to zero in

each $\mathcal{G}(W_iW)$. The separation axiom in $\mathcal{F}$ then yields that $s_n^{-1}x = 0$, hence $x = 0$. So $\mathcal{G}$ is satisfying the separation axiom.
Consider a finite cover $\{W \to W_iW\}$ and look at $x_i \in \mathcal{G}(W_iW)$ such that x_i and x_j are mapping to the same element in $\mathcal{G}(W_iWW_jW)$ for every i and j. Since S_n is the last letter of W, W_iW, W_iWW_jW for all i and j, all sections are $Q_{S_n}(R)$-modules and again we may choose $s_n \in S_n$ such that $s_n^{-1}x_i \in F_0\mathcal{G}(W_iW) = \mathcal{F}(W_iW)$, say $s_n^{-1}x_i \in \mathcal{F}_d(W_iW)$ for some $d \leq 0$ (i.e. $x_i \in F_{d+\deg(s_n)}\mathcal{G}(W_iW)$) and we may assume d is smallest possible. By the gluing property of $\mathcal{F}$ there is a $y \in \mathcal{F}(W)$ such that y maps to $s_n^{-1}x_i$ under each restriction morphism; in fact since the gluing property of $\mathcal{F}$ respects the $\mathcal{F}_m$, we have $y \in \mathcal{F}_d(W)$. Then s_ny maps to x_i under each restriction from W to W_iW, and $s_ny \in F_{d+\deg(s_n)}\mathcal{G}(W)$.
We still have to check the axioms for the trivial word $W = 1$. Since $G(R)$ is a domain, we have a total quotient ring $Q(R)$ and we can consider $\mathcal{G}(W)$ as an R-submodule of $Q(R) \otimes_R \mathcal{G}(1)$. If elements $x_i \in \mathcal{G}(W_i)$ are such that x_i and x_j map to the same element in $\mathcal{G}(W_iW_j)$ for every i and j, then since the cover is finite, they become equal in $\mathcal{G}(W) \subset Q(R) \otimes_R \mathcal{G}(1)$ for some word W. This implies that the x_i are in the intersection of the $\mathcal{G}(W_i)$. But since $\{W_i\}$ is a global cover, this intersection is just $\mathcal{G}(1)$. So we established the filtered gluing property of $\mathcal{G}$. □

Corollary 7.4.16. *Let $\Omega \supset M$ be a glider representation such that $G(\Omega)$ is a faithful $G(R)$-module (assumptions on $R, G(R)$ as before). The presheaves $\mathcal{O}_M$ and $\mathcal{O}_\Omega$, $\mathcal{O}_M = F_0\mathcal{O}_\Omega$ defined before Definition 7.4.14 are both sheaves if one is.*

Example 7.4.17. For FR commutative, say with $F_0R = K$ a field, we have the classical Zariski topology $X = \mathrm{Spec(R)}$ with structure sheaf $\mathcal{O}_X$. If $G(R)$ is a domain, then $\mathcal{F}\mathcal{O}_X$ is a filtered sheaf. The proof of the previous proposition reduces in this case to considering a finite cover $X = \bigcup_i \mathbb{X}(f_i)$, with non-trivial Ore sets $< f_i >$. For any $g \in R$, we have a cover $\mathbb{X}(g) = \cup_i \mathbb{X}(gf_i)$. Since $\mathbb{X}(gf_i) \subset \mathbb{X}(g)$ (if and only if $gf_i \in \sqrt{(g)}$) implies that Ω_{gf_i} is an R_g-module, we can lower the degree of $x_i \in \Omega_{gf_i}$ by multiplying with g^{-1}.

Definition 7.4.18. A sheaf $\mathcal{F} = \mathcal{F}_0\mathcal{G}$ of $\mathcal{O}_R$-glider representations is quasi-coherent if there exists a global cover of Ore sets $\{T_i, i \in I\}$ together with $F^+Q_{T_i}(R)$ glider representations $\Omega_i \supset M_i$ such that for any morphism

$V \to W$ in $\underline{\mathcal{W}}(R)$ we obtain a commutative diagram of filtered morphisms:

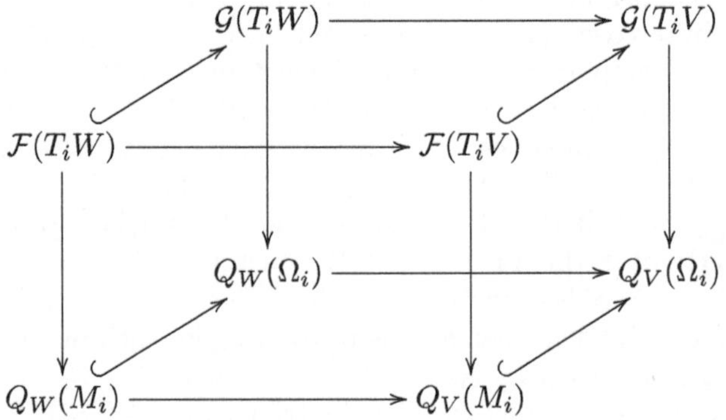

(7.6)

in which the vertical maps in the front plane are isomorphisms of glider representations and $\mathcal{G}$ is a strict sheaf in the sense that the restriction morphisms are strict filtered maps.

Observe that by Proposition 7.4.15, an isomorphism of sheaves between $\mathcal{F}(T_i-)$ and $\mathcal{O}_{M_i}$ implies an isomorphism of sheaves between $\mathcal{G}(T_i-)$ and $\mathcal{O}_{\Omega_i}$. Hence, the vertical maps in the back plane are strict isomorphisms of filtered $FQ_{WT_i}(R)$-modules. Serre's global section theorem for quasi-coherent sheaves of $\mathcal{O}_R$-modules (see [69, Theorem 2.1.4]) remains valid in the glider case:

Theorem 7.4.19. *Let $\mathcal{F} \subset \mathcal{G}$ be a quasi-coherent sheaf of $\mathcal{O}_R$ glider representations with $G\mathcal{F}(W)$ faithful over $G(R)$ for all W. If $\mathcal{F}(1)$ denotes the global section F^+R-glider representation, then $\mathcal{F}$ is sheaf isomorphic to the structure sheaf $\mathcal{O}_{\mathcal{F}(1)}$.*

Proof. Proposition 7.4.15 shows that $\mathcal{F}$ being a sheaf is equivalent to $\mathcal{G}$ being a sheaf of filtered $\mathcal{O}_R$-modules. By the global section theorem for quasi-coherent sheaves of modules (cf. [72]), we have that $\mathcal{G} \cong \mathcal{O}_{\mathcal{G}(1)}$. The latter sheaf is a sheaf of filtered $\mathcal{O}_R$-modules by the quotient filtration. Denote this filtration by $F\mathcal{G}$. The original filtration on $\mathcal{G}$ is denoted by $f\mathcal{G}$. Since $\mathcal{G}$ is a strict sheaf, $\mathcal{G}(1) \to \mathcal{G}(W) = Q_W(\mathcal{G}(1))$ is strict filtered for every $W \in \underline{\mathcal{W}}(R)$. Let $x \in f_m\mathcal{G}(W)$, say $W = S_1 \dots S_n$. There are $s_i \in S_i$ such that $s_1 \dots s_n x \in \mathcal{G}(1)$. Moreover, since $\mathcal{G}$ is a sheaf of filtered $\mathcal{O}_R$-modules, $\mathcal{G}(W)$ is in particular a filtered FR-module and therefore $s_1 \dots s_n x \in f_{m+d}\mathcal{G}(W)$ where $d = \deg(s_1 \dots s_n)$. Hence $s_1 \dots s_n x \in f_{m+d}\mathcal{G}(W) \cap \mathcal{G}(1) = f_{m+d}\mathcal{G}(1)$ by strictness. For $W = 1$, the quotient filtration corresponds with the original filtration: $F\mathcal{O}_{\mathcal{G}(1)}(1) = F\mathcal{G}(1) = f\mathcal{G}(1)$. Hence $s_1 \dots s_n x \in F_{m+d}\mathcal{G}(1)$ and by definition of the quotient filtration this means that $x \in F_m\mathcal{O}_{\mathcal{G}(1)}(W)$. This shows that $f_m\mathcal{G}(W) \subset F_m\mathcal{G}(W)$. By the faithfulness assumption, multiplying by $s_1 \dots s_n$ does not lower the degree for $f\mathcal{G}$. Hence the converse $F_m\mathcal{G}(W) \subset f_m\mathcal{G}(W)$ is obvious. We conclude that $\mathcal{G} \cong \mathcal{O}_{\mathcal{G}(1)}$ is an isomorphism of filtered $\mathcal{O}_R$-modules. Reducing to the degree 0 part yields $\mathcal{F} \cong \mathcal{O}_{\mathcal{F}(1)}$ as desired. □

7.5 Applications to commutative algebraic geometry

As we already mentioned in the introductory part of this chapter, we intend to apply the developed techniques amongst others for rings of differential operators (see remarks before Theorem 7.2.12) and in classical algebraic geometry. For the latter, we lay out a few examples here.

Example 7.5.1. Consider the normalization of the cusp $C : X^3 = Y^2$, given by the embedding of coordinate rings

$$\mathbb{C}[X^2, X^3] \subset \mathbb{C}[X],$$

where the standard filtration from the introduction is considered by choosing X as an $\mathbb{C}[X^2, X^3]$-ring generator for $\mathbb{C}[X]$. For this filtration the associated graded ring is isomorphic to $\mathbb{C}[X^2, X^3][\epsilon]/(\epsilon^2)$. We have a glider representation

$$\Omega = \mathbb{C}[X] \supset M = \mathbb{C}[X^2, X^3] \supset \mathbb{C}X^2 + (X^4) \supset (X^5) \supset \dots$$

and one easily sees that the ideal generated by ϵ sits in $\mathcal{A}\mathrm{nn}(\Omega)$. If we localize the glider at $S = S_X$, then we obtain a glider inside $\mathbb{C}[X, X^{-1}]$ which has non-zero body. Indeed, for any $n \geq 2$ and $\gamma \geq 0$ we have that $F_\gamma(X^n)X^4 = (X^{4+n}) \subset F_{-n-1}\Omega \subset F_{\gamma-n-1}\Omega$, so $X^4 \in F_{-n-1}(S_X^{-1}\Omega)$, so X^4 sits in the body.

Example 7.5.2. Let K be an algebraically closed field. Consider

$$V = \mathrm{Spec}(\mathrm{K[X, Y, T]/(XY - T)}) \quad \text{and} \quad \mathrm{W = Spec(K[T])}.$$

Consider the inclusion of coordinate rings $K[W] \hookrightarrow K[V]$. If we choose $\overline{X}, \overline{Y}$ to be $K[W]$-ring generators for $K[V]$, we get a standard filtration on $K[V]$ with associated graded $G(K[V]) = K[T][\overline{X}, \overline{Y}]/(\overline{XY})$. Consider the glider representation

$$\Omega = K[V] \supset M = K[T, \overline{Y}] \supset (\overline{Y}) \supset (\overline{Y})^2 \supset \dots$$

which is bodyless. Its associated graded is

$$G(M) = K[T][\overline{Y}],$$

with $\deg(T) = 0, \deg(\overline{Y}) = -1$. We have that $\mathcal{A}\mathrm{nn}(\Omega) = \mathrm{gen}(\kappa_{S_{\overline{Y}}})$, the torsion theories that contain $\overline{Y}$. Hence the strong characteristic variety is a proper subset of $G(K[V])$-tors. On the Zariski open $\mathbb{X}(\overline{Y})$, we have that $\widetilde{M}(\mathbb{X}(\overline{Y})) = Q_{\kappa_{S_{\overline{Y}}}}(M)$ has non-zero body. Indeed, let $m, n > 0$. Then $F_\gamma(Y^n) = 0$ if $\gamma < n$ so assume that $\gamma \geq n$. Then we have

$$F_\gamma(Y^n)Y^m \subset M_{\gamma+m} \subset M_n = F_{-n}\Omega \subset F_{\gamma-n}\Omega.$$

It follows that $Y^m \in F_{-n}Q_{\kappa S_{\overline{Y}}}(M)$ and we have shown that

$$(\overline{Y}) \subset B(Q_{\kappa S_{\overline{Y}}}(M)).$$

Observe that in the proof of Theorem 7.2.12 we indicated that $I \in \mathcal{L}(\kappa)$ is independent to define the degree. We needed the κ-separatedness for this, which does not hold in this situation. Obviously, we can interchange the role of X and Y.

The previous examples indicate a link with singularities or smoothness of morphisms between schemes. Recall that a morphism $f : X \to Y$ of schemes of finite type over some field k is smooth if for every closed point $x \in X$, the induced map on the Zariski tangent spaces $T_f = (df)^* : T_x \to T_{f(x)}$ is surjective.

Proposition 7.5.3. *Let K be algebraically closed and let $K[W] \subset K[V]$ be an embedding of coordinate rings corresponding to a dominant polynomial map $f : V \to W$, where W and V are embedded in $\mathbb{A}^m$, resp. $\mathbb{A}^n$ with $m < n$. If f is not smooth, there exists a standard filtration on $K[V]$ with degree zero part $K[W]$ and a glider $K[V] \supset M$ with proper strong characteristic variety.*

Proof. Let $X_1, \ldots, X_m$ be the variables in $\mathbb{A}^m$ and extend them to $X_1, \ldots, X_n$ for $\mathbb{A}^n$. We may assume that f is not smooth in the origin, so the map

$$df : \frac{P \cap K[W]}{(P \cap K[W])^2} \to \frac{P}{P^2}$$

is not injective, where $P = (\overline{X_1}, \ldots, \overline{X_n})$. W.l.o.g. we may assume that $\overline{X_1}$ in $P \cap K[W]$ is such that $\overline{X_1} \in P^2 \setminus K[W]$. Hence, we can write $\overline{X_1} = f(\underline{X})g(\underline{X})$, where f, g are two polynomials with zero constant coefficient. Write $K[V] = K[W][f, g, a_1, \ldots, a_k]$ as a ring extension and use these generators to define a standard filtration on $K[V]$. We have a glider representation

$$K[V] \supset M = K[W][a_1, \ldots, a_k][f] \supset (f) \supset (f)^2 \supset \ldots$$

The associated graded is $G(M) = K[W][a_1, \ldots, a_k][\overline{f}]$, where $\deg(\overline{f}) = -1$. It is clear that the torsion theory $\overline{S}_{\overline{f}} \notin \xi(K[V])$ and we are done. □

The above (standard) filtrations appearing in algebraic geometry are not the only interesting ones. Admittedly, we laid out the sheaf theory for almost commutative rings, such that all filtered localizations were separated and induced the original filtration. This is however not necessary. Indeed, on the one hand we already discussed the existence of the strong characteristic variety related to separatedness of the filtered localization. On the other hand, there are interesting situations where a filtered localization does exist but is not inducing. Of course, the situations we have in mind are the first type of filtrations mentioned in the introduction and hinted at in Example 7.2.14.

Let R be a ring with given $\mathbb{Z}$-filtration

$$\ldots \subset 0 \subset 0 \subset S = F_0R \subset R_1 = F_1R \subset \ldots \subset R = F_nR \subset R \subset R \subset \ldots$$

where each R_i is a subring of R. Consider a kernel functor κ on R-mod and write $\pi : R \twoheadrightarrow \overline{R} = R/\kappa(R)$ for the canonical epimorphism, putting $F_i\overline{R} = \pi(R_i)$. Since the associated graded $G(R)$ is certainly not a domain, we can not hope for κ-separatedness. Nonetheless, we can still define a filtration on the localization

$$F_dQ_\kappa(R) = \{x \in Q_\kappa(R) \mid \exists I \in \mathcal{L}(\kappa) \text{ such that } F_\gamma Ix \subset F_{\gamma+d}\overline{R},\ \forall\gamma \in \mathbb{Z}\}.$$

Proposition 7.5.4. *With notations as above: $FQ_\kappa(R)$ is a separated (exhaustive) filtration of the ring $Q_\kappa(R)$ making the canonical localization morphism $j_\kappa : R \to Q_\kappa(R)$ into a filtered morphism, which need not be strict.*

Proof. We only have to show that for $y \in F_pQ_\kappa(R), z \in F_qQ_\kappa(R)$, we have that $yz \in F_{p+q}Q_\kappa(R)$, the other properties of an exhaustive filtration are easily checked. Say $I \in \mathcal{L}(\kappa)$ is such that $F_\gamma Iy \subset F_{\gamma+p}\overline{R}$ for all $\gamma \in \mathbb{Z}$, $J \in \mathcal{L}(\kappa)$ is such that $F_\mu Jz \subset F_{\mu+q}\overline{R}$ for all $\mu \in \mathbb{Z}$. Let $I_1 \in \mathcal{L}(\kappa)$ be the set $\{x \in R,\ xy \in \overline{R}\}$. Now look at $(\overline{J} : y) = \{x \in R,\ xy \in \overline{J}\}$, then $(\overline{J} : y) \subset I_1$ and for every $i \in I_1$ we have that $(J : iy) = \{r \in R,\ riy \in \overline{J}\} \in \mathcal{L}(\kappa)$. We also have that $(J : iy)iy \subset \overline{J}$ hence $(J : iy)i \subset (\overline{J} : y)$, or $(J : iy) \subset ((\overline{J} : y) : i)$. By idempotency of κ it follows that $(\overline{J} : y) \in \mathcal{L}(\kappa)$ and hence $H = (\overline{J} : y) \cap I$ in $\mathcal{L}(\kappa)$. $F_\gamma H \subset F_\gamma I$ for all γ thus $F_\gamma Hy \subset F_{\gamma+p}\overline{J}$ and thus

$$F_\gamma Hyz \subset F_{\gamma+p}\overline{J}z \subset F_{\gamma+p+q}\overline{R}.$$

In other words $yz \in F_{p+q}Q_\kappa(R)$.
It remains to verify that $FQ_\kappa(R)$ is separated, so assume that $z \in \cap_d F_dQ_\kappa(R)$. This means that for any d there is an $I \in \mathcal{L}(\kappa)$ such that $F_\gamma Iz \subset F_{\gamma+d}\overline{R}$, in particular for $d < -n$. We have for $\gamma = n$ that $Iz \subset F_{n+d}\overline{R} \subset F_{-1}\overline{R} = 0$ with $I \in \mathcal{L}(\kappa)$. But $z \in Q_\kappa(R)$ then entails that $z = 0$ and $FQ_\kappa(R)$ is indeed separated. □

In order for the localization morphism $j_\kappa : R \to Q_\kappa(R)$ to be strict, we would have to have that $\overline{R} \cap F_mQ_\kappa(R) = F_m\overline{R}$. However, $F_\gamma Ix \subset F_{\gamma+m}\overline{R}$ is possible for $x \notin F_m\overline{R}$. For example, take $x \in F_{m+1}\overline{R} \setminus F_m\overline{R}$ (such m exists by separability) with $\max\{\gamma, m+1\} < \gamma + m$, then $F_\gamma Ix \subset F_{\max\{\gamma,m+1\}}\overline{R} \subset F_{\gamma+m}\overline{R}$ with $x \notin F_m\overline{R}$.

In the above situation we now consider an R-module M with discrete filtration of finite length $m+n$ say, starting at $-m$ and reaching M at $F_nM = M$. Then $\overline{M} = M/\kappa(M)$ is discrete too and hence $F\overline{M}$ is separated as the induced filtration by FM. Define

$$F_dQ_\kappa(M) = \{q \in Q_\kappa(M) \mid \exists I \in \mathcal{L}(\kappa) \text{ such that } F_\gamma Iq \subset F_{\gamma+d}\overline{M}\}.$$

Proposition 7.5.5. *With notation as above: $FQ_\kappa(M)$ is a separated (exhaustive) filtration and $Q_\kappa(M)$ is a filtered $Q_\kappa(R)$-module with respect to the quotient filtration on $Q_\kappa(R)$ as in Proposition 7.5.4.*

Proof. We just point out the separability; the other properties are easily checked just as in Proposition 7.5.4. So look at $q \in \cap_d F_d Q_\kappa(M)$. For $d = -n - m - 1$, $\forall\gamma$, $F_\gamma I q \subset F_{\gamma-n-m-1}\overline{M}$ for some $I \in \mathcal{L}(\kappa)$. For $\gamma = n$, we obtain that $Iq \subset F_{-m-1}\overline{M} = 0$, hence $q = 0$. □

Remark 7.5.6. We actually prove in the above propositions that $Q_\kappa(R)$ and $Q_\kappa(M)$ are both discrete filtered ($F_{-n-1}Q_\kappa(R) = 0, F_{-n-m-1}Q_\kappa(M) = 0$).

Again looking at a finite ring-filtered ring R, we now consider a glider representation $\Omega \supset M$ of finite essential length and zero body, say $M_m = 0$, $M_{m-1} \neq 0$. Then the filtration we introduced on RM starts at $F_{-m}RM = 0, F_{-m+1}RM = M_{m-1}$ and reaches RM at F_nRM, so it is discrete of length $m+n$. Put $\Omega = RM$ and consider $Q_\kappa(\Omega)$ with the quotient filtration $FQ_\kappa(\Omega)$ as defined above and we define $Q_\kappa(M) = F_0Q_\kappa(\Omega)$ as earlier in the chapter. Then $Q_\kappa(M)$ is a glider representation of finite essential length for $Q_\kappa(R)$ in $Q_\kappa(\Omega)$.

Now we specify the nature of κ with respect to $S = F_0R$. We assume that κ induces a localization κ_S on S via $I_1 \in \mathcal{L}(\kappa_S)$ if $I_1 = I \cap S$ for some $I \in \mathcal{L}(\kappa)$ and for any $J \in \mathcal{L}(\kappa)$, $R(J \cap S) \in \mathcal{L}(\kappa)$. We also assume that S is (left) Noetherian (in particular this applies to a tower of coordinate rings $K[W] \subset \ldots \subset K[V]$). Under this assumption, κ_S and also κ are of finite type, that is, $\mathcal{L}(\kappa_S)$ and $\mathcal{L}(\kappa)$ have a filterbasis of finitely generated left ideals. We say that the glider $\Omega \supset M$ is (κ, κ_S)-orthogonal if $\Omega/F_d\Omega$ is κ_S-torsion free for all d.

Proposition 7.5.7. *If M is a glider of finite essential length over the finite ring-filtered ring R, $\Omega = RM \supset M$ and κ, κ_S are symmetric kernel functors on R-mod, resp. S-mod such that M is (κ, κ_S)-orthogonal, then the separated filtration on $Q_\kappa(\Omega)$ with $F_0Q_\kappa(\Omega) = Q_\kappa(M)$ induces the filtration $F\overline{\Omega}$ on $\overline{\Omega}$. Similarly, if R is (κ, κ_S)-orthogonal then $FQ_\kappa(R)$ induces $F\overline{R}$, i.e. $j : R \to Q_\kappa(R)$ is strict filtered.*

Proof. We write the proof for M, Ω; the proof for R is similar. Take $\overline{q} \in F_dQ_\kappa(\Omega) \cap \overline{\Omega}$, i.e. there is an $I \in \mathcal{L}(\kappa)$ such that $F_\gamma I\overline{q} \subset F_{\gamma+d}\overline{\Omega}$ for all γ. In particular for $\gamma = 0$, we have that $(I \cap S)\overline{q} \subset F_d\overline{\Omega}$ with $I \cap S \in \mathcal{L}(\kappa_S)$. Pick a representative $q \in \Omega$ for $\overline{q}$, then $(I \cap S)q \subset F_d\Omega + \kappa_S(\Omega)$. Since S is Noetherian, $I \cap S$ is $\sum_{j=1}^t Si_j$ for some finite $i_j \in I$. Then $i_jq \in F_d\Omega + \kappa(\Omega)$, say $i_jq = f_{d,j} + t_j$. Take $J_i \in \mathcal{L}(\kappa)$ such that $J_it_i = 0$, then $(S \cap J_i)i_jq \subset F_d\Omega$ and it follows that $(\sum(S \cap J_i)i_j)q \subset F_d\Omega$. Since κ_S is symmetric, $\sum(S \cap J_i)i_j$ in $\mathcal{L}(\kappa_S)$. So if $\Omega/F_d\Omega$ is κ_S-torsion free, $q \in F_d\Omega$ and $\overline{q} \in F_d\overline{\Omega}$. Thus $\overline{\Omega} \hookrightarrow Q_\kappa(\Omega)$ is strict and $j : \Omega \to Q_\kappa(\Omega)$ is strict too. In particular, the fragment structure of $F_0Q_\kappa(\Omega) = Q_\kappa(M)$ induces the chain of the fragment M. □

Corollary 7.5.8. *Let $K[W] \subset \ldots \subset K[V_i] \subset \ldots \subset K[V] = K[V_n]$ be a tower of coordinate rings for a chain of varieties $V \to \cdots \to V_i \to \cdots \to W$ and $K(V_i) \cap K[V] = K[V_i]$, say all $K[V_i]$ are Noetherian domains, then every κ inducing κ_W as before is such that $K[V]$ is (κ, κ_W)-orthogonal; hence the discrete quotient filtration on $Q_\kappa(K[V])$ induces the finite ring filtration on $K[V]$.*

Proof. Let $z \in K[V]$ be such that $(I \cap K[W])z \subset K[V_i]$ for some $i < n$. Then for some $a \in I \cap K[W]$ we have that $az \in K[V_i]$ or that $z \in K(V_i)$. Hence $z \in K(V_i) \cap K[V] = K[V_i]$. □

Remark 7.5.9. The condition concerning the function fields of the varieties rules out any birationalities in the chain of varieties.

7.6 Projective aspects of gliders

From examples at the end of the foregoing section it is clear that relations between modules over rings of sections over Zariski opens in Spec(C) may be expressed in terms of fragments. It is not hard to see that this may be extended to covers by affine subvarieties of some abstract variety or scheme, e.g. the scheme Proj(R) for some positively graded ring R. Further generalization to non-commutative Proj(A) may be obtained and then we look at the micro-structure sheaf of a Zariski filtered ring; in fact it is the non-commutative projective scheme of the Rees ring of the considered filtration that has to be studied. Then fragment or glider theory will be connected to so called quantum non-commutative geometry of Proj as introduced in [61] and [69]. Those more geometric considerations are the topic of future research, but the projective aspects of glider theory are already apparent via links to filtrations over suitably defined rings, i.e. the so called "modulizers" of the glider. We present the initial theory here and will establish the relation with some localizations, themselves representing algebraic descriptions of a global section functor when interpreted geometrically, as encountered in the foregeoing section.

Let us start by setting the scene by introducing an example we call normalizing birationality. The non-commutativity version of this is a research topic in its own right. The relation between the Witten algebra $W_q(\mathfrak{sl}_2)$ and the Woronowicz quantum enveloping algebra $V_q^w(\mathfrak{sl}_1)$ mentioned in example g) in Section 3.3. is a typical case of this. Let k be a field and A, B k-algebras. Consider a central element t of A and a normalizing element z of B (that means $Bz = zB$), such that there exists a k-algebra isomorphism $A[t^{-1}] \cong B[z^{-1}]$ given by $\Psi : A[t^{-1}] \to B[z^{-1}]$. We may identify B and $\Psi^{-1}(B) \subset A[t^{-1}]$ so that we may assume $B \subset A[t^{-1}]$ and $B[z^{-1}] = A[t^{-1}]$ where z is still normalizing in B. On $A[t^{-1}]$ we have the t-adic filtration $F(t^{-1})$ given by:

$$A[t^{-1}] \supset \ldots \supset At^{-i} \supset \ldots \supset At^{-1} \supset A \supset At \supset At^2 \supset \ldots$$

This induces on B the filtration $F(t^{-1}) \cap B$, the induced filtration. It is readily verified that A is now an $F(t^{-1})$-fragment of $A[t^{-1}]$ and hence also an $(F(t^{-1}) \cap B)$-fragment for B. In formally a similar way we may define $F(z^{-1})$ on $B[z^{-1}]$ and $F(z^{-1}) \cap A$. These filtrations are compatible in the way expressed in the following diagram

$$\begin{array}{ccccccc}
B[z^{-1}] = A[t^{-1}] \supset \ldots \supset & At^{-i} & \supset \ldots \supset At^{-1} \supset & A & \supset At \supset & \ldots \supset At^i \supset & \ldots \\
\cup & \cup & & \cup & & \cup & \\
\vdots & \vdots & & \vdots & & \vdots & \\
\cup & \cup & & \cup & & \cup & \\
Bz^{-j} \supset \ldots \supset & Bz^{-j} \cap At^{-i} \supset \ldots & \ldots \supset & Bz^{-j} \cap A & \supset \ldots \supset & \ldots \supset Bz^{-j} \cap At^i \supset & \ldots \\
\cup & \cup & & \cup & & \cup & \\
\vdots & \vdots & & \vdots & & \vdots & \\
\cup & \cup & & \cup & & \cup & \\
B \supset \ldots \supset & B \cap At^{-i} \supset \ldots & \ldots \supset & B \cap A & \supset \ldots \supset & \ldots \supset \cap At^i \supset & \ldots \\
\cup & \cup & & \cup & & \cup & \\
\vdots & \vdots & & \vdots & & \vdots & \\
\cup & \cup & & \cup & & \cup & \\
Bz^n \supset \ldots \supset & Bz^n \cap At^{-i} \supset \ldots & \ldots \supset & Bz^n \cap A & \supset \ldots \supset & \ldots \supset Bz^n \cap At^i \supset & \ldots \\
\cup & \cup & & \cup & & \cup & \\
\vdots & \vdots & & \vdots & & \vdots &
\end{array}$$

In this diagram A and B may be replaced by modules ${}_AM$, resp. ${}_BN$, such that $A[t^{-1}] \otimes_A {}_AM \cong B[z^{-1}] \otimes_B {}_BN$, with filtrations induced in the obvious ways by the t-adic, resp. z-adic filtration. In some practical examples we actually have $z^{-1} \in A$ and $t^{-1} \in B$ and then $AB = BA = A[t^{-1}] = B[z^{-1}]$. This is for example the case if $A = K[t]$, $B = K[t^{-1}]$ appearing as a cover for $\mathbb{P}_1(k) \cong \mathrm{Proj}(\mathrm{k}[\mathrm{X}, \mathrm{Y}])$, but also in more complicated examples, like the normalizing birationality for $W_q(\mathfrak{sl}_2)$ and $V_q^w(\mathfrak{sl}_1)$ mentioned before. In the latter case we really only have a normalizing non-central element which entails interpretations in terms of some twisting automorphisms.

7.6.1 Glider modulizers and rings of definition

Let R be a positively filtered ring with filtration FR, $S = F_0R$. Consider an FR-glider representation $\Omega \supset M$ given by its chain: $M_0 = M \supset M_1 \supset \ldots$. We generalize the theory started in [53] for gliders here; in loc. cit. it was developed for natural fragments only. Define for $d \in \mathbb{Z}$,

$$F_d^*(R) = \{x \in R \mid xM_{n+d} \subset M_n, \text{ for } n \in \mathbb{N} \text{ such that } n + d \geq 0\}.$$

Proposition 7.6.1. *With notation and assumptions as above:*

1. *If we define F_+^* by $F_{+,d}^*R = F_d^*(R)$ for positive d, then F_+^*R is a positive filtration on the ring R such that $FR \subset F_+^*R$ and M is an F_+^*R-glider representation with respect to Ω and the same chain: $M_0 = M \supset M_1 \supset \ldots$;*

2. *Put $F^*_{-,d}R = F^*_d(R)$ for negative d. Then F^*_- defines a filtration on $S^* = F^*_0(R) = \{x \in R \mid xM_n \subset M_n$ for $n \geq 0\}$ such that M is a filtered left S^*-module with filtration given by the chain $M_0 = M \supset M_1 \supset \dots$;*

3. *The ascending chain F^*R makes R into a filtered right S^*-module via the filtration F^*_- on S^*.*

Proof. 1. For $n, m \in \mathbb{N}$ look at $x \in F^*_n(R), y \in F^*_m(R)$, then for every $l \in \mathbb{N}$ we have: $xM_{l+n} \subset M_l, yM_{l+m} \subset M_l$. Thus

$$xyM_{m+n+l} \subset xM_{l+n} \subset M_l,$$

hence $xy \in F^*_{n+m}(R)$. It is obvious that $1 \in F^*_0(R) = S^*$ and also that $F_nR \subset F^*_{+,n}(R)$ for every $n \in \mathbb{N}$. Thus F^*_+R is indeed a positive filtration on R dominating the filtration FR.
Next, for $n, m \in \mathbb{N}$ we have that $F^*_{+,n}RM_m \subset M_{m-n}$ if $m \geq n$; thus the chain $M = M_0 \supset M_1 \supset \dots$ also defines a glider structure on M for F^*_+R. Note that all M_i are S^*-modules by definition of $S^* = F^*_0R$.
2. Look at $x \in F^*_{-d}(R), y \in F^*_{-e}(R)$ with $e, d \geq 0$. For $n \geq d + e$ we have : $xyM_{n-d-e} \subset xM_{n-d} \subset M_n$; thus we obtain that $xy \in F^*_{-(d+e)}R$, so F^*_-R defines a filtration on the ring S^*, it is a negative filtration. Now defining FM by putting $F_{-n}M = M_n$ for $n \geq 0$ we derive from $F^*_{-d}RM_n \subset M_{n+d}$ for all $n \geq 0$ that FM is a filtration of the left S^*-module M over F^*_-R.
3. For $i, d \geq 0$ and $x \in F^*_iR, y \in F^*_{-d}R$ we distinguish two cases. First, for $i \geq d \geq 0$ we have that $xyM_{i-d+n} \subset xM_{i+n} \subset M_n$ for all $n \in \mathbb{N}$. Secondly, for $d \geq i \geq 0$ and $n + i - d \geq 0$ we have: $xyM_{n+i-d} \subset xM_{n+i} \subset M_n$. So in either case we obtain $xy \in F^*_{i-d}R$, thus $F^*_iRF^*_{-d}R \subset F^*_{i-d}R$ and this makes R into a filtered right S^*-module with respect to $F^*_-S^*$. Obviously R is also a left filtered F^*_+R-module. □

Observation 7.6.2. Note that F^*R does not make R into a filtered ring even if $F^*_0R = S^*$ is a filtered ring with respect to F^*_-R and R is filtered as a ring by F^*_+R. The obstruction for F^*R to be a ring filtration is that in the foregoing proof the multiplication yxM_{n+i-d} is not necessarily defined unless $n \geq d$ but in fact we have to deal with $n \geq d - i$. This asymmetry assigns a sign to each side, that is $-$ on the right and $+$ on the left! If R is commutative however, F^*R makes R into a filtered ring. Indeed the missing condition $F^*_{-d}RF^*_iR \subset F^*_{i-d}R$ for $d \geq 0, i \geq 0$ now follows from the proof given because the order of multiplication may be reversed. The $+$ and $-$ cannot necessarily be gathered on the same side in case R is non-commutative. Here, surprisingly, some projective aspect of the theory will show up. First we add a note on the separatedness of F^*_-R.

Proposition 7.6.3. *With notations as before:*

$$\cap_d F^*_dR = \{x \in S^* \mid xM \subset B(M)\},$$

a left ideal of R contained in S^*.

Proof. If $x \in \cap_d F^*_d R$, then $xM_0 = M_{n-n} \subset M_n$ because $x \in F^*_{-n}R$, for every $n \geq 0$, thus $xM \subset B(M)$ follows. Conversely if $zM \subset B(M)$ then $z \in F^*_d R$ for every d is obvious. □

By reducing to the spirit $M/B(M) \subset \Omega/B(M)$ we have defined for the glider $M/B(M)$ the F^*R with $\cap_d F^*_d R$ being the annihilator $\mathrm{ann}_{S^*}\mathrm{M}$, hence the F^*_- is a separated filtration on S^* if and only if M is a faithful S^*-module.

For settling the problem about F^*R not being a ring filtration on R we now define a weaker version of F^*R as follows:

$$F^{(*)}_d R = \{x \in R \,|\, xM_{n+d} \subset M_n \text{ for all } n \geq n_0 \text{ for some } n_0 \in \mathbb{N} \text{ and } n+d \geq 0\}.$$

Note that the $n_0 \in \mathbb{N}$ is depending on the d considered. The $F^{(*)}_d R$ build an ascending chain: $\ldots \subset F^{(*)}_d R \subset F^{(*)}_{d+1} R \subset \ldots$. Indeed, if $x \in F^{(*)}_d R$ then $xM_{n+d} \subset M_n$ for $n \geq n_0, n+d \geq 0$. Since $M_{n+d+1} \subset M_{n+d}$ we also have $xM_{n+d+1} \subset M_n$ for $n \geq n_0$ and $n+d \geq 0$ so $n+d+1 \geq 0$ too, or $x \in F^{(*)}_{d+1} R$. So we established that the n_0 chosen for x in $F^{(*)}_d R$ actually works for x in all larger $F^{(*)}_m R$.

Proposition 7.6.4. *The chain $F^{(*)}R$ makes R into a filtered ring.*

Proof. Look at $x \in F^{(*)}_d R, y \in F^{(*)}_e R$. Then we have $xM_{n+d} \subset M_n$ for $n \geq n_0, yM_{m+e} \subset M_m$ for $m \geq m_0$ for some $n_0, m_0 \in \mathbb{N}$. Now look at xyM_{d+e+n}, then first $yM_{d+e+n} \subset M_{d+n}$ for some $d+n \geq m_0$, thus $n \geq m_0 - d$, secondly $xM_{d+n} \subset M_n$ for $n \geq n_0$. Hence for $n \geq \max\{m_0 - d, n_0\}$ we now have that $xyM_{d+e+n} \subset M_n$ or $xy \in F^{(*)}_{d+e} R$. As obviously $1 \in F^{(*)}_0 R$ we established that $F^{(*)}R$ is a ring filtration of R. □

Observation 7.6.5. Now we know $F^{(*)}R$ is a ring filtration but now M need not be an $S^{(*)}$-module. We call S^* the modulizer of M with respect to FR and $S^{(*)}$ is the ring of definition of M. Modulizers generalize to some extent the idea of an idealizer of some subset of a ring. They have an independent ring theoretical interest and may deserve further research. For example, let S and R be Noetherian rings and $M \subset \Omega$ a finitely generated S-module. If Ω is faithful as an R-module, is then S^* a Noetherian ring? In fact without the faithfulness assumption the answer is negative; there is an easy counterexample over a ring of differential operators mentioned to us by K. Goodearl.

Corollary 7.6.6. *1. If $\widetilde{S^*}$ is the Rees ring of S^* with respect to $F^*_- S^*$ then the Rees right module $\widetilde{R^*}$ with respect to F^*R is a graded right $\widetilde{S^*}$-module. On the other hand $\widetilde{R^*}$ is also the Rees left module with respect to $F^*_+ R$ hence a graded $\widetilde{R^*_+}$-module where $\widetilde{R^*_+}$ is in fact a ring and it is the Rees ring of R with respect to the filtration $F^*_+ R$;*

2. *Put $G(R) = G_{FR}(R), G^*(S^*) = G_{F^*_- S^*}(S^*), G^*(R) = G_{F^*R}(R)$. Then it is clear that $G^*(S^*) = G^*(R)_{\leq 0} = \oplus_{d \leq 0} G^*(R)_d$. Moreover $G^*(R)_{\geq 0}$ is the associated graded ring with respect to $F^*_+ R$ and $G^*(R)$ is a graded right $G^*(S^*)$-module as well as a graded left $G^*(R)_{\geq 0}$-module. In other words, the chain F^*R defines on $G^*(R)$ the structure of a graded $(G^*(R)_{\geq 0} - G^*(R)_{\leq 0})$-bimodule;*

3. *To M we may associate the S-module $g(M) = \oplus_{i \geq 0} M_i/M_{i+1}$ which is in a canonical way an $fG(R)$-fragment where $fG(R)$ is the gradation filtration of $G(R)$. Now the second part of Proposition 7.6.1 entails that $g(M) = G^*(M)$ is the associated graded $G^*(S^*)$-module corresponding to the filtration of M as a left S^*-module. Obviously $G^*(M)$ is filtered with respect to the grading filtration $G^*(R)_{\leq 0}$; moreover $g(M) = G^*(M)$ is an $fG^*(R)_{\geq 0}$-fragment with $f_p G^*(R)_{\geq 0} = \oplus_{0 \leq m \leq p} G^*(R)_m$ (see Lemma 1.9.1).*

From the foregoing it is clear that fragmented structures combined with other filtered or graded structures do appear naturally in the study of gliders (and fragments). For example this also happens in fragments studied in connection with the valuation theory (or non-commutative generalizations thereof).

7.6.2 Closures of filtrations, rings and gliders

Consider a Noetherian positively graded K-algebra A, $A = K \oplus A_1 \oplus A_2 \oplus \cdots$, with gr-maximal ideal $A_+ = \oplus_{i>0} A_i$. Then a graded A-module has finite length if and only if it can be annihilated by a power of A_+. The Gabriel filter $\mathcal{L}(\kappa_+)$ defined as the set of left ideals containing a power of A_+ defines a kernel functor (torsion theory) κ_+ and a quotient category of the category of graded A-modules with respect to the localizing subcategory of those graded A-modules of finite length, the latter quotient category is (A, κ_+)-gr, cfr. [55]. Objects of (A, κ_+)-gr having a finitely generated representative in A-gr form a full subcategory $(A, \kappa_+)^f$-gr and one may define $\mathrm{Proj(A)} = (\mathrm{A}, \kappa_+)^{\mathrm{f}}$-gr and study some projective geometry for the non-commutative algebra A, cfr. [5], [61], [69]. Here we avoid for the moment the use of localizations at κ_+ but we define corresponding closure operations intrinsically. Everything could be interpreted in terms of Proj, in some cases i.e. for κ_+, but we will not insist on this and view the projective aspects as encoded in the κ-closure operations we define.

As before we consider a glider $M \subset \Omega$ with respect to FR and $F^*R, F^{(*)}R$ as defined earlier. For $n \in \mathbb{Z}$ we define

$$\mathrm{cl}_\kappa(\mathrm{F}^*_\mathrm{n}\mathrm{R}) = \{\mathrm{x} \in \mathrm{R} \mid \mathrm{F}^*_\gamma(\mathrm{I})\mathrm{x} \subset \mathrm{F}^*_{\mathrm{n}+\gamma}\mathrm{R} \text{ for all } \gamma \geq \gamma_0 \geq 0 \text{ for some } \gamma_0 \in \mathbb{N} \text{ and some } \mathrm{I} \in \mathcal{L}(\kappa)\},$$

where we wrote $F^*_\gamma(I) I \cap F^*_\gamma R$ and κ any kernel functor on R-mod. In particular for κ being trivial, that is $\mathcal{L}(\kappa) = \{R\}$ we have

$$\mathrm{cl}(\mathrm{F}^*_\mathrm{n}\mathrm{R}) = \{\mathrm{x} \in \mathrm{R} \mid \mathrm{F}^*_\gamma \mathrm{R}\mathrm{x} \subset \mathrm{F}^*_{\mathrm{n}+\gamma}\mathrm{R} \text{ for all } \gamma \geq \gamma_0 \geq 0 \text{ for some } \gamma_0 \in \mathbb{N}\}.$$

For Ω and $n \geq 0$ we define now:

$$F_n^*(\Omega) = \{\omega \in \Omega \mid F_{-n-t}^* R\omega \subset M_t \text{ for } t \geq 0\}.$$

For $n \leq 0$ we take $F_n^*\Omega = M_{-n}$. Then it is clear that Ω is a filtered S^*-module with filtration given by the chain $F^*\Omega$. Indeed, if $\omega \in F_n^*\Omega$ with $n \geq 0$ then $\omega \in M_{-n}$ and for $d \geq 0, F_{-d}^* R\omega \subset M_{-n+d}$ because of Proposition 7.6.1.2. If however $n \geq 0$ then look at $F_{-d}^* R\omega$ and $F_{-n+d-t}^* RF_{-d}^* R\omega$. We have to check that this is in M_t. First if $-n+d-t \leq 0$ then $F_{-n+d-t}^* RF_{-d}^*$ is in $F_{-n-t}^* R$ for $n, t \geq 0$ hence $F_{-n-t}^* R\omega \subset M_t$ for $t \geq 0$ as $\omega \in F_n^*\Omega$ yields the claim. Secondly, if $-n+d-t \geq 0$ then from 3. in Proposition 7.6.1 we retain that $F_{-n+d-t}^* RF_{-d}^* R \subset F_{-n-t}^* R$ and the conclusion follows as in the first case.

For any kernel functor κ on R-mod we now define:

$$\mathrm{cl}_\kappa(\mathrm{M_n}) = \{\omega \in \Omega \mid \mathrm{F}_\gamma^*(\mathrm{I})\omega \subset \mathrm{F}_{\gamma-\mathrm{n}}^*\Omega \text{ for } \gamma \geq \gamma_0 \text{ and some } \mathrm{I} \in \mathcal{L}(\kappa)\},$$

$$\mathrm{cl}_\kappa(\mathrm{F_n^*}\Omega) = \{\omega \in \Omega \mid \mathrm{F}_\gamma^*(\mathrm{I})\omega \subset \mathrm{F}_{\gamma+\mathrm{n}}^*\Omega \text{ for } \gamma \geq \gamma_0 \text{ and some } \mathrm{I} \in \mathcal{L}(\kappa)\}.$$

For the trivial Gabriel filter $\mathcal{L}(\kappa) = \{R\}$ this leads to:

$$\mathrm{cl}(\mathrm{F_n^*R}) = \{\mathrm{x} \in \mathrm{R} \mid \mathrm{F}_\gamma^*\mathrm{Rx} \subset \mathrm{F}_{\mathrm{n}+\gamma}^*\mathrm{R} \text{ for } \gamma \geq \gamma_0 \geq 0 \text{ for some } \gamma_0 \in \mathbb{N}\},$$

$$\mathrm{cl}(\mathrm{M_n}) = \{\omega \in \Omega \mid \mathrm{F}_\gamma^*\mathrm{R}\omega \subset \mathrm{F}_{\gamma-\mathrm{n}}^*\Omega \text{ for some } \gamma_0 \in \mathbb{N}, \text{ all } \gamma \geq \gamma_0\},$$

$$F_n^{(*)}\Omega =: \mathrm{cl}(\mathrm{F_n^*}\Omega) = \{\omega \in \Omega \,\big|\, \mathrm{F}_\gamma^*\mathrm{R}\omega \subset \mathrm{F}_{\gamma+\mathrm{n}}^*\Omega \text{ for some } \gamma_0 \in \mathbb{N}, \text{ all } \gamma \geq \gamma_0\},$$

$$\mathrm{cl}(\mathrm{S}^*) = \mathrm{cl}(\mathrm{F_0^*R}) = \{\mathrm{x} \in \mathrm{R} \mid \mathrm{F}_\gamma^*\mathrm{Rx} \subset \mathrm{F}_\gamma^*\mathrm{R} \text{ for all } \gamma \geq \gamma_0 \geq 0 \text{ for some } \gamma_0 \in \mathbb{N}\}.$$

Proposition 7.6.7. *With notation as above we obtain:*

1. *For* $d \in \mathbb{Z}, F_d^{(*)}R \subset \mathrm{cl}_\kappa(\mathrm{F_d^*R})$, *in particular we have* $S^{(*)} \subset \mathrm{cl}(\mathrm{S}^*) \subset \mathrm{cl}_\kappa(\mathrm{S}^*)$. *Moreover* $\mathrm{cl}_\kappa(\mathrm{S}^*)$ *is a filtered ring with respect to* $\mathrm{cl}_\kappa(\mathrm{F}_-^*)$ *on* $\mathrm{cl}_\kappa(\mathrm{S}^*)$.
2. *The chain* $\mathrm{cl}_\kappa\mathrm{F}^*$ *makes* R *into a filtered ring and into a* $\mathrm{cl}_\kappa\mathrm{F}_-^*$*-filtered left* $\mathrm{cl}_\kappa(\mathrm{S}^*)$*-module.*

Proof. 1. Take $y \in F_d^{(*)}R$, say $yM_{n+d} \subset M_n$ for all $n \geq n_0$ and $n+d \geq 0$. Then $F_{n_0}^* RyM_{t+n_0+d} \subset M_t$ for all $t \geq 0$, hence $F_{n_0}^* Ry \subset F_{n_0+d}^* R$ with $n_0 + d \geq 0$. We may repeat this argument for any $n_1 \geq n_0$ which leads to $F_{n_1}^* Ry \subset F_{n_1+d}^* R$. Since $R \in \mathcal{L}(\kappa)$ the foregoing yields $y \in \mathrm{cl}_\kappa(\mathrm{F_d^*R})$, thus we have $F_d^{(*)}R \subset \mathrm{cl}_\kappa(\mathrm{F_d^*R})$ for every d. Looking at $d = 0$ we find

$$S^{(*)} \subset \mathrm{cl}(\mathrm{S}^*) \subset \mathrm{cl}_\kappa(\mathrm{S}^*),$$

where the first inclusion follows by taking κ to be trivial in the foregoing and the second inclusion is obvious since $R \in \mathcal{L}(\kappa)$. Next look at $\gamma, \delta \geq 0$ and choose $x \in \mathrm{cl}_\kappa(\mathrm{F}_{-\gamma}^*\mathrm{R}), \mathrm{y} \in \mathrm{cl}_\kappa(\mathrm{F}_{-\delta}^*\mathrm{R})$. Then for some $I \in \mathcal{L}(\kappa)$ we have

$F^*_n(I)x \subset F^*_{n-\gamma}R$ for $n \geq n_0 \geq 0$ and for some $J \in \mathcal{L}(\kappa), F_m(J)y \subset F^*_{m-\delta}R$ for all $m \geq m_0 \geq 0$. Since $\mathcal{L}(\kappa)$ is a Gabriel filter there is a $J' \in \mathcal{L}(\kappa)$ such that $J'x \subset J$ and then look at $I \cap J' \in \mathcal{L}(\kappa)$. For $n \geq \gamma + n_0 + m_0$ we easily calculate:

$$F^*_n(I \cap J')x \subset F^*_{n-\gamma}R \cap J = F^*_{n-\gamma}J,$$

hence

$$F^*_n(I \cap J')xy \subset F^*_{n-\gamma}Jy \subset F^*_{n-\gamma-\delta}R,$$

because $n - \gamma \geq n_0 + m_0 \geq m_0$. Thus this means that $xy \in \mathrm{cl}_\kappa(\mathrm{F}^*_{-\gamma-\delta}\mathrm{R}*)$ or that $\mathrm{cl}_\kappa(\mathrm{S}^*)$ is a filtered ring with respect to the $\mathrm{cl}_\kappa \mathrm{F}^*_-$.
2. A proof formally similar to the above establishes that R is a filtered ring with respect to $\mathrm{cl}_\kappa \mathrm{F}^*$; indeed the fact that $F^*_\gamma(H)xy \subset F^*_{\gamma+\alpha+\beta}R$ for some $H \in \mathcal{L}(\kappa)$, where we took $x \in \mathrm{cl}_\kappa(\mathrm{F}^*_\alpha \mathrm{R}), \mathrm{y} \in \mathrm{cl}_\kappa(\mathrm{F}^*_\beta \mathrm{R})$, has to be established only for large enough γ, so this allows us to take γ large enough to have that $\gamma + \alpha \geq n_0 + m_0$ so that all operations encountered are indeed well-defined. □

Proposition 7.6.8. *With notation as before we obtain: the chain* $\mathrm{cl}_\kappa(\mathrm{F}^*\Omega)$ *defines on* Ω *the structure of a filtered* R*-module with respect to* $\mathrm{cl}_\kappa \mathrm{F}^*$. *Moreover:*

$$\mathrm{cl}_\kappa(\mathrm{M}) = \mathrm{cl}_\kappa(\mathrm{F}^*_0\Omega) = \{\omega \in \Omega \mid \mathrm{F}^*_\gamma(\mathrm{I})\omega \subset \mathrm{F}^*_\gamma\Omega \text{ for } \gamma \geq \gamma_0 \text{ some } \gamma_0 \text{ and some } \mathrm{I} \in \mathcal{L}(\kappa)\}$$

is a filtered submodule of Ω *with respect to* $\mathrm{cl}_\kappa \mathrm{F}^*_-$ *and therefore* $\mathrm{cl}_\kappa(\mathrm{M})$ *is a* $\mathrm{cl}_\kappa \mathrm{F}^*_+$*-fragment.*

Proof. Take $\omega \in \mathrm{cl}_\kappa(\mathrm{F}^*_\mathrm{n}\Omega), \mathrm{r} \in \mathrm{cl}_\kappa(\mathrm{F}^*_\mathrm{m}\mathrm{R})$, say we have: $F_\gamma * (I)\omega \subset F^*_{\gamma+r}\Omega$ for $\gamma \geq \gamma_0$ and for some $I \in \mathcal{L}(\kappa)$, $F_\delta(H)r \subset F^*_{m-\delta}R$ for $\delta \geq \delta_0$ and some $H \in \mathcal{L}(\kappa)$. Taking now $\delta \geq \delta_0$ and $m + \delta \geq \gamma_0$ we may take $J \in \mathcal{L}(\kappa)$ such that $Jr \subset I$ and calculate for $\delta \geq \delta_0, \delta \geq \gamma_0 - m$:

$$F^*_\delta(H \cap J)r \subset F^*_{m+\delta}R \cap I = F^*_{m+\delta}(I),$$

and

$$F^*_\delta(H \cap J)r\omega \subset F^*_{m+\delta}(I)\omega \subset F^*_{m+\delta+n}\Omega.$$

Consequently we arrive at $r\omega \in \mathrm{cl}_\kappa(\mathrm{F}^*_{\mathrm{m+n}}\Omega)$. It is clear that $\mathrm{cl}_\kappa(\mathrm{M})$ is $\mathrm{cl}_\kappa(\mathrm{F}^*_0\Omega)$ and $\mathrm{cl}_\kappa(\mathrm{M_n}) = \mathrm{cl}_\kappa(\mathrm{F}^*_{-\mathrm{n}}\Omega)$ for all $n \geq 0$. Hence $\mathrm{cl}_\kappa(\mathrm{M})$ is certainly a filtered $\mathrm{cl}_\kappa(\mathrm{S}^*)$-module with respect to the filtration $\mathrm{cl}_\kappa \mathrm{F}^*_-$ on $\mathrm{cl}_\kappa(\mathrm{S}^*)$. The final statement follows directly from it because $\mathrm{cl}_\kappa(\mathrm{M})$ is the negative part of the filtered R-module Ω with respect to the ring filtration $\mathrm{cl}_\kappa \mathrm{F}^*\mathrm{R}$ on R. □

Remark 7.6.9. The appearance of κ allows us to relate this theory to the geometric theory in Section 7.2 to Section 7.4. In fact, then one can look at κ coming from some $\widetilde{\kappa}$ on $\widetilde{R^*}$-gr where $\widetilde{R^*}$ is the Rees ring with respect to the positive filtration F^*_+R of R, which also yields a $\overline{\kappa}$ on $G(R)$-gr, the associated graded ring $G(R)$ stemming from F^*_+R. We do not go into the ramifications of considering κ which are induced from a kernel functor on S-mod (or S^*-mod) and investigate some relations with respect to microlocalizations and quantum sections. A first development of microlocalization and quantum sections for gliders (it was done first for natural fragments in [53] but it can easily be generalized to glider representations) is in loc. cit. We leave all this for further research here.

Let us finish this section by some special result for natural gliders.

Proposition 7.6.10. *With notation and assumptions as before but now assuming that M is a natural glider, then we have* $\mathrm{clF}^* = \mathrm{F}^{(*)}$ *and* $F^{(*)}\Omega$ *makes* Ω *into an* $F^{(*)}$*-filtered module such that* $\mathrm{cl(M)} = \mathrm{F}_0^{(*)}\Omega$ *is an* $F_+^{(*)}R$*-glider.*

Proof. It will follow directly from Proposition 7.6.8 if we establish that $\mathrm{cl(F^*)} = \mathrm{F}^{(*)}$. We have already seen that $F^{(*)} \subset \mathrm{cl(F^*)}$, so look at $x \in \mathrm{cl(F_n^*R)}$. This means that for all $\gamma \geq \gamma_0$, for some γ_0, $F_\gamma^* Rx \subset F_{n+\gamma}^* R$. Consider $xM_{\gamma+n}$, then we have:

$$F_\gamma^* RxM_{\gamma+n} \subset F_{n+\gamma}^* RM_{\gamma+n}$$

for $\gamma \geq \gamma_0$. If we also have that $\gamma + n \geq 0$ then $F_\gamma^* RxM_{\gamma+n} \subset M$. Since M is natural this implies that $xM_{\gamma+n} \subset M_\gamma$ because $F_\gamma^* RxM_{\gamma+n} \subset M$ also entails that $F_\gamma RxM_{\gamma+n} \subset M$ and thus $xM_{\gamma+n} \subset M_\gamma$. But $xM_{\gamma+n} \subset M_\gamma$ for large γ yields $x \in F_n^{(*)}R$. □

7.6.3 Sections of affine type and good filtrations

In this section we consider a positively filtered ring R with $S = F_0R$ and $G(R) = \bigoplus_{n\in\mathbb{N}} F_nR/F_{n-1}R$ for its associated graded ring. We consider a multiplicatively closed set X in R such that $o \notin X, 1 \in X$ and $\sigma(X)$ is an Ore set in $G(R)$. Recall that $\sigma : R \to G(R)$ is the principal symbol map, that is $\sigma(x) = x \mod F_{n-1}R$ if $x \in F_nR - F_{n-1}R$ for some $n \in \mathbb{N}$. We have $0 \notin \sigma(X)$ since FR is separated and therefore σ is multiplicative on X, i.e. for $x_1, x_2 \in X$ we have: $\sigma(x_1x_2) = \sigma(x_1)\sigma(x_2)$. The set $X_{sat} = \{x \in R \mid \sigma(x) \in \sigma(X)\}$ is again a multiplicatively closed set in R and it is the maximal one with respect to $\sigma(Y) = \sigma(X)$. If $G(R)$ is a Noetherian ring then X_{sat} is also an Ore set in R (since then FR is a Zariskian filtration cfr [46]). In this case, the microlocalization of R at X, say $Q_X^\mu(R)$, equals the completion of the localization at X_{sat}, $(X_{sat}^{-1}R)^\wedge$. We will focus on natural FR-fragments M with $B(M) = 0$. It is clear at first glance that several cases may yield different results, for example: if $X \subset S$ or $X \cap S = \{1\}$, or neither of these hold.

Standing assumptions: We consider R positively filtered by FR and such that $G(R)$ is a Noetherian ring (thus we know that then FR is a Zariskian filtration and all F_nR are finitely generated S-modules, cf. [46]). Let X be an Ore set in R such that $\sigma(X)$ is an Ore set in $G(R)$ not concentrated in degree zero (i.e. $X \not\subset S$) and such that $G(R)$ is $\sigma(X)$-torsion free. M is a natural FR-fragment; $g(M)$ is as in Corollary 7.6.6.3. In this situation we may define $F^*\Omega$ as before

Proposition 7.6.7. Note that even when $g(M)$ is $\sigma(X)$-torsion free there need be no reason why $G^*(\Omega)$ should be $\sigma(X)$-torsion free too. There would be no harm in assuming that $g(M)$ and $G_F(R)$ are $\sigma(X)$-torsion free since one may reduce to that case by harmless assumptions. But since F^*R and $F^*\Omega$ depend on M in a less stringent way, it is not a priori clear one may reduce to the situation where $G^*(\Omega)$ is $\sigma(X)$-torsion free.

Proposition 7.6.11. *1. If $G(R)$ is $\sigma(X)$-torsion free then R is X-torsion free; moreover if $g(M)$ is $\sigma(X)$-torsion free (that means no nonzero $\overline{m}_i$ in $g(M)_i$ can be annihilated by an $\overline{s}_\gamma \in \sigma(X)$ for which multiplication $\overline{s}_\gamma\overline{m}_i$ is defined), then $t_X(M)$ given by*

$$t_X(M)_i = \{m \in M_i \mid xm = 0 \text{ for some } x \in X\}$$

is such that $t_X(M) \subset B(M)$ and $t_X(M)$ is a strict subfragment of M.

2. With standing assumptions and assuming that FR is a standard filtration, that is $R = S[F_1R]$ and $F_nR = (F_1R)^n$ for every $n \geq 1$. Define $F_n\Omega$ by $F_n\Omega = M_{-n}$ for $n \leq 0$ and $F_n\Omega = F_nRM \subset \Omega$ for $n > 0$. Then $F\Omega$ is an FR-filtration on Ω with $F_0\Omega = M$ and $G(\Omega)_{\leq 0} = g(M)$.

Proof. 1. For $r \in R, x \in X$ we have $\sigma(r) \neq 0, \sigma(x) \neq 0$ if $r \neq 0$, then $\sigma(x)\sigma(r) \neq 0$ if $G(R)$ is $\sigma(X)$-torsion free but this yields $\sigma(xr) = \sigma(x)\sigma(r)$ by definition of σ. Then $xr \neq 0$ or we arrive at R being X-torsion free. If $m \not\in B(M)$ then $m \in M_i - M_{i+1}$ for some $i \in \mathbb{N}$, so if $xm = 0$ for some $x \in X$ then $\sigma(x)\sigma(m) = 0$ because if not $\sigma(xm) = \sigma(x)\sigma(m) \neq 0$ would contradict $xm = 0$. Therefore if $g(M)$ is $\sigma(X)$-torsion free then $xm = 0$ leads to $m \in B(M)$ and thus $t_X(M) = t_X(B(M))$ is an R-module. That $t_X(M)$ with chain as given in the proposition is a strict subfragment of M is obvious, and if M is natural then so is $t_X(M)$.
2. Obvious verification using $F_nR = (F_1R)^n$. □

If M is a natural fragment, $M \subset \Omega = RM$, then we may view $S(M) = M/B(M)$ as a natural fragment in $\Omega/B(M)$ (an R-module) and also $RS(M) = \Omega/B(M)$. These remarks and the proposition above then lead to the following.

Corollary 7.6.12. *In the situation of Proposition 7.6.11.1 we have $g(M/t_X(M)) = g(M), B(M/t_X(M)) = B(M)/t_X(B(M))$ and $S(M/t_X(M)) = S(M)$. Moreover $g(S(M)) = g(M)$ and $S(M)$ is X-torsion free if $g(M)$ is $\sigma(X)$-torsion free. For natural M we may then look at $S(M)$ as being natural again.*

Remark 7.6.13. We will restrict attention (without loss of generality essentially) to natural fragments with $B(M) = 0$, thus $S(M) = M$. Assuming that $g(M)$ is $\sigma(X)$-torsion free then leads to $t_X(M) \subset B(M) = 0$, so M is X-torsion free. We have $M \subset \Omega = RM$ and since $t_X(\Omega) \cap M = 0$ we have an embedding $M \subset \overline{\Omega} = \Omega/t_X(\Omega)$ with $R.M = \overline{\Omega}$ where $R.M$ is taken in $\overline{\Omega}$.

Standing assumptions: We assume the same conditions as stated before Proposition 7.6.11 plus that FR is a standard filtration ($R = S[F_1R]$ and $F_nR = (F_1R)^n$ for $n \geq 0$). Moreover we assume that $\Omega = RM$ is X-torsion free; this is somewhat more general than $B(M) = 0$ (see previous remark) but in fact we also may take $B(M) = 0$. Note that this does not imply that $G(\Omega)$, with respect to $F\Omega$, or in fact $G^*(\Omega)$, with respect to $F^*\Omega$ introduced earlier, is $\sigma(X)$-torsion free!

First observe that now we do not know whether $F\Omega \subset F^*\Omega$. Recall that for $n \leq 0$ we defined $F_n^*\Omega = M_{-n}$ and for $n > 0$

$$F_n^*(\Omega) = \{\omega \in \Omega \mid F_{-n-t}^* R\omega \subset M_t \text{ for all } t \geq 0\}.$$

However, we can alter F^*R and $F^*\Omega$ slightly and obtain interesting new structures with better properties in fact. Define $F^\circ R$ by putting

$$F_d^\circ R = \{x \in R \mid xF_{n-d}\Omega \subset F_n\Omega \text{ for all } n\}.$$

Then

$$S^\circ = F_0^\circ R = \{x \in R \mid xF_n\Omega \subset F_n\Omega \text{ for all } n\},$$

so in particular $x \in S^\circ$ yields $xM_d \subset M_d$ by taking $n = -d$ in the foregoing. Also from $xF_{n-d}\Omega \subset F_n\Omega$ for $-n = m \geq 0$ and $n - d \leq 0$ thus $m + d \geq 0$, it follows that $xM_{m+d} \subset M_m$ for $m + d \geq 0$ and this just means $x \in F_d^*R$. So we established $F^\circ R \subset F^*R$. It is now easy to arrive at the following.

Proposition 7.6.14. *R with $F^\circ R$ is a filtered ring and $FR \subset F^\circ R \subset F^*R$.*

Proof. Since $F_dR \subset F_d^\circ R$ is clear, the remarks preceding the proposition yield the inclusions $FR \subset F^\circ R \subset F^*R$. To check that $F^\circ R$ makes R into a filtered ring, take $x \in F_n^\circ R, y \in F_y^\circ R$ and look at $xyF_{t-n-m}\Omega$. Then $yF_{t-n-m}\Omega \subset F_{t-n}\Omega$ because $y \in F_m^\circ R$ and this is for all t. Further, $xF_{t-n}\Omega \subset F_t\Omega$ for all t because $x \in F_n^\circ R$. Hence $xyF_{t-n-m}\Omega \subset F_t\Omega$ for all t and this yields that $xy \in F_{n+m}^\circ R$. It is obvious that $1 \in S^\circ$ and thus R is a filtered ring with respect to $F^\circ R$. □

We drop the assumption that M is natural in the following proposition, so we prove it for gliders.

Proposition 7.6.15. *If M is an FR-glider, then it is an F°-glider.*

Proof. Look at $F_d^\circ RM_i$ for $d \leq i$. If $d \geq 0$ then from $F_d^\circ R \subset F_d^*R$ we obtain $F_d^\circ RM_i \subset F_d^*RM_i \subset M_{i-d}$ because M is a fragment fpr F_+^*R as observed in Proposition 7.6.1. If $d \leq 0$ then $xF_{n-d}\Omega \subset F_n\Omega$ for all n, if $x \in F_d^\circ R$. For $n-d \leq 0$, thus also $n \leq 0$, say $n = -m$ with $m \geq 0$, this yields $xM_{d+m} \subset M_m$ for $d + m \geq 0$; thus M is an $F^\circ R$-fragment and an $F^\circ R$-glider if M was a glider (say with $\Omega = RM$). □

Now for a glider $M \subset \Omega$ with respect to FR we define for $n \in \mathbb{Z}$:

$$F_n^\circ\Omega = \{\omega \in \Omega \mid F_{-n-t}^\circ R\omega \subset M_t \text{ for all } t \geq 0\}.$$

Since $F^\circ R$ is an ascending chain, $F^\circ\Omega$ is an ascending chain and all $F_n^\circ\Omega$ are obviously S°-modules.

Proposition 7.6.16. *The chain $F^\circ\Omega$ makes the R-module Ω into a filtered $F^\circ R$-module such that for $n \leq 0$ we have $F_n^\circ\Omega = M_{-n}$, thus the fragment chain for M appears also as the negative part of the filtration $F^\circ\Omega$.*

Proof. Take $x \in F_d^\circ R, \omega \in F_n^\circ\Omega$, then for all $t \geq 0$ we have $F_{-n-t}^\circ R\omega \subset M_t$ and we want to check that $x\omega \in F_{n+d}^\circ\Omega$, i.e. that $F_{-n-d-t}^\circ Rx\omega \subset M_t$ for all $t \geq 0$. Since we have that $F^\circ R$ is a ring filtration on R we obtain $F_{-n-d-t}^\circ Rx \subset F_{-n-t}^\circ$ and then also $F_{-n-t}^\circ R\omega \subset M_t$ for all $t \geq 0$. Consequently the claim $F_{-n-d-t}^\circ Rx\omega \subset M_t$ for all $t \geq 0$ is verified and thus $x\omega \in F_{n+d}^\circ\Omega$.
We have $F_0^\circ\Omega = \{\omega \in \Omega \mid F_{-t}^\circ R\omega \subset M_t \text{ for all } t \geq 0\}$. In particular if $\omega \in F_0^\circ\Omega$ then taking $t = 0$ yields $F_0^\circ R\omega \subset M$ and $1 \in F_0^\circ R$ entails $\omega \in M$ and so $F_0^\circ\Omega \subset M$. Conversely, if $y \in F_{-t}^\circ R$, say $yF_{n+t}\Omega \subset F_n\Omega$ for all n, then for $n = -t$ we get $yM \subset F_{-t}\Omega = M_t$ and thus $F_{-t}^\circ RM \subset M_t$ or $M \subset F_0^\circ\Omega$ and finally $M = F_0^\circ\Omega$. Now for $n < 0$ and $\omega \in F_n^\circ\Omega$ we have $F_{-n-t}^\circ R\omega \subset M_t$ for all $t \geq 0$; thus for $t = -n > 0$ we obtain $F_0^\circ R\omega \subset M_{-n}$ or $F_n^\circ\Omega \subset M_{-n}$ for $n < 0$ as $1 \in F_0^\circ R$. On the other hand, M_{-n} satisfies:

$$F_{-n-t}^\circ RM_{-n} = F_{-n-t}^\circ RF_n\Omega \subset M_n,$$

because M is an $F^\circ R$-fragment as established in Proposition 7.6.15. It follows that $M_{-n} \subset F_n^\circ\Omega$ for $n \leq 0$ and the claim $F_n^\circ\Omega = M_{-n}$ for $n \leq 0$ has been established. □

The advantage of $F^\circ R$ over F^*R is that one does not need to go to a closure $F^{(*)}R$ in order to get a ring filtration on R.

Remark 7.6.17. The filtrations $F\Omega$ and $F^\circ\Omega$ are separated if and only if $B(M) = 0$. But for $F^\circ R$ we easily see that $\cap_d F_d^\circ R = t(\Omega) = \{x \in R \mid x\Omega = 0\}$ in that case, so $F^\circ R$ is separated if and only if $t(\Omega) = 0$ or Ω is a faithful R-module.

We now return to the standing assumptions, hence assuming that M is natural with $RM = \Omega$ and Ω taken to be X-torsion free but now not assuming $B(M) = 0$. Define

$$Q_X^f(M)_d = \{z \in Q_X(\Omega) \mid s_i z \in M_{d-i} \text{ for some } s_i \in X, d \geq i\}.$$

Lemma 7.6.18. *With the standing assumptions as above:*

$$\begin{aligned} Q_X^f(M) &:= \bigcup_{d\geq 0}\{z \in Q_X(\Omega) \mid s_{(i)}z \in M_{d-i} \text{ for some } s_i \in X, d \geq i\} \\ &= \bigcup_{d\geq 0}\{z \in Q_X(\Omega) \mid F_k(Rs_{(i)})z \subset M_{d-k}, d \geq k \geq i\} \\ &= \bigcup_{d\geq 0}\{z \in Q_X(\Omega) \mid F_k(Rs_{(i)})z \subset F_{k-d}\Omega \text{ for } k \in \mathbb{N}, d \geq i\} \end{aligned}$$

where by $r_{(i)} \in R$ we indicate that $\sigma(r_{(i)}) \in G(R)_i$, i.e. that $r_{(i)} \in F_iR - F_{i-1}R$.

Proof. By definition $F_k(Rs_{(i)}) = F_kR \cap Rs_{(i)}$ and since $G(R)$ is assumed to be $\sigma(X)$-torsion free it follows that $F_kR \cap Rs_{(i)} = F_{k-i}Rs_{(i)}$ where $F_{k-i}R = 0$ for $k < i$. Now if $s_{(i)}z \in M_{d-i}$ then $F_{k-i}Rs_{(i)}z$ in $M_{d-i+i-k} = M_{d-k}$ and thus we get $F_k(Rs_{(i)})z \subset M_{d-k}$. On the other hand if $F_k(Rs_{(i)})z \subset M_{d-k}$ for all admissible $k \in \mathbb{N}$ then $F_{k-i}Rs_{(i)}z \subset M_{d-k}$ yields $s_{(i)}z \in M_{d-k+k-i} = M_{d-i}$ because M is a natural fragment, hence the second equality follows. The third equality derives from the fact that for $k > d : F_k(Rs_{(i)})z = F_{k-d}RF_{d-i}Rs_{(i)}z \subset F_{k-d}RM = F_{k-d}\Omega$ (FR is standard!) while for $k \leq d$ the statement $F_k(Rs_{(i)})z \subset F_{k-d}\Omega$ reduces to $F_k(Rs_{(i)})z \subset M_{d-k}$ as long as $k \geq i$ and it reduces to $z \in M_{d-k}$ otherwise. □

Observation 7.6.19. We have $Q_X(B(M)) \subset Q_X^f(M) \subset Q_X(\Omega)$.

Lemma 7.6.20. *With the standing assumptions we obtain: if $z \in Q_X^f(M) - Q_X(B(M))$ then there is an $s_{(i)} \in X$ and some $d \in \mathbb{N}$ such that $F_jRs_{(i)}z \subset M_{d-i-j}$ for all $d - i \geq j \geq 0$ but $F_{j_0}Rs_{(i)}z \not\subset M_{d-i-j_0+1}$ for some $j_0, d - i \geq j_0 \geq 0$.*

Proof. First there is a d' such that for some $s_{(i)} \in X$ we have $s_{(i)}z$ in $M_{d'-i}$. If for this d' no j_0 with the desired properties exists then $F_jRs_{(i)}z \subset M_{d'+1-i-j}$ for all j such that $d' - i \geq j \geq 0$, hence $s_{(i)}z \in M_{d'+1-i}$ (by naturality of M). If there is a j_1 such that $d' + 1 - i \geq j_1 \geq 0$ such that $F_{j_1}Rs_{(i)}z \not\subset M_{d'+2-i-j}$ then $d = d' + 1$ is the integer searched for. If there is no such j_1 we arrive at $F_jRs_{(i)}z \subset M_{d'+2-i-j}$ for all j such that $d' + 1 - i \geq j \geq 0$; hence we obtain $s_{(i)}z \in M_{d'+2-i}$ and so on until we arrive at a $d = d' + k$ such that desired properties hold or else we cannot find such a d but then this means that $s_{(i)}z \in \cap_d M_{d-1} = B(M)$. The latter contradicts the assumption $z \notin Q_X(B(M))$. □

Corollary 7.6.21. *If $z \in Q_X^f(M) - Q_X(B(M))$ then there is a $d \in \mathbb{N}$ such that for some $s_{(i)} \in X$ we have $s_{(i)}z \in M_{d-i}$ but $s_{(i)}z \notin M_{d-i+1}$.*

Lemma 7.6.22. *With assumptions as before and supposing now also that $G(\Omega)$ is $\sigma(X)$-torsion free, then for $z \in Q_X^f(M) - Q_X(B(M))$ the d obtained in $\mathbb{N}$ in the foregoing corollary depends on z only and not on the $s_{(i)} \in X$ used.*

Proof. Assume $s_{(i)} \in X$ is such that $s_{(i)}z \in M_{d-i} - M_{d-i+1}$ and $s_{(k)} \in X$ is such that $s_{(k)}z \in M_{d+\nu-k} - M_{d+\nu-k+1}$ for some fixed $\nu > 0$. Since X is an Ore set there is an $s_{(\mu)} \in X$ such that $s_{(\mu)}s_{(i)} = rs_{(k)}$ for some $r \in R$. Now we have that $\sigma(s_{(\mu)}s_{(i)}) = \sigma(s_{(\mu)})\sigma(s_{(i)})$ because the latter is non-zero. Moreover since $G(R)$ has no $\sigma(X)$-torsion, no element of $G(R)$ can be annihilated on the right by an element of $\sigma(X)$ because of the second Ore condition for $\sigma(X)$ in $G(R)$. Hence it follows that $r \in F_{i+\mu-k}R - F_{i+\mu-k-1}R$ so we write it as $r = r_{(i+\mu-k)}$. Thus

$$s_{(mu)}s_{(i)}z = r_{(i+\mu-k)}s_{(k)}z \in F_{-d-\nu+i+\mu}\Omega,$$

and $m = s_{(i)}z$ is such that $s_{(\mu)}z$ is contained in $F_{-d-\nu+i+\mu}\Omega$ with $\nu > 0$ and $m \in M_{d-i} = F_{i-d}\Omega$. By the definition of the $G(R)$-structure on $G(\Omega)$ we have: $\sigma(s_{(\mu)})\sigma(m) = s_{(\mu)}m$ modulo $F_{-d+i+\mu-1}\Omega$ and now $\nu > 0$ yields $s_{(\mu)}m$ in $F_{i+\mu-d-1}\Omega$, hence $\sigma(s_{(\mu)})\sigma(m) = 0$. Now the fact that $G(\Omega)$ is supposed to be $\sigma(X)$-torsion free; the latter would lead to $\sigma(m) = 0$ but that contradicts $m = s_{(i)}z \notin B(M)$ because $z \notin Q_X(B(M))$. We have established that $\nu > 0$ leads to a contradiction. Of course the same proof applies in the case $\nu < 0$ up to interchanging the role of $s_{(i)}$ and $s_{(k)}$, respectively d and $d - \nu$. That d is therefore uniquely determined by z not depending on the $s_{(i)}, s_{(k)} \in X$ used thus follows. □

Remark 7.6.23.

1. It was necessary to assume $G(\Omega)$ is $\sigma(X)$-torsion free because we cannot control the degrees μ of the $s_{(\mu)}$ appearing in the proof; i.e. the multiplication of $s_{(k)}z$ and $r_{(i+\mu-k)}$ may not be possible in the fragment M.

2. It is possible to force the existence of $s_{(\mu)}$ in low degree μ. The necessary condition to make the proof work assuming only $g(M)$ to be $\sigma(X)$-torsion free is the following: for $f_{(i)} \in R, s_{(k)} \in X$ and $d \geq i, k$ there is an $s_{(\mu)} \in F_{d-i}R$ such that $s_{(\mu)}f_{(i)} = r_{(i+\mu-k)}s_{(k)}$. One easily sees that X must then be of a very restricted type; that is for every $s_{(k)} \in X$ there exists $s_{(0)} \in X, t_{(0)} \in S$ such that $s_{(0)}s_{(k)} = t_{(0)}s_{(1)}^k$ for some fixed $s_{(1)} \in X$. In other words, we then have to consider only X generated by a single degree one element up to elements from $F_0R = S$. In this situation we say that X is monogenic. The restriction to monogenic X is strict but in geometric contexts they appear on basic affine coverrings!

We now define $Q_X(M)_d = \{z \in Q_X(\Omega) \mid s_{(\mu)}z \in F_{\mu-d}\Omega \text{ for some } s_{(\mu)} \in X\}$ and $Q_X(M) = \bigcup_d Q_X(M)_d$. Obviously we have $Q^f_X(M)_d \subset Q_X(M)_d$.

Proposition 7.6.24. *The analogues of Lemma 7.6.18 (where now $d \geq i$ is no longer assumed), Lemma 7.6.20 and Lemma 7.6.22 may be established for $Q_X(M)_d$ and $Q_X(M)$.*

Proof. Very similar to the proofs given before, the simple verification is left to the reader. □

Proposition 7.6.25. *In the situation of Lemma 7.6.22, $Q_X(M)$ is an FR-fragment; if X is monogenic then $Q^f_X(M)$ is an FR-fragment too.*

Proof. If $a, b \in Q_X(M)_d$ for some $d \in \mathbb{N}$ then for some $s_{(i)}, s_{(k)} \in X$ we have $s_{(i)}a \in F_{i-d}\Omega, s_{(k)}b \in F_{k-d}\Omega$. Now let $s_{(\mu)} \in X$ be chosen such that $s_{(\mu)}s_{(i)} = r_{(i+\mu-k)}s_{(k)}$. Then we obtain

$$s_{(\mu)}s_{(i)}(a + b) = s_{(\mu)}s_{(i)}a + r_{(i+\mu-k)}s_{(k)}b,$$

where $s_{(\mu)}s_{(i)}a \in F_{\mu+i-d}\Omega$ and $r_{(i+\mu-k)}s_{(k)}b \in F_{\mu+i-d}\Omega$. Putting $s_{(i+\mu)} = s_{(\mu)}s_{(i)}$ we arrive at $s_{(i+\mu)}(a+b) \in F_{i+\mu-d}\Omega$, hence $a+b \in Q_X(M)_d$. Now consider $x \in Q_X(M)_d$ and $j \leq d, f_{(j)} \in F_jR$. Pick an $s_{(\mu)} \in X$ such that $s_{(\mu)}f_{(j)} = r_{(j+\mu-i)}s_{(i)}$. We calculate $s_{(\mu)}f_{(j)}x \in F_{j+\mu-d}\Omega$, hence $f_{(j)}x \in Q_X(M)_{d-j}$. If f is also in $F_{(j')}R$ with $j' < j$ then we must have $fx \in Q_X(M)_{d-j'} \subset Q_X(M)_{d-j}$, hence $F_jRx \subset Q_X(M)_{d-j}$ as desired. The proof of the second statement in case of monogenic X is completely similar and it is modified such that μ can be chosen such that only $F_d\Omega$ with $d \leq 0$ appears. □

Corollary 7.6.26. *By uniqueness for d, as above, for any z in $Q_X(M) - Q_X(B(M))$ it follows that $B(Q_X(M)) = Q_X(B(M))$ is the localization at X of the R-module $B(M)$. Also if $B(M) = 0$ then $B(Q_X(M)) = 0$.*

The fragment $Q_X(M)$ is also filtered as a fragment with respect to some filtrations defined on $Q_X(R)$, as in the following.

Proposition 7.6.27. *The fragmentation chain of $Q_X(\Omega)$ makes it into a filtered $F_0Q_X(R)$-module with respect to the negative filtration F^- defined by $F_n^-(F_0Q_X(R)) = F_{-n}Q_X(R)$ for $n \geq 0$ where $FQ_X(R)$ is the localized filtration. Moreover $Q_X(M)$ is a fragment with respect to the positive F^+ defined by $F_n^+Q_X(R) = F_nQ_X(R), n \geq 0$, for the localized filtration $FQ_X(R)$.*

Proof. By the definitions of the localized filtrations (see Section 7.1, Section 7.2, Section 7.3) we have that $r \in F^- + n$ exactly when there is an $s_{(i)} \in X$ such that $s_{(i)}r \in F_{i-n}R, i \geq n$. Look at $z \in Q_X(M)_d$ and at $rz \in Q_X(\Omega)$. For some $s_{(j)} \in X$ we have $s_{(j)}z \in F_{j-d}\Omega$. Take some $s_{(\mu)} \in X$ such that $s_{(\mu)}(s_{(i)}r) = r_{(\mu+i-n-j)}s_{(j)}$; note that we automatically may assume that n is minimal with respect to $r \in F_n^-$ so that $s_{(i)}r \notin F_{i-n-1}R$ and so the selection of degrees in the foregoing equation is indeed allowed, in fact enforced. Now: $r_{(\mu+i-n-j)}s_{(j)}z = s_{(\mu)}(s_{(i)}r)z$ and so by putting $s_{(\mu+i)} = s_{(\mu)}s_{(i)} \in X$ and $s_{(\mu+i)}rz \in F_{(\mu+i)-n-d}\Omega$, we arrive at: $rz \in Q_X(M)_{n+d}$.
The second statement follows in a similar way, only by modifying the argument by replacing $-n$ by n and look at rz for $r \in F_n^+, z \in Q_X(M)_d$ with $n \leq d$. The transcription of the argument is just a simple exercise. □

Remark 7.6.28. Even when X is monogenic it seems that $Q_X^f(M)$ need not be a fragment with respect to F^+, nor filtered with respect to F^-. The difference between $Q_X^f(M)$ and $Q_X(M)$ resides in the fact that elements of $Q_X^f(M)_d$ arise as fractions $s_{(i)}^{-1}\omega_{(j)}$ with $j \leq 0$ and $j - i = -d$, whereas elements in $Q_X(M)_d$ arise as $s_{(\mu)}^{-1}\omega_{(k)}$ with $k - \mu = -d$ but $k > 0$ allowed. Hence $Q_X^f(M)$ collects those elements that can be reached by denominators of "small enough" degree. In general when X is not monogenic, $Q_X^f(M)$ is just a set which we will refer to as the "fragment heart" of $Q_X(M)$. As we said before, monogenic X are however encountered often in the theory of Zariskian filtered rings R with commutative positively graded $G(R)$ when these are quantum-localized at Ore sets X with $\sigma(X) = \langle \overline{f} \rangle$ for some homogeneous $\overline{f} \in G(R)$, even $\overline{f} \in G(R)_1$.

In order to apply directly some techniques of filtered rings and modules with respect to microlocalizations, we will henceforth assume that $B(M) = 0$. In that case Ω is a separated filtered R-module by $F\Omega$, the condition that $g(M)$ or $G(\Omega)$ is $\sigma(X)$-torsion free automatically entails that M, resp. Ω, is X-torsion free; moreover $\bigcap_d Q_X^f(M)_d = Q_X(B(M)) = 0$ and also $\bigcap_d Q_X(M)_d = 0$, hence $Q_X(M)$ is separated and exhaustively filtered by its fragmentation chain with respect to F^- on $F_0Q_X(R)$.

By general theory of microlocalizations we know that $Q^{\mu}_{\sigma(X)}(R) = (Q_{X_{sat}}(R))^{\wedge}$ where $\wedge$ denotes the completion with respect to the localized filtration on $Q_{X_{sat}}(R)$, cf. [46], [6]. In case, as we consider above, X is itself an Ore set we obtain: $Q_X(R)^{\wedge} = Q_{X_{sat}}(R)^{\wedge}$ where on the left $\wedge$ stands for completion in the localized filtration on $Q_X(R)$, using the same symbol $\wedge$ does not cause any ambiguity here since the ring is explicitly mentioned. Note that $Q_X(R) \subsetneq Q_{X_{sat}}(R)$ is possible but their completions do coincide. Moreover note also:

$$\begin{array}{ccccc} G(Q_X(R)) & = & Q_{\sigma(X)}(G(R)) & = & G(Q_{X_{sat}}(R)) \\ \| & & & & \| \\ G(Q_X(R)^{\wedge}) & = & G(Q^{\mu}_{\sigma(X)}(R)) & = & G(Q_{X_{sat}}(R)^{\wedge}) \end{array}$$

Quantum localization at $\sigma(X)$ is defined as $F_0Q^{\mu}_{\sigma(X)}(R)$ and we will denote this by $Q^{q}_{\sigma(X)}(R)$, the ring of quantum sections at X. The same ring may be obtained by completing $F_0Q_X(R)$ at F^-. Similar results transfer immediately to filtered R-modules with good filtration (which are necessarily finitely generated).

Lemma 7.6.29. *The filtration $F\Omega$ is good if and only if it is finitely generated as a fragment.*

Proof. Assume M is finitely generated as a fragment; then there exists a finite set of $m_i \in M_{n_i}, i = 1, \ldots, t$, such that for all $n \in \mathbb{N}$ we have: $M_d = \sum_{i=1}^t F_{n_i - d}Rm_i$, hence $M = \sum_{i=1}^t F_{n_i}Rm_i$. Since $F_\lambda\Omega$ for $\lambda \geq 0$ is $F_\lambda RM$, we obtain $F_\lambda\Omega = \sum_{i=1}^t F_{\lambda+n_i}Rm_i$ for all λ. This states that $F\Omega$ is a good filtration because $m_i \in F_{-n_i}\Omega, i = 1, \ldots, t$. Conversely, suppose $\Omega = R\omega_1 + \ldots + R\omega_t$ such that for all λ, $F_\lambda\Omega = \sum_{i=1}^t F_{\lambda-n_i}R\omega_i$ with $\omega_i \in F_{n_i}\Omega$. Let $\omega_{i_1}, \ldots, \omega_{i_n}$ be the subset of the ω_i for which $n_i \leq 0$. For negative λ we then see that in $F_\lambda\Omega = M_{-\lambda} = \sum_{j=1}^n F_{\lambda - n_{i_j}}R\omega_{i_j}$ because when n_i is positive $\lambda - n_i \leq 0$ then yields $F_{\lambda-n_i}R = 0$ as FR is positive. Hence the set $\omega_{i_j}, j = 1, \ldots, n$ is a finite generating set for M as a fragment. □

Note that for a finitely generated fragment the chain $\ldots \subset M_d \subset \ldots \subset M$ is finite and M is a finitely generated S-module because every M_d is a finitely generated S-module (FR is a standard filtration!).

Proposition 7.6.30. *1. If M is a finitely generated fragment, $Q^q_X(\Omega) = Q_X(M)^\wedge$ where $\wedge$ stands for completion with respect to F^- on $F_0Q_X(R)$. Then $Q_X(M)^\wedge$ is a fragment with respect to the positive filtration defined on $Q^\mu_X(R)$, moreover we have*

$$Q_X(M)^\wedge = F_0(Q^\mu_X(R) \otimes_R \Omega).$$

2. If M is not finitely generated but assuming that $FQ_X(\Omega)$ is good with respect to the localized $FQ_X(R)$ then the claims of 1. do hold.

Proof. 1. First observe that $Q_X(M) = F_0Q_X(\Omega)$. Indeed the inclusion $Q_X(M) \subset F_0Q_X(\Omega)$ is obvious and conversely if $y \in F_0Q_X(\Omega)$ then $s_{(i)}y$ in $F_i\Omega = F_iRM(i \geq 0)$, for some $s_{(i)} \in X$. By the standing $\sigma(X)$-torsion free assumption on $G(\Omega)$ we then must have: $s_{(i)}y \in F_iRM$ yields $y \in Q_X(M)$. By definition of $Q^q_X(\Omega)$ the first statement is clear. So we have $Q_X(M)^\wedge = F_0Q^\mu_X(\Omega)$ where $Q^\mu_X(\Omega)$ has the microlocalized filtration on it; therefore it is obviously a fragment with respect to $F^+Q^\mu_X(R)$. Since $F\Omega$ is good, it is well-known (cf. [6]) that $Q^\mu_X(\Omega) = Q^\mu_X(R) \otimes_R \Omega$.
2. We make the following modification in the proof of 1. above. Replace R by $Q_X(R)$, FR by $FQ_X(R)$, S by $F_0Q_X(R)$, M by $Q_X(M)$, the fragment chain of M_d by the chain of $Q_X(M)_d$, Ω by $Q_X(\Omega)$. According to definition of $Q_X(M)$, we now define $Q'_X(Q_X(M)) \subset Q_X(Q_X(\Omega)) = Q_X(\Omega)$ by the following rule: $z \in Q'_X(Q_X(M))_d$ if for some $s_{(i)} \in X$ (now viewed in $Q_X(R)$ and thus consisting of units) we have $s_{(i)}z \in F_{i-d}Q_X(\Omega)$. Hence for some $s_{(j)} \in X$ we have $s_{(j)}(s_{(i)}z) \in F_{j+i-d}\Omega$ but this yields $z \in Q_X(M)_d$. Therefore we arrive at $Q'_X(Q_X(M)) = Q_X(M)$ (since $Q_X(M)$ is not a real localization of M this short proof was necessary!). Now because $FQ_X(\Omega)$ is good with respect to $FQ_X(R)$ we obtain:

$$Q^\mu_X(M) = Q^\mu_X(Q_X(\Omega)) = Q^\mu_X(Q_X(R) \otimes Q_X(\Omega)).$$

Since X is invertible in $Q_X(R)$, the microlocalization of $Q_X(R)$ with respect to X is just the completion of $Q_X(R)$ with respect to $FQ_X(R)$; hence $Q^\mu_X(Q_X(R)) = Q^\mu_X(R)$, and now $Q_X(\Omega) = Q_X(R) \otimes_R \Omega$ thus $Q^\mu_X(\Omega) = Q^\mu_X(R) \otimes_R \Omega$. All claims of 1. are now easily verified. □

The condition $F\Omega$ is good is rather restrictive on M because FR is a positive filtration. But $Q_X(R)$ is not positively filtered by $FQ_X(R)$ so the condition for $FQ_X(\Omega)$ to be good with respect to $FQ_X(R)$ is less restrictive on $F_0Q_X(\Omega) = Q_X(M)$. This can be seen in the trivial case where we take $M = \Omega$ and $R = S$ with trivial filtration FR. Then $X = R - \{0\}$ and let us look at a commutative domain R with field of fractions K. Then $F\Omega$ is good only when Ω is finitely generated as an R-module. Moreover the condition $Q_X(\Omega) = K \otimes_R \Omega$ is K-finitely generated does not imply a strong finiteness condition for $_R\Omega$.

Another easy example is the following.

Example 7.6.31. Let $S = O_v$ be a discrete valuation ring in the field $K = R$ and take X the multiplicative set $\langle \pi^{-1} \rangle$ where $m_v = (\pi)$ is the maximal ideal of O_v. Define $F_n K = O_v \pi^{-n}$ for $n \geq 0$, of course $Q_X(K) = K$ but the localized filtration $F^\times K$ is now the $\mathbb{Z}$-filtration corresponding to the valuation v, i.e. $F_n^\times K = \pi^{-n} O_v$ for $n \in \mathbb{Z}$. An FR-fragment is nothing but an m_v-adically filtered O_v-module M and it is always natural with $\Omega = K \otimes_{O_v} M$. Hence $F\Omega$ is the filtration $F_n \Omega = \pi^{-n} M$ for $n \in \mathbb{N}$ and clearly $Q_X(\Omega) = \Omega$, $F\Omega$ coinciding with the localized filtration. Now $z \in Q_X^f(M)_d$ means $\pi^{-i} z \in M_{d-i} \subset \pi^{d-i} M$ with $d \geq i$, hence $z \in \pi^d M$ and $Q_X^f(M) = M$ follows. On the other hand $y \in Q_X(M)_d$ means $\pi^{-i} y \in F_{i-d}\Omega$ for some $i \in \mathbb{N}$, i.e. $\pi^{-i} y \in \pi^{d-i} M$ or $y \in \pi^d M$ and again $Q_X(M) = M$. Furthermore $Q_X^\mu(K) = K^{\wedge_v}$ where $\wedge_v$ is the v-adic completion and therefore $Q_X^q(K) = O_v^\wedge$, the complete valuation ring of $K^{\wedge_v}$. The completion of M in the m_v-adic topology is $Q_X^q(\Omega)$ and this is $(F_0(K \otimes_{O_v} M))^\wedge$ indeed.

We are now ready to go on with the theory of fragments and gliders over nice Zariskian filtered rings like Weyl algebras and other rings of differential and pseudo-differential operators. This is the topic of research in progress.

Chapter 8

Glider Brauer-Severi varieties

In the second chapter of part III we investigate glider theory for unbounded filtrations on central simple algebras. Such algebras stem from classical representation theory of finite groups and they were studied in detail in the theory of the Brauer group of a field. Using descent theory for example, there can be found a strong relation to non-Abelian cohomology in some projective linear group. Brauer-Severi varieties are geometric objects associated with a central simple algebra (csa) and these also relate to the non-Abelian cohomology as before, see [4], [64], [21]. Brauer-Severi (BS-)varieties found interesting applications in the geometric theory, for example in the Artin-Mumford example of unirational non-rational varieties, and also in some approaches to the Mercurjev-Suslin theorem on the co-torsion of the second K-group. Now the Brauer-Severi variety of a csa A over a field K is defined as the variety of irreducible left ideals of A, its variety structure coming from an obvious embedding as a closed subset of some Grassmann variety.

The gliders we are interested in in this chapter lie inside a csa A over a field K, that is, having $\Omega \subset A$. In fact, the question we want to investigate is whether such gliders that are irreducible define something like a BS-variety. The gliders in A are defined with respect to a filtration FA having a ring of filtration-degree zero, B say, inducing a filtration on K with ring of filtration degree zero, R say. We work in the situation where $KB = A$ and usually we assume R is a Noetherian ring (in future work we would like to deal with Noetherian integrally closed domains; then B is an order over R in A). The definition of the GBS(A), the glider Brauer-Severi variety, in terms of irreducible subgliders of A with respect to some filtration FA (with some extra properties usually), does indeed lead to some geometric structure. A first observation is that GBS(K) is not trivial, unlike the BS(K), so we study the GBS of a field first and, perhaps a surprise, we find it is the product of the Riemann surface $R(K)$ of K with the integers $\mathbb{Z}$. The latter is the effect of some shift which is always possible on gliders; we can get rid of this factor $\mathbb{Z}$ by introducing glider classes under shifts, but we do not do that here. The Riemann surface of a field is the set of all discrete valuation rings of the field and it may be given a nice topology of Zariski-type and viewed as some geometric object but not a variety. One of the telling results of the theory about GBS(A) is that the relative GBS of A introduced in Section 8.3 turns out to be BS(A) $\times$ GBS(K), so the product of the BS(A) and the Riemann surface $R(K)$ with a further factor $\mathbb{Z}$. In a sense this thus yields some geometric structure in terms of the BS of A and the Riemann surface of K which is not a variety but still nicely described. The theory of orders and maximal orders enters the glider theory naturally here. For example, we also have a version of the Brandt groupoid appearing. We could go to rings K instead of fields and start from Azumaya algebras and BS-schemes, but this is left for work in progress.

The structure of the chapter is as follows: In Section 8.1 we show some results about filtrations on fields and central simple algebras. For example, in Proposition 8.1.2 we make a connection between irreducible glider representations and the filtration being strong. Section 8.2 starts with the definition of GBS(A), the glider Brauer-Severi variety of a csa A. However, the rest of the section is entirely devoted to the calculation of the GBS of a field K. We conclude with Theorem 8.2.10, which shows how the Riemann surface $R(K)$ enters the game. Subsequently, in Section 8.3 we deal with the GBS of central simple algebras A over a field K. In fact, we introduce the relative $\mathrm{GBS^K(A)}$, in which we restrict to filtrations FA that induce separated, exhaustive and unbounded filtrations FK. The main result of this section is Corollary 8.3.6 which shows that $\mathrm{GBS^K(A)}$ equals $\mathrm{BS(A)} \times \mathrm{GBS(K)}$. In Section 8.4 we define a tensor product with a field extension L/K, which in the case of a strong filtration on a csa A over K gives rise to a map $\mathrm{GBS_F(A)} \to \mathrm{GBS_f(A \otimes_K L)}$. These observations then allow us to deduce that the relative glider Brauer-Severi variety $\mathrm{GBS^K(A)}$ is a twist of the relative glider Brauer-Severi variety of a matrix algebra $M_n(L)$ for a suitable field extension L/K. Throughout these sections, we indicate links with the theory of (maximal) orders. This then inspires the construction of the Brandt groupoid of normal glider ideals in a csa A. This is carried out in Section 8.5. Finally, since glider representations can be defined for Γ-filtrations with Γ any totally ordered group, we include a final Section 8.6 in which we work with $\Gamma = \mathbb{Z}^2$ with lexicographical order. We establish Theorem 8.6.4 which shows that all rank 2 valuation rings in a field K enter the scene. The content of this chapter is based on [19].

8.1 Some results on separated, exhaustive filtrations on central simple algebras

In this section K is a field and A a central simple K-algebra. As in the previous chapter, we work with $\Gamma = \mathbb{Z}$, except in Section 8.6. We always assume that the filtrations on A considered in the entire chapter are separated, exhaustive and unbounded. The latter means that all $F_nA \neq 0$ and there exists no $m \in \mathbb{Z}$ such that $F_mA = A$. We call the filtration FA strong if $F_nAF_{-n}A = F_0A = F_{-n}AF_nA$ for all $n \geq 0$. In fact, FA is strong if and only if $F_1AF_{-1}A = F_0A = F_{-1}AF_1A$.

When fixing such a filtration FA, one can consider left (or right) FA-glider representations $\Omega \supset M \supset M_1 \supset M_2 \supset \ldots$ We recall that a subglider N of M is said to be trivial if either:

T_1. There is an $n \in \mathbb{N}$ such that $N_n = B(N)$ but $M_n \neq B(M)$.
T_2. There is an $n \in \mathbb{N}$ such that $N_n = 0$ but $M_n \neq 0$.
T_3. There exists a monotone increasing map $\alpha : \mathbb{N} \to \mathbb{N}$ such that $N_n = M_{\alpha(n)}$ and $\alpha(n) - m \geq \alpha(n-m)$ for all $m \leq n$ in $\mathbb{N}$.

The monotone increasing map α for type T_3 is in fact strict monotone. Indeed, $1 \leq n+1$ implies that $\alpha(n+1) - 1 \geq \alpha(n)$.

As a particular example of a left (and right) FA-glider, we have the negative part of the filtration:

$$F^-B : \quad B = F_0A \supset F_{-1}A \supset F_{-2}A \supset \ldots$$

Let us investigate when this glider is in fact irreducible.

For ease of notation, we will write F_n rather than F_nA. First of all, if $F_0 = F_{-1}$, then for all $n > 0$ we would have $F_n = F_nF_0 = F_nF_{-1} \subset F_{n-1}$. It follows that F^+A is the trivial chain and in order to be exhaustive $F_0 = A$ and then FA is just the trivial chain, whence it is not separated. Thus $F_{-1} \subsetneq F_0$. Also, there are no idempotent elements in the negative part. Indeed, suppose on the contrary that $F_{-n}F_{-n} = F_{-n}$, then for all $k > 0$ it holds $F_{-n} = F_{-n}^k \subset F_{-nk}$. This would entail that F_{-n} is in the core of the filtration.

Remark 8.1.1. By the assumption on the filtrations being separated, exhaustive and unbounded, we know that the left glider ideal F^-B is of infinite essential length. If we would drop the left boundedness condition then F^-B being of finite essential length, say d, would entail that FA is a positive filtration. Indeed, it would follow that $F_{-d} \supsetneq F_{-d-1} = 0$ because the filtration is separated. By irreducibility of F^-B, the subglider

$$F_dF_{-d} \supset F_{d-1}F_{-d} \supset F_1F_{-d} \supset F_{-d} \supsetneq 0 \supset \ldots$$

must be of type T_3 and hence $F_0 = F_dF_{-d}$. It follows that $F_{-1} = F_0F_{-1} = F_dF_{-d}F_{-1} \subset F_dF_{-d-1} = 0$, whence FA is indeed a positive filtration.

The filtration FA induces on K a filtration FK defined by $F_nK = F_n \cap K$. Due to the assumption on all filtrations considered, the filtration on K is obviously also separated and exhaustive. The subring F_0K must be proper, for otherwise all F_nA are K-vector spaces and since A is finite dimensional, this would entail left and right boundedness of FA. We suppose moreover that the induced filtration FK on K has non-trivial negative part. Since K has no zero divisors, this is equivalent to saying that $F_{-n}K \neq 0$ for all $n \geq 0$.

Separatedness of FA entails the existence of a smallest integer $e \geq 1$ such that $F_{-e} \supsetneq F_{-e-1}$, i.e. $G(A)_{-e} = F_{-e}/F_{-e-1} \neq 0$. By a strong e-step filtration FA we mean that for all $n \in \mathbb{Z}$

$$F_{ne}A = F_{ne+1}A = \ldots = F_{(n+1)e-1}A,$$

and that for all $n, m \in \mathbb{Z}$ it holds $F_{ne}AF_{me}A = F_{(n+m)e}A$.

Proposition 8.1.2. *Let FA be a filtration such that F^-B is a left irreducible FA-glider, then the filtration is a strong e-step filtration, with e defined as above.*

Proof. To begin with, consider the subglider

$$\begin{array}{ccccccccccc} F_0 & \supset & F_{-1} & \supset & \ldots & \supset & F_{-e} & \supsetneq & F_{-e-1} & \supset & \ldots \\ \cup & & \| & & & & \cup & & \cup & & \\ F_eF_{-e} & \supset & F_{e-1}F_{-e} & \supset & \ldots & \supset & F_0F_{-e} & \supset & F_{-1}F_{-e} & \supset & \ldots \end{array}$$

For obvious reasons it cannot be trivial of type T_1 or T_2. Because $F_0F_{-e} = F_{-e} \supsetneq F_{-e-1}$, triviality of type T_3 implies that $F_eF_{-e} = F_0$. A similar argument shows that $F_{-e}F_e = F_0$ as well. Irreducibility of F^-B implies furthermore that $F_{-e}F_{-e} = F_{-2e-d}$ for some $d \geq 0$. But we also have

$$F_{-e} = F_0F_{-e} = F_eF_{-e}F_{-e} = F_eF_{-2e-d} \subset F_{-e-d},$$

and by definition of e it follows that $d = 0$, i.e. $F_{-e}F_{-e} = F_{-2e}$. Suppose now that $F_{-2e} = F_{-3e}$, then we would have

$$F_{-e} = F_0F_{-e} = F_eF_{-e}F_{-e} = F_eF_{-2e} = F_eF_{-3e} \subset F_{-2e},$$

a contradiction. Hence $F_{-2e} \supsetneq F_{-3e}$. This allows us to show that $F_{-e}^3 = F_{-3e}$. Indeed, irreducibility of F^-B implies that $F_{-e}^3 = F_{-3e-d}$ for some $d \geq 0$. We also have

$$F_{-2e} = F_0F_{-2e} = F_eF_{-e}F_{-2e} = F_eF_{-3e-d} \subset F_{-2e-d}.$$

If $d > 0$, then we would have that $F_{-2e} \subset F_{-2e-d} \subset F_{-3e} \subset F_{-2e}$, a contradiction. By induction one then shows that for all $n \geq 0$, $F_{-ne} \supsetneq F_{-(n+1)e}$ and $F_{-e}^n = F_{-ne}$.
Consider now the subglider

$$F_eF_{-2e} \supset F_{e-1}F_{-2e} \supset \ldots \supset F_0F_{-2e} \supset F_{-1}F_{-2e} \supset \ldots$$

of

$$F_{-e} \supsetneq F_{-e-1} \supset \ldots \supset F_{-2e} \supset F_{-2e-1} \supset \ldots$$

Since $F_eF_{-2e} = F_eF_{-e}F_{-e} = F_{-e}$ and $F_{-1}F_{-2e} = F_{-e}F_{-2e} = F_{-3e} \subsetneq F_{-2e}$ and triviality of type T_3 implies that the associated monotone increasing map is the identity on $\{0, 1 \ldots, e\}$. In particular, it follows that

$$F_{-e-1} = F_{e-1}F_{-2e} = F_{e-1}F_{-e}F_{-e} \subset F_{-1}F_{-e} = F_{-e}F_{-e} = F_{-2e},$$

whence

$$F_{-e-1} = F_{-e-2} = \cdots = F_{-2e}.$$

If $F_{-2e} = F_{-2e-1}$, then $F_{-e} = F_eF_{-2e} = F_eF_{-2e-1} \subset F_{-e-1}$, a contradiction. Using induction one then shows that

$$F_{-ne} = F_{-ne+e-1} \quad \text{for all } n \geq 0.$$

Using similar arguments, one shows by induction that $F_{ne}F_{-ne} = F_0$ for all $n > 0$. We also have

$$F_{e-1} = F_{e-1}F_0 = F_{e-1}F_{-e}F_e \subset F_{-1}F_e = F_{-e}F_e = F_0.$$

Both results then allow us to show that

$$F_{ne} = F_{ne+e-1} \quad \text{for all } n \geq 0.$$

We can conclude if we show that $F_{ne} = F_e^n$ for all $n \geq 2$. Since we have

$$F_{ne} = F_{ne}F_0 = F_{ne}F_{-e}F_e \subset F_{(n-1)e}F_e,$$

this follows easily using induction. □

In case $A = K$ is a field, we know by [45, Theorem 2.6] that $F_0K = R$ is a discrete valuation ring and the associated valuation filtration is FK if and only if the associated graded $G(K) = \oplus_{n\in\mathbb{Z}}F_nK/F_{n-1}K$ is a domain. In [46, Theorem 3.10], the authors moreover show that quasi-simple rings A with exhaustive and separated filtration FA and $G(A)$ a semiprime, Noetherian P.I. ring actually have for $G(A)$ a prime ring. So for filtrations FK on K, $G(K)$ being semiprime implies $G(K)$ to be prime, whence a domain.

Proposition 8.1.3. *Let FK be a filtration such that F^-R is an irreducible FK-glider. If $e = 1$, then $R = F_0K$ is a discrete valuation ring and the filtration corresponds to the valuation filtration.*

Proof. Suppose that $G(K)$ is not a domain, whence not semiprime. Since FK is strong, the associated graded ring $G(K)$ is strongly graded. From [26] we know that $G(K)$-gr $\cong F_0K/F_{-1}K$-mod is an equivalence of categories, whence there exists $a \in F_0K - F_{-1}K$ and $n > 1$ minimal such that $a^n \in F_{-1}K$. The induced filtration on F_0Ka defines a subglider, which must be trivial of type T_3. Because $a \in F_0Aa - F_{-1}A$, this entails that $F_0Aa = F_0A$. In particular, a is then invertible in F_0A, say $ba = 1$ for $b \in F_0A$. But then $a^{n-1} = 1a^{n-1} = ba^n \in F_{-1}A$, contradicting the minimality of n. Hence $G(K)$ is a domain. It follows that $F_0K = B$ is a DVR with unique maximal ideal M. We have that $F_{-1}K = M^n$ for some $n \geq 1$, but because $F_0K/F_{-1}K$ is a domain, $n = 1$, i.e. $F_{-1}K = M$. If F^vK denotes the valuation filtration, then

$$F_1K = F_1KF_0K = F_1KMF_1^vK = F_1KF_{-1}KF_1^vK = F_1^vK.$$

And since both filtrations are determined by their degree -1 and degree 1 part, both filtrations agree. □

Proposition 8.1.4. *Let A be a simple Artinian ring with filtration FA that induces FK with non-trivial negative part. If F^-B is an irreducible FA-glider then F_0K is a discrete valuation ring in K. If FK is a strong f-step filtration then $F_{nf}A = F_{nf}KF_0A$ for all $n \in \mathbb{Z}$.*

Proof. As in the beginning of the proof of Proposition 8.1.3 we can show that $G(A)$ must be a prime ring by using [45, Theorem 3.10]. Indeed, suppose $G(A)$ is not semiprime. Then there exists an ideal $J \lhd F_0A$ such that $J^n \subset F_{-1}A$ for some $n > 1$ and assume n is minimal with this property. The triviality of the subglider $J^{n-1} \supset F_{-1}J^{n-1} \supset F_{-2}J^{n-1} \supset \ldots$ entails that $F_0 = J^{n-1}$ but then it follows that $J \subset F_{-1}$, a contradiction. The center of a prime ring is an integral domain, whence so is $G(K) \subset Z(G(A))$. Hence FK is the filtration associated with a valuation ring O_v on K when considered over $f\mathbb{Z}$ for some $f > 0$. In particular, f is minimal with the property that $F_{-f-1}K \subsetneq F_{-f}K$ and $G(K)$ is $f\mathbb{Z}$-strongly graded. Define the glider $M_j = F_{-f}KF_{f-j}A \subset F_{-j}A$. For $i \leq j$ we indeed have

$$F_iAM_j \subset= F_{-f}KF_iAF_{f-j}A \subset F_{-f}KF_{f+i-j}A = M_{j-i}.$$

Irreducibility of F^-B implies that the subglider $M \supset M_1 \supset \ldots$ is trivial of type T_3. If the strict monotone increasing map α is not the identity, then $F_{-f}KF_0A = F_{-f-d}A$ for some $d > 0$. Since $1 \in F_0A$, this would entail

$$F_{-f}K \subset F_{-f-d}A \cap K = F_{-f-d}K.$$

By definition of f this yields $d = 0$, i.e. $F_{-f}F_0A = F_{-f}A$. By Proposition 8.1.2 we know that FA is a strong e-step filtration. Since GK is an f-step filtration, e divides f. Hence

$$F_{-nf}KF_0A = F_{-nf}A \quad \forall n \geq 0.$$

From $F_{-f}KF_0A = F_{-f}A$ it follows that $F_{-f}KF_fA = F_0A$. Consequently

$$F_fKF_0A = F_fKF_{-f}KF_fA = F_0KF_fA = F_fA,$$

and then

$$F_{nf}KF_0A = F_{nf}A \quad \forall n \geq 0$$

follows. □

In Section 3.1 completely irreducible gliders were introduced and it was shown that a simple Artinian ring A has a filtration FA with subring $F_0A = A'$ making the negative part into a completely irreducible FA-glider if and only if A is a skew field, A' is a discrete valuation ring and FA is the corresponding valuation filtration, Theorem 3.1.16. The approach here leads to a generalization, as we do not require the notion of completely irreducibility.

Proposition 8.1.5. *Let A be a central simple K-algebra with filtration FA inducing a filtration FK with non-trivial negative part and such that F^-B is an irreducible left glider. Then A is a skew field.*

Proof. Suppose that A is not a skew field. So there exists a proper left ideal Av. Since FA is exhaustive and separated, $v \in F_n - F_{n-1}$ for some $n \in \mathbb{Z}$. The subglider $F_{-n}v \supset F_{-n-1}v \supset \ldots$ of $F^{-}B$ must be trivial. If it is trivial if type T_1, then $F_m v = 0$ for some $m \in \mathbb{Z}$. Since $0 \neq F_m \subset \mathrm{Ann}_A(Av)$, it follows that $F_0 = F_{-m}F_m \subset \mathrm{Ann}_A(Av)$, a contradiction. If it is trivial of type T_2, then $F_m v = F_{m+1}v$ for some $m \in \mathbb{Z}$. By Proposition 8.1.4 it follows that $Av = F_0 v = F_{-n}v$ for all $n \in \mathbb{Z}$. In particular, $v \in B(FA) = 0$, a contradiction. So the subglider must be trivial of type T_3, but in this case, $F_{-n}v = F_{-m}$ for some $m \in \mathbb{N}$. Hence $F_{m-n}v = F_0$, which shows that $Av = A$, contradiction. □

Proposition 8.1.4 also reveals a link with the theory of orders in simple algebras. Denote $R = F_0K$. We recall that a full R-lattice is a finitely generated R-torsion free module M in A such that $KM = A$. By definition, an R-order C in A is a subring of A which is also a full R-lattice. If C is not contained in any proper R-order D, we call C a maximal order. We refer the reader to [58] for a detailed treatment of the theory of maximal orders.

Corollary 8.1.6. *In the situation of Proposition 8.1.4, $B = F_0A$ is an R-order in A.*

Proof. B is prime since $BK = A$ is a csa and $Z(B) = F_0K = R$ is a DVR hence a Noetherian ring. Moreover, B is a P.I. ring as a subring of a P.I. ring A, hence by a result of Formanek, see [30, Theorem 2]: B is a finitely generated $Z(B)$-module □

Suppose that $R = F_0K$ is a Dedekind domain and B is a maximal R-order in A. A prime ideal of B is by definition a proper two-sided ideal P in B such that $KP = A$ and such that for every pair of two-sided ideals S, T in B and containing P, we have

$$ST \subset P \Rightarrow S \subset P \text{ or } T \subset P.$$

In fact, the prime ideals of B coincide with the maximal two-sided ideals of B and if P is such a prime ideal, then $P \cap R \in \mathrm{Spec(R)}$. Vice-versa, for $p \in \mathrm{Spec(R)}$, $P = B \cap \mathrm{rad}(\mathrm{B_p})$ is a prime ideal of B, and this yields a one-to-one correspondence. Also, the product of prime ideals in B is commutative and every two-sided ideal of B can be written uniquely as a product of prime ideals. Since R is Dedekind, there are only a finite number of prime (hence maximal) ideals $p_1, \ldots, p_n$. If P_i corresponds to p_i, that is, if $P_i = B \cap \mathrm{rad}(\mathrm{B_{p_i}})$ and $p_i = P_i \cap R$, then

$$p_i B = P_i^{e_i}$$

for some $e_i > 0$, and we call e_i the ramification index at P_i. In particular, for $m = ke_i + j$ with $0 < j \leq e_i$ it follows that:

$$P_i^m \cap R = p_i^{k+1}. \tag{8.1}$$

Lemma 8.1.7. *Let FK be an filtration, then $F_{-1}K \subset J(F_0K)$.*

Proof. Let $N \in \mathrm{Max}(\mathrm{F_0K})$ and let $x \in N$. Since FK is exhaustive there exists $n > 0$ such that $x^{-1} \in F_nK - F_{n-1}K$. Hence $F_{-1}^n x^{-1} \subset F_0$ and it follows that $F_{-1}^n \subset (x) \subset N$. Since N is prime, it follows that $F_{-1} \subset N$ and the result follows. □

Lemma 8.1.8. *Let R be Dedekind with quotient field K and B be a maximal R-order in A. If $P \neq Q$ are prime ideals of B, then*

$$PQ \cap K = (P \cap K) \cap (Q \cap K) = (P \cap R) \cap (Q \cap R).$$

Proof. This follows since $PQ \cap K \subset (P \cap R) \cap (Q \cap R)$, by the correspondence of prime ideals and the fact that every two-sided ideal can be written uniquely as the product of prime ideals. □

We denote the ceil function by $\lceil \cdot \rceil$.

Lemma 8.1.9. *Let $e > 0$ and $k, l \in \mathbb{Z}$. We have the inequality*

$$\left\lceil \frac{k}{e} \right\rceil + \left\lceil \frac{l}{e} \right\rceil \geq \left\lceil \frac{k+l}{e} \right\rceil,$$

and we have strict inequality if and only if

$$k = k_e e + i,\ 0 < i \leq e - 1,$$

$$l = l_e e + j,\ 0 < j \leq e - 1$$

and $2 \leq i + j \leq e$.

Proof. By writing $k = k_e e + i, l = l_e e + j$ with $0 \leq i, j \leq e - 1$, we have

$$\left\lceil \frac{k}{e} \right\rceil = k_e + 1 - \delta_{0,i}, \quad \left\lceil \frac{l}{e} \right\rceil = l_e + 1 - \delta_{0,j}.$$

The statements now follow easily. □

Theorem 8.1.10. *Let A be a csa over K with filtration FA such that $F_0A = B$ is a maximal $R = F_0K$-order, with R a Dedekind domain. Then FA is strong if and only if*

$$F_{-1}A = P_1^{k_1} \ldots P_n^{k_n},$$

with $e_i | k_i$ for all $1 \leq i \leq n$, where e_i denotes the ramification index at P_i.

Proof. By the theory of maximal orders, $F_{-1}A = P_1^{k_1} \ldots P_n^{k_n}$ for some $k_i \geq 0$. By Lemma 8.1.8 and (8.1) we obtain that

$$F_{-1}K = p_1^{\left\lceil \frac{k_1}{e_1} \right\rceil} \ldots p_n^{\left\lceil \frac{k_n}{e_n} \right\rceil}.$$

If $e_i | k_i$ for all $1 \leq i \leq n$, then $m \left\lceil \frac{k_i}{e_i} \right\rceil = \left\lceil \frac{mk_i}{e_i} \right\rceil$ for all $1 \leq i \leq n$ and $m \in \mathbb{Z}$. It follows that FK is strong. Conversely, suppose that FK is strong, but that

$\left\lceil \frac{k_1}{e_1} \right\rceil > \frac{k_1}{e_1}$, i.e. $k_1 = le_1 + j$ for some $0 < j < e_1$. Since FK is strong, we have for every m

$$m\left\lceil \frac{k_1}{e_1} \right\rceil = \left\lceil \frac{mk_1}{e_1} \right\rceil,$$

i.e. $(m-1)e_1 < mj < me_1 - m$. It follows that for every $m \in \mathbb{Z}$

$$\frac{m-1}{m}e_1 < j \le e_1 - 1,$$

which is a contradiction. □

8.2 The glider Brauer-Severi variety of a field

Let A be a central simple algebra of degree n^2 over a field K. For every $1 \le r \le n$ one may define the (generalized) Brauer-Severi variety $\mathrm{BS_r(A)}$ as the variety of left ideals of reduced dimension r in A, see [42]. Such an ideal is represented by a non-zero rn-vector $u_1 \wedge \ldots \wedge u_{rn} \in \bigwedge^{rn} A$. If $(e_i)_{1\le i\le n^2}$ denotes a basis of A, then the rn-dimensional subspace represented by $u_1 \wedge \ldots \wedge u_{rn} \in \bigwedge^{rn} A$ is a left ideal of reduced dimension r if and only if it is preserved under left multiplication by $e_1, \ldots, e_{n^2}$. When $r = 1$, we obtain the classical Brauer-Severi variety $\mathrm{BS(A)}$ and we see that

$$\mathrm{BS(A)} = \{\mathrm{L} \le \mathrm{A} \text{ irreducible left ideal}\}.$$

If we consider the set of all left ideals of A then we obtain

$$\sqcup_{r=1}^{n} \mathrm{BS_r(A)}.$$

Definition 8.2.1. Let FR be a filtered ring with subring $F_0R = S$. We call $L \subset R$ a left glider ideal if L is a left FR-glider; in particular L is a filtered left F^-S-module. Analogously, one introduces right glider ideals and glider ideals:

Definition 8.2.2. Let A be a K-algebra. For a filtration FA on A, we define the glider Brauer-Severi variety associated to the filtration FA

$$\mathrm{GBS_F(A)} = \{\text{irreducible left FA}-\text{glider ideals of A}\}.$$

And more generally, we define the glider Brauer-Severi variety

$$\mathrm{GBS(A)} = \bigcup_{\text{FA filtration}} \mathrm{GBS_F(A)},$$

where the union is taken over all separated, exhaustive and unbounded filtrations FA.

The study of irreducible gliders is highly dependent on the type of filtrations one is working with. For example, when working with right bounded filtrations, we showed Lemma 3.2.3. In particular, irreducible gliders for these type of filtrations are of finite essential length d and if they have zero body, their structure is determined by the simple F_0A-module M_d. For the unbounded separated, exhaustive filtrations on central simple algebras, the question of what irreducible gliders look like has not been answered up to now. The following lemma will be useful for tackling this problem

Lemma 8.2.3. *If M is irreducible of infinite essential length, then for all $i \geq 0$ either $M_i = M_{i+1}$ or M_i/M_{i+1} is an irreducible left F_0A-module.*

Proof. Suppose that $M_i > N > M_{i+1}$ is a proper left F_0A-module. Then

$$\begin{array}{ccccccc} M_i & \supsetneq & M_{i+1} & \supset & M_{i+2} & \supset & \dots \\ \cup & & \cup & & \cup & & \\ N & \supset & F_{-1}N & \supset & F_{-2}N & \supset & \dots \end{array}$$

would be a non-trivial subglider of the irreducible glider $M_i \supset M_{i+1} \supset \dots$. □

Throughout the rest of this section, we focus on $A = K$ and determine its glider Brauer-Severi variety GBS(K). For unbounded filtrations, it appears that irreducible gliders have infinite essential length. Indeed, K being a field implies that an irreducible left glider ideal M has zero body. So if el(M) = d, then $F_{-d-1}KM = 0$, a contradiction. Suppose that $M \in \mathrm{GBS}_{\mathrm{F}}(\mathrm{K})$. Lemma 8.2.3 entails that there exists $n \in \mathbb{Z}, i \in \mathbb{N}$ such that $M_i = F_nK$. Indeed, consider on $KM = K$ the filtration given by $F_{-n}KM = M_n, F_n(KM) = \sum_{j \geq 0} F_{j+n}M_j$ for $n \geq 0$. Suppose first that $1 \in M_i \setminus M_{i+1}$. Then $F_iK \subset M$. The submodule

$$\frac{F_iK}{F_iK \cap M_1} \leq \frac{M}{M_1}$$

cannot be zero, whence $F_iK = M$. If $1 \in F_n(KM) - F_{n-1}(KM)$ for some $n > 0$, then we can write

$$1 = \sum f_{j+n} m_j$$

for some $f_{j+n} \in F_{j+n}K, m_j \in M_j$. In this case, $F_{-n}K \subset M$, and since $B(M) = 0$, there exists a maximal $i \geq 0$ such that $F_{-n}K \subset M_i$. By irreducibility as F_0K-module of M_i/M_{i+1} it follows that $F_{-n}K = M_i$.

To continue, we again use $B(M) = 0$ for the existence of an $i_1 \geq i$ maximal such that $M_{i_1} \supset F_{n-1}K$. And by Lemma 8.2.3 we actually have equality. This leads to a sequence $i \leq i_1 \leq i_2 \leq \dots$ such that

$$M_{i_k} = F_{n-k}$$

and for all $0 \leq l \leq i_{k+1} - i_k$ we obtain the equalities

$$M_{i_k+l} = F_{i_{k+1}-l}F_{n-k}.$$

In particular, we have that

$$l_{F_0K}(M_{i_k}/M_{i_{k+1}}) \leq i_{k+1} - i_k.$$

In particular, if FK is a strong filtration, then $i_{k+1} = i_k + 1$ for all k. It follows that there exists an $n \in \mathbb{Z}$ such that M equals

$$(F_nK)_* : \quad F_nK \supset F_{n-1}K \supset F_{n-2}K \supset \ldots$$

Example 8.2.4. Let $O_v \subset K$ be a discrete valuation ring and $F = F^v$ the v-adic filtration on the field of fractions K. Let $L \subset \Omega \in \mathrm{GBS_{F^v}(K)}$. Suppose that $L_n = L_{n+1}$ for some $n \geq 0$, then $F_1L_n = F_1L_{n+1} \subset L_n \subset F_1L_n$, whence $F_1L_n = L_n$ and it follows that $F_nL_n = L_n = K$. But then it follows that $K = F_{-1}L_{n+1} \subset L_{n+2}$ and L is just the glider

$$K \supset K \supset K \supset \ldots,$$

which has the non-trivial subglider $O_v \supset F_{-1} \supset \ldots$. In particular, it follows that $L_0 \supsetneq L_1$. Let $l \in L_0 - L_1$. Then $O_vl \supset F_{-1}l \supset F_{-2}l \supset \ldots$ is a trivial subglider of type T_3, and because $l \notin L_1$, it follows that $L = O_vl$. In particular, L is a fractional ideal and since O_v is a local Dedekind domain, $L = F_n$ for some $n \in \mathbb{Z}$. In fact, one shows that all L_i are fractional ideals and so $L_1 = F_{n+m}$ for some $m < 0$. If $m < -1$, then

$$\Omega \supset F_{n-1} \supset L_1 \supset L_2 \supset L_3 \supset \ldots$$

would be a non-trivial subglider. Hence $m = 1$ and actually one can show that $L_i = F_{-n-i}$ for all $i \geq 0$. Hence we have shown that

$$\mathrm{GBS_{F^v}(K)} = \{(\mathrm{F^v_n})_* \mid \mathrm{n} \in \mathbb{Z}\},$$

where F^v_n has the chain $F^v_n \supset F^v_{n-1} \supset F^v_{n-2} \supset \ldots$

Example 8.2.5. Let p, q be two distinct prime numbers. Consider the filtration on $\mathbb{Q}$ defined by $F_0 = \mathbb{Z}_S$ where S is the multiplicatively closed set generated by all prime numbers except for p and q. The negative part is the (pq)-adic filtration and for $i > 0$ we set

$$F_i = \sum_{k,l \leq i} \mathbb{Z}_S \frac{1}{p^k q^l}.$$

This is a strong filtration. We have that $\Omega = \mathbb{Q}$ whence the M-chain must be

$$F_n \supset F_{n-1} \supset F_{n-2} \supset \ldots$$

starting at some $n \in \mathbb{Z}$, i.e. the glider is $(F_n)_\gamma$. This is however not irreducible since the consecutive quotients are not simple $\mathbb{Z}_S$-modules. Indeed, for any $m \in \mathbb{N}$

$$(pq)^n \supsetneq (p^{n+1}q^n) \supsetneq (pq)^{n+1}.$$

These examples indicate that for strong filtrations FK on fields, non-emptiness of $\mathrm{GBS_F(K)}$ is equivalent to F_0K being a discrete valuation ring. Indeed, we have the following:

Proposition 8.2.6. *Let K be a field with strong filtration FK. Then $\mathrm{GBS_F(K)} \neq \emptyset$ if and only if $F_0K = O_v$ is a DVR and $F = F^v$ is the associated v-adic filtration.*

Proof. Suppose that $\mathrm{GBS_F(K)} \neq \emptyset$ and let $M \in \mathrm{GBS_F(K)}$. Suppose first that $F_0 = O_v$ is a DVR. There exists $a > 0$ such that $F_{-1} = (\pi^a)$ if π is a uniformizing parameter of F_0. By the structure of the elements in the $\mathrm{GBS_F(K)}$ for strong filtrations, there exists an $n \in \mathbb{Z}$ such that $M = F_n$. If $n \geq 0$, then

$$F_0 \supset F_{-1} \supset F_{-2} \supset \dots$$

is also irreducible and Proposition 8.1.3 entails that the filtration corresponds to the valuation filtration. If $-n > 0$ then $M = F_n = (\pi^{an})$. Since FK is strong, $M_1 = F_{n-1} = (\pi^{a(n+1)})$. Because $M/M_1 \cong (\pi^a)$ is simple, $a = 1$ and it follows that $FK = F^vK$. Conversely, suppose that F_0 is not a DVR. In particular, when $M = F_nK$ as before, the n must be strictly smaller than 0 by Proposition 8.1.3. Let $y \in F_0 \setminus F_{-1}$. Then $y^{-1} \in \dot{F}_m$ for some $m \geq 0$. We want to show that $m = 0$. The glider

$$y^{-1}F_{-1}^{m+n} \supset y^{-1}F_{-1}^{n+m+1} \supset \dots$$

is a subglider of

$$F_{-1}^n \supset F_{-1}^{n+1} \supset \dots,$$

whence must be trivial of type T_3. Hence there exists $r \geq 0$ such that $y^{-1}F_{-1}^{m+n} = F_{-1}^{n+r}$. It follows that $y^{-1} \in F_{-1}^{n+r}F_{-1}^{m+n} = F_{m+n-n-r} = F_{m-r}$, whence $m - r \geq m$ or $r \leq 0$. So $r = 0$ and $y^{-1}F_{-1}^{m+n} = F_{-1}^n$ from which $y^{-1}F_{-1}^m = F_0$ follows. In particular we obtain $1 = y^{-1}x$ for some $x \in F_{-1}^m$. However, x must be y and it follows that $y \in F_{-1}^m$. Finally, we started with $y \in \dot{F}_0$; m must indeed be equal to 0. This shows that F_0 is local with maximal ideal F_{-1}. However, since F_0 is not a DVR, F_{-1} is not principal. Since FK is strong, there exists $x \in F_{-1} - F_{-2}$. The subglider

$$\begin{array}{ccccc} F_{-n} & \supset & F_{-n-1} & \supset & \dots \\ \cup & & \cup & & \\ xF_{-n+1} & \supset & xF_{-n} & \supset & \dots \end{array}$$

must be trivial of type T_3, so $xF_{-n+1} = F_{-n-r}$ for some $r \geq 0$, or $(x) = F_{-r-1}$. It follows that $r = 0$, which contradicts the principality of F_{-1}. Hence $\mathrm{GBS_F(K)} = \emptyset$.

□

Together with Example 8.2.4 we obtain that when running over all unbounded, separated, exhaustive strong filtrations FK on K, we have

$$\bigcup_{FK} \mathrm{GBS_F(K)} = \mathrm{R(K)} \times \mathbb{Z}, \tag{8.2}$$

where $R(K)$ denotes the Riemann surface of K. Filtrations that are not strong can also have non-zero glider Brauer-Severi variety. Indeed, one can for example consider a DVR $R = O_v$ with maximal ideal $M = (x)$. The positive part on K is just defined by the standard filtration generated by $F_1K = R + Rx^{-1}$. And for the negative part, one takes

$$R \supset M^2 \supset M^3 \supset \dots$$

In fact $\mathrm{GBS_F(K)} = \mathrm{GBS_{F^v}(K)}$.

Proposition 8.2.7. *Let K be a field with filtration FK. If $\mathrm{GBS_F(K)} \neq \emptyset$, then there exists $n \geq 0$ such that for all $m \geq n$, F_{-m} is a principal ideal of F_0.*

Proof. By the structure of elements of the glider Brauer-Severi variety, we know that $M = F_l F_m$ for some $l \geq 0$ and $m \in \mathbb{Z}$. In particular, we know that an F_{-n} for $n \geq 0$ appears as an M_i. Take $a \in M_i$, then

$$\begin{array}{ccccccccccc} F_{-n} & \supset & F_{i_1-1}F_{-n-1} & \supset & \dots & \supset & F_1F_{-n-1} & \supset & F_{-n-1} & \supset & \dots \\ \cup & & \cup & & & & \cup & & \cup & & \\ (a) & \supset & (a)F_{-1} & \supset & \dots & \supset & (a)F_{-i_1+1} & \supset & (a)F_{-i_1} & \supset & \dots \end{array}$$

is a subglider of type T_3. In particular, there exists $k_a \geq 0$ and $0 \leq l_a \leq i_{k_a}$ such that $(a) = F_{l_a}F_{-n-k_a}$. If $(a) \subset F_{-n}$ is proper, then there exists $a_1 \in F_{-n} - (a)$, and there exist $k_{a_1} \geq 0$ and $0 \leq l_{a_1} \leq i_{k_{a_1}}$ such that $(a_1) = F_{l_{a_1}}F_{-n-k_{a_1}}$. It must hold that $k_{a_1} \leq k_a$ and if equality holds, then $l_{a_1} > l_a$. In particular $(a) \subsetneq (a_1) \subset F_{-n}$. If the last inclusion is proper, then we can continue this argument, which must stop by the restrictions on the k_{a_j}, l_{a_j}. This shows that F_{-n} is indeed principal. The glider starting from F_{-n-m} for any $m \geq 0$ remains irreducible, and the reasoning above shows that F_{-n-m} is again principal. □

Lemma 8.2.8. *Let FK be filtration on a field K. If $M \in \mathrm{GBS_F(K)}$ with $M = F_0K = R$, then the negative chain $F_{-n}R = M_n$ defines a ring filtration on R.*

Proof. Let $i > 0$ and consider the subglider

$$\begin{array}{ccccccccccc} M & \supset & M_1 & \supset & \dots & \supset & M_i & \supset & M_{i+1} & \supset & \dots \\ \cup & & \cup & & & & \cup & & \cup & & \\ MM_i & \supset & M_1M_i & \supset & \dots & \supset & M_iM_i & \supset & M_{i+1}M_i & \supset & \dots \end{array}$$

Since K is a field it cannot be of type T_2. Suppose that $M_nM_i = M_{n+m}M_i$ for some n and all $m > 0$. It would follow that $F_mKM_nM_i = F_mKM_{n+m}M_i \subset$

M_nM_i. Exhaustivity of FK implies that M_nM_i is a K-vector space in K, i.e. $M_nM_i = K$, a contradiction. The subglider is thus of type T_3. Because $MM_i = M_i$, $\alpha(0) \geq i$, which implies that for every $j > 0$, $\alpha(j) > i+j$, which amounts to saying that

$$M_jM_i = M_{\alpha(j)} \subset M_{i+j},$$

or

$$F_{-j}RF_{-i}R \subset F_{-i-j}R,$$

proving the claim. □

Proposition 8.2.9. *Let FK be a non-strong filtration such that* $\mathrm{GBS_F(K)} \neq \emptyset$ *and $F_0K = R$. Then there exists a strong e-step filtration F^sK on K such that $F^+K = F^{s,+}K$ and F^-R is a trivial subglider of type T_3 of $F^{s,-}K$.*

Proof. Let $M \in \mathrm{GBS_F(K)}$. Up to considering the irreducible glider $M_n \supset M_{n+1} \supset \ldots$, there exists an $m \in \mathbb{Z}$ such that $M = F_mK$. Suppose that $m < 0$. Let $x \in F_{-m}K$, then since FK is exhaustive, $x^{-1} \in F_nK$ for some $n > 0$. Since we are working in a field K, the glider

$$xM \supset xM_1 \supset xM_2 \supset \ldots$$

is also irreducible. And because $1 \in xM$, we have $R \subset xM$ and it follows that there appears an F_rK as an xM_j for some $r \geq 0$, hence also $F_0K = R$ appears. In particular, whenever $\mathrm{GBS_F(K)} \neq \emptyset$, we can find an irreducible glider starting with $M = F_0K = R$. By the structure of irreducible gliders, we know that there exists $n > 0$ such that $M_n = F_{-1}K$, which leads to $F_0K = F_nKF_{-1}K$. So actually $n > 1$. The previous lemma entails that the M-chain defines a negative ring filtration on R. Define the filtration F^sK on K by

$$F^s_nK = F_nK, \quad F^s_{-n}K = M_n, \quad n \geq 0.$$

Proposition 8.1.2 entails that F^sK is a strong e-step filtration. Since n is the smallest number such that $F_nKF_{-1}K = R$, it follows that $e \leq n < 2e$. And since, $F_{n-1}KF_{-1}K < R$, n actually equals e. □

The e in the above proposition is determined by the positive filtration F^+K; i.e. it is the smallest number $m \geq 1$ such that $F_mK \supsetneq F_{m-1}K$.

Theorem 8.2.10. *Let K be a field, then*

$$\mathrm{GBS(K)} = \mathrm{R(K)} \times \mathbb{Z}.$$

Proof. Let FK be a filtration. If $\mathrm{GBS_F(K)} \neq \emptyset$, then $\mathrm{GBS_F(K)} = \mathrm{GBS_{F^s}(K)}$, so the result follows by (8.2).

□

8.3 The relative glider Brauer-Severi variety for a central simple algebra

In this section we determine the relative glider Brauer-Severi variety GBS(A) of a central simple algebra A over a field K, meaning that we run over all filtrations FA extending some fixed filtration FK, i.e. satisfying $F_nA \cap K = F_nK$ for all $n \in \mathbb{Z}$. Before we put some additional conditions on FK, we prove the following.

Lemma 8.3.1. *Let FA be a filtration extending FK. If $M \in \mathrm{GBS_F(A)}$ and M is not a left A-module, then* $\mathrm{el(M)} = +\infty$.

Proof. Suppose that $\mathrm{el(M)} = \mathrm{d} < +\infty$. It follows that $F_{-d-1}M \subset B(M)$. However, since $0 \neq F_{-d-1}K \subset F_{-d_1}$ contains invertible elements, it follows that $M \subset B(M) \subset M$. This entails that the glider is $M \supset M \supset M \supset \dots$. If $M = B(M) = Av$ is a left ideal of A, then $F_0Av \supset F_{-1}Av \supset F_{-2}Av \supset \dots$ is a non-trivial subglider, a contradiction. □

Corollary 8.3.2. *In the situation of the previous lemma, if $N = M_i$ is a left A-module for some $i \in \mathbb{N}$, then $M_i = N = M$.*

Proof. The previous lemma entails $\mathrm{el(M)} = +\infty$, whence we may assume that $M_i \supsetneq M_{i+1}$. The subglider

$$\begin{array}{ccccccccc} M & \supset & \dots & \supset & M_i & \supsetneq & M_{i+1} & \supset & \dots \\ \cup & & & & \cup & & \cup & & \\ N & \supset & \dots & \supset & N & \supsetneq & M_{i+1} & \supset & \dots \end{array}$$

must be trivial of type T_3 and α the identity map. □

Proposition 8.3.3. *Suppose we are in the situation of Lemma 8.3.1. If $N \subsetneq M$ is a left A-module, then $N \subset B(M)$.*

Proof. Corollary 8.3.2 entails $N \neq M_i$ for all $i \geq 0$. Irreducibility of M entails that $N \supset N \cap M_1 \supset N \cap M_2 \supset \dots$ must be trivial and it cannot be of type T_3. Since A is semisimple, triviality of type T_2 would entail that $AM = N \oplus V$ for some left A-module V and $M = N \oplus W$ for $W \subset V$ a left F_0A-submodule. Since $M_n \subset W$ it follows that $W \cap M \supset W \cap M_1 \supset \dots$ would be a non-trivial subglider. Hence the subglider is trivial of type T_1, i.e. there exists $n > 0$ such that $N \cap M_n = N \cap M_{n+m}$ for all $m \geq 0$. Let $x \in N$ and $k \geq n \in \mathbb{N}$. Then $F_{-k}N \subset N \cap M_k$. Since $F_{-k}K \neq 0$, take $s_k \in F_{-k}K$. If $t_k s_k = 1$ with $t_k \in F_{\deg(t_k)}K$, then

$$x = t_k s_k x \in t_k(N \cap M_k) = t_k(N \cap M_{k+\deg(t_k)}) \subset M_k,$$

which shows that $N \subset M_k$ for all $k \geq n$, and $N \subset B(M)$ follows. □

Proposition 8.3.4. *Let A be a csa over K with filtration FA inducing FK with non-trivial negative part. If $M \in \mathrm{GBS_F(A)}$ then $B(M) = 0$ and AM is an irreducible principal left A-module.*

Proof. A csa is left Noetherian, so $AM/B(M)$ is finitely generated. We first show that $AM/B(M)$ is a principal left A-module. Suppose that $AM/B(M) = A\overline{v} + A\overline{w}$ is generated by two elements. The proof for more generators is an easy modification. Let v, w be lifts for $\overline{v}, \overline{w}$ respectively. If $F_n v \subset M$ for all $n \in \mathbb{Z}$, then $Av \subsetneq M$ and $Av \subset B(M)$ by Proposition 8.3.3, which would lead to $\overline{v} = 0$. Reasoning similarly for Aw we have shown that there exist n, m maximal such that $F_n v \subset M$ and $F_m w \subset M$. Since $\overline{v} \neq 0$ there exists $i \geq 0$ minimal such that $F_n v \subset M_i$ and $F_n v \not\subset M_{i+1}$. In particular, $M_{i+1} \subsetneq M_i$ and Lemma 8.2.3 entails that $M_i = F_n v$. Similarly, we arrive at the equality $F_m w = M_j$ for some j. Without loss of generality, we may assume $i \leq j$, i.e. $F_n v \subset F_m w$. However, since $F_n K \subset F_n$ contains invertible elements, it follows that $v \in Aw$, which entails $AM/B(M)$ is principal. Since A is semisimple, $\Omega = AM = B(M) \oplus Av$ for some left A-module Av. However, the irreducible left glider M then becomes a strong glider direct sum - i.e. the sum is direct on every level -

$$B(M)\dot{\oplus}\Big(N \supset N_1 \supset \dots\Big),$$

$N = F_n v$ for some $n \in \mathbb{Z}$. Irreducibility then shows that $B(M)$ must be zero and Av must be an irreducible left ideal. □

The previous proposition shows that $AM = Av$ for some $Av \in \mathrm{BS(A)}$, the classical Brauer-Severi variety. One can perform similar reasoning to ours for fields to show that there exists $n \in \mathbb{Z}, i \in \mathbb{N}$ such that $M_i = F_n Av$. If we further assume that FA is strong, then M becomes the glider

$$(F_m Av)_* : \quad F_m Av \supset F_{m-1}Av \supset F_{m-2}Av \supset \dots \tag{8.3}$$

On the field, we put a valuation filtration corresponding to a DVR $F_0K = O_v = R$. Since FA extends FK, we have for $n \geq 0$, $F_{-n}KF_nA \subset F_0A$ and since FK is strong, it follows that $F_nA \subset F_nKF_0A$. The other inclusion is obvious, so $F_nA = F_nKF_0A$ for all $n \in \mathbb{Z}$. This shows that FA is strong as well, and the irreducible left glider ideals are of the form (8.3). We denote the relative glider Brauer-Severi variety with regard to FK by $\mathrm{GBS^{FK}(A)}$.

Proposition 8.3.5. *Let A be a csa with FA extending FK a valuation filtration with F_0K a DVR and put $F_0A = B$. Let M be non-empty in* $\mathrm{GBS_F(A)} = \mathrm{GBS^{FK}(A)}$*; then $AM = KM = Av$ with $Av \in \mathrm{BS(A)}$, v any generator of an A-module. Then there is an $n \in \mathbb{N}$ such that $M_i = F_{n-i}Av$ and $M_i = F_{n-i}Av \cap Kv = F_{n-i}Kv$ is an irreducible FK-glider in Kv. Thus the $F_{n-i}Kv$ define an element of* $\mathrm{GBS_F(K)}$. *As a consequence we get* $\mathrm{GBS^{FK}(A)} = \mathrm{BS(A)} \times \mathbb{Z}$ *(as sets).*

Proof. We have $AM = KM = Av$ by Proposition 8.3.4, and also $B(M) = 0$ and $Av \in \mathrm{BS(A)}$. Since $v \in KM, \lambda v \in M$ for some $\lambda \in K^*$ we may assume v is chosen so that $v \in M$. We have a subglider

$$\begin{array}{ccccccccc} M = M_0 & \supset & M_1 & \supset & \dots & \supset & M_n & \supset & \dots \\ \cup & & \cup & & & & \cup & & \\ Bv & \supset & F_{-1}Av & \supset & \dots & \supset & F_{-n}Av & \supset & \dots \end{array}$$

Since FK is strong, FA is strong too and it follows for n that $F_nKF_0A = F_nA$. Since $B(M) = 0, B(Bv) = 0$ too and so no $F_{-n}Av \supset F_{-n}Kv$ is zero; thus the glider BV above has to be trivial of type T_3, say $Bv = M_n$ for some $n \geq 0$. Then $M = F_nAM_n = F_nKM_n$ yields $M_i = F_{n-i}Av$ for $i \geq 0$. Since FK is a valuation filtration, we know that $F_nK \supset \ldots \supset F_{n-i}K \supset \ldots$, is an irreducible FK-glider hence an element of $\mathrm{GBS_F(K)} \subset \mathrm{GBS(K)}$. But then $K \cong Kv$ yields that $F_nKv \supset F_{n-1}Kv \supset \ldots$, is also an irreducible glider in Kv. Now consider the chain

$$Kv \supset M_i \cap Kv \supset F_{n-i}Kv.$$

Then $B(M_i \cap Kv)$ contains $BF_{n-i}Kv = F_{n-i}Av = M_i$, thus $B(M_i \cap Kv) = M_i$. If for some M_i we would have $M_i \cap Kv \supsetneq F_{n-i}Kv$ then since FK is a valuation filtration $M_i \cap Kv = F_mKv$ with $m > n - i$ and then $M_i = B(M_i \cap Kv) = BF_mKv = F_mAv \supset BF_{n-i}Kv = M_i$ with $m > n - i$. Thus $F_{-1}KM_i = M_i$ and since FK is strong $F_{-1}K \subset J(F_0K)$. Now B is prime since BK is a csa and $Z(B) = F_0K$ is a DVR hence a Noetherian ring. The same reasoning as in the proof of Corollary 8.1.6 shows that B is a finitely generated $Z(B)$-module, and thus $F_{-1}KM_i = M_i$ with $M_i = BF_{n-i}Kv$ also a finitely generated F_0K-module, which yields $M_i = 0$, a contradiction. Hence for all $M_i \neq 0$ we thus get $M_i \cap Kv = F_{n-i}Kv$. If we choose another generator w for Av, thus $Av = Aw$ then w in some $F_mA - F_{m-1}A$ and the change from v to w in the foregoing is just coming down to a shift by $F_{m-n}K$ in the glider of K we found; consequently the Av define the element of $BS(A)$, the FA corresponds to the valuation F_0K and the choice of generator for $KM = Av$ yields a shift of the K-glider $F_0K \supset F_{-1}Kv \supset \ldots \supset F_{-n}K \supset \ldots$. Since we may replace v by any λv with $\lambda \in K^*$, it is clear that all shifts over $\mathbb{Z}$ appear and thus we finally arrive at

$$\mathrm{GBS^{FK}(A) = BS(A) \times \mathbb{Z}}.$$

as sets. □

By the previous section we know that we obtain the whole of GBS(K) by running over all valuation filtrations on K. Hence, we define the relative glider Brauer-Severi variety

$$\mathrm{GBS^K(A)} = \bigcup_{\text{FK valuation filtration}} \mathrm{GBS^{FK}(A)}.$$

Corollary 8.3.6. *We have a bijection as sets:*

$$\mathrm{GBS^K(A) = BS(A) \times R(K) \times \mathbb{Z} = BS(A) \times GBS(K)}.$$

Corollary 8.1.6 indicated a link between irreducible gliders and the theory of orders. This indication also manifests itself in the relative glider Brauer-Severi varieties of central simple algebras when FK is a valuation filtration. Before we exploit this, let us recall a few facts about maximal orders over a DVR R.

For a detailed overview, the reader is referred to [58]. A maximal R-order in a central simple algebra $M_n(D)$ is conjugate to $M_n(\Lambda)$ where Λ is the unique maximal R-order in D, a central division algebra. In fact, Λ is the integral closure of R in D. The 1- and 2-sided ideals in Λ are of the form $\pi^i\Lambda$. It follows that maximal R-orders in $M_n(D)$ are also principal ideal rings and there exists a uniformizer Π generating these ideals.

Proposition 8.3.7. *Let B be a maximal R-order in A with filtration FA extending FK. If $\mathrm{GBS_F(A)} \neq \emptyset$ then the ramification index e at $F_{-1}A = P = \Pi B$ is 1.*

Proof. If Π, π are uniformizers for B, R respectively, then $\pi B = \Pi^e B$. If $M \in \mathrm{GBS_F(A)}$, then there exists $m \in \mathbb{Z}$, $Av \in P(A)$ such that $M = F_m KBv = \pi^m Bv = \Pi^{me} Bv$ and $M_1 = F_{m-1}KBv = \pi^{m-1}Bv = \Pi^{(m-1)e}Bv$. The quotient M/M_1 must be a simple left B-module, hence $e = 1$. □

Example 8.3.8. Consider $R = \mathbb{Z}_{(2)}[i] \subset K = \mathbb{Q}(i)$. Then $M_n(R)$ is a maximal R-order in $A = M_n(K)$, with ramification index 2 at the prime ideal $P = M_n((1+i)R)$: $P^2 = M_n(2R) = (2)M_n(R)$. Let $Av \leq_l A$ correspond to $v \in \mathbb{P}^n$, i.e. to

$$\left\{ \begin{pmatrix} (\sum_{i=1}^n a_{i1})v \\ (\sum_{i=1}^n a_{i2})v \\ \cdots \\ (\sum_{i=1}^n a_{in})v \end{pmatrix} \middle| \, a_{ij} \in K \right\}.$$

From this it follows readily that $M_n(R)v/M_n(2R)v \cong M_n(R/2R)\overline{v}$ is not irreducible as left $M_n(R)$-module.

There is an obvious way to also reach the maximal orders with ramification index strictly bigger than 1. Indeed, we can allow the valuation filtration FK to have a higher step size $f > 1$. If FA is a filtration extending FK, then we only have that $F_{nf}A = F_{nf}KF_0A$ for all $n \in \mathbb{Z}$. The filtration FA is therefore not strong in general. If we do impose the strongness condition, we get a generalization of Proposition 8.3.7.

Proposition 8.3.9. *Let FK be the strong f-step filtration associated with $R = O^v$. Let FA be a strong filtration extending FK such that $F_0A = B$ is a maximal R-order. Then $\mathrm{GBS_F(A)} \neq \emptyset$ implies $e = f$ is the ramification index at $F_{-1}A = P$.*

Proof. Let p denote the maximal ideal of R. The filtration FA being strong implies that $M = (F_nA)_\gamma v \in \mathrm{GBS_F(A)}$, for some $n \in \mathbb{Z}$ and we may assume that $M = F_{-mf}Av$ for some $m \geq 0$. We have $M = F_{-mf}KBv = p^m Bv = P^{me}v = F_{-1}^{me}v$ and in a similar way we have $M_f = P^{(m+1)e}v$. Strongness of FA furthermore implies that $F_jAv \neq F_{j-1}Av$ for all $j \geq n$. Together with

$$P^{(m+1)e}v = M_f = F_{-f}AM = P^f P^{me}v,$$

this shows that $e = f$. □

Proposition 8.3.10. *Let A be a csa with strong filtration FA extending FK an f-step valuation filtration with F_0K a DVR and put $F_0A = B$. Let M be non-empty in* $\mathrm{GBS_F(A)}$*; then $AM = KM = Av$ with $Av \in \mathrm{BS(A)}$, v any generator of an A-module. Then there exists $0 \leq d < f$ and $m \in \mathbb{N}$ such that M is the chain*

$$\begin{array}{ccccccccccccccccc} F_dAF_{mf}Kv & \supset & F_{d-1}AF_{mf}Kv & \supset \dots \supset & F_1AF_{mf}Kv & \supset & F_{mf}Kv & \supset & F_{f-1}AF_{(m-1)f}Kv & \supset \dots \supset & F_1AF_{(m-1)f}Kv & \supset & F_{(m-2)}Kv & \supset \dots \\ \| & & \| & & \| & & \| & & \| & & \| & & \| & \\ M & \supset & M_1 & \supset \dots \supset & M_{d-1} & \supset & M_d & \supset & M_{d+1} & \supset \dots \supset & M_{d+f-1} & \supset & M_{d+f} & \supset \dots \end{array}$$

Moreover, the chain $M_d \supset M_{d+f} \supset M_{d+2f} \supset \dots$ defines an element in $\mathrm{GBS_F(K)}$.

Proof. Essentially just modify the proof of Proposition 8.3.5. □

We observe that unlike the case for $f = 1$, strong filtrations FA and $F'A$ extending an f-step valuation filtration FK have different $\mathrm{GBS_F(A)} \neq \mathrm{GBS_{F'}(A)}$ if $F_0A \neq F_0'A$. In particular, $\mathrm{GBS^{F^fK}}$ contains as a subset

$$\mathrm{BS(A)} \times \mathrm{A(R,f)} \times \mathbb{Z},$$

where $A(R,f)$ denotes the set of all maximal R-orders in A with ramification index f. When considering all discrete valuations on K, for every point of the Riemann surface $R(K)$ we obtain differences corresponding to the maximal orders of ramification index f lying over it. It remains a question whether there exist filtrations FA extending an f-step valuation filtration with non-empty $\mathrm{GBS_F(A)}$ and which are not strong.

8.4 Tensor product with a field extension

Let A be a csa over K with separated, exhaustive filtration FA inducing a filtration FK on K. Let L/K be a field extension and choose a filtration FL on L inducing FK on K. We define a filtration on $A \otimes_K L$ by putting

$$F_q(A \otimes_K L) = \sum_{k \leq q} F_kA \otimes F_{q-k}L,$$

where the tensor product is taken over K. Note that F_kA and $F_{q-k}L$ are not necessarily K-vector spaces, but we consider them inside K, resp. L. The following table shows how the $F_q(A \otimes_K L)$ are formed.

F_{-2}	F_{-1}	F_0	F_1	F_2
				$F_2A \otimes F_0L$
			$F_1A \otimes F_0L$	$F_1A \otimes F_1L$
		$F_0A \otimes F_0L$	$F_0A \otimes F_1L$	$F_0A \otimes F_2L$
	$F_{-1}A \otimes F_0L$	$F_{-1}A \otimes F_1L$	$F_{-1}A \otimes F_2L$	$F_{-1}A \otimes F_3L$
$F_{-2}A \otimes F_0L$	$F_{-2}A \otimes F_1L$	$F_{-2}A \otimes F_2L$	$F_{-2}A \otimes F_3L$	$F_{-2}A \otimes F_4L$
$\vdots$	$\vdots$	$\vdots$	$\vdots$	$\vdots$

Let $\Omega \supset M$ be a (left) FA-glider. Inside $\Omega \otimes_K L$ we want to define a glider representation for $F(A \otimes_K L)$, which we denote by $M \otimes L$. We define

$$(M \otimes L)_p = \sum_{i \geq p} M_i \otimes F_{i-p}L. \tag{8.4}$$

We depict $M \otimes L$ in the following table to show how the chain is formed.

0	1	2
$M_0 \otimes F_0L$		
$M_1 \otimes F_1L$	$M_1 \otimes F_0L$	
$M_2 \otimes F_2L$	$M_2 \otimes F_1L$	$M_2 \otimes F_0L$

for $a \otimes s \in F_j(A \otimes_K L)$ and $m \otimes t \in (M \otimes L)_i$ with $j \leq i$, we define

$$(a \otimes s) \cdot (m \otimes t) = (am \otimes st). \tag{8.5}$$

Lemma 8.4.1. *The left $A \otimes_K L$-module $\Omega \otimes_K L$ with chain and partial actions defined by* (8.4) *and* (8.5) *yields a left $F(A \otimes_K L)$-glider representation.*

Proof. Let $j \leq i$, then

$$\begin{aligned}\Big(\sum_{k\leq j} F_kA \otimes F_{j-k}L\Big)\Big(\sum_{l \geq i} M_l \otimes F_{l-i}L\Big) &\subset \sum_{k \leq j}\sum_{l \geq i} F_kAM_l \otimes F_{j-k}LF_{l-i}L \\ &\subset \sum_{k \leq j}\sum_{l \geq i} M_{l-k} \otimes F_{l-k+j-i},\end{aligned}$$

which is a subset of $\sum_{h \geq l-k} M_h \otimes F_{h+j-i} = (M \otimes L)_{i-j}$. □

In case we start from a strong filtration FK on K, the extended filtration on L satisfies $F_nL = F_nKF_0L$ and the filtration on $A \otimes_K L$ becomes $FA \otimes F_0L$ and for the glider filtration we obtain the chain

$$M \otimes F_0L \supset M_1 \otimes F_0L \supset M_2 \otimes F_0L \supset \dots$$

Proposition 8.4.2. *When FK is strong, we obtain a map*

$$\mathrm{GBS_F(A)} \to \mathrm{GBS_F(A \otimes_K L)}.$$

Proof. Take $M \in \mathrm{GBS_F(A)}$ and consider an $F(A \otimes_K L)$-subglider $N \supset N_1 \supset \dots$ of $M \otimes F_0L \supset M_1 \otimes F_0L \supset \dots$. By the previous section, we know that $\Omega = Av$ is a simple left ideal, whence so is $Av \otimes_K L$. If $N = 0$; then there is nothing to prove. If $N \neq 0$, then it follows that $(A \otimes_K L)N = (A \otimes_K L)(M \otimes_K F_0L)$. In particular, there exist $m \in M, l \in F_0L, a_i \in A, l_i \in F_0L, n_i \in N$ such that

$$m \otimes l = \sum_{i=1}^{t} (a_i \otimes l_i)n_i.$$

Because FA is unbounded below, we can find elements $b \in A$ such that $ba_i \in F_0A$ and $bm \in M$. It follows that

$$bm \otimes l = \sum_{i=1}^{t} (ba_i \otimes l_i)n_i \in N.$$

In particular, we know that N contains a monomial, say $m \otimes l \in M \otimes F_0 L \cap N$. In fact, for every $i \geq 0$, N_i contains a monomial $m_i \otimes l$. Consider now the $FA \otimes F_0 K$-subglider

$$\begin{array}{ccccccc} M \otimes F_0 Kl & \supset & M_1 \otimes F_0 Kl & \supset & M_2 \otimes F_0 Kl & \supset & \dots \\ \cup & & \cup & & \cup & & \\ N \cap (M \otimes F_0 Kl) & \supset & N_1 \cap (M_1 \otimes F_0 Kl) & \supset & N_2 \cap (M_2 \otimes F_0 Kl) & \supset & \dots \end{array}$$

Since $FA \otimes F_0 K \cong FA$ is an isomorphism of filtered rings and since K is central in A, the glider $M \otimes F_0 Kl \supset M_1 \otimes F_0 Kl \supset \dots$ is isomorphic to $M \supset M_1 \supset \dots$, whence is irreducible. If $N \cap (M \otimes F_0 Kl)$ is trivial of type T_1, then there exists $n \geq 0$ such that for all $m \geq 0$

$$N_n \cap (M_n \otimes F_0 Kl) = N_{n+m} \cap (M_{n+m} \cap F_0 Kl) = B(N \cap (M \otimes F_0 Kl)).$$

In this case, $N_n \cap (M_n \otimes F_0 Kl) = B(N \cap (M \otimes F_0 Kl) \subset B(M \otimes F_0 Kl)$ would be a non-zero left A-module, which contradicts Proposition 8.3.4. Since every N_i contains a monomial of the form $m_i \otimes l$, the subglider cannot be trivial of type T_2, so it is of type T_3 with monotone increasing map α. Since we can choose any $l \in F_0 L$ above, we can assume $l \in F_0 L^\times$. Now, suppose that the monomial $m \otimes l \in N$ is such that $m \in M_s - M_{s+1}$. Since FA is strong, there exists $c \in F_s A - F_{s-1} A$, whence $(cm \otimes l) \in (M \otimes F_0 L) \cap N$ and not in $M_1 \otimes F_0 L$. From this it follows that $\alpha = id$. Because l is invertible in $F_0 L$, it follows that

$$M_m \otimes F_0 Ll = M_m \otimes F_0 L \subset N_m \subset M_m \otimes F_0 L,$$

for every $m \geq 0$. □

Example 8.4.3. Let FK be the valuation filtration for $F_0 K = O_v$ a DVR and let O_w be a valuation extension of O_v with fraction field L/K and consider on L the w-adic filtration. If FA extends FK, then $F(A \otimes_K L)$ extends FL, whence also FK. Since we have $F_n(A \otimes_K L) = F_n A \otimes O_w = F_n K(F_0 A \otimes O_w)$, Proposition 8.4.2 yields a map

$$\mathrm{GBS^{FK}(A)} \to \mathrm{GBS^{FK}(A \otimes_K L)},$$

which sends $M = (F_n Ka)_*$ to $(F_n Kb)_*$ if $Aa \otimes_K L = (A \otimes_K L)b \in \mathrm{BS(A \otimes_K L)}$. By choosing a valuation extension for every element in $R(K)$, we obtain a map

$$\mathrm{BS(A) \times GBS(K) = GBS^K(A) \to GBS^K(A \otimes_K L) = BS(A \otimes_K L) \times GBS(K)}.$$

This shows that the relative glider Brauer-Severi variety $\mathrm{GBS^K(A)}$ is a twist of the relative glider Brauer-Severi variety of matrix algebra $M_n(L)$ for a suitable field extension L/K.

8.5 The glider Brandt groupoid

Let K be a field with filtration FK such that $R = F_0K$ is a Dedekind domain. Consider a central simple K-algebra A. Classically, one defines the Brandt groupoid as the set of all normal ideals M in A with proper multiplication, see [58, Section 19-22]. Such a normal ideal M is a finitely generated R-torsion free R-module such that $KM = A$. We will consider finitely generated R-torsion free FK-gliders

$$M \supset M_1 \supset M_2 \supset \dots$$

inside A and such that $KM = A$. Define the following subrings of A

$$O_l(M_i) = \{x \in A \mid xM_i \subset M_i\},$$

$$O_l^{gl}(M) = \{x \in A \,|\, xM_i \subset M_i \ \forall i \geq 0\}.$$

Suppose that M is generated by $\{(m_i, d_i), 1 \leq i \leq t\}$, i.e.

$$M_n = \sum_{i=1}^{t} F_{d_i - n}Km_i$$

for all $n \geq 0$. Moreover, Lemma 8.1.7 and Lemma 2.1.5 entail that $m_i \in M_{d_i} - M_{d_i+1}$.

Lemma 8.5.1. *We have the equalities*

$$O_l^{gl}(M) = \bigcap_{i=1}^{t} O_l(M_{d_i}) = \bigcap_{i\geq 0} O_l(M_i).$$

Proof. Suppose that $x \in \bigcap_{i=1}^{t} O_l(M_{d_i})$ and let $i \geq 0$. We have that

$$\begin{aligned} xM_0 &= x\sum_{k=1}^{t} F_{d_k - i}Km_k = \sum_{k=1}^{t} F_{d_k - i}Kxm_k \\ &\subset \sum_{k=1}^{t} F_{d_k - i}KM_{d_k} \subset M_i, \end{aligned}$$

whence $x \in O_l^{gl}(M)$. The result now follows since we have the inclusions

$$\bigcap_{i=1}^{t} O_l(M_{d_i}) \subset O_l^{gl}(M) \subset \bigcap_{i\geq 0} O_l(M_i) \subset \bigcap_{i=1}^{t} O_l(M_{d_i}). \qquad \square$$

We denote $B = B^l = O_l^{gl}(M)$ and define a filtration FA on A by putting $F_0A = B$ and $F_nA = F_nKB$.

Proposition 8.5.2. *The chain $M \supset M_1 \supset \ldots$ is a left FA-glider ideal.*

Proof. The previous lemma shows that all M_i are left B-modules. For $i \leq j$ we have

$$F_iAM_j = BF_iKM_j \subset BM_{j-i} \subset M_{j-i}. \quad \square$$

Analogously, one introduces $O_r^{gl}(M)$, $O_r(M_i)$ and one obtains a subring B^r yielding another filtration F^rA on A such that M becomes a right F^rA-glider ideal as well.

Suppose now that we have two finitely generated R-torsion free FK-gliders in A, say M and N. For $i \geq 0$, define

$$(M \cdot N)_i = \{\sum_j' m_j n_j \mid \mathrm{gr}(m_j) + \mathrm{gr}(n_j) \geq i\}.$$

It is clear that $(M \cdot N)_i$ is an R-module.

Lemma 8.5.3. *We have the equality*

$$(M \cdot N)_i = \sum_{k=0}^{i} M_k N_{i-k}.$$

Proof. Let $x \in (M \cdot N)_i$ and suppose that $m_i n_i$ appears in the expression of x as a finite sum. If $\mathrm{gr}(m_i) \geq i$, then $m_i n_i \in M_k N$. If $0 \leq \mathrm{gr}(m_i) < i$ then from $\mathrm{gr}(n_i) \geq i - \mathrm{gr}(m_i)$ it follows that $n_i \in N_{i-\mathrm{gr}(m_i)}$. The other inclusion is trivial. $\square$

Suppose that M and N are generated by $\{(m_i, d_i),\ 1 \leq i \leq t\}$ and $\{(n_j, e_j),\ 1 \leq j \leq s\}$ respectively. The previously lemma then entails that $M \cdot N$ is finitely generated by $\{(m_i n_j, d_i + e_j), 1 \leq i \leq t, 1 \leq j \leq s\}$. Indeed,

$$\begin{aligned}
(M \cdot N)_l &= \sum_{k=0}^{l} M_k N_{l-k} \\
&= \sum_{k=0}^{l} \sum_{i=1}^{t} \sum_{j=1}^{s} F_{d_i - k} K F_{e_j - l + k} K m_i n_j \\
&\subset \sum_{i=1}^{t} \sum_{j=1}^{s} F_{d_i + e_j - l} K m_i n_j \subset (M \cdot N)_l.
\end{aligned}$$

Our next goal is to define the inverse glider

$$M^{-1} \supset (M^{-1})_1 \supset (M^{-1})_2 \supset \ldots$$

To this extent, we put

$$M^{-1} = \{x \in A \mid MxM \subset M\},$$

as in the classical situation of normal ideals. For $i > 0$ we then define

$$(M^{-1})_i = \{x \in A | \, MxM \subset M_i\}.$$

Proposition 8.5.4. *M^{-1} together with the chain defined by the $(M^{-1})_i$ is a finitely generated R-torsion free FK-glider.*

Proof. Let $i \leq j, x \in (M^{-1})_j$, then $MF_iKxM \subset F_iKM_j \subset M_{j-i}$ shows that $F_iK(M^{-1})_j \subset (M^{-1})_{j-i}$ and M^{-1} is indeed an FK-glider. R-torsion freeness is obvious. Next, we show that M^{-1} is a finitely generated R-module. To this extent, let $a \in A$. The R-module MaM is finitely generated and since $KM = A$ it follows that there exists $r \in R$ such that $MraM = rMaM \subset M$. Hence $ra \in M^{-1}$ and it follows that $KM^{-1} = A$. We have an embedding as R-modules

$$MM^{-1} \hookrightarrow \mathrm{Hom}_R(M, M), \ w \mapsto (m \mapsto wm).$$

Indeed, suppose that $w, w' \in MM^{-1}$ define the same morphism; then $wM = w'M$ element-wise and since $MK = A$, also $wA = w'A$ element-wise. In particular, $w = w'$. By choosing generators $u_1, \ldots, u_n$ for M as R-module, we obtain a surjective morphism $M_n(R) \twoheadrightarrow \mathrm{Hom}_R(M, M)$. Since $M_n(R)$ is a finitely generated R-module, so is $\mathrm{Hom}_R(M, M)$. Because R is Noetherian, it follows that MM^{-1} is also finitely generated. The morphism

$$M^{-1} \to \mathrm{Hom}_R(M, MM^{-1}), \ w \mapsto (m \mapsto mw)$$

is again injective. The R-module $\mathrm{Hom}_R(M, MM^{-1})$ is the image of $M_{d,n}(R)$ by choosing generators $y_1, \ldots, y_d$ in MM^{-1} and we arrive at M^{-1} being finitely generated as R-module. Suppose now that $M^{-1} = \sum_{i=1}^{t} Rx_i$, then we define $j_i \geq 0$ to be maximal such that $Mx_iM \subset M_{j_i}$. It follows that for all $n \geq 0$, $\sum_{i=1}^{t} F_{j_i - n}Kx_i \subset (M^{-1})_n$. By enlarging one of the indices F_{j_i-n+d} the sum on the left sits inside $(M^{-1})_{n-d}$ which shows that the above inclusion is in fact an equality. This shows that M^{-1} as an FK-glider is generated by $\{(x_i, j_i), \ 1 \leq i \leq t\}$. □

By definition, we have

$$((M^{-1})^{-1})_i = \{x \in A \, | \, M^{-1}xM^{-1} \subset (M^{-1})_i\}.$$

Proposition 8.5.5. *If B is a maximal order, then $M = (M^{-1})^{-1}$ as FK-gliders.*

Proof. Let x be such that $M^{-1}xM^{-1} \subset (M^{-1})_i$. This means that

$$MM^{-1}xM^{-1}M \subset M_i.$$

Since B is a maximal order and M is a normal ideal in A, the classical theory yields that $MM^{-1} = B$ and $M^{-1}M = B^r$. In particular, $1 \in B \cap B^r$, which shows that $x \in M_i$. Conversely, let $x \in M_i$. We have to show that $MM^{-1}xM^{-1}M \subset M_i$. This follows since M_i is a $B - B^r$-bimodule. □

From now on, we assume B is a maximal order.

Corollary 8.5.6. *$M \cdot M^{-1}$ is a left FA-glider ideal.*

Proof. From the deductions made after Lemma 8.5.3, we know that $M \cdot M^{-1}$ is a finitely generated FK-glider. Since $F_nA = F_nKB$ and all M_i are left B-modules, $M \cdot M^{-1}$ is indeed a left FA-glider. □

Remark 8.5.7. The same results hold on the right, that is, with regards to B^r and $M^{-1} \cdot M$.

Let us recall the definition of a groupoid G from [58] as being a collection of elements, certain of whose products are defined and lie in G, such that

1. For each $a_{ij} \in G$ there exist unique elements $e_i, e_j \in G$ such that $e_ia_{ij} = a_{ij} = a_{ij}e_j$, where all indicated products are defined. Further, $e_ie_i = e_i, e_je_j = e_j$. We call e_i the left unit of a_{ij} and e_j the right unit of a_{ij};

2. $a_{ij}b_{kl}$ is defined if and only if $j = k$, that is, if and only if the right unit of a_{ij} equals the left unit of b_{kl};

3. If ab and bc are defined, so are $(ab)c$ and $a(bc)$ and these are equal;

4. For each $a_{ij} \in G$, there exists an $a_{ij}^{-1} \in G$ with left unit e_j, right unit e_i, such that
$$a_{ij} \cdot a_{ij}^{-1} = e_i, \quad a_{ij}^{-1} \cdot a_{ij} = e_j.$$

5. Given any pair of units $e, e' \in G$, there is an element $a_{ij} \in G$ with left unit e, right unit e'.

We call a finitely generated R-torsion free FK-glider M inside A such that $KM = A$ and such that both $B^l = B^l(M)$ and $B^r = B^r(M)$ are R-orders a normal FK-glider ideal in A. Let M, N be two normal FK-glider ideals in A. We have to be careful, however, with the notion of left and right units, as there can exist multiple elements e satisfying $e \cdot M = M$. Indeed,

Lemma 8.5.8. *Let FA be a filtration extending FK with $F_0A = B$ and negative part F^-B. Let $M \supset M_1 \supset \dots$ be a left FK-glider.*

1. *If M is a left FA-glider, then $F^-B \cdot M = M$;*

2. *If $F^-B \cdot M = M$ then M is a left filtered F^-B-module for the chain $F_{-n}M = M_n$;*

3. *M is an idempotent, i.e. $M \cdot M = M$ if and only if*
$$M \supset M_1 \supset M_2 \supset \dots$$
defines a negative algebra filtration on the ring $M \subset A$.

Proof. (1) and (3) are straightforward. We prove (2). Let $d, n \geq 0$. Then

$$F_{-d}BM_n = F_{-d}BM_{n-d+d} \subset (F^-B \cdot M)_{n-d} = M_{n-d}. \quad \square$$

For two idempotents F^-B and F^-C we put $F^-B \leq F^-C$ if and only if $F^-B \cdot F^-C = F^-C$.

Proposition 8.5.9. *Let M be a normal FK-glider ideal in A. The set*

$$\{e \text{ normal glider ideal} \mid e \cdot e = e \text{ and } e \cdot M = M\}$$

has a unique maximal element $E^l = E^l(M)$, which we call the left unit of M. The same result holds for right multiplication, leading to a right unit $E^r = E^r(M)$.

Proof. The glider ideal M yields a subring $B = B^l(M)$ and filtration FA on A such that M is a left FA-glider. The theory of glider modulizers, see Subsection 7.6.1, shows that there exists a subring $B^* \supset B$ with chain $F^{-,*}B^*$ which yields a negative ring filtration and such that M with filtration $F_{-n}M = M_n$ is a left filtered $F^{-,*}B^*$-module. We recall that the negative part of the chain is defined by

$$F^*_{-d}B = \{x \in B \mid xM_{n-d} \subset M_n \text{ for } n \in \mathbb{N} \text{ such that } n - d \geq 0\},$$

for $d \geq 0$. This entails that

$$F^{-,*}B^* \cdot M = M.$$

In fact, by definition of B, B^* actually equals B. Now suppose that M is also a left fA-glider for some exhaustive, separated filtration fA on A. Let $x \in f_{-d}A$, then for all $n \geq d$ it holds

$$xM_{n-d} \subset M_n,$$

which entails that $x \in F^{-d,*}B$. This shows that $f^-A \cdot F^{-,*}B = F^{-,*}B$ and $E = F^{-,*}B$. $\square$

The previous proposition leads to calling the multiplication $M \cdot N$ proper if $E^r(M) = E^l(N)$. We will show that the collection of normal FK-glider ideals in A with proper multiplication is in fact a groupoid.

Proposition 8.5.10. *We have the equalities*

$$E^l(M) = F^{-,*}B^l = M \cdot M^{-1},$$

and

$$E^r(M) = F^{-,*}B^r = M^{-1} \cdot M.$$

Proof. Let $x \in F^*_{-d}A$, i.e. $xM_{n-d} \subset M_n$ for all $n \geq d$. Then

$$x \in xB = xMM^{-1} = xM_{d-d}M^{-1} \subset M_dM^{-1} \subset (M \cdot M^{-1})_d.$$

This shows that $F^{-,*}B \subset M \cdot M^{-1}$. Lemma 8.5.8 then entails that

$$M \cdot M^{-1} \subset (M \cdot M^{-1}) \cdot (M \cdot M^{-1}).$$

Let $i \geq 0$; then we have

$$\begin{aligned}
((M \cdot M^{-1}) \cdot (M \cdot M^{-1}))_i &= \sum_{k=0}^{i}\sum_{s=0}^{k}\sum_{t=0}^{i-k} M_s(M^{-1})_{k-s}M_t(M^{-1})_{i-k-t} \\
&\subset \sum_{k=0}^{i}\sum_{s=0}^{k}\sum_{t=0}^{i-k} M_sB^r(M^{-1})_{i-k-t} \\
&\subset \sum_{k=0}^{i}\sum_{s=0}^{k}\sum_{t+k=0}^{i} M_s(M^{-1})_{i-k-t} \\
&\subset \sum_{s=0}^{k} M_s(M^{-1})_{i-s} = (M \cdot M^{-1})_i.
\end{aligned}$$

It follows that $M \cdot M^{-1}$ is an idempotent, hence equals $E^l(M)$ by Proposition 8.5.9. The proof for $E^r(M)$ is analogous. □

Proposition 8.5.11. *Let FK be a separated, exhaustive filtration on a field K and let A be a central simple K-algebra. The collection of all normal FK-glider ideals in A with proper multiplication $M \cdot N$ and units being the idempotent elements forms a groupoid.*

Proof. Let M, N and V be normal FK-glider ideals such that $M \cdot N$ and $N \cdot V$ are defined. For $i \geq 0$ we have by definition

$$\begin{aligned}
((M \cdot N) \cdot V)_i &= \sum_{k=0}^{i}\sum_{j=0}^{k} M_jN_{k-j}V_{i-k} \\
&= \sum_{j=0}^{i}\sum_{k=j}^{i} M_jN_{k-j}V_{i-k} \\
&= \sum_{j=0}^{i}\sum_{l=0}^{i-j} M_jN_lV_{i-j-l} \\
&= (M \cdot (N \cdot V))_i.
\end{aligned}$$

Properties (1) and (4) from the definition of a groupoid follow from Proposition 8.5.10. For property (5), the element $F^-B \cdot F^-C$ with F^-B, F^-C being two units, i.e. negative algebra filtrations on subrings B and C of A, does the trick.

□

Example 8.5.12. When FK is strong, it follows from M being finitely generated, that $M_i = F_{-i}KM$ for all $i \geq 0$. This shows that $B = O_l(M) = O_l(M_i)$ for all $i \geq i$. One then shows that $M_i(M^{-1})_j \subset F_{-i-j}A$. Since the filtration FA is also strong, one subsequently shows that

$$(M \cdot M^{-1})_i = F_{-i}A.$$

Thus, when FK is strong and B is a maximal order, then $M \cdot M^{-1}$ is the left FA-glider

$$B \supset F_{-1}A \supset F_{-2}A \supset \ldots$$

and $F^{-}B = F^{-,*}B = M \cdot M^{-1}$ and similar for B^r.

8.6 Higher rank valuations

In this section we consider Γ-filtered rings where $\Gamma = \mathbb{Z}^n$ equipped with the lexicographical order, that is

$$(a_1, \ldots, a_n) < (b_1, \ldots, b_n) \Leftrightarrow \begin{cases} \text{the first non} - \text{zero element of} \\ b_1 - a_1, \ldots, b_n - a_n \text{ is positive.} \end{cases}$$

In fact, we will restrict to $n = 2$. Let us recall from [48] how one obtains higher rank (commutative) valuation rings. To this extent, let K be a field with valuation, and call R its valuation ring. If M denotes the unique maximal ideal of R, then we may assume that the residue field $K' = R/M$ is also equipped with a valuation. If we denote its valuation ring by

$$\overline{R}' \subset K',$$

then the subset of R given by

$$R' := \{x \in R \,|\, x + M \in \overline{R}'\},$$

is a valuation ring. The value groups of R, R' and $\overline{R}'$ are related by a short exact sequence

$$0 \to \Gamma_{\overline{R}'} \to \Gamma_{R'} \to \Gamma_R \to 0,$$

where the morphisms preserve the ordering.
We call R' the composite of the valuation rings R and $\overline{R}'$.
From the short exact sequence above we see that $\mathbb{Z} \cong \Gamma_{\overline{R}'}$ is an isolated subgroup of $\Gamma_{R'}$, which means that if $x \in \Gamma_{\overline{R}'}$, then so is $y \in \Gamma_{\overline{R}'}$ for all $y \leq x$. In [62, Chapter 1 Section 2] it is shown that on the factor group $\Gamma_{R'}/\Gamma_{\overline{R}'}$ one can put a total ordering and one checks this ordering corresponds to the ordering on Γ_R. This shows that $\Gamma_{\overline{R}'}$ is a discrete group of rank 2 and it follows from [62, p. 8] that $\Gamma_{\overline{R}'}$ is order isomorphic to $\mathbb{Z}^2$ with lexicographical ordering.

Example 8.6.1. Let $K = k((y))((x))$ be the fraction field of $R = k((y))[[x]]$. The x-adic valuation yields a discrete valuation on K with valuation ring R and value group $\mathbb{Z}$. On the residue field $K' = k((y))$ we put the y-adic valuation with valuation ring $k[[y]]$. The composite is the valuation ring $R' = k[[y]] + xk((y))[[x]]$ and the short exact sequence

$$0 \to \mathbb{Z} \to \Gamma_{R'} \to \mathbb{Z} \to 0$$

splits, so that $\Gamma_{R'} = \mathbb{Z}^2$. One checks that the total ordering is given by the lexicographical ordering. The negative part of the associated valuation filtration on $R' = F_{(0,0)}K$ is given by

$$F_{(-i,-j)}K = x^i(y^j k[[y]] + xk((y))[[x]]).$$

From now on, all filtrations FK are assumed to be $\mathbb{Z}^2$-filtrations, where we may assume the ordering on $\mathbb{Z}^2$ to be the lexicographical one by the observations preceding the previous example. We recall the standing assumption that FK is separated, exhaustive and unbounded. In order to generalize Theorem 8.2.10 for fields K equipped with $\mathbb{Z}^2$-filtrations, we first observe that the additive chain

$$f^hK: \quad \ldots \subset \bigcup_{n\in\mathbb{Z}} F_{(-1,n)}K \subset \bigcup_{n\in\mathbb{Z}} F_{(0,n)}K \subset \bigcup_{n\in\mathbb{Z}} F_{(1,n)}K \subset \ldots$$

defines a separated, exhaustive and unbounded $\mathbb{Z}$-filtration on K. We also have a filtration on $R = \bigcup_{n\in\mathbb{Z}} F_{(0,n)}$ given by

$$f^vR: \quad \ldots F_{(0,-2)}K \subset F_{(0,-1)}K \subset F_{(0,0)}K \subset F_{(0,1)}K \subset F_{(0,2)}K \subset \ldots$$

Let M be an FK-glider. We will depict this by

$$\begin{array}{ccccccc} \vdots & & \vdots & & \vdots & & \ddots \\ \cap & & \cap & & \cap & & \\ M_{(0,2)} & \supset & M_{(1,2)} & \supset & M_{(2,2)} & \supset & \cdots \\ \cap & & \cap & & \cap & & \\ M_{(0,1)} & \supset & M_{(1,1)} & \supset & M_{(2,1)} & \supset & \cdots \\ \cap & & \cap & & \cap & & \\ M_{(0,0)} & \supset & M_{(1,0)} & \supset & M_{(2,0)} & \supset & \cdots \end{array}$$

For $m \geq 0$, the chain

$$M_{(m,*)}: \quad M_{(m,0)} \supset M_{(m,1)} \supset M_{(m,2)} \supset \ldots$$

is an f^vR-glider, whence its body $B(M_{(m,*)})$ is an R-module.

Lemma 8.6.2. *Let* $M \in \mathrm{GBS_F(K)}$*, then the chain*

$$B^v(M): \quad B(M_{(0,*)}) \supset B(M_{(1,*)}) \supset B(M_{(2,*)}) \supset \ldots$$

is an irreducible f^hK*-glider, i.e.* $B^v(M) \in \mathrm{GBS}_{\mathrm{f^hK}}(\mathrm{K})$.

Proof. That $B^v(M)$ is an f^hK-glider is obvious. Suppose that $T \supset T_1 \supset \dots$ is a subglider. There is an associated FK-subglider of M, given by

$$\begin{array}{ccccccc} \vdots & & \vdots & & \vdots & & \ddots \\ \cap & & \cap & & \cap & & \\ T & \supset & T_1 & \supset & T_2 & \supset & \cdots \\ \cap & & \cap & & \cap & & \\ T & \supset & T_1 & \supset & T_2 & \supset & \cdots \\ \cap & & \cap & & \cap & & \\ T & \supset & T_1 & \supset & T_2 & \supset & \cdots \end{array}$$

so it must be trivial. It is clear that triviality of type T_1 or T_2 easily yields that $T \supset T_1 \supset \dots$ is trivial of the same type. So suppose that $T \supset T_1 \supset \dots$ is not trivial of type T_1 or T_2. It follows that there exists a monotone increasing map $\alpha : \mathbb{N}^2 \to \mathbb{N}^2$ such that $N_{(m,n)} = M_{\alpha(m,n)}$. We define a map $\beta : \mathbb{N} \to \mathbb{N}$ by

$$\beta(n) := \pi_1(\sup_m \alpha(n, m)),$$

where $\pi_1 : \mathbb{N}^2 \to \mathbb{N}$ is the projection on the first component. Observe that our assumption that the subglider is not of type T_1 or T_2 ensures that the supremum is indeed finite. One checks that β is monotone increasing and satisfies $T_n = B(M_{(\beta(n),*)})$. This shows that $T \supset T_1 \supset \dots$ is trivial of type T_3. □

Since we have that $B^v(M)_0 \subset f_0^hK$, we can invoke Proposition 8.2.9 to refine the negative part of f^hK to obtain a strong e-step filtration f^sK such that $B^v(M) \in \mathrm{GBS}_{f^s}(\mathrm{K})$. We know that the negative part of f^hK is a trivial f^sK-subfragment of type T_3 of the negative part of f^sK. Suppose that $\alpha : \mathbb{N} \to \mathbb{N}$ yields such a relation, i.e. $f^v_{-n}K = f^s_{-\alpha(n)}K$. We refine the negative part of FK, i.e. when the first component is negative by putting

$$F'_{(-\alpha(n),m)}K = F_{(-n,m)}K,$$

and

$$F'_{(-i,m)}K = f^s_{-i}K \cap F_{(-n-1,m)}K$$

if $\alpha(n) < i < \alpha(n+1)$ for some n. Since we only altered the horizontal direction, $M \in \mathrm{GBS}_{\mathrm{F'K}}(\mathrm{K})$, so we may replace $F'K$ by FK without harm. Suppose now that $e > 1$. In particular, we have that $f^s_{-1}K = f^s_{-2}K$. It follows that

$$f_{-1}K \supset \cap_{n \in \mathbb{Z}} F_{(-1,n)}K \supset f_{-2},$$

whence $f_{-1}K = F_{(-1,n)} = f_{-2}K$ for all $n \in \mathbb{Z}$. Next, since $f_0^sK.f^s_{-1}K = f^s_{-1}K$ we obtain

$$F_{(0,n)}KF_{(-1,0)}K = F_{(-1,0)}K,$$

for all $n \in \mathbb{Z}$ and since we are working in a field, it follows that $f_0^s K = F_{(0,n)}K$ for all $n \in \mathbb{Z}$. After performing similar reasonings, one deduces that FK is trivially filtered in the vertical direction, and so FK is essentially a $\mathbb{Z}$-filtration, which we can exclude. Hence $e = 1$ and we may assume that $f^h K = f^s K$.

Proposition 8.2.6 entails that $f_0^s K = f_0^h K = R$ is a DVR and $f^h K$ is the associated valuation filtration. One checks that

$$gK': \quad \ldots \subset \frac{F_{(0,-2)}K}{F_{(-1,-2)}K} \subset \frac{F_{(0,-1)}K}{F_{(-1,-1)}K} \subset \frac{F_{(0,0)}K}{F_{(-1,0)}K} \subset \frac{F_{(0,1)}K}{F_{(-1,1)}K} \subset \frac{F_{(0,2)}K}{F_{(-1,2)}K} \subset \ldots$$

defines a separated, exhaustive filtration on the residue field $K' = f_0^h K / f_{-1}^h K$.

Lemma 8.6.3. *The descending chain*

$$M^{res}: \quad \frac{M_{(0,0)}}{M_{(1,0)}} \supset \frac{M_{(0,1)}}{M_{(1,1)}} \supset \frac{M_{(0,2)}}{M_{(1,2)}} \supset \ldots$$

is an irreducible gK'-glider.

Proof. Suppose that $\overline{N} \supset \overline{N}_1 \supset \overline{N}_2 \supset \ldots$ is a non-trivial gK'-subglider. Without loss of generality, we may assume that $\overline{N}$ corresponds to an $F_{(0,0)}K$-submodule

$$M_{(0,0)} \supsetneq N \supsetneq M_{(1,0)}.$$

However, $F_{(-i,-j)}KN$ would define a non-trivial FK-subglider of M. □

Observe that we do not know whether $M^{res} \in \mathrm{GBS_{gK'}(K')}$ as we do not know whether $M_{(1,0)} \subset f_{-1}^h K$. However, we can consider for any $s \geq 0$, the descending chain

$$M_s^{res}: \quad \frac{M_{(s,0)}}{M_{(s+1,0)}} \supset \frac{M_{(s,1)}}{M_{(s+1,1)}} \supset \frac{M_{(s,2)}}{M_{(s+1,2)}} \supset \ldots$$

One shows analogously that they are all irreducible gK'-gliders. Hence, they must be all isomorphic since they are contained in one another. Choosing s large enough such that $M_{(s+1,1)} \subset f_{-1}^h K$ then shows that $\mathrm{GBS_{gK'}(K')} \neq \emptyset$. Hence, we can again invoke Proposition 8.2.6 to deduce that $g_0 K'$ is a DVR and $g'K$ is the strong f-step valuation filtration for some $f \geq 1$.

We denote by $\mathrm{GBS_{\mathbb{Z}^2}(K)}$ the union of all $\mathrm{GBS_F(K)}$ where FK is a $\mathbb{Z}^2$-filtration. Let K/k be a field extension; then $R(K/k, 2)$ denotes the Riemann surface of all rank 2 valuations containing k, i.e. which are trivial on k. We have the generalization of Theorem 8.2.10.

Theorem 8.6.4. *Let K/k be a field of transcendence degree* $\mathrm{tdeg_k(K)} \geq 2$. *We have an isomorphism as sets*

$$\mathrm{GBS_{\mathbb{Z}^2}(K) = R(K/k, 2) \times \mathbb{Z}^2}.$$

Proof. We showed that if $M \in \mathrm{GBS}_{\mathrm{F}}(\mathrm{K})$, then $M \in \mathrm{GBS}_{\mathrm{F^v}}(\mathrm{K})$ where $F^v K$ is the valuation filtration of a valuation of rank 2. From the structure of irreducible gliders for $\mathbb{Z}$-filtrations on fields we know that there exists $m \in \mathbb{Z}$ such that

$$\bigcap_i M_{(0,i)} = \bigcup_{j \in \mathbb{Z}} F^v_{(m,j)} K \supset M_{(1,0)}.$$

Since $M_{(1,0)}$ is an $F^v_{(0,0)}K$-module, there exists $n \in \mathbb{Z}$ such that

$$M_{(1,0)} = F^v_{(m,n)} K.$$

It then follows that

$$M_{j,i} = F^v_{m-1-j,n-i} K$$

for all $i, j \geq 0$. □

Chapter 9

Odds and ends

9.1 The road to Hopf gliders

In this section we introduce the notion of a cofragment or coglider over a positive coalgebra FC. For background on coalgebras and Hopf algebras we refer to [1],[49].

Let us recall some notions on filtered coalgebras and Hopf algebras.

Definition 9.1.1. A k-coalgebra C together with an ascending chain of k-linear subspaces

$$\ldots \subset F_{-1}C \subset F_0C \subset F_1C \subset \ldots$$

is a filtered k-coalgebra if for $n \in \mathbb{Z}$:

$$\Delta(F_nC) \subset \sum_{i+j=n} F_iC \otimes F_jC.$$

An interesting coalgebra filtration is for example the coradical filtration on a coalgebra, which is positive. Observe that for positive filtrations, i.e. $F_{-n}C = 0$ for $n > 0$ we have

$$\Delta(F_nC) \subset \sum_{i=0}^{n} F_iC \otimes F_{n-i}C.$$

It follows in particular that every F_nC is a subcoalgebra of C.

Definition 9.1.2. Let FC be a k-coalgebra filtration. A filtered (left) comodule FM is a (left) C-comodule (M, ρ) with ascending chain of k-linear subspaces

$$\ldots \subset F_{-1}M \subset F_0M \subset F_1M \subset \ldots$$

such that for all $s \in \mathbb{Z}$:

$$\rho(F_sM) \subset \sum_{p+q=s} F_pC \otimes F_qM.$$

Definition 9.1.3. Let B be a k-bialgebra. If there is an ascending chain of k-linear subspaces

$$\ldots \subset F_{-1}B \subset F_0B \subset F_1B \subset \ldots$$

such that FB is simultaneously a k-algebra and k-coalgebra filtration, then B is called a filtered k-bialgebra.

Definition 9.1.4. Let H be a k-Hopf algebra with bialgebra filtration FH. If $S(F_iH) \subset F_iH$ for all $i \in \mathbb{Z}$, then we call H a filtered k-Hopf algebra.

Hopf filtrations were studied for example in [29].

What should be the right definition of a cofragment? On the module and comodule level there exist nice dualities. For example, the dual vector space C^* of a coalgebra C is equipped with an algebra structure. We recall that its multiplication is given by

$$m_{C^*} = \Delta^* \circ i : C^* \otimes C^* \to (C \otimes C)^* \to C^*.$$

Specifically, the multiplication is given by

$$(c^* * d^*)(c) = \langle c^*, c_{(1)} \rangle \langle d^*, c_{(2)} \rangle.$$

Conversely, if we have a K-algebra A such that A is finitely generated and projective as a K-module, then we can turn A^* into a coalgebra. The comultiplication is defined by

$$\Delta_{A^*} = i^{-1} \circ m_A^* : A^* \to (A \otimes A)^* \to A^* \otimes A^*$$

and the counit

$$\epsilon_{A^*}(a^*) = \langle a^*, 1 \rangle$$

Let $\{(e_i, e_i^*)\}$ be a dual basis of A. Then the comultiplication is described explicitly by

$$\Delta(a^*) = \sum_{i,j} \langle a^*, e_i e_j \rangle e_i^* \otimes e_j^*.$$

If $e_i \in \dot{F}_k A$ we will denote this by e_i^k.

We obtain a nice duality, because there exists a functor $F : \mathcal{M}^C \to {}_{C^*}\mathcal{M}$ from the category of right C-comodules and left C^*-modules, which is an equivalence of categories if C^* is finitely generated and projective as a K-module. How the functor and its inverse are defined will be recalled below. A first property we would like to have is a generalization to fragments and cofragments. Hence, we must find a way to associate with a coalgebra filtration an algebra filtration on the dual.

The above functor F sends objects M to M, but we could also consider the dual vector space M^*. We have the following

- Let C be a coalgebra and let M be a left C-comodule; then the dual M^* has a natural left C^*-module structure;

- Let A be an algebra (finitely generated projective) and let M be a left A-module; then the dual M^* has a natural left A^*-module structure.

Hence, we also would like to define the dual of a (co)fragment, such that the above properties generalize.

Let FC be a positive coalgebra filtration and consider the sequence of projections

$$C \twoheadrightarrow C/F_0C \twoheadrightarrow C/F_1C \twoheadrightarrow \cdots \twoheadrightarrow C/F_nC \twoheadrightarrow \cdots$$

The functor $\mathrm{Hom}_k(-,k)$ is contravariant and yields the following chain

$$C^* \supset (C/F_0C)^* \supset (C/F_1C)^* \supset \ldots \supset (C/F_nC)^* \supset \ldots$$

Let $\pi_i : C \to C/F_iC$ denote the projection; then an element $f_i \in (C/F_iC)^*$ is embedded in C^* by $f_i \circ \pi_i$.

Proposition 9.1.5. *The algebra C^* with descending chain of vector spaces $(C/F_iC)^*$ yields a negative algebra filtration F^-C^*, where $F_{-j}C^* = (C/F_{j-1}C)^*$.*

Proof. Let $f \in F_{-i}C^*, g \in F_{-j}C^*$. In order for $f * g$ to be in $F_{-i-j}C^*$ we have that show that $f * g$ factorizes through $F_{i+j-1}C$. So let $c \in F_{i+j-1}C$, then $\Delta(c) \in \sum_{p+q=i+j-1} F_pC \otimes F_qC$. Hence either $p < i$ or $q < j$. This shows that

$$\langle f * g, c\rangle = \langle f, c_{(1)}\rangle\langle g, c_{(2)}\rangle = 0,$$

as f factorizes through $C/F_{i-1}C$ and g through $C/F_{j-1}C$. □

Example 9.1.6. Let C be a finite dimensional coalgebra with coradical filtration, i.e. $F_0C = R$ is the coradical of C. We have that $R^\perp = J(C^*)$ and more generally $(F_nC)^\perp = \Lambda^{n+1}R^\perp = J(C^*)^{n+1}$. Hence, an element c^* sits in $F_{-n}C^* = (C/F_{n-1}C)^*$ exactly when it sits in $F_{n-1}C^\perp = J(C^*)^n$. So the dual filtration of the coradical filtration is the $J(C^*)$-adic filtration.

Dually, for a positive algebra filtration FA, with A finitely generated and projective as a k-module, we have

Proposition 9.1.7. *The coalgebra A^* with descending chain of vector spaces $(A/F_iA)^*$ yields a negative coalgebra filtration F^-A^*, where $F_{-n}A^* = (A/F_{n-1}A)^*$.*

Proof. The dual basis element $(e_i^k)^*$ factorizes through $A/F_{k-1}A$ but not through A/F_kA, hence $(e_i^k)^* \in F_{-k}A^*$ Let $f \in F_{-n}A^* = (A/F_{n-1}A)^*$. Since f factorizes through $A/F_{n-1}A$ we have that $\langle f, e_i^k e_j^l\rangle = 0$ if $k + l \leq n - 1$. Hence

$$\Delta(f) \in \sum_{k+l>n} F_{-k}A^* \otimes F_{-l}A^*.$$

If $k+l>n$, then write $0<\delta=k+l-n=\delta_1+\delta_2$ such that $\delta_1\leq k$ and $\delta_2\leq l$. It follows that

$$F_{-k}A^*\otimes F_{-l}A^*\subset F_{-k+\delta_1}A^*\otimes F_{-l+\delta_2}A^*,$$

whence

$$\Delta(f)\in\sum_{k+l=n}F_{-k}A^*\otimes F_{-l}A^*.\quad\square$$

Example 9.1.8. Let G be a finite group with descending chain of normal subgroups $1\triangleleft G_1\triangleleft\ldots\triangleleft G_n=G$. There is an associated algebra filtration on the group algebra KG given by $F_{-i}KG=0, F_iKG=KG_i$ for $i\geq 0$ and $G_i=G$ for $i\geq n$. On the dual coalgebra $(KG)^*$ we obtain a negative coalgebra filtration

$$\ldots\subset 0\subset(K(G/G_{n-1}))^*\subset\ldots\subset(K(G/G_1))^*\subset(KG)^*.$$

For $g\in G$ we denote by $v_g: KG\to K$ the projection onto the g-component. If $\overline{g}^i\in G/G_i$ then

$$v_{\overline{g}^i}=\sum_{h\in G_i}v_{gh}\in(KG)^*.$$

We verify that

$$\begin{aligned}\Delta(v_{\overline{g}^i}) &= \sum_{h\in G_i}\sum_{x\in G}v_x\otimes v_{x^{-1}gh}\\ &= \sum_{x\in G}v_x\otimes v_{\overline{x^{-1}g}^i}\in F_0(KG)^*\otimes F_{-i}(KG)^*.\end{aligned}$$

Example 9.1.9. Let C be the divided power coalgebra, i.e. C is the free K-module with basis $\{c_m\,|\,m\in\mathbb{N}\}$ and comultiplication, resp. counit defined by

$$\Delta(c_m)=\sum_{i=0}^{m}c_i\otimes c_{m-i},\ \text{resp. }\epsilon(c_m)=\delta_{0,m}.$$

The dual algebra C^* is isomorphic to the formal power series in one variable $K[[X]]$. The connecting isomorphism φ is given by

$$\varphi(f)=\sum_{n=0}^{\infty}f(c_n)X^n.$$

The algebra $K[[X]]$ has positive standard filtration defined by the generator X. The associated negative coalgebra filtration on $C^{**}\cap C$ is then given by

$$C\supset K\langle c_i\,|\,i\geq 1\rangle\supset K\langle c_i\,|\,i\geq 2\rangle\supset\ldots\supset K\langle c_i\,|\,i\geq n\rangle\supset\ldots.$$

With the above constructions in mind, we define

Definition 9.1.10. Let FC be a positive k-coalgebra filtration. A (left) FC-cofragment M is a k-vector space with a descending chain of subspaces such that for all $i \geq 0$ there exists a k-linear map

$$\rho_i : M_i \to \sum_{l \geq 0} F_l C \otimes M_{l+i},$$

such that for all $(\rho_i)_{|M_j} = \rho_j$ for all $i \leq j$ and such that $(\Delta \otimes M) \circ \rho = (C \otimes \rho) \circ \rho$.

The definition of a right FC-comodule is analogous.

Remark 9.1.11. Recall that for an FA-fragment $M \supset M_1 \supset \ldots$ all the M_i are $F_0 A$-modules, but not A-modules in general. Special fragments arise when M is embedded in an A-module Ω and the partial actions are induced from the action of Ω. For cofragments however, all the M_i are already C-comodules, so a cofragment is already a coglider.

Let FA be a positive algebra filtration and $\Omega \supset M \supset M_1 \supset \ldots \supset M_n \supset 0 \supset \ldots$ be a left FA-glider of essential length n. On Ω^* we have the descending chain

$$\Omega^* \supset (\Omega/M_n)^* \supset \ldots \supset (\Omega/M)^* \supset 0 \supset \ldots$$

If A is finitely generated and projective as a k-module, then by duality, Ω^* is a right A^*-comodule where the coaction is given by

$$i(\rho(g))(\omega \otimes a) = g(a \cdot \omega).$$

Proposition 9.1.12. *Let FA be a positive algebra filtration and $\Omega \supset M \supset M_1 \supset \ldots \supset M_n \supset 0 \supset \ldots$ be a left FA-glider of finite essential length and zero body. The chain $F_{-i}\Omega^* = (\Omega/M_{n+1-i})^*$ for $1 \leq i \leq n+1$:*

$$\ldots \subset 0 \subset (\Omega/M)^* \subset (\Omega/M_1)^* \subset \ldots \subset (\Omega/M_n)^* \subset \Omega^*$$

is a negative filtered comodule with regard to the negative coalgebra filtration $F^- A^*$.

Proof. Let $f \in F_{-i}\Omega^* = (\Omega/M_{n+1-i})^*$. We have to show that

$$\begin{aligned} \rho(f) &\in \sum_{p+q=-i, p,q \leq 0} F_p A^* \otimes F_q \Omega^* \\ &= \sum_{p+q=-i, p,q \leq 0} (A/F_{-p-1}A)^* \otimes (\Omega/M_{n+1+q})^*. \end{aligned}$$

Because $i(\rho(f))(a \otimes m) = f(m \cdot a)$ and because $M_{n+1+q} \cdot F_{-p-1}A \subset M_{n-i+2} \subset M_{n-i}$ we see that $i(\rho(f))$ vanishes on $\sum_{p+q=-i,p,q \leq 0} F_{-p-1}A \otimes M_{n+1+q}$. □

Now, if FC is a filtered coalgebra and $M \supset M_1 \supset \ldots \supset M_n \supset 0 \supset \ldots$ a right FC-cofragment, then M has a left C^*-module structure given by

$$c^* \cdot m = \langle c^*, m_{[1]} \rangle m_{[0]}.$$

Proposition 9.1.13. *Let FC be a positive coalgebra filtration and $M \supset M_1 \supset \ldots \supset M_n \supset 0 \supset \ldots$ be a right FC-cofragment. The chain $F_{-i}M^* = (M/M_{n+1-i})^*$ for $0 \leq i \leq n+1$:*

$$\ldots \subset 0 \subset (M/M_1)^* \subset \ldots \subset (M/M_n)^* \subset M^*$$

is a negative filtered module with regard to the negative algebra filtration F^-C^*.

Proof. The C^*-action on M^* is given by

$$c^* \cdot m^*(m) = \langle c^*, m_{[1]}\rangle\langle m^*, m_{[0]}\rangle.$$

Let $f \in F_{-i}C^* = (C/F_{i-1}C)^*$ and $g \in F_{-j}M^* = (M/M_{n+1-j})^*$. We have to show that $f \cdot g \in F_{-i-j}M^*$. For $m \in M_{n+1-i-j}$ it holds that $\rho(m) \in \sum_{l\geq 0} M_{l+n+1-i-j} \otimes F_lC$. If $l < i$, then $\langle f, m_{[1]}\rangle = 0$, if $l \geq i$, then $l+n+1-i-j \geq n+1-j$ and $\langle g, m_{[0]}\rangle$ is zero in this case. □

Proposition 9.1.12 and Proposition 9.1.13 cover the duality between fragments and cofragments. We also have the following:

Proposition 9.1.14. *Let FC be a positive coalgebra filtration and $M \supset M_1 \supset \ldots$ a right FC-cofragment. The negative chain $F_{-i}M = M_i$ is a negative filtered F^-C^*-module.*

Proof. Let $f \in F_{-i}C^*, m \in F_{-j}M = M_j$. Since f vanishes on $F_{i-1}C$, it follows that the degree of $m_{[0]}$ appearing in

$$f \cdot m = \langle f, m_{[1]}\rangle m_{[0]}$$

is bigger than $i+j$, hence $f \cdot m \in M_{i+j} = F_{-i-j}M$. □

Conversely, let F^-M be a left filtered F^-C^*-module. Pick a finite dual basis $\{(c_i, c_i^*) \mid i = 1, \ldots, n\}$ for C. By our construction, if $c_i \in F_{\delta_i}C$, then $c_i^* \in F_{-\delta_i}C^*$. The coaction on M is given by

$$\rho(m) = \sum_{i=1}^{n} c_i^* \cdot m \otimes c_i.$$

Denote $M_j = F_{-j}M$, then we have the following:

Proposition 9.1.15. *The chain $M \supset M_1 \supset \ldots$ is a right FC-cofragment.*

Proof. Let $m \in M_j = F_{-j}\Omega$. Then

$$\rho(m) \in \sum_{i=1}^{n} F_{-j-\delta_i}M \otimes F_{\delta_i}C \subset \sum_{l\geq 0} M_{l+j} \otimes F_lC. \quad \square$$

For these reasons we claim that our definition of a cofragment is plausible. One can go further and define as follows:

Definition 9.1.16. Let FB be a positive k-bialgebra filtration. A descending chain of k-vector spaces $M \supset M_1 \supset \ldots \supset M_n \supset \ldots$ with an FB-fragment and FB-cofragment structure is called an FB-bifragment if for all $m \in M_i, b \in B_j$ with $j \leq i$ we have that

$$\rho_{j-i}(bm) = b_{(1)}m_{[0]} \otimes b_{(2)}m_{[1]} \in \sum_{l \geq 0} F_l B \otimes M_{l+j-i}.$$

Suppose that $b_{(1)} \in F_l B, m_{[0]} \in F_m B$, then $b_{(2)} \in F_{j-l}B, m_{[1]} \in M_{m+i}$. Hence

$$b_{(1)}m_{[0]} \in F_{l+m}, \quad b_{(2)}m_{[1]} \in M_{m+l+i-j},$$

whence the condition that $b_{(1)}m_{[0]} \otimes b_{(2)}m_{[1]} \in \sum_{l \geq 0} F_l B \otimes M_{l+j-i}$ is automatically satisfied. By definition, a bifragment $M \supset M_1 \supset \ldots \supset M_n \supset \ldots$ has M a B-comodule. In general, M is not a B-module, so certainly not a B-bimodule.

Proposition 9.1.17. *Let FB be a right bounded k-bialgebra filtration with $F_n B = B$, n minimal as such. If $M \supset \ldots \supset M_n \supset 0 \supset \ldots$ is an FB-bifragment, irreducible as fragment, then M is a bimodule.*

Proof. Irreducibility implies that $M = BM_n$, so suppose that $m = \sum b_i m_i \in M$. Let $c \in B$ then $cm = \sum cb_i m_i$ with $cb_i \in F_n B, m_i \in M_n$. Because M is a bifragment we have that

$$\begin{aligned} \rho(cm) &= \sum \rho(cb_i m_i) = \sum (cb_i)_{(1)}(m_i)_{[0]} \otimes (cb_i)_{(2)}(m_i)_{[1]} \\ &= \sum c_{(1)}(b_i)_{(1)}(m_i)_{[0]} \otimes c_{(2)}(b_i)_{(2)}(m_i)_{[1]} \\ &= c \cdot (b_i)_{(1)}(m_i)_{[0]} \otimes (b_i)_{(2)}(m_i)_{[1]} = c \cdot \rho(m). \end{aligned}$$

Hence we see that the B-coaction ρ on M is B-linear, or that M is a bimodule. □

By the fundamental theorem, we have that $M \cong B \otimes M^{coH}$. In particular, if B and M are finite dimensional, then $\dim_k(B)$ divides $\dim_k(M)$.

Example 9.1.18. Consider the k-Hopf algebra filtration $k \subset kG$ with G a finite group, k algebraically closed of characteristic 0. If $M \supset Ka \supset 0 \supset \ldots$ satisfies the conditions of the previous lemma, then $|G|$ divides $\dim_k(M)$. By Theorem 4.1.12 we know that $M \cong \oplus_{i=1}^n V_i^{\oplus m_i}$ where the V_i are non-isomorphic simple G-representations of dimension n_i. Moreover, $m_i \leq n_i$ for all i. Hence $\dim_k(M) = \sum_i n_i m_i \leq \sum n_i^2 = |G|$. It follows that $M \cong kG$.

Lemma 9.1.19. *Let $M \supset \ldots \supset M_n \supset \ldots$ be a cofragment, then the chain $F_{-n}M = M_n, n \geq 0$ and $F_n M = M, n \geq 0$ defines a filtered F^+C-comodule.*

Proof. By definition, $\rho_0(M) \subset \sum_{l\geq 0} F_l C \otimes M_l \subset C \otimes M$, so M is a left C-comodule. For $n > 0$, we have

$$\begin{aligned}\rho(F_n M) = \rho(M) &\subset \sum_{l\geq 0} F_l C \otimes M_l \\ &\subset \sum_{l=0}^{n-1} F_l C \otimes F_{n-l} M + \sum_{l\geq n} F_l C \otimes M_{l-n}. \quad \square\end{aligned}$$

Definition 9.1.20. We call $B(M) = \bigcap_i M_i$ the body of the cofragment.

A subcofragment N of M is said to be trivial if either:

T_1. There is an $i \in \mathbb{N}$ such that $N_i = B(N)$ but $M_i \neq B(M)$;
T_3. There exists a monotone increasing map $\alpha : \mathbb{N} \to \mathbb{N}$ such that $N_i = M_{\alpha(i)}$.

Definition 9.1.21. A cofragment M is said to be cosimple if it has no non-trivial subcofragments.

The body of a cofragment is a C-subcomodule.

Proposition 9.1.22. *Let FC be a positive coalgebra filtration and $M \supset M_1 \supset \ldots$ a cosimple cofragment. Then M_i/M_{i+1} is a cosimple C-comodule for every $i \geq 0$.*

Proof. Suppose that M_i/M_{i+1} is not cosimple. Hence there exists a proper subcomodule $M_i \supsetneq N_i \supsetneq M_{i+1}$ and the subcofragment

$$\begin{array}{ccccccccccccc} M & \supset & M_1 & \supset & \ldots & \supset & M_{i-1} & \supset & M_i & \supset & M_{i+1} & \supset & \ldots \\ \cup & & \cup & & & & \cup & & \cup & & \cup & & \\ N_i & \supset & N_i & \supset & \ldots & \supset & N_i & \supset & N_i & \supset & M_{i+1} & \supset & \ldots \end{array}$$

is non-trivial. $\square$

9.1.1 Application: iterated one-point extensions of finite dimensional elementary algebras

Let k be an algebraically closed field of characteristic zero. A quiver Q consists of a set Q_0 of vertices and a set Q_1 of arrows between these vertices. For an arrow $a \in Q_1$ we define its head $h(a) = e_1 \in Q_0$ and its tail $t(a) = e_0 \in Q_0$, i.e. a is an arrow from e_0 to e_1. A quiver representation V associates a vector space V_i to every vertex e_i and to any arrow $a : e_i \to e_j$ a k-linear map

$$\varphi_a : V_i \to V_j.$$

To Q there is also associated the path algebra kQ which as a vector space is spanned by all possible paths in Q. A path is a composition of arrows, and here we make the following convention: If $a : e_0 \to e_1$ and $b : e_1 \to e_2$ are arrows in Q then ba is the path from e_0 to e_2, i.e. ba is read as $b \circ a$. The multiplication on kQ is then given by concatenation of paths.

From representation theory, see e.g. [7], we know that any finite-dimensional k-algebra A is Morita equivalent to an elementary or basic algebra A^b, which is by definition an algebra C such that $C/J(C) = k \times \cdots \times k$. The associated basic algebra A^b is isomorphic to $\mathrm{End}(\mathrm{P_1} \oplus \ldots \oplus \mathrm{P_n})$, where $\{P_1, \ldots, P_n\}$ is a complete set of indecomposable projective modules. In fact, A^b is isomorphic to the path algebra $kQ(A)$ modulo some relations $\langle \rho \rangle$. The quiver $Q(A)$ is referred to as the quiver of A and is computed in the following way: one picks a complete set $S_1, \ldots, S_r$ of simple modules with projective covers $P_i = Ae_i$ of A^b. The $\{e_i\}$ form a complete set of orthogonal primitive idempotents of A^b. If $J = J(A)$ denotes the Jacobson radical of A, then let $\epsilon_i = e_i + J \in A/J$. The quiver $Q(A)$ has vertices $f_1, \ldots, f_r$ corresponding to these simple modules and the number of arrows between f_i and f_j equals the dimension $\dim_k(\epsilon_j(J/J^2)\epsilon_i)$. Moreover, one can show that this dimension equals $\dim_k \mathrm{Ext}^1_A(S_i, S_j)$ ([10, Definition 4.1.6]). To prove the isomorphism, one defines an algebra morphism $kQ(A) \to A^b$ as follows: the vertex f_i associated with the idempotent e_i is sent to e_i. For any i, j one picks a k-basis $\{\overline{(y_{ij})_s}\}$ of $\epsilon_j(J/J^2)\epsilon_i$ and lifts this to a set $\{(y_{ij})_s\}$, i.e. $(y_{ij})_s + J^2 = \overline{(y_{ij})_s}$ for each i, j, s. The set of arrows is formed by these $\{\overline{(y_{ij})_s}\}$ and the algebra map sends $\overline{(y_{ij})_s}$ to $(y_{ij})_s$. Last but not least, the ideal of relations $\langle \rho \rangle$ is zero exactly when the algebra A^b is hereditary.

In all that follows, A will be a finite dimensional basic hereditary algebra; that is, A is isomorphic to the path algebra kQ of a finite quiver Q with no loops. We will make no distinction between A and kQ.

From [59, Section 2.5] we recall that a one-point extension of a finite-dimensional algebra A is defined by an A-module R and given by the algebra

$$B = A[R] = \begin{pmatrix} A & R \\ 0 & k \end{pmatrix}.$$

Indeed, we obtain one additional simple module $S = \begin{pmatrix} 0 & 0 \\ 0 & k \end{pmatrix}$, hence one additional vertex, and we have an equality

$$\dim_k \mathrm{Ext}^1_B(S, S_i) = \dim_k \mathrm{Hom}_A(R, S_i),$$

so R determines the number of arrows from the additional vertex, which is moreover always a source. In order to determine when $A[R]$ remains hereditary, we recall that the category Rep(Q) of quiver representations for some quiver Q is equivalent to the category of left kQ-modules. We summarize that the equivalence sends a quiver representation $(\{V_i\}_{i \in Q_0}, \{\varphi_a\}_{a \in Q_1})$ to the kQ-module $\overline{V} = \oplus_{i \in Q_0} V_i$ in which the action is defined as follows: if $p = a_0 a_1 \ldots a_n$

is a path in Q, i.e. $h(a_i) = t(a_{i-1})$ or a_i ends where a_{i-1} starts, and $v = \sum_{i \in Q_0} v_i \in \overline{V}$ one defines

$$p \cdot v = \varphi_{a_0} \circ \varphi_{a_1} \circ \ldots \circ \varphi_{a_n}(v_{t(a_n)}).$$

The projective modules P_i of kQ are of the form kQe_i for $e_i \in Q_0$, i.e. the span of all the paths starting at vertex e_i. Even more, they are projective indecomposable and are the projective covers of the simple modules S_i, which correspond to the quiver representations having only a one-dimensional vector space k at vertex e_i.

Lemma 9.1.23. *Let A be a finite dimensional algebra such that the quiver $Q = Q(A)$ has no relations. The quiver of the one-point extension $B = A[P_i]$ has no relations.*

Proof. Since $\dim_k \mathrm{Hom}_A(P_i, S_i) = 1$, we get an additional arrow $\gamma : e \to e_i$ from the extra vertex e to e_i. Under the algebra morphism from $kQ(B)$ to B this arrow is mapped to $\gamma \mapsto \begin{pmatrix} 0 & e_i \\ 0 & 0 \end{pmatrix}$. The kernel of the algebra map $kQ(B) \to B$ is contained in R^2, where R is the arrow ideal. So suppose that $c = \alpha + \beta\gamma \in R^2$ maps to zero. In here $\alpha \in R_A^2, \beta \in R_A$. Under the algebra map, c is sent to

$$\begin{pmatrix} \alpha & 0 \\ 0 & 0 \end{pmatrix} + \begin{pmatrix} \beta & 0 \\ 0 & 0 \end{pmatrix} \begin{pmatrix} 0 & e_i \\ 0 & 0 \end{pmatrix} = 0.$$

It follows immediately that $\alpha = 0$. If we write $\beta = \sum \lambda_i a_i$ with a_i paths in $Q(A)$ starting in e_i and $\lambda_i \in k$, then $\beta \cdot e_i = 0$ implies that $\beta = 0$. It follows that the kernel is zero, which proves the claim. □

As a corollary, we can take for R any direct sum of indecomposable projectives. Still there are other possibilities for R, but they are related to the indecomposable projective ones.

Lemma 9.1.24. *Let A be a finite dimensional algebra such that the quiver $Q = Q(A)$ has no relations. If R is an indecomposable module such that the quiver of the one-point extension has no relations, then for every vertex $e_i \in Q_0(A)$ such that R projects onto S_i we have that $P_i \subset R$.*

Proof. By the lifting property, there exists an A-linear map $f : P_i \to R$ such that the diagram

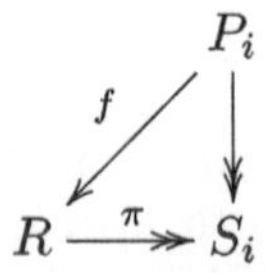

commutes. The simple module S_i is isomorphic to ke_i; in fact we may assume it is equal to it, and $P_i = Ae_i$. If $f(e_i) = 0$, then f would be zero since it is A-linear, contradicting the lifting property. Hence $0 \neq f(e_1) \in R$. Assume that

$af(e_i) = f(ae_i) = 0$ but $ae_1 \neq 0$ for some $a \in A$, i.e. a is a linear combination of paths starting at e_i. Because there are no relations on $Q(B)$, the algebra map $g : kQ(B) \to B$ is an isomorphism, hence we find an element $\gamma \in kQ(B)$ which maps to

$$\begin{pmatrix} 0 & f(e_i) \\ 0 & 0 \end{pmatrix}.$$

But then we would have that

$$g(a\gamma) = \begin{pmatrix} a & 0 \\ 0 & 0 \end{pmatrix} \begin{pmatrix} 0 & f(e_i) \\ 0 & 0 \end{pmatrix} = 0,$$

a contradiction. Thus f is a monomorphism.

□

Example 9.1.25. Consider the following quiver representation $V = (V_i, \varphi_j)$ where the numbers inside the nodes represent the dimension of the V_i.

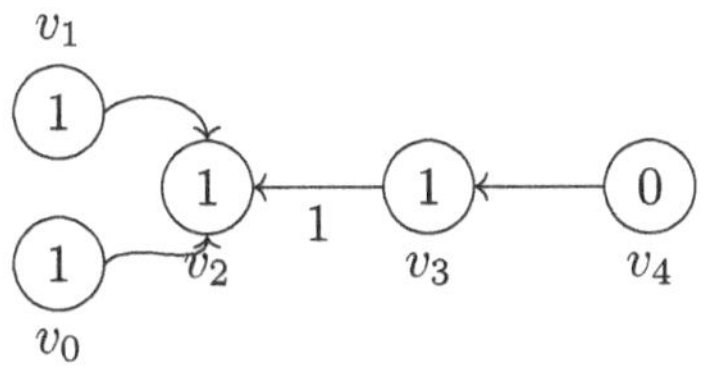

We see that the associated kQ-representation $\overline{V}$ projects onto S_0, S_1 and S_3 and indeed contains P_0, P_1 and P_3 but $\overline{V}$ itself is not projective.

If a module R projects onto a simple S_i, then we call $n = \dim_k \operatorname{Hom}_A(R, S_i)$ its S_i-multiplicity.

Proposition 9.1.26. *Let R be as in the previous lemma and suppose that $\{S_i, i \in I\}$ are all the simple representations on which R project with multiplicity n_i. Then the one-point extensions $A[R]$ and $A[\oplus_{i\in I} P_i^{n_i}]$ are Morita equivalent.*

Proof. We have that

$$\begin{aligned} \dim_k \operatorname{Ext}^1_{A[R]}(S, S_i) &= \dim_k \operatorname{Hom}_A(R, S_i) = n_i \\ &= \dim_k \operatorname{Hom}_A(\oplus_{j\in I} P_j^{n_j}, S_i) \\ &= \dim_k \operatorname{Ext}^1_{A[\oplus_{i\in I} P_i^{n_i}]}(S, S_i), \end{aligned}$$

from which it follows that both quivers have the same number of vertices and arrows. In both cases we do not have any relations, hence the one-point extensions are Morita equivalent. □

Theorem 9.1.27. *Let A be a finite dimensional algebra such that the quiver $Q = Q(A)$ has no relations and let R be an A-module. The quiver of the one-point extension $B = A[R]$ has no relations if and only if B is Morita equivalent to a one-point extension $A[R']$ with $R' = \bigoplus_i P_i$ a direct sum of projective indecomposables.*

Proof. Decompose $R = \bigoplus_i V_i$ into indecomposable modules. If all V_i are projective then we are done. Suppose that V_1 is not projective; then Lemma 9.1.24 and Proposition 9.1.26 show that $A[R]$ is Morita equivalent to $A[P_1 \oplus \ldots \oplus P_{n_1} \oplus V_2 \oplus \ldots \oplus V_m]$ where the P_j for $1 \leq j \leq n_1$ are the projective covers of the simple representations determined by Lemma 9.1.24. Then we use induction on the number of non-projective components V_i. □

Example 9.1.28. As an example we obtain the one-point extensions of the ADE-type quivers to the extended Dynkin quivers. Indeed, this follows from [59, Section 3.6] since we can take for $R = \bigoplus P_i^{d_i}$ with $d_i = D_i\chi(\underline{\dim}(M))$ in which M is a sincere, directing wing module and $D_i\chi$ is the i-th partial derivative of the the quadratic form χ associated to the quiver.

By Theorem 9.1.27 we may assume that R is a direct sum of projective modules. There is a canonical projection

$$B \xrightarrow{\pi} A\,,$$

which after dualizing yields the inclusion of coalgebras $A^* \subset A[R]^*$. Since we are dealing with finite quivers and do not allow any loops, the coalgebra $A^* = (kQ)^*$ is just the path coalgeba $k[Q]$ of the quiver Q. In general, for any quiver Q the path coalgebra $k[Q]$ has as vector space, the same vector space as for the path algebra, but the comultiplication is defined by

$$\Delta(c) = \sum_{c=ab} a \otimes b,$$

where a, b, c are paths in Q. The counit is given by $\epsilon(c) = \delta_{0,|c|}$, where $|c|$ denotes the length of the path c. The group-like elements correspond to the vertices and the arrows (paths of length 1) are primitive. Path coalgebras were first introduced in [23] and in [22] it is shown that the categories $\mathcal{M}^{k[Q]}$ of right $k[Q]$-comodules and InRep(Q) of locally finite quiver representations are equivalent. Again, since we deal with finite quivers we arrive at the equivalence

$$\mathcal{M}^{k[Q]} \cong \mathrm{Rep(Q)} \cong_{\mathrm{kQ}} \mathcal{M}. \tag{9.1}$$

If (V_i, φ_j) is a quiver representation, then the associated right comodule is $V = \oplus_i V_i$ and if $x \in V_i$, then $\rho(x) = \sum \varphi_a(x) \otimes a$, where a runs through the set of paths in Q starting at e_i.

In any coalgebra filtration, the group-like elements must be contained in the degree zero part, so we will consider the following positive coalgebra filtration

$$A^* + ke \subset A[R]^* \subset A[R]^* \subset \ldots$$

Remark 9.1.29. The algebra A is by definition not a subalgebra of $A[R]$, since both algebras have a different unit. However, we could consider

$$A + ke \subset A[R]$$

as algebra filtration. We choose to consider the approach via the dual coalgebras, since the projection $A[R] \twoheadrightarrow A$ is more natural and since for cofragments we do not need to worry about the action of unit elements.

Let us construct some cofragments for this coalgebra filtration. We denote the quiver of A by Q and the one of B by Q'. The additional vertex is denoted by $e_R = e \in Q'_0$. We denote the set of paths of length ≥ 1 of Q, resp. Q' by $\mathcal{P}$, resp. $\mathcal{P}'$. For $p \in \mathcal{P}$, we denote $\epsilon_p : k[Q] \to k$ the functional defined by $\epsilon_p(q) = \delta_{p,q}$ for $q \in \mathcal{P}$. Let M be a right $A[R]^*$-comodule. Since the category $\mathcal{M}^{A[R]^*}$ is Krull-Schmidt ([22, Corollary 2.2]) and by the equivalence with quiver representations, we can write

$$M = \bigoplus_{e_i \in Q'_0} M_{e_i}.$$

We define

$$M_1 = \langle x \in M \mid \exists p \in \mathcal{P}' \setminus \mathcal{P}, y \in M_e \text{ such that } x = (1 \otimes \epsilon_p)\rho(y) \rangle.$$

Proposition 9.1.30. *The k-vector space M_1 is an A^*-comodule.*

Proof. Let $\{x_i, i \in I\}$ be a basis for M_1 with $x_i \in M_{e_{j_i}}$ for some $e_{j_i} \in Q_0$ and let $y_i \in M, p_i \in \mathcal{P}'$ be such that $x_i = (1 \otimes \epsilon_{p_i})\rho(y_i)$. Take $x = x_i, y = y_i, p = p_i$. We can write

$$\rho(y) = x \otimes p + \sum_{q \neq p} y_q \otimes q.$$

Applying $(I_M \otimes \Delta)$ and $(\rho \otimes I_{A[R]^*})$ to both sides yields the equality

$$\rho(x) \otimes p + \sum_{q \neq p} \rho(y_q) \otimes q = x \otimes \Delta(p) + \sum_{q \neq p} y_q \otimes \Delta(q).$$

Let $r \in \mathcal{P}'$ and apply $1 \otimes \epsilon_r \otimes \epsilon_p$ to the above equality. We obtain

$$(1 \otimes \epsilon_r)\rho(x) = y_{rp} \in M_1,$$

which is possibly not zero if and only if $rp \in \mathcal{P}'$. If r is a vertex, then we have

$$(1 \otimes \epsilon_r)\rho(x) = \delta_{r,h(p)} x \in M_1.$$

This shows that $\rho(M_1) \subset M_1 \otimes A[R]^*$. However, in order that $(1 \otimes \epsilon_r)\rho(x) \neq 0$ it is necessary that either $rp \neq 0$ in $A[R]^*$ or that $r = h(p)$. Since $A[R]$ is a one-point extension, $r \in \mathcal{P}$ or $r = h(p) \in A^*$. Hence M_1 is in fact an A^*-subcomodule. □

By construction, we have that

$$\rho(M) \subset M \otimes F_0 A[R]^* + M_1 \otimes A[R]^*.$$

It also follows that $M_1 = 0$ if $M_e = 0$.

Example 9.1.31. Let Q' be a quiver of type D_4 with orientation

$$\begin{array}{ccccc} & & e_0 & & \\ & & \downarrow & & \\ e & \longrightarrow & e_1 & \longrightarrow & e_2 \end{array}$$

A is the path algebra of the A_3-quiver with vertices e_0, e_1, e_2 and $B = A[P_1]$. Let M be the B^*-comodule determined by the dimension vector $\begin{smallmatrix} 1 \\ 110 \end{smallmatrix}$. Then M_1 is the A^*-comodule determined by dimension vector $\begin{smallmatrix} 0 \\ 010 \end{smallmatrix}$.

Now, let us consider an iterated one-point extension $A[R][S]$. We denote the third quiver by Q'' and the set of paths of length ≥ 1 in Q'' by $\mathcal{P}''$. We would like to have that the chain of coalgebras

$$A^* + \sum_{e \in Q_0''} ke \subset A[R]^* \subset A[R][S]^*$$

is a positive coalgebra filtration. Let e_R, resp. e_S denote the additional vertex of Q', resp. Q''. Then if $\mathrm{Hom}_{A[R]}(S, S_{e_R}) \neq 0$, then we have an arrow $q : e_S \to e_R$. Let $p : e_R \to e_i$ be a path in Q', then

$$\Delta(pq) = e_i \otimes pq + p \otimes q + pq \otimes e_S$$

does not satisfy the conditions of a coalgebra filtration. Hence $pq = 0$, and $A[R][S]$ is not hereditary!

We distinguish two cases.

Case 1: $A[R][S] \cong A[S][R]$ or R and S are both A-modules. In this case the above situation does not occur and we can define

$$M_1 = \langle x \in M \mid \exists p \in \mathcal{P}' \setminus \mathcal{P}, y \in M_{e_R} \text{ such that } x = (1 \otimes \epsilon_p)\rho(y) \rangle,$$

and

$$M_2 = \langle x \in M \mid \exists p \in \mathcal{P}'' \setminus \mathcal{P}, y \in M_{e_S} \text{ such that } x = (1 \otimes \epsilon_p)\rho(y) \rangle.$$

By Proposition 9.1.30, both M_1 and M_2 are A^*-comodules and we can form the cofragment

$$M \supset M_1 + M_2 \supset M_i \supset 0 \supset \dots \quad (i = 1, 2).$$

Case 2: S is an $A[R]$-module, but not an A-module. In this case, every path pq with $q : e_S \to e_R$ and $p : e_R \to e_i$, $e_i \in Q_0$ must be zero. Since S is not an A-module and e_R is a source vertex in Q', we know by (9.1) that

$\mathrm{Hom}_{A[R]}(S, S_{e_R}) \neq 0$ and so there must be at least one arrow $a : e_S \to e_R$ in Q''. A priori, we define

$$M_1 = \langle x \in M \mid \exists p \in \mathcal{P}' \setminus \mathcal{P}, y \in M_{e_R} \text{ such that } x = (1 \otimes \epsilon_p)\rho(y)\rangle,$$

$$M_2 = \langle x \in M \mid \exists p \in \mathcal{P}'' \setminus \mathcal{P}', y \in M_{e_S} \text{ such that } x = (1 \otimes \epsilon_p)\rho(y)\rangle.$$

By Proposition 9.1.30, M_1 is an A^*-comodule and M_2 is an $A[R]^*$-comodule. Next we define

$$M_3 = \langle x \in M_2 \mid \exists p \in \mathcal{P}' \setminus \mathcal{P}, y \in M_{e_R} \text{ such that } x = (1 \otimes \epsilon_p)\rho(y)\rangle,$$

which is an A^*-comodule. We obtain the cofragment

$$M \supset M_1 + M_2 \supset M_2 \supset M_3 \supset 0 \supset \dots$$

Lemma 9.1.32. *If the associated Q''-quiver representation (M_i, φ_j) is such that $M_{e_R} = \bigcup_{a:e_S \to e_R} \mathrm{Im}(\varphi_a)$, then $M_1 \subset M_2$.*

Proof. Let $x \in M_1$ with path p starting in e_R and $y \in M_{e_R}$ such that $x = (1 \otimes \epsilon_p)\rho(y)$. Let $y = \varphi_a(z)$. Then $x = (1 \otimes \epsilon_{pa})\rho(z) \in M_2$. □

In the situation of the lemma, we obtain the cofragment

$$M \supset M_2 \supset M_2 \supset M_3 \supset 0 \supset \dots$$

Example 9.1.33. Consider the following quiver representation

$$M_{e_S} \xrightarrow{\begin{pmatrix} 1 \\ 0 \end{pmatrix}} M_{e_R} \xrightarrow{\begin{pmatrix} 1 & 0 \\ 0 & 1 \end{pmatrix}} M_0$$

Let $M_{e_S} = km_S, M_{e_R} = \langle m_R^1, m_R^2\rangle, M_0 = \langle m^1, m^2\rangle$ be bases establishing the above linear maps. Then we have that

$$M_1 = \langle m^1, m^2\rangle \not\subset M_2 = \langle m^1, m_R^1\rangle.$$

The associated cofragment is

$$M \supset \langle m_R^1, m^1, m^2\rangle \supset \langle m^1, m_R^1\rangle \supset km^1 \supset 0 \supset \dots$$

9.2 Glider representations for the Weyl algebras

In [8] the authors classify the simple modules for $\mathbb{Z}$-graded rings of the following type: $A = \bigoplus_{i\in\mathbb{Z}} A_i$ with $A_i = A_0 v_i$ a free left A_0-module of rank 1 and $v_0 = 1$. The multiplication in these rings is given by

$$\alpha v_i \beta v_j = \alpha\sigma^i(\beta)c(i,j)v_{i+j},$$

for $\alpha, \beta \in A_0, i, j \in \mathbb{Z}$ and where σ is an automorphism of A_0 and c is a 2-cocycle

$$c : \mathbb{Z} \times \mathbb{Z} \to Z(A_0).$$

As a particular example, they obtain a classification of the simple modules for generalized Weyl algebras (GWA) introduced by Bavula. They are $\mathbb{Z}$-graded rings generated by $D = A_0$ and two indeterminates $X = v_1$ and $Y = v_{-1}$ satisfying the relations

$$X\alpha = \sigma(\alpha)X \quad \text{and} \quad Y\alpha = \sigma^{-1}(\alpha)Y, \ \forall \alpha \in D,$$
$$YX = a, \ XY = \sigma(a),$$

for an automorphism σ of D and some $a \in Z(D)$. We denote this GWA by $A = D(\sigma, a)$. One easily checks that the first Weyl algebra $A_1 = A_1(K)$; K some field of characteristic zero, is just $K[H](\sigma, H)$ where σ is the automorphism of $K[H]$ defined by $\sigma(H) = H - 1$. More generally, Bavula introduced generalized Weyl algebras of degree n which are $\mathbb{Z}^n$-graded rings and as particular examples one obtains the nth Weyl algebra. In [9] the authors then classify the simple holonomic modules over the second Weyl algebra A_2.

In example f) from Section 3.3 we know that even the first Weyl algebra carries different ring filtrations and fragments again appear naturally. It is moreover possible to define ring filtrations FA_2 of the second Weyl algebra such that $F_0A_2 = A_1$, so the question arises as to whether our fragment theory can reveal new information. Let us first give an overview of how the simple A_1-modules can be classified.

We denote by $k = K(H)$ the field of fractions of $D = K[H]$, which is obtained by localizing D at $S = D \setminus \{0\}$. The localization B of the Weyl algebra $A_1 = K[H](\sigma, H)$ is isomorphic to the skew Laurent polynomial ring

$$B \cong k[X, X^{-1}; \sigma],$$

an isomorphism given by $X \mapsto X$, $Y \mapsto HX^{-1}$, $H \mapsto H$. We denote by $\hat{A}_1$ the set of simple A_1-modules (up to isomorphism) and let $M \in \hat{A}_1$. The module M has a D-torsion submodule tor(M); hence after localizing we have that $S^{-1}M = 0$ (we say that M is D-torsion) or that $S^{-1}M \neq 0$ (we say that M is D-torsion free). We arrive at the following distinction

$$\hat{A}_1 = \hat{A}_1(D - \text{torsion}) \bigsqcup \hat{A}_1(D - \text{torsion free}).$$

The class of simple D-torsion modules correspond with the simple modules that are D-weight modules, that is, that are semisimple when considered as D-modules. It can be shown that the weights of such modules lie completely in one orbit $\mathcal{O}$ of the action of cyclic subgroup $\langle\sigma\rangle$ on the set of all maximal ideals Max(K[H]) and in general there is a difference between linear and cyclic orbits. For the first Weyl algebra, however, all orbits are linear. Furthermore, there is an equivalence relation on the set of all maximal ideals that lie in a linear orbit: two maximal ideals $P \sim Q$ are equivalent if $P = \sigma^i(Q)$ for some $i \in \mathbb{Z}$. In our setting, $P = (H - \lambda) \sim Q = (H - \mu)$ if and only if $\lambda = \mu + i$ for some $i \in \mathbb{Z}$. Theorem 3.2 from [8] entails that $\hat{A}_1(D - \text{torsion})$ is given by

Max(K[H])/ $\sim$	$\xrightarrow{\cong}$	$\hat{A}_1(D - \text{torsion})$
$[\lambda] \neq 0 \in K/\mathbb{Z}$	$\mapsto$	$A_1/A_1(H-\lambda) \cong K_\lambda[X,Y]/(XY - \lambda - 1)(H \cdot 1 = \lambda)$,
$[1]\ (\mathbb{Z}_0^+)$	$\mapsto$	$A_1/A_1(H-1, Y) \cong K[X](H \cdot 1 = 1)$,
$[0]\ (\mathbb{Z}^-)$	$\mapsto$	$A_1/A_1(H, X) \cong K[Y](H \cdot 1 = 0)$

For the other type of simple A_1-modules, there appears to be a bijection between $\hat{A}_1(D - \text{torsion free})$ and $\hat{B}(A_1 - \text{socle})$, which is the set of simple B-modules that are socle when considered as A_1-modules. The latter means that the A_1-socle, i.e. the sum of the simple A_1-submodules, is not zero. The authors further show that the D-torsion free simple A_1-modules are of the form $M_f = A_1/A_1 \cap Bf$ where $Bf \leq B$ is a maximal left ideal. Subsequently, one has to determine when such an irreducible element $f \in B$ is such that M_f is simple and socle. In case of the first Weyl algebra, all modules M_f for $f \in B$ irreducible are A_1-socle. In order for M_f to also be simple, it suffices to pick $f \in B$ l-normal. In order to define l-normal elements one first has to introduce a relation $<$ on S: we say that $\alpha < \beta$ if and only if there are no maximal ideals P and Q which belong to the same linear orbit, contain α and β resp. and $P = \sigma^i(Q)$ for some $i \in \mathbb{N}$. The l-normal elements can then be defined as elements of the form $f = Y^m\beta_{-m}(H) + \cdots + \beta_0(H) \in A_1, \beta_i(H) \in K[H], i = -m, \ldots, 0, m > 0$ such that $\beta_0(H) < \beta_{-m}(H)$ and $\beta_0(H) < a = H$. In case of the Weyl algebra, we can write

$$\beta_{-i}(H) = \Pi_j(H - \lambda_j^i).$$

The conditions $\beta_0(H) < H$ and $\beta_0(H) < \beta_{-m}(H)$ are equivalent with

$$\forall j : \lambda_j^0 \notin \mathbb{Z}^+,$$

and

$$\forall j\ \forall i : \lambda_i^0 \notin \lambda_j^m + \mathbb{Z}^+.$$

In what follows, we will interchangeably use the isomorphisms

$$A_1 = K\langle X, Y\rangle/(XY - YX - 1) \cong K[X][Y; \frac{\partial}{\partial_X}] \cong K[H][\sigma, H],$$

where the second ring is the Ore extension and the last ring is a GWA. Both the gradation filtration $F^g A_1(K)$ $(\deg(X) = -1, \deg(Y) = 1)$ and the operator filtration $F^o A_1(K)$ are standard filtrations, with $F_0^g A_1(K)$ the Ore extension $C = K[X][XY; X\frac{\partial}{\partial_X}]$ and $F_0^o A_1(K) = K[X]$. If we only consider the positive part of the grading filtration, we obtain the inclusion

$$F^{o,+}A_1(K) \subset F^{g,+}A_1(K).$$

For certain simple $A_1(K)$-modules M we can construct a (standard) fragment structure for either $F^{o,+}A_1(K)$ or $F^{g,+}A_1(K)$. For example, for

$$M = A_1/A_1(YX - \lambda), \lambda \in K \setminus \mathbb{Z},$$

the descending chain

$$M \supset (X)M \supset (X)^2M \supset \dots$$

defines a standard fragment for the operator filtration. For the other simple D-torsion modules, there does not exist such a standard fragment structure. For D-torsion free modules, one can consider for example the l-normal element $f = Y(H - \frac{1}{3}) + (H - \frac{1}{2})$. Then

$$C(XY - \tfrac{3}{2})M_f \supset C(XY - \tfrac{4}{3})(XY - \tfrac{3}{2})M_f \supset C(XY - \tfrac{4}{3})^2(XY - \tfrac{3}{2})M_f \supset \dots$$

is a fragment for the grading filtration. It is however not standard.

Holonomic modules arose by the observation that the Gelfand-Kirillov dimension GKdim(M) of any A_n-module is bounded by $n \leq \mathrm{GKdim(M)} \leq 2\mathrm{n}$. There is also a notion of the Gelfand-Kirillov dimension of fragments or glider representations, which is the topic of the next and final section. It is our aim to develop the Gelfand-Kirillov dimension further and to apply it to the short introduction to glider representations for the first Weyl algebra above.

9.3 Gelfand-Kirillov dimension of fragments

In this final section we propose the definition of the Gelfand-Kirillov dimension of fragments and show a few preliminary results. Let us first recall the definition of GK dimension of algebras and modules. We refer to [43] for a detailed overview.

Let A be a K-algebra and V a finite dimensional subspace of A containing 1. We denote by $K[V]$ the K-algebra generated by V in A and we call V a subframe of A. If $V = Kv_1 + \ldots + Kv_d$ we denote by V^n the set of monomials of length n in the $v_1, \ldots, v_d$ and $F_n^V(A) = K + V + V^2 + \ldots + V^n$. We obtain a monotone increasing function $d_V(n) = \dim_K(F_n^V(A))$ and we define the Gelfand-Kirillov dimension of A to be

$$\mathrm{GKdim(A)} = \sup_{\mathrm{V}} \overline{\lim} \ \log_{\mathrm{n}}(\mathrm{d_V(n)}),$$

where the supremum is taken over all subframes of A containing 1, $\overline{\lim}$ denotes the limsup and $\log_n(d_V(n))$ is defined to be $\log(d_V(n))/\log(n)$. It can be shown that if A is finitely generated by $\{a_1, \ldots, a_d\}$ and if $V = K1 + Ka_1 + \ldots + Ka_d$, then $\mathrm{GKdim(A)} = \overline{\lim} \log_n(d_V(n))$.

The GK dimension also applies to (left) modules: Let M be a left A-module, F a finite dimensional subspace of M - we say that F is a subframe of M - and V a finite dimensional subspace of A containing 1. We have the monotone increasing function $d_{V,F}(n) = \dim_k(V^n F)$ and we define

$$\mathrm{GKdim}({}_A\mathrm{M}) = \sup_{F,V} \overline{\lim} \log_n(d_{V,F}(n)),$$

where the supremum is taken over all subframes of A and M. To define a suitable GK dimension for fragments we would have to incorporate both the chain structures from the algebra filtration FA and from the chain of F_0A-modules itself. To this extent, we observe that for any subframe N of a fragment M, there exists an $e(N) \geq 0$ such that $N_{e(N)+1} = N \cap M_{e(N)+1} = 0$ (unless $N \cap B(M) \neq 0$ of course).

Definition 9.3.1. Let FA be a positive algebra filtration with subalgebra $F_0A = B$ and let $M \supset M_1 \supset \ldots \supset M_n \supset \ldots$ be a (left) FA-fragment. The Gelfand-Kirillov dimension gkdim(M) is defined by

$$\mathrm{gkdim(M)} = \sup_{V,N} \overline{\lim} \log_{\min n, e(N)} \dim_K \Big(\sum_{m=0}^{e(N)} F_m(V^n) N_m \Big),$$

where the supremum is taken over all subframes V, N of A, resp. M.

Lemma 9.3.2. *Let FA be a positive K-algebra filtration with $F_0A = B$ and let $M \supset M_1 \supset \ldots \supset M_n \supset \ldots$ be a (left) FA-fragment. Then*

$$\mathrm{GKdim}({}_A\mathrm{B(M)}) \leq \mathrm{gkdim(M)}.$$

Proof. For $N \subset B(M)$ a finite dimensional subspace, we have that $e(N) = +\infty$ and for any subframe V of A we have that $\sum_{m=0}^{+\infty} F_m(V^n) N_m \subset V^n N$ for all $n >> 0$. □

Proposition 9.3.3. *Let FA be a positive K-algebra filtration with $F_0A = B$ and let $\Omega \supset M \supset M_1 \supset \ldots \supset M_n \supset \ldots$ be a (left) FA-glider representation. We have the inequalities*

$$\mathrm{GKdim}({}_B\mathrm{M}) \leq \mathrm{gkdim(M)} \leq \mathrm{GKdim}({}_A\mathrm{RM}) \leq \mathrm{GKdim}({}_A\Omega) \leq \mathrm{GKdim(A)}.$$

Proof. The third and last inequalities are well-known properties of the GK dimensions of modules. Hence we only have to prove the first two. For V, N subframes of A, resp. M we have that

$$\sum_{m=0}^{e(N)} F_m(V^n) N_m \subset V^n N,$$

from which the second inequality easily follows. For the first one, let V be a subframe of $B = F_0A$; then for all $m, n \geq 0, F_m(V^n) = V^n$, whence

$$V^nN = F_0(V^n)N_0 = \sum_{m=0}^{e(N)} F_m(V^n)N_m,$$

and the first inequality follows. □

Example 9.3.4. Let $A = A_1(K)$ be the first Weyl algebra with the operator filtration F^oA. Let $\Omega \supset M \supset M_1 \supset \ldots$ be a left FA-glider such that $\Omega = AM$ is a holonomic module, i.e. $\mathrm{GKdim}(_{\mathrm{A}}\mathrm{AM}) = 1$. Proposition 9.3.3 entails that

$$\mathrm{GKdim}(_{\mathrm{K[X]}}\mathrm{M}) \leq \mathrm{gkdim(M)} \leq \mathrm{GKdim}(_{\mathrm{A}}\mathrm{AM}) = 1.$$

It follows that the Gelfand-Kirillov dimension of the $K[X]$-module M is either 1 or 0. Suppose we are in the latter case, that is, M is finite dimensional. From [8] we know what the holonomic modules look like. In particular, the $K[XY]$-torsion modules are enlisted by

$\mathrm{Max(K[H])}/\sim$	$\xrightarrow{\cong}$	$\hat{A}_1(D - \text{torsion})$
$[\lambda] \neq 0 \in K/\mathbb{Z}$	$\mapsto$	$A_1/A_1(H - \lambda) \cong K_\lambda[X, Y]/(XY - \lambda - 1)(H \cdot 1 = \lambda),$
$[1]\ (\mathbb{Z}_0^+)$	$\mapsto$	$A_1/A_1(H - 1, Y) \cong K[X](H \cdot 1 = 1),$
$[0]\ (\mathbb{Z}^-)$	$\mapsto$	$A_1/A_1(H, X) \cong K[Y](H \cdot 1 = 0)$

The second case is easily seen to be excluded. Suppose that $AM \cong K_\lambda[X, Y]/(XY - \lambda - 1)$ for some $\lambda \in K/\mathbb{Z}$ not zero and let $m = P(X) + Q(Y) \in M$. By multiplying with a suitable power of X, we have that $X \cdot m \in K[X]$, from which it follows that $K[X]m$ is not finite dimensional. Hence it follows that $\Omega = K[Y]$ and $M \subset \Omega$ is a finite dimensional subspace. It is clear that $\mathrm{gkdim(M)} = 0 < 1$.

The other type of holonomic modules, that is, the $K[XY]$-torsion free modules, are of the form $M_f = A_1/(A_1 \cap Bf)$, where $B \cong K[X, X^{-1}; \sigma]$ and $f \in B$ is l-normal. By definition of l-normal elements, it follows that $M \subset A_1M = M_f$ must contain a polynomial $P(X)$ in order for $A_1M = M_f$ to hold, hence $K[X]P(X) \subset M$ is not finite dimensional, and it follows that $\mathrm{gkdim(M)} = 1$.

From the example above, it follows that the Gelfand-Kirillov dimension of a glider with Ω holonomic is always 1, except for one particular case. To investigate the behavior for $A = A_2(K)$ the second Weyl algebra, we recall that

$$A_2 = A_1 \otimes A_1.$$

In general, let FA, FA' be positively filtered K-algebras. Then we can define a filtration on the tensor product $A \otimes A'$ as follows:

$$F_n(A \otimes A') = \sum_{i+j \leq n} F_iA \otimes F_jA'.$$

In case the first Weyl algebra A_1 is equipped with the operator filtration as before, then one checks that the associated filtration on the second Weyl algebra is just the operator filtration again:

$$F(A_1 \otimes A_1) = F^o A_2.$$

A notion of a tensor product of fragments already appeared in [52], but we need a slightly different construction.

Proposition 9.3.5. *Let FA, FA' be two positively filtered K-algebras, $M \supset M_1 \supset \ldots$ an FA-fragment and $M' \supset M'_1 \supset \ldots$ an FA'-fragment. The chain of $F_0A \otimes F_0A'$-modules*

$$M \otimes M' \supset M_1 \otimes M'_1 \supset \ldots \supset M_n \otimes M'_n \supset \ldots$$

is an $F(A \otimes A')$-fragment.

Proof. Let $j \leq i$ and $m_i \otimes m'_i \in (M \otimes M')_i$. Let $a_s \otimes a'_t$ be a monomial in $F_j(A \otimes A')$, so $s + t \leq j$. By definition, $(a_s \otimes a'_t) \cdot (m_i \otimes m'_i) = a_s m_i \otimes a'_t m'_i \in M_{i-s} \otimes M'_{i-j} \subset M_{i-j} \otimes M'_{i-j} = (M \otimes M')_{i-j}$ (both filtrations FA, FA' are positive!). □

Lemma 9.3.6. *Let V, V' be subframes of A, resp. A'. Then for all $n, m \geq 0$ it holds that*

$$F_m((V \otimes V')^n) \subset F_m(V^n) \otimes F_m((V')^n).$$

Proof. By definition

$$\begin{aligned} F_m((V \otimes V')^n) &= \Big(\sum_{i+j\leq m} F_iA \otimes F_jA'\Big) \cap (V \otimes V')^n \\ &\subset \Big(\sum_{i+j\leq m} F_iA \otimes F_jA'\Big) \cap (V^n \otimes (V')^n) \\ &= \sum_{i+j\leq m} F_i(V^n) \otimes F_j((V')^n) \\ &\subset F_m(V^n) \otimes F_m((V')^n). \quad \square \end{aligned}$$

Proposition 9.3.7. *Let FA, FA' be two positively filtered K-algebras, $M \supset M_1 \supset \ldots$ an FA-fragment and $M' \supset M'_1 \supset \ldots$ an FA'-fragment. If both fragments are of infinite essential length and if $B(M), B(M')$ are both non-zero of both zero, then*

$$\mathrm{gkdim}(\mathrm{M} \otimes \mathrm{M}') \geq \max\{\mathrm{gkdim}(\mathrm{M}), \mathrm{gkdim}(\mathrm{M}')\}.$$

Proof. Let V, N be subframes of A, M respectively. If $e(N) < \infty$, pick any $m' \in M'_{e(N)}$; then $T = N \otimes Km'$ is a subframe of $M \otimes M'$ such that $e(N) = e(T)$. We have

$$\begin{aligned}\dim_K\Big(\sum_{m=0}^{e(N)} F_m(V^n)N_m\Big) &= \dim_K\Big(\sum_{m=0}^{e(N)} F_m(V^n)\otimes F_m(K^n)N_m \otimes Km'\Big)\\ &= \dim_K\Big(\sum_{m=0}^{e(T)} F_m((V\otimes K)^n)T_m\Big).\end{aligned}$$

If $N \cap B(M) \neq \emptyset$, then take $m' \in B(M')$. We have shown the inequality $\mathrm{gkdim(M)} \le \mathrm{gkdim(M \otimes M')}$. By symmetry, $\mathrm{gkdim(M')} \le \mathrm{gkdim(M \otimes M')}$ also holds. □

Proposition 9.3.8. *Let $0 \to N \to M \to L \to 0$ be a (strict) short exact sequence of fragments. Then*

$$\mathrm{gkdim(M)} \ge \max\{\mathrm{gkdim(N)}, \mathrm{gkdim(L)}\}.$$

Proof. This follows from the definition of gkdim(M) as in the module case. □

Corollary 9.3.9. *For any fragment M we have that* $\mathrm{gkdim(M)} \ge \mathrm{gkdim(B(M))}$.

Observe that we take the GK dimension of the body as a fragment with trivial chain.

Classically, one can use filtrations to obtain lower bounds of the GK dimension of a module. For example, when the algebra carries a filtration FA and a left A-module M has a filtration FM making it into a filtered FA-module, then

$$\mathrm{GKdim(G(M))} \le \mathrm{GKdim(M)}.$$

For any subframe N of M, we set $G(N) = \bigoplus_{n\in\mathbb{Z}} N_i/N_{i+1}$, where $N_i = N \cap M_i$. For fragments M, we denoted the associated graded fragment by $g(M)$, and for subframes N of M we denote $\bigoplus_{n\ge 0} N_n/N_{n+1}$ by $g(N)$. We have a similar result for fragments.

Proposition 9.3.10. *Let $M \supset M_1 \supset \dots$ be an FA-fragment. Then* $\mathrm{gkdim(g(M))} \le \mathrm{gkdim(M)}$.

Proof. We denote the grading filtrations on the associated graded levels with a small f. Let W be a subframe of $G(A)$, T a subframe of $g(M)$. There exist subframes V, resp. N of A, resp. M such that $W \subset G(V)$ and $T \subset g(N)$. For $n, m \ge 0$ we have that

$$f_m(G(V^n)) = \bigoplus_{l\le m} F_lV^n/F_{l-1}V^n = G(F_mV^n),$$

and

$$g(N)_m = \bigoplus_{p \geq m} N_p/N_{p+1} = g(N_m).$$

Hence

$$\begin{aligned}
\sum_{m=0}^{e(T)} f_m(W^n)T_m &\subset \sum_{m=0}^{e(T)} f_m(G(V)^n)g(N)_m \\
&\subset \sum_{m=0}^{e(T)} f_m(G(V^n))g(N)_m \\
&= \sum_{m=0}^{e(T)} G(F_m V^n)g(N_m) \subset \sum_{m=0}^{e(T)} g(F_m V^n N_m).
\end{aligned}$$

Since $\sum_{m=0}^{e(T)} F_m V^n N_m$ is finite dimensional and since $e(T) \leq e(N)$, we arrive at

$$\dim_K(\sum_{m=0}^{e(T)} f_m(W^n)T_m) \leq \dim_K(\sum_{m=0}^{e(N)} F_m V^n N_m).$$

This shows the desired inequality. □

We say that a fragment $M \supset M_1 \supset \ldots$ is finite if $\dim_K(M_i/M_{i+1}) < \infty$ for all $i \geq 0$. For finite fragments, the GK dimension contains some information on the chain of the fragment.

Proposition 9.3.11. *Let $M \supset M_1 \supset \ldots$ be a finite fragment. Then*

$$\mathrm{gkdim(M)} \geq \sup_{\mathrm{n} \geq 2} \frac{\mathrm{dim_K(M_n/M_{n+1})}}{\mathrm{log(n)}}.$$

Proof. Let $n \geq 2$. As a subframe N of M we can lift K-basis elements of M_n/M_{n+1}. Then

$$\dim_K(M_n/M_{n+1}) = \dim_K(KN) = \dim_K(F_n K N_n) = \dim_K(\sum_{m=0}^{n} F_m K N_m),$$

which shows that

$$\mathrm{gkdim(M)} \geq \frac{\mathrm{dim_K(M_n/M_{n+1})}}{\mathrm{log(n)}}. \quad \square$$

Bibliography

[1] E. Abe. *Hopf Algebras, Translated from the Japanese by Hisae Kinoshita and Hiroko Tanaka.* Cambridge University Press, Cambridge - New York, 1980.

[2] S.A. Amitsur. Groups with representations of bounded degree. ii. *Illinois J. Math.*, 5:198–205, 1961.

[3] F.W. Anderson and K.R. Fuller. *Rings and Categories of Modules.* Number 13 in Graduate Texts in Mathematics. Springer-Verlag, Second edition, 1992.

[4] M. Artin. Brauer-Severi varieties (Notes by A. Verschoren). In *F. Van Oystaeyen, A. Verschoren (Eds.), Brauer Groups in Ring Theory and Algebraic Geometry, Proceedings of a conference held at Antwerp 1981*, volume 917 of *Lecture Notes Math.*, pages 194 – 210. Springer, Berlin, Heidelberg, New York, 1982.

[5] M. Artin. Geometry of quantum planes. *Contemp. Math.*, 124:1–15, 1992.

[6] M.J. Asensio, M. Van den Bergh, and F. Van Oystaeyen. A new algebraic approach to microlocalization of filtered rings. *Trans. Amer. Math. Soc.*, 316(2):537–553, 1989.

[7] I. Assem, D. Simson, and A. Skowroński. *Elements of the Representation Theory of Associative Algebras. Vol. 1.* Number 65 in London Mathematical Society Student Texts. Cambridge University Press, Cambridge, Second edition, 2006.

[8] V. Bavula and F. Van Oystaeyen. The simple modules of certain generalized crossed products. *J. Algebra*, 194(2):521–566, 1997.

[9] V. Bavula and F. Van Oystaeyen. Simple holonomic modules over the second Weyl algebra A_2. *Adv. Math.*, 150(1):80–116, 2000.

[10] D. Benson. *Representations and Cohomology. I. Basic Representation Theory of Finite Groups and Associative Algebras.* Number 30 in Cambridge Studies in Advanced Mathematics. Cambridge University Press, Cambridge, Second edition, 1998.

[11] I.N. Bernstein, I.M. Gelfand, and S.I. Gelfand. Structure of representations generated by highest weights. *Funktsional. Anal. i Prilozhen*, 5(1):1–9, 1971.

[12] J.-E. Björk. *Rings of Differential Operators*. Number 21 in Math Library. North Holland, Amsterdam, 1979.

[13] A. Braverman and D. Gaitsgory. Poincaré-Birkhoff-Witt theorem for quadratic algebras of Koszul type. *J. Algebra*, 181(2):315–328, 1996.

[14] F. Caenepeel. Glider representations of chains of semisimple Lie algebras. *Comm. Algebra*, 46(11):4985–5005, 2018.

[15] F. Caenepeel and G. Janssens. On distinguishing isocategorical groups with glider representation rings. Available on the ArXiv, 2019.

[16] F. Caenepeel and F. Van Oystaeyen. Clifford Theory for glider representations. *Algebr. Represent. Theory*, 19(6):1477–1493, 2016.

[17] F. Caenepeel and F. Van Oystaeyen. Glider representations of group algebra filtrations of nilpotent groups. *Algebr. Represent. Theory*, 21(3):529–550, 2018.

[18] F. Caenepeel and F. Van Oystaeyen. Generalized characters for glider representations of groups. *Algebr. Represent. Theory*, 2019.

[19] F. Caenepeel and F. Van Oystaeyen. Glider Brauer-Severi varieties of central simple algebras. arXiv:1906.03411, 2019.

[20] F. Caenepeel and F. Van Oystaeyen. Localization and sheaves of glider representations. *J. Pure Appl. Algebra*, 223(6):2635–2672, 2019.

[21] F. Châtelet. Variations sur un thème de H. Poincaré. *Ann. Sch. E.N.S.*, 61:249–300, 1944.

[22] W. Chin, M. Kleiner, and D. Quinn. Almost split sequences for comodules. *J. Algebra*, 249(1):1–19, 2002.

[23] W. Chin and S. Montgomery. Basic coalgebras. In *Modular Interfaces*, Volume 4 of *AMS/IP Stud. Adv. Math.*, pages 41 – 47. Amer. Math. Soc., Providence, RI, 1997.

[24] A.H. Clifford. Representations induced in an invariant subgroup. *Ann. of Math.*, 38(2):533–550, 1937.

[25] D.H. Collingwood and W.M. McGovern. *Nilpotent Orbits in Semisimple Lie Algebras*. Van Nostrand Reinhold Mathematics Series. Van Nostrand Reinhold Co., New York, 1993.

[26] E.C. Dade. Group graded rings and modules. *Math. Zeitschrift*, 174(3):241–262, 1980.

[27] P. Etingof and S. Gelaki. Isocategorical groups. *Internat. Math. Res. Notices*, 2:59–76, 2001.

[28] P. Etingof, O. Golberg, S. Hensel, T. Liu, A. Schwendner, D. Vaintrob, and E. Yudovina. *Introduction to Representation Theory.* Number 59 in Student Mathematical Library. Amer. Math. Soc., Providence, RI, 2011.

[29] A.S. Farahat and F. Van Oystaeyen. Hopf filtrations and Larson-type orders in Hopf algebras. *J. Algebra*, 267(2):756–772, 2003.

[30] E. Formanek. Noetherian PI-rings. *Comm. Algebra*, 1:79–86, 1974.

[31] W. Fulton and J. Harris. *Representation Theory: a First Course.* Number 129 in Graduate Texts in Mathematics. Readings in Mathematics, Springer-Verlag, New York, 1991.

[32] P.X. Gallagher. Group characters and normal Hall subgroups. *Nagoya Math. J.*, 21:223–230, 1962.

[33] J. Golan. *Localization of Noncommutative Rings.* Number 30 in Lect. Notes Pure Appl. Math. Marcel Dekker, New York, 1975.

[34] O. Goldman. Rings and modules of quotients. *J. Algebra*, 13:10–47, 1969.

[35] I. Goyvaerts and E. Meir. The Witt group of a braided monoidal category. arXiv:1412.3417, 2014.

[36] R. Hartshorne. *Algebraic Geometry.* Number 52 in Graduate Texts in Mathematics. Springer-Verlag, New York - Heidelberg, 1977.

[37] J.E. Humphreys. *Introduction to Lie Algebras and Representation Theory.* Number 9 in Graduate Texts in Mathematics. Springer-Verlag, New York - Berlin, 1972.

[38] J.E. Humphreys. *Representations of Semisimple Lie Algebras in the BGG Category $\mathcal{O}$.* Number 94 in Graduate Studies in Mathematics. Amer. Math. Soc., Providence, RI, 2008.

[39] M.I. Isaacs. *Character Theory of Finite Groups.* Number 69 in Pure and Applied Mathematics. Academic Press [Harcourt Brace Jovanovich, Publishers], New York - London, 1976.

[40] G. James and M. Liebeck. *Representations and Characters of Groups.* Cambridge University Press, New York, Second edition, 2001.

[41] E. Jespers, G. Leal, and A. Paques. Central idempotents in the rational group algebra of a finite nilpotent group. *J. Algebra Appl.*, 2(1):57–62, 2003.

[42] M-A. Knus, A Merkurjev, M. Rost, and J.-P. Tignol. *The Book of Involutions.* Number 44 in American Mathematical Society Colloquium Publications. Amer. Math. Soc., Providence, RI, 1998.

[43] G.R. Krause and T. H. Lenagan. *Growth of Algebras and Gelfand-Kirillov Dimension.* Number 22 in Graduate Studies in Mathematics. Amer. Math. Soc., Providence, RI, revised edition, 2000.

[44] T.Y. Lam. *A First Course in Noncommutative Rings.* Number 131 in Graduate Studies in Mathematics. Springer-Verlag, New York, Second edition, 1991.

[45] H. Li and F. Van Oystaeyen. Filtrations on simple Artinian rings. *J. Algebra*, 132(2):361–376, 1990.

[46] H. Li and F. Van Oystaeyen. *Zariskian Filtrations.* Number 2 in K-Monographs Math. Kluwer, Dordrecht, 1995.

[47] G. Mackey. On induced representations of groups. *Amer. J. Math.*, 73:576–592, 1951.

[48] H. Matsumura. *Commutative Ring Theory.* Number 8 in Cambridge Studies in Advanced Mathematics. Cambridge University Press, Cambridge, Second edition, 1989. Translated from the Japanese by M. Reid.

[49] S. Montgomery. *Hopf Algebras and their Actions on Rings.* Number 82 in CBMS Regional Conference Series in Mathematics. Amer. Math. Soc., Providence, RI, 1993. Published for the Conference Board of the Mathematical Sciences, Washington DC.

[50] D. Mumford. *The Red Book of Varieties and Schemes.* Number 1358 in Lecture Notes in Mathematics. Springer-Verlag, Berlin, Second, expanded edition, 1999. Includes the Michigan lectures (1974) on curves and their Jacobians. With contributions by Enrico Arbarello.

[51] D.C. Murdoch and F. Van Oystaeyen. Noncommutative localization and sheaves. *J. Algebra*, 35:500–515, 1975.

[52] S. Nawal and F. Van Oystaeyen. An Introduction of Fragmented Structures over Filtered Rings. *Comm. Algebra*, 23:975–993, 1995.

[53] S. Nawal and F. Van Oystaeyen. Projective aspects of fragments over filtered rings. *Comm. Algebra*, 24:165–185, 1996.

[54] C. Năstăsescu and F. Van Oystaeyen. *Graded and Filtered Rings and Modules.* Number 758 in Lecture Notes in Mathematics. Springer, Berlin, 1979.

[55] C. Năstăsescu and F. Van Oystaeyen. *Graded Ring Theory.* Number 28 in North-Holland Mathematical Library. North-Holland Publishing Co., Amsterdam - New York, 1982.

[56] C. Năstăsescu and F. Van Oystaeyen. *Methods of Graded Rings*. Number 1838 in Lecture Notes in Mathematics. Springer-Verlag, Berlin, 2004.

[57] J. Okniński. *Semigroup Algebras*. Number 138 in Monographs and Textbooks in Pure and Appl. Mathematics. Marcel Dekker, New York, 1990.

[58] I. Reiner. *Maximal Orders*. Number 5 in London Mathematical Society Monographs. Academic Press, London - New York, 1975.

[59] C.M. Ringel. *Tame Algebras and Integral Quadratic Forms*. Number 1099 in Lecture Notes in Mathematics. Springer-Verlag, Berlin, 1984.

[60] A. Rosenberg. *Noncommutative Geometry*. Reidel Publ. G., Amsterdam, 1997.

[61] R. Sallam and F. Van Oystaeyen. A Microstructure Sheaf and Quantum Sections over a Projective Scheme. *J. Algebra*, 158(1):201–225, 1993.

[62] O. Schilling. *The Theory of Valuations*. Number 4 in Mathematical Surveys. Amer. Math. Soc., New York, 1950.

[63] J.P. Serre. Faisceaux algbriques coherents. *Ann. of Math.*, 61:197–278, 1955.

[64] J.P. Serre. *Corps Locaux*. Number VIII in Publications de l'Institut de Mathématique de l'Université de Nancago. Hermann, Paris, Second edition, 1968.

[65] J.P. Serre. *Linear Representations of Finite Groups*. Number 42 in Graduate Texts in Mathematics. Springer-Verlag, New York - Heidelberg, 1977. Translated from the second French edition by Leonard L. Scott.

[66] B. Stenström. *Rings of Quotients*. Number 217 in Grundlehren Math. Wissenschaften. Springer, Berlin, 1975.

[67] F. Van Oystaeyen. *Prime Spectra in Non-commutative Algebra*. Number 444 in Lecture Notes in Mathematics. Springer, Berlin-New York, 1975.

[68] F. Van Oystaeyen. Compatibility of kernel functors and localization functors. *Bull. Soc. Math. Belg.*, 28(2):131–137, 1976.

[69] F. Van Oystaeyen. *Algebraic Geometry for Associative Algebras*. Number 232 in Monographs and Textbooks in Pure and Applied Mathematics. Marcel Dekker, Inc., New York, 2000.

[70] F. Van Oystaeyen. *Virtual Topology and Functor Geometry*. Number 256 in Lecture Notes in Pure and Applied Mathematics. Chapman & Hall/CRC, Boca Raton, FL, 2008.

[71] F. Van Oystaeyen and A. Verschoren. *Reflectors and Localization. Application to Sheaf Theory.* Number 41 in Lecture Notes in Pure and Applied Mathematics. Marcel Dekker, Inc., New York, 1979.

[72] F. Van Oystaeyen and L. Willaert. Grothendieck topology, coherent sheaves and Serre's theorem for schematic algebras. *J. Pure Appl. Algebra*, 104(1):109–122, 1995.

[73] D.N. Verma. *Structure of certain induced representations of complex semisimple Lie algebras.* PhD thesis, Yale University, 1966.

Index